Experimental Meson Spectroscopy-1980
(Sixth International Conference, Brookhaven)

AIP Conference Proceedings
Series Editor: Hugh C. Wolfe
Number 67
Particles and Fields Subseries No. 21

Experimental Meson Spectroscopy-1980
(Sixth International Conference, Brookhaven)

Editors
S. U. Chung
Brookhaven National Laboratory
and
S. J. Lindenbaum
Brookhaven National Laboratory
and City College of New York

American Institute of Physics
New York   1981

L.C. Catalog Card No. 80-71123
ISBN 0-88318-166-5
DOE CONF- 8004102

Preface

The Sixth International Conference on Experimental Meson
Spectroscopy (EMS-80) was held at Brookhaven National Laboratory,
Upton, New York, on April 25-26, 1980. Approximately two hundred
physicists participated. They represented most of the experimental
groups and many theorists active in this field throughout the world.

Meson Spectroscopy has been changing with the advent of quantum
chromodynamics which has explained many of its gross features. Nevertheless many of the crucial points of soft QCD still remain to be
tested in meson-spectroscopy experiments.

The Conference Program Committee consisted of G. Goldhaber (LBL),
R. Jaffe (MIT), L. Montanet (CERN) and P. Söding (DESY), L.L. Wang
(BNL), A.B. Wicklund (ANL) and S.U. Chung, Chairman (BNL). The
Organizing Committee consisted of P.D. Grannis (SUNY, Stony Brook),
I. Muzinich (BNL), F.E. Paige, Jr. (BNL), J.S. Russ (Carnegie-Mellon),
R. Weinstein (Northeastern), E.H. Willen (BNL), and S.J. Lindenbaum,
Chairman (BNL). The Program and Organizing Committees worked together
in an integrated and harmonious fashion which produced an excellent
program and a smoothly run and enjoyable conference. The excellent
job done by our speakers and chairmen (see Table of Contents) is also
much appreciated.

Miss Sharon Smith, the Conference Secretary, organized the
Secretariat and did an admirable job of organizing the multitude of
details required to make a conference run. Mrs. Rae Greenberg was
an essential member of the Conference Secretariat; Mrs. Violet Bezler
assisted the Conference Secretaries. The editors were also assisted
by Miss Sharon Smith in preparing the proceedings for publication.

Our after-dinner speaker at the Banquet, Dr. Jane Setlow (BNL),
Chairman of the NIH Recombinant DNA Molecule Program Advisory Committee,
gave us a very stimulating and thought-provoking talk on "Recombinant
DNA, Hope or Hazard." The Scientific Secretaries were Chang S. Chan,
Venky Ganapathi, Henry Glass, Jagdish Prasad and Lee Spencer, and we
wish to thank them for their efforts. We also wish to acknowledge
the cooperation of Mrs. Dorothy Schroeder of the BNL Directors Office,

the BNL Staff Services Division and the Instrumentation Division. Last but not least we wish to thank all the participants for their numerous contributions to the conference both in public and much more frequently in private discussions.

We are looking forward to hosting the next EMS Conference in about a couple of years.

S.U. Chung

Program Committee Chairman
and Proceedings Co-editor

S.J. Lindenbaum

Organizing Committee Chairman
and Proceedings Co-editor

INDEX OF SPEAKERS

(* = Manuscript not received)

Page

Bebek, C.J. 365

Bloom, E.D. 312

Cashmore, R.J. 1

Chen, M. 421

Crittenden, R. .... *

Dickey, J.O. 583

Donoghue, J.F...104

Eichten, E..387

Ferrer, A. 123

Glashow, S. .... *

Goldhaber, G. 223

Green, D.R. 152

Hey, A.J.G..194

Knies, G. 513

Kumar, B.R. 69

Lee-Franzini, J. 375

Littenberg, L.S. 415

Lynch, H.L..543

Nakamura, K. 217

Nef, C. 55

O'Neill, Jr., L.H. 559

Prentice, J.D. 297

Protopopescu, S. 170

Ratcliff, B.N. 37

Ruddick, K. 94

Scharre, D.L. 329

Schröder, H. 356

Smith, G.A. 186

Wang, L.L.C. 403

Wiss, J.E. 257

Witherell, M.S. 285

TABLE OF CONTENTS

Page

Preface and Acknowledgements

Index of Speakers

April 25, First Morning Session -- Chairman: D.W.G.S. Leith

Resonances in $\pi\pi\pi$ and $K\pi\pi$ Systems
R.J. Cashmore 1

$\bar{K}p$ Interactions at 11 GeV/c: New Results on
Strange Meson Systems
Blair N. Ratcliff 37

New Results on $K\pi$, $K\bar{K}$ and $\bar{\Lambda}p$ Mesonic States
W.E. Cleland, A. Delfosse, P.A. Dorsaz,
M.N. Kienzle-Focacci, G. Mancarella,
M. Martin, P. Muhlemann, C. Nef, T. Pal,
J. Rutschmann, J.L. Gloor, H. Zeidler,
A.D. Martin 55
(Presented by C. Nef)

April 25, Second Morning Session -- Chairman: R. Jaffe

Heavy Vector Mesons
B.R. Kumar 69

New Results on Radiative Meson Decays
B. Collick, S. Heppelmann, T. Joyce, Y. Makdisi,
M. Marshak, E. Peterson, K. Ruddick, M. Shupe,
D. Berg, C. Chandlee, S. Cihangir, T. Ferbel,
J. Huston, T. Jensen, F. Lobkowicz, M. McLaughlin,
T. Oshima, P. Slattery, P. Thompson, J. Biel,
A. Jonckheere, P.F. Koehler, C.A. Nelson 94
(Presented by K. Ruddick)

Dynamics of Light Hadrons - Mostly Glueballs
John F. Donoghue 104

April 25, First Afternoon Session -- Chairman: S. Ozaki

Review of $\bar{p}p$ Production Experiments at CERN
A. Ferrer 123

Search for Narrow $\bar{p}p$ States in 10 GeV/c $\pi^{+}p$
Interactions
D.R. Green 152

April 25, First Afternoon Session (continued)

Search for Narrow $\bar{p}p$ States in the Reaction
$\pi^- p \rightarrow p\pi^- \bar{p}p$ at 16 GeV/c
    S.U. Chung, A. Etkin, R. Fernow, K. Foley,
    J.H. Goldman, H. Kirk, J. Kopp, A. Lesnik,
    W. Love, T. Morris, S. Ozaki, E. Platner,
    S.D. Protopopescu, A. Saulys, D. Weygand,
    C.D. Wheeler, E. Willen, J. Bensinger,
    W. Morris, S.J. Lindenbaum, M.A. Kramer,
    Z. Bar Yam, J. Dowd, W. Kern, M. Winik,
    J. Button-Shafer, S. Dhar, R.L. Lichti 170
        (Presented by S. Protopopescu)

Review of $\bar{N}N$ Formation Experiments
    Gerald A. Smith.....................................186

Theories of Baryonium Exotics and Multiquark
Systems
    Anthony J.G. Hey 194

Search for the Narrow S-Resonance in a Measurement
of $\bar{p}p$ Total Cross Section Using Wire Chambers
    H. Aihara, J. Chiba, H. Fujii, T. Fujii,
    H. Iwasaki, T. Kamae, K. Nakamura, T. Sumiyoshi,
    Y. Takada, T. Takeda, M. Yamauchi, H. Fukuma,
    T. Takeshita 217
        (Presented by K. Nakamura)

April 25, Second Afternoon Session -- Chairman: K. Strauch

Charmed Mesons from e^+e^- Annihilation
    Gerson Goldhaber 223

D. Meson Production by Photons and Neutrinos
    James E. Wiss 257

Charm Production in Hadronic Interactions
    M.S. Witherell......................................285

Measurements of Charmed Meson Lifetimes
    J.D. Prentice 297

April 26, First Morning Session -- Chairman: J. Ballam

Radiative Transitions to an $\eta_c(2980)$ Candidate
State and the Observation of Hadronic Decays of
This State
    Elliott D. Bloom 312

April 26, First Morning Session (continued)

Radiative Transitions from the ψ(3095) to
Ordinary Hadrons
D.L. Scharre 329

New Results on the T-System from DORIS
H. Schröder 356

Results on the T Region from CLEO at CESR
C.J. Bebek 365

Results on the T Region from the Columbia-
Stony Brook Experiment at CESR
J. Lee-Fransini... 375

April 26, Second Morning Session -- Chairman: L. Trueman

Recent Theoretical Developments for Heavy
Quark-Antiquark Systems
E. Eichten 387

Flavor Mixing and Quark Decay
Ling-Lie Chau Wang 403

A Search for Narrow States Produced in the
Reaction $\pi^- p \to n + \gamma$'s at 13 GeV/c
I-H. Chiang, R.A. Johnson, B. Kwan,
T.F. Kycia, K.K. Li, L.S. Littenberg,
A. Wijangco, L.A. Garren, J.J. Thaler,
G.E. Hogan, K.T. McDonald, A.J.S. Smith................ 415
(Presented by L.S. Littenberg)

April 26, Afternoon Session -- Chairman: R. Weinstein

Recent Results of Mark J
The AACHEN/DESY/MIT/NIKHEF/PEKING
Collaboration.. 421
(Presented by M. Chen)

Recent Results on $\gamma\gamma$ and e^+e^- Annihilation
from the Pluto Experiment
Gerhard Knies 513

An Overview of QCD in Relation to e^+e^- Annihilation
and Brief View of Single Particle Spectra
Harvey L. Lynch 543

Recent Results from the JADE Collaboration at
PETRA on e^+e^- Annihilation to Multihadrons
L.H. O'Neill, Jr... 559

April 26, Afternoon Session (continued)

    D and E Mesons and Possible KKK Enhancements
        C. Bromberg, J. Dickey, G. Fox, R. Gomez,
        W. Kropac, J. Pine, S. Stampke, H. Haggerty,
        E. Malamud, R. Abrams, R. Delzenero,
        H. Goldberg, F. Lopez, S. Margulies,
        D. McLeod, J. Solomon, A. Dzierba,
        F. Fredericksen, R. Heinz, J. Krider,
        H. Martin, D. Petersen 583
          (Presented by Jean O. Dickey)

List of Participants 600

Complete Author Index 606

RESONANCES IN $\pi\pi\pi$ AND $K\pi\pi$ SYSTEMS

R. J. Cashmore
Nuclear Physics Laboratory, Keble Road, Oxford OX1 3RH, UK.

ABSTRACT

Recent data on $\pi\pi\pi$ and $K\pi\pi$ systems have led to substantial improvements in our knowledge of the light quark meson spectrum. In particular there is now unambiguous evidence for the Al, the H meson is observed and an almost consistent picture of the axial vector mesons exists. Only the D and E meson decays lead to possible difficulties. The π' and K' are observed in new experiments. The A3 and L_1 mesons are also observed so that the L=2 $q\bar{q}$ super multiplet is gradually being substantiated. Various other evidence exists for radial excitations - A3', Al' and Q', A2' - although there is clear need for confirmation.

1. INTRODUCTION

In this review I will deal with the meson resonances which can be and are observed in the $\pi\pi\pi$ and $K\pi\pi$ systems. Historically these systems have provided, almost exclusively, the information on unnatural spin parity resonances and have been vital in producing the current view of meson resonances. In order to guide us through the multitude of states I will discuss I want to introduce the spectrum associated with the $q\bar{q}$ model of mesons. This is not because I necessarily believe that this is the correct model for mesons, although it is very successful, but rather that it will give us a framework and allow any peculiarities to be conspicuous. One of the vital questions in meson spectroscopy is whether multiquark states or glueballs exist. These will lead to extra states over those predicted with the $q\bar{q}$ model and hence it is essential to identify whether or not a new state can be accommodated in the $q\bar{q}$ spectrum.

In Fig.1 the $q\bar{q}$ spectrum, assuming a SHO potential, is given together with the well identified states. The quantum numbers of the mesons are given by

$$\underline{J} = \underline{L} + \underline{S}$$

$$P = (-1)^{L+1}$$

$$C = (-1)^{L+S}$$

where L is the internal orbital $q\bar{q}$ angular momentum and S the $q\bar{q}$ spin state. In general there are 4 SU(3) nonets associated with each L value (except for the case where L=0). Because of the assumption of a SHO potential, there is a dynamical degeneracy resulting in mass values given by

$$M \; \alpha \; (2n+L)K = NK$$

ISSN:0094-243X/81/67001-36$1.50 Copyright 1981 American Institute of Physics

where n is the radial degree of excitation and K some constant. If a different potential is assumed then the same multiplet structures will exist but their masses will not be related in the same way (e.g. a Coulomb potential would lead to degeneracy of 2s and 2p states). Thus in studying the meson resonances, we should not expect to find new multiplets in exactly predicted positions. Indeed, the positions of the new states will provide a guide to the potential which exists (this is, of course, providing potential models are appropriate to give an adequate description of the states which is rather dubious if the quarks are relativistic).

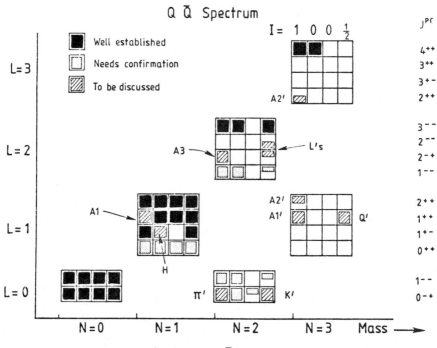

Fig.1 The q$\bar{\text{q}}$ spectrum

Many of the states of Fig. 1 are accessible to $\pi\pi\pi$ and $K\pi\pi$ systems and those I will discuss are indicated. In particular, I will concentrate on the following topics:

(i)   L=1 q$\bar{\text{q}}$ multiplet – A1, H and H', Q's

(ii) N=2, L=0 q$\bar{\text{q}}$ multiplet – radial excitations of the π and K i.e. the π' and K'

(iii) N=2, L=2 q$\bar{\text{q}}$ multiplet – K***(3$^-$), A3, L's and vector mesons

(iv) Higher mass q$\bar{\text{q}}$ multiplets – possible 1$^+$, 2$^-$ and 2$^+$ states

We will find that there has been a dramatic improvement in our understanding over the last few years due principally to the existence of new high statistics experiments. Only with correspondingly better experiments in the future will we be able to further refine our knowledge of the meson spectrum associated with the light quarks.

2. THE NEW EXPERIMENTAL INFORMATION

The high statistics data I will discuss have originated from the three experiments shown in Table I, where I also include the name by which I will refer to them.

Table I   New Experimental Data

Reaction	Momentum (GeV/c)	No of events	Ref.	Name
$\pi^- p \to \pi^- \pi^- \pi^+ p$	63 93	~600K	1	
$K^- p \to K^- \pi^- \pi^+ p$	63	~200K	2	ACCMOR
$K^- p \to K^- \pi^- \pi^+ \pi^\circ p$	63	~2K	2	WA3
$\to (K\omega) p$				
$\pi^- p \to \pi^+ \pi^- \pi^\circ n$	8.5	~50K	3	CEX
$K^- p \to \bar{K}^\circ \pi^+ \pi^- n$	6.0	~5K	4	MPS

All experiments are performed with sophisticated spectrometers implying that acceptance corrections have to be made. However, in general the acceptances are large so that little uncertainty is associated with them. The ACCMOR experiment measures diffractive production at the SPS (the K_ω reaction is included since it is the first high statistics measurement of this reaction and bears directly on the Q mesons), while the two charge exchange reactions were measured at Argonne and Brookhaven. I will concentrate mainly on the ACCMOR and CEX experiments as they provide the most illuminating results.

3. PARTIAL WAVE ANALYSES OF $\pi\pi\pi$ AND $K\pi\pi$ FINAL STATES

To extract any physics from these systems a partial wave analysis has to be performed. The results I will describe use the isobar model analyses based on the programs originating form SLAC/LBL[5] and Illinois[6]. The model and nomenclature are summarized in Figure 2. In performing the analyses, the experiments used the amplitudes summarized in Table II.

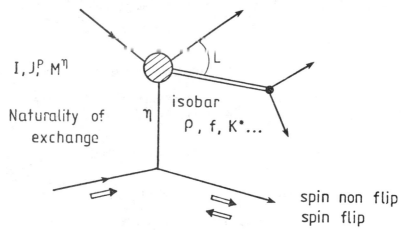

$$I, J^P_i M^\eta$$

Naturality of exchange

isobar

$\rho, f, K^*...$

spin non flip
spin flip

Fig.2 The Isobar Model

Table II   Amplitude Analyses

Experiment	Amplitudes	Total Number
ACCMOR	$\pi\pi\pi$; I=1 K$\pi\pi$: I=1/2 Natural Parity Exchange (η=+); Spin non flip	1
MPS	K$\pi\pi$: I=1/2 Natural and Unnatural Parity Exchange Spin non flip	2
CEX	$\pi\pi\pi$: I=0,1,2 Natural and Unnatural Parity Exchange; Spin flip and non flip	12

One of the major reasons for performing the charge exchange experiment is to avoid the Deck process inherent in diffractive reactions. However, the increase in complexity is enormous - one set of amplitudes is replaced by 12 sets - and it is quite an achievement to have obtained any results at all in this environment.

The type of result obtained is shown in Figures 3 and 4, where the prominent A1 and A2 features are observed in the diffractive reaction and the ω, A2 and ω(1670) are seen in the CEX. It is the study of this type of distribution and the relative phases between the various waves that will reveal the presence of any resonant structures.

4.   THE L=1 q$\bar{q}$ MULTIPLET

(i)   JPC = 2++

As can be seen from Figs.3 and 4 the A2 and, similarly, the K**(1430) are easily observed. ACCMOR(1),(2) measurements give the following parameters:

A2:   M=1317±2 MeV   Γ=96±9 MeV

K**(1430)   M=1432 MeV   Γ=110 MeV $\frac{\Gamma(K^*\pi)}{\Gamma(K\rho)}$=2.3

The $K^*\pi/K\rho$ ratio is consistent with SU(3) predictions.

The major importance of these waves is that they provide a "calibration" and a reference for the other waves.

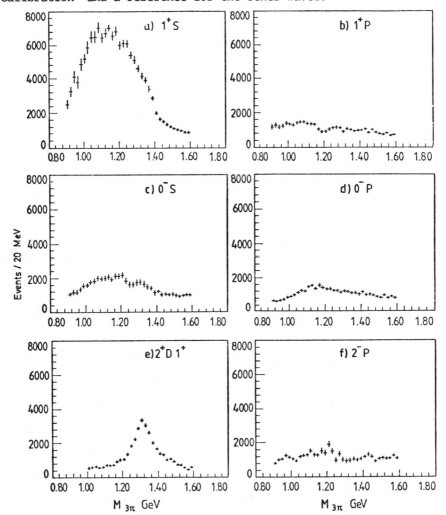

Fig.3   Individual partial wave contributions in the final state $\pi^-\pi^-\pi^+$ in $\pi^-p \to \pi^-\pi^-\pi^+p$ collisions at 63 and 94 Gev/c. The A_2 resonance and A1 enhancement are obvious.

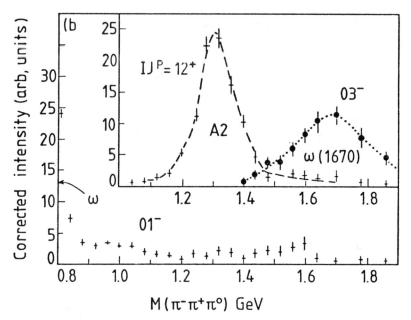

Fig.4 The I=0 J^P=1⁻, 3⁻ and I=1 J^P=2⁺ contributions in the final state $\pi^-\pi^+\pi^0$ from the reaction $\pi^-p \to \pi^+\pi^-\pi^0 n$ at 8.45 GeV/c. The ω(780), A_2(1300) and ω(1670) are clearly visible.

(ii) J^{PC} = 1⁺⁺, 1⁺⁻

Under these J^{PC} values I have to deal with the A1, H and H', the Q's and the question of whether a consistent picture of these states is present.

(a)   A1 Resonance J^{PC} = 1⁺⁺

Before looking at the new data, it is worth reminding ourselves of the recent history of the A1.

Historical survey: A number of 'observations' of the A1 exist in the literature, the most common being in diffractive reactions, although there are discussions of the A1 in backward production and τ decay.

Diffractive production: The production of the A1 resonance in reactions such as

$$\pi^-p \to \pi^-\pi^-\pi^+p$$

is confused by the presence of Deck processes of the type indicated in Figure 5b, which populate low 3π masses. Attempts by a variety of authors[7],[8],[9] to fit the 1⁺ $\pi\rho$ mass spectra and phases all lead to the same conclusions. It is impossible to fit with either a resonance alone (not enough phase variation in the data) or Deck alone (mass spectrum drops far too sharply in the data) and thus it is essential to include both resonant (Fig.5a) and Deck (Fig.5b) contributions and certainly a rescattering of the type shown in

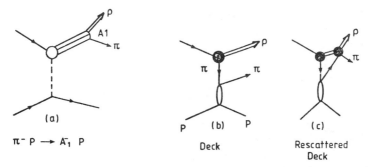

Fig.5   Resonance, Deck and Rescattering contributions to the
reaction $\pi^-p \to \pi^-\pi^-\pi^+p$

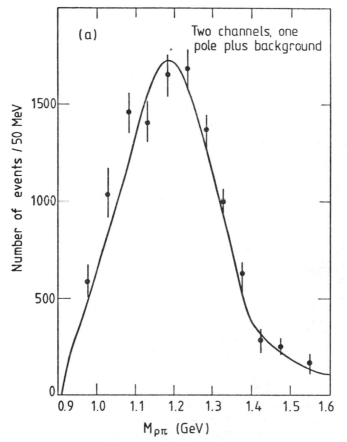

Fig.6   A fit (Ref.8) to the $J^P=1^+$ ($\pi^-\pi^-\pi^+$) spectrum
in 40 GeV/c $\pi^-p \to \pi^-\pi^-\pi^+p$

Fig.5c.   However, rescatering prescriptions differ and the Al
resonance parameters differ, depending on the exact quantity quoted
by the authors, e.g. K matrix pole, T matrix pole, Breit-Wigner mass

8

and width, second sheet pole. The fits are almost indistinguishable, an example being given in Fig.6, and there is unanimous agreement on both the existence of the A1 and the importance of rescattering. The A1 resonance parameters lie in the region

$$M: \quad 1185 \rightarrow 1450 \text{ MeV}$$

$$\Gamma: \quad 200 \rightarrow 300 \text{ MeV}$$

Backward production: The production of the A1 has been claimed[10] in reactions such as

$$K^-p \rightarrow \pi^+\pi^-\pi^+\Sigma^-$$

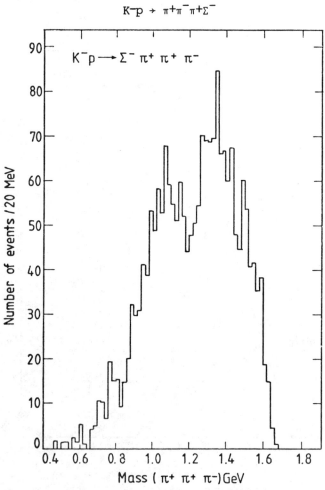

Fig.7 Backward production of the $\pi^-\pi^+\pi^+$ system in $K^-p \rightarrow \Sigma^-\pi^+\pi^+\pi^-$ at 4.2 GeV. The shoulder in the region of 1.08 GeV is suggested as evidence for a low mass A1.

A bump or shoulder is observed in the mass spectrum at ~1.04 GeV as shown in Fig.7, while the partial wave analysis indicates an increase in the $J^P=1^+$ wave in this vicinity. On this basis a resonance is claimed with a mass ~1.04 GeV.

In comparison with the diffractive reaction, the data is sparse at the very least and the partial wave analysis has to be performed in very large mass bins. It is then impossible to identify unambiguously a $J^P=1^+$ effects with this narrowish effect in the data. Furthermore, background amplitudes (e.g. of the multi Regge type) can and almost certainly will be present and it is then naive not to expect complicated interference effects of the sort present in diffractive processes. That a shoulder is present in the data is almost certainly true, but to immediately associate this with a pure Al resonance is not reasonable. An interference of background and a higher mass resonance could without doubt produce such a result.
Tau decay: The 3π decay of the τ lepton[11]

$$\tau^- \to \nu\pi^-\pi^-\pi^+$$

shown in Fig.8 is another potential source of information on the Al meson. Unfortunately, the statistics are even smaller (~40 events in total in any one experiment) so that the 3π mass spectrum provides little constraint on the Al mass and width. The spectrum is consistent with Al masses in the region 1100-1200 MeV and with a width of ~400 MeV, i.e. a value not inconsistent with those obtained from diffractive analyses. Only with the advent of much greater statistics will this reaction make any real contribution to the total Al picture.

In summary, even before the recent data, there was no doubt that the Al existed with a mass somewhere in the broad range 1180-1400 MeV and that it was necessary to make careful models of reactions in which it might be produced.

New ACCMOR and CEX data: I will first deal with the ACCMOR diffractive data[1]. In Fig.9 the dramatic variation of the $1^+\pi\rho$ s-wave is shown for different t selections. At high t the Deck process has effectively disappeared, revealing an almost pure Al resonance with a mass ~1280 MeV. To further substantiate this, the phases of the $J^P=2^+$ wave, the A2, relative to all others can be studied. In Fig.10 it is possible to see that the resonant A2 possesses too little phase variation against the 1^+S, whereas against the 0^-P wave it is just right. This suggests that the 1^+S wave is moving forward in phase in the A2 region and this can be demonstrated as in Fig.11a, where the (1^+S-2^+D) phase is plotted with the expected A2 Breit-Wigner phase variation subtracted out. This implies a broad Al resonance in the region 1200 to 1300 MeV. The next step is to make a quantitative fit to the data using the model of Fig.5 and the formulation of Bowler[7]. The fits are clearly an excellent description of the data and result in Al parameters

$$M \sim 1280 \quad 40 \text{ MeV}$$

$$\Gamma \sim 300 \quad 30 \text{ MeV}$$

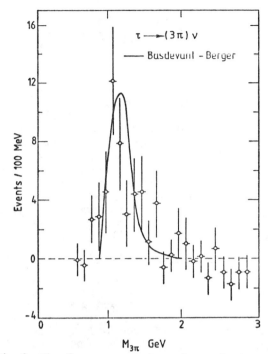

Fig.8 The 3π mass spectrum in $\tau \rightarrow \nu 3\pi$ together with a fit (Ref.11).

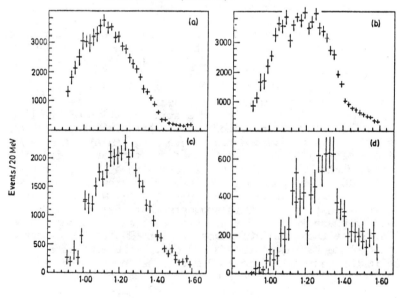

Fig.9 The variation of the 1^+ partial wave intensity as a function of t (a) $|t|<.05$ (b) $.05 <|t|<0.7$ (c) $.16<|t|<.3$ (d) $.3<|t|<.7$.

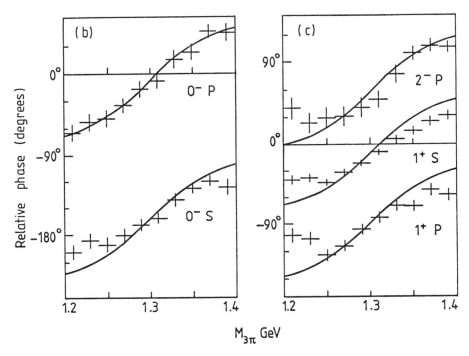

Fig.10 The phase of the JP=2+D wave measured with respect
to the other waves in the vicinity of M(3π)=1.30 GeV.

The CEX data pertinent to the A1 are shown in Figs.12 and 13,
where the intensities and phase variations of the natural parity
exchange spin flip amplitudes are plotted. The JPMn=1+1+ $_{\pi\rho}$ s–wave
shows a substantial bump together with a phase variation relative to
the I=2 wave, indicative of a resonance. Preliminary interpretation
might suggest a resonance with a mass of ∼1150 MeV which is broad.
However, if fits are made with the model of Bowler including a modest
background amplitude (as is almost certainly present), then much
higher mass values still reproduce the data well. An example of such
a fit with the A1 mass at ∼1200 MeV and a width ∼300 MeV is shown in
Fig.14.

Thus the compatibility in interpretations of the diffractive and
CEX data hinges on the presence of the background amplitudes. In
fact, the measured JP=1+ mass spectra are not grossly different, as
can be seen in Fig.15, and thus it is not surprising that consistent
views can be obtained. Of course, it is also possible to introduce a
further background phase into the $_{\pi\rho}$ scattering amplitude (e.g. from
some potential scattering or even more bizarre source such as a $q^2\bar{q}^2$
state so that the $_{\pi\rho}$ scattering is not entirely due to a resonance)
leading to good descriptions of both the diffractive and CEX data.

12

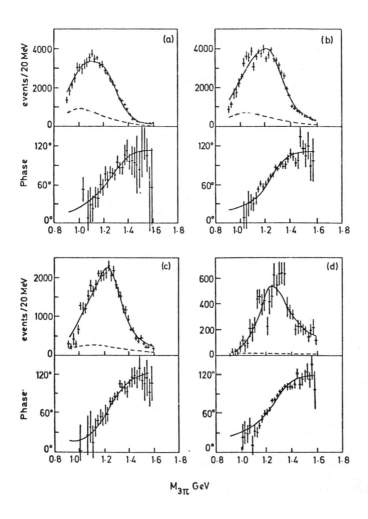

Fig.11 Fits to the $J^P=1^+S$ intensity and phase (measured with respect to the 2^+D production phase) as a function of t (see Fig.9).

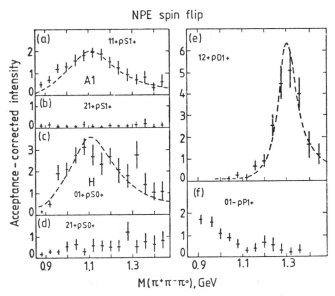

Fig.12 The natural parity exchange (NPE) spin flip
cross-section in $\pi^-p \to \pi^-\pi^+\pi^0 n$.

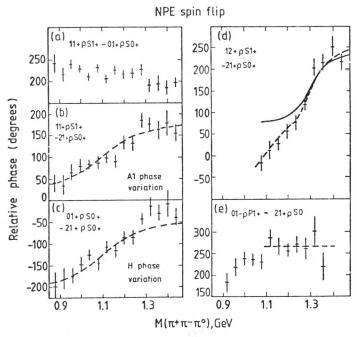

Fig.13 The phases of the natural parity exchange spin
flip amplitudes in $\pi^-p \to \pi^-\pi^+\pi^0 n$.

14

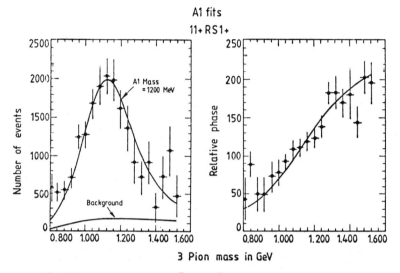

A1 fits

11+ RS1+

Number of events

A1 Mass = 1200 MeV

Background

Relative phase

3 Pion mass in GeV

Fig.14   Fits to the $J^P M \eta = 1^+ 1^+$ partial wave using resonance and background.

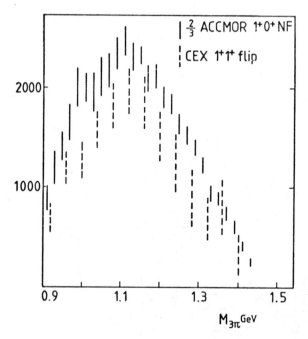

$|\frac{2}{3}$ ACCMOR 1^+0^+ NF

$|$ CEX 1^+1^+ flip

$M_{3\pi}$ GeV

Fig.15   A comparison of the diffractive $J^P M \eta = 1^+ 0^+$ (Ref.1) and charge exchange $J^P M \eta = 1^+ 1^+$ (Ref.3) partial wave intensities. The normalization is arbitrary.

The introduction of these various backgrounds, either in the production process (DECK etc.) or the $\pi\rho$ scattering, all lead to an increase in Al mass bringing it closer to the expected value from the ACCMOR high t data. It is the combination of the CEX and diffractive data which eventually leads to a mass in the range ∼1230-1270 MeV and a width ∼300 MeV.

Conclusions: All the evidence clearly points to the existence of the Al. However to extract its parameters requires a detail model including background amplitudes, rescattering and possibly a $\pi\rho$ background phase shift. When such prescriptions are used the resulting parameters are

$$M_{Al} \sim 1230\text{-}1270 \text{ MeV}$$

$$\Gamma_{Al} \sim 300 \text{ MeV}$$

These results are clearly compatible with the earlier diffractive experiments and τ decay. To understand whether there is any problem with the sparse backward production data would require a much higher statistics experiment allowing a more sophisticated analysis.

(b)   H and H' resonances   I=0 $J^{PC}=1^{+-}$
These states could both decay into the $(3\pi)^{\circ}$ final state. In Figs.12 and 13 the intensity and phase variations of the spin flip I=0 $J^P M \eta = 1^+0^+{}_{\pi\rho}$ s-wave are indicative of a resonance. This resonance is further supported by the properties of the spin non flip I=0 $1^+0^+{}_{\pi\rho}$ wave shown in Fig.16. There is a clear resonance like bump and when the A2 phase is measured relative to this partial wave there appears to be insufficient phase variation to accommodate the A2 Breit Wigner.
Thus it would appear that there is the first clear evidence for the H meson. If the bumps are literally interpreted then the following parameters for the H would be obtained

$$M_H \quad 1135 \pm 30 \text{ MeV}$$

$$\Gamma_H \quad 270 \pm 50 \text{ MeV}$$

However, from the experience of the analysis of the charged 3π data, if some backgrounds are introduced (they are expected from multi Regge diagram) then the mass of this state will almost certainly rise leading to a value closer to that of the Al. Furthermore a better fit will probably be obtained to the spin flip H intensity of Fig.12.
At present there is no clear evidence which suggests the existence of a second I=0 state, the H'.

16

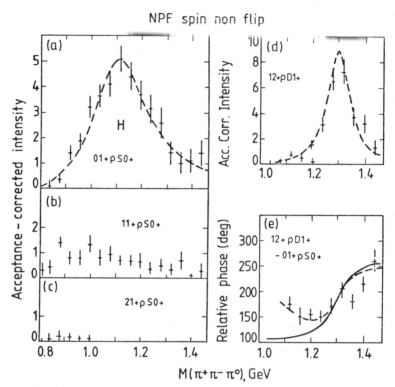

H Resonance

NPE spin non flip

Fig.16 The partial wave intensities and phases of
the natural parity exchange spin non flip
contributions in the reaction $\pi^- p \to \pi^- \pi^+ \pi^0 n$

(c) <u>The Q resonances I=1/2 S=±1</u>
     Before introducing the new data I would again like to review the
recent history of the Q mesons.

History: The Q_1 and Q_2 resonances were finally unambiguously
<u>identified</u>(12) at SLAC in the reactions

$$K^\pm \ p \ \to \ K \pm \pi^+ \pi^- p$$

in 1975. It was found that Q1 decayed mainly to K_ρ and Q_2 to πK^* the
intensities and phase variations of the partial waves reflecting
these properties. This means that the physical Q_1 and Q2 must be
mixtures of the Q_A and Q_B states, which are the strange members of
the Al and B octets, i.e.

$$Q_1 = \cos\theta \ Q_A + \sin\theta \ Q_B$$

$$Q_2 = -\sin\theta \ Q_A + \cos\theta \ Q_B$$

where the mixing angle θ lies in the range 45-55°.

These observations have subsequently been confirmed by many experiments.

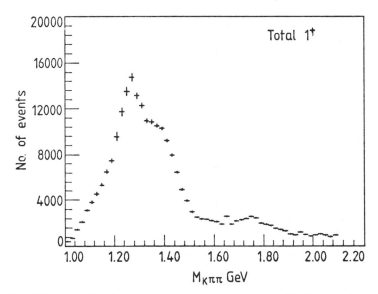

Fig.17 The JP=1+ intensity in the K⁻π⁻π+ final state in
K⁻p→K⁻π⁻π+p at 63 GeV/c (Ref.2).

New Data: Additional information has emerged from both the MPS experiment and more importantly the ACCMOR experiment.

The MPS results show that Q_2 is produced by natural parity exchange in the charge exchange reaction which corroborates the SLAC observation of a large C=-1 contribution in diffractive Q_2 production. This has always been a puzzle and remains one since the Q_2 only weakly couples to the Kρ system the ρ being the natural candidate for both this I=1 and C=-1 exchange.

Turning to the ACCMOR experiment the total 1+ intensity is shown in Fig.17. The peak and shoulder associated with Q_1 and Q_2 are clearly visible and perhaps there is a hint of some excess in the region of 1750 MeV which I will return to in a later section. The K*π and Kρ S waves have very similar properties to the SLAC experiment but a new feature, the K*π D-wave, is observed for the first time. The intensity and phase (relative to the 1+0+πK*S wave) of the 1+0+ πK* D-wave are shown in Fig.18. The peak occurs at ∼1.30 GeV and there is a very dramatic phase variation. This observation is reassuring since the B meson is observed to have a πω D wave decay implying that Q_B should also have a πK* D wave decay. (The πρD wave decay of the A1 is not observed indicating that Q_A will have no πK* D wave decay). If Q_1 and Q_2 are approximately equal mixtures (θ∼45°) then they should each have a πK* D wave decay leading to yet a different interference pattern in this wave in the Q region. The curves of Fig.18 correspond to such a model and give excellent agreement with the data.

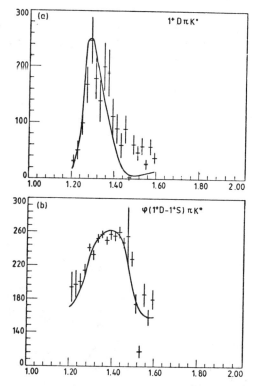

Fig.18   The 1^+0^+ $_\pi K^*D$ intensity and phase (with respect
to the dominant 1^+0^+ $_\pi K^*S$ wave) in $K^-_{\pi}{}^-_{\pi}{}^+$

Fitting the Q data ($_\pi K^*$, K_ρ, K_ϵ, $_{\pi\kappa}$, $(_\pi K^*)_D$ partial waves in
both M=0 and M=1) with a model including rescattering (of the type of
Fig.5) leads to the resonance parameters of Table III, both Q's being
absolutely essential for successful fits.   These parameters are
essentially independent of the model used.   It is interesting to note

Table III   Q resonances

Resonance	M (GeV)	Γ (GeV)	Mixing Angle
Q_1	1.27 ± .01	0.08 ± .01	$\theta=55-65°$
Q_2	1.41 ± .01	0.20 ± .03	

that these data lead to a smaller Q_1 width, closer in value to that
quoted for the Q_1 meson[13] observed in $\bar{p}p$ annihilations.
The other new data on the Q's comes from the ACCMOR observations
of the K_ω final state, a high resolution shower counter having been
introduce into the detector to measure the γ rays.   The performance
of the spectrometer in identifying the ω is summarized in Fig.19
where a clear K_ω signal can be easily identified.   In Fig.20 the

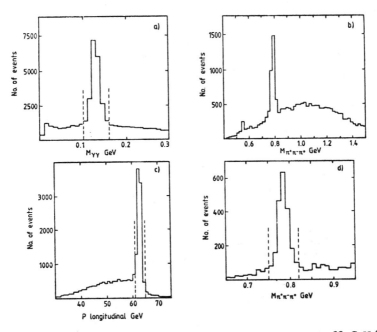

Fig.19 The selection of the reaction $K^-p \to K^-\omega^0 p$ at 63 GeV/c.

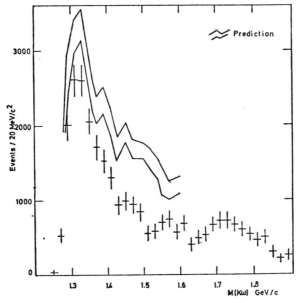

Fig.20 A comparison of the $K\rho$ and $K\omega$ spectrum in the region
of 1.3 GeV. The shapes are clearly very similar indicating
the presence of a large $Q_1 \to K\omega$ decay.

corrected $K\omega$ mass spectrum is shown together with the predicted spectrum assuming that $K\omega$ is equivalent to K_ρ^0 (the $K\omega$ angular distributions are consistent with dominant $J^P=1^+$ partial waves). Clearly the assumption is good although the relative normalization is not yet perfect(?). Since the π_β spectrum is dominated by the Q_1 this implies a much larger Q_1 to $K\omega$ coupling than Q_2 to $K\omega$.

Conclusions: The Q_1 and Q_2 are well established resonances and the recent diffractive data not only confirms their presence and the interpretation but also brings consistency between different observations. There is little change in production mechanism in moving from the SLAC energy of 13 GeV/c to the ACCMOR energy of 63 GeV/c.

(d)   The Axial Vector Mesons – A consistent picture?
        The only missing member of the 1^+ nonets is now the H' and it is appropriate to question whether all of the information on these states is consistent. The only sensible way of doing this is examine the SU(3) relations that should exist and emphasize any problems that occur. The following comments are very preliminary and will, I am sure, change in detail with more careful analysis.

Q_1,Q_2, A and B;   The masses, coupling constants and mixing angle for the Q's can be used to predict the widths of the A_1 and B, and the D/S ratio in $B_{\to\pi\omega}$ decay. These results are summarized in Table IV and clearly represent a considerable success.

Table IV   Predictions from the Q_1,Q_2 system

Quantity	Prediction	Measurement
$\Gamma(A_1 \to \rho\pi)$	270 ± 45 MeV	300 ± 30 MeV
$\Gamma(B \to \omega\pi)$	137 ± 30 MeV	128 ± 10 MeV
(D/S) $B_{\to\pi\omega}$	.40 ± .10	.29 ± .05

B nonet: From the Q_B and B mass values it is possible to predict the mass of the unmixed octet member H_8

$$M_{H_8} \sim 1.40 \text{ GeV}$$

If the H observed in the CEX experiment has a mass of 1.23 GeV (more or less degenerate with the B) and is magically mixed (i.e. same quark content as the B) then it is possible to obtain the result that

$$M_H = 1.23 \text{ GeV} \qquad \Gamma(H \to \rho\pi) \sim 360 \text{ MeV}$$

$$M_{H'} = 1.48 \text{ GeV} \qquad \Gamma(H' \to K^*\bar{K}) \sim 165 \text{ MeV}$$

where the H' being composed only of strange quarks automatically decays to $\bar{K}K^*$ final states. The width of the H is certainly not inconsistent with the observed properties. Clearly observation of the H' would finally tie down the properties of this nonet.

A nonet: Using the Q_A and A_1 masses it is possible to calculate the mixing angle leading to the observed D and E mesons

$$|D\rangle = \cos_\theta |8\rangle + \sin_\theta |1\rangle$$

$$|E\rangle = -\sin_\theta |8\rangle + \cos_\theta |1\rangle$$

The result is a value of $\theta = 52° \pm 15°$ which does not correspond to magic mixing ($\theta = 37°$). This then allows a calculation of the E decay into $\bar{K}K^*$ resulting in

$$\Gamma \ (E \rightarrow K^*\bar{K} + K\bar{K}^*) \sim 48 \pm 10 \ \text{MeV} \quad \theta = 52°$$

to be compared with the result in the case of 'magic mixing' of

$$\Gamma \ (E \rightarrow K^*\bar{K} + K\bar{K}^*) \sim 85 \pm 15 \ \text{MeV} \quad \text{'Magic mixing'} \ \theta = 37°$$

Recent measurements[14] of the E width give a value

$$\Gamma \text{total}(E) \lesssim 60 \ \text{MeV}$$

At first sight this may be reassuring. However the Q_1 is observed to have a substantial $\kappa\pi$ decay mode which can be used to predict the $D \rightarrow \delta\pi$ and $E \rightarrow \delta\pi$ decays giving contributions to the D and E total widths. From the Q data it is found that

$$\Gamma(D \rightarrow \delta\pi) + \Gamma(E \rightarrow \delta\pi) \gtrsim 200 \ \text{MeV}$$

and for $\theta = 52°$ $\Gamma(E \rightarrow \delta\pi) \simeq 45$ MeV (assuming the δ is a $q\bar{q}$ state. If the δ is a $q^2\bar{q}^2$ state this value is probably even larger). This clearly exceeds the E total width and agreement for the E can only be obtained by invoking a larger mixing angle ($\theta \sim 70°$). However a large width for the D results. We are then left with a situation in which the D is almost purely octet and E purely singlet and a width for the D which exceeds it observed small total width. Thus there appear to be some difficulties, if one accepts the $Q_1 \rightarrow \pi\kappa$ decay where the κ itself is a rather peculiar object. Observations of the E meson, but not the D meson, in radiative decays of the ψ[15] may be related to these problems. There is still clearly confusion over the D and E properties, particularly the decays.

Conclusions: In general there is a pleasing SU(3) picture of the axial vector mesons with only the H' remaining to be found. There are however some problems in the D and E decays and this could be an indication of some new physics in this region e.g. the possibility of glue balls.

(iii) $J^{PC} = 0^{++}$
The 0^+ states are not accessible to the $\pi\pi\pi$ and $K\pi\pi$ systems.

(iv) Conclusions
     The L=1 multiplet now appears to be complete except for the H'
state. Some doubt still remains over the scalar mesons and the
decays of the D and E. I have spent a substantial amount of time on
the A₁ and its other axial vector meson counterparts. There should
now be little doubt as to its existence and the general state of
these resonances.

5.   THE L=0 $q\bar{q}$ RADIAL EXCITATIONS

(i) $JPC = 1^{--}$
     The only new data with any bearing on these states comes from
the MPS experiment. In Fig.21 I show the 1^- $_\pi K^*$ and K_ρ waves. The
analysis clearly indicates some structure in the intensities and a
forward phase variation which the authors would like to associate
with two vector mesons at ~1450 MeV and ~1800 MeV. However the data
is rather sparse and the acceptance is certainly becoming poor at the
high masses. The evidence is suggestive but not compelling and
should provide the encouragement to repeat the experiment with much
higher statistics.
     It is important to resolve this question and see if either of
these states can be associated with the possible 1^- K* in K$_\pi$
scattering in the vicinity of 1650 MeV. Two vector states are
required in this mass range, one to be associated with the L=0 $q\bar{q}$
radial excitation and the other with the L=2 $q\bar{q}$ system.

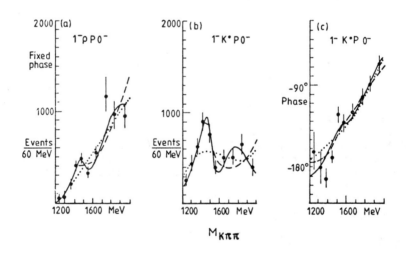

Fig.21   The JP=1- K$\pi\pi$ waves in the reaction
         K⁻p→K̄⁰ π⁺π⁻n (Ref.4).

(ii) $\underline{J^{PC} = 0^{-+}}$ π' and K'

π': In Fig.22 are the 0^- $\pi\varepsilon$ and 0^- $\pi\rho$ partial waves from the ACCMOR diffractive data. The phases are measured relative to the A$_2$ production phase i.e. with the A2 Breit Wigner phase variation subtracted from the A2 total phase. Both waves suggest the existence of a broad resonance which was already apparent in the 0^- phases in Fig.10. To extract parameters clearly requires a model similar to that used for the Al. Such fits(16) including a π', Deck and rescattering lead to the following resonance parameters

$$M \sim 1273$$

$$\Gamma_{tot} \sim 510 \qquad \Gamma(\rho\pi) \sim 165 \text{ MeV} \qquad \Gamma(\varepsilon\pi) \sim 345 \text{ MeV}$$

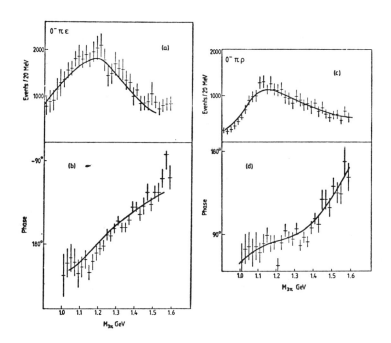

Fig.22   The $J^P=0^-$ waves in the $\pi^-\pi^-\pi^+$ final state. The curves are due to Longacre.

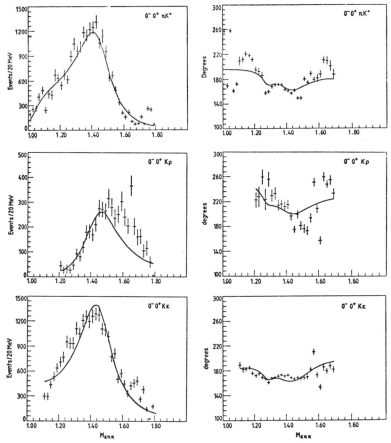

Fig.23 The $J^P=0^-$ waves in the $K^-\pi^-\pi^+$ final state (Ref.2).

<u>K'</u>: The existence of a K' was suggested in the SLAC $K\pi\pi$ experiment(17). In Fig.23 the 0^- πK^*, $K\rho$ and $K\epsilon$ waves derived from the ACCMOR data are presented, the phases being measured relative to the $1^+0^+\pi K^*$ wave which will itself possess substantial forward phase variation due to the presence of the Q_2 resonance. This approximately constant relative phase is thus indicative of a substantial forward phase variation in the πK^* and $K\epsilon$ waves. Furthermore a rapid drop in intensity is observed in the πK^* and ϵK waves at masses ~1500 MeV, reminiscent of the type of structure observed in 1^+ waves (Al). It is impossible to obtain any representation of this data using Deck amplitudes alone - the presence of a resonance is essential. Detail fits, as for the π', indicate the following parameters

$$M \sim 1450 \text{ MeV} \qquad \Gamma \sim 250 \text{ MeV}$$

These parameter values, particularly the width, should not be taken too seriously as the fits are far from good and some variation will clearly be possible.

Conclusions: In both the 3_π and $K_{\pi\pi}$ systems the 0^- waves qualitatively indicate the presence of resonances. It is impossible to describe these waves with Deck amplitudes alone and one is thus forced to a quantitative need for the π' and K'. Both resonances are clearly broad with masses

$$M_{\pi'} \sim 1270 \qquad M_{K'} \sim 1450$$

The observation of these states is important since they can only be assigned to a radial excitation in the $q\bar{q}$ model.

(iii) Conclusions
The L=0 $q\bar{q}$ radial excitation is nearly complete providing one accepts the variety of vector meson states that are reported. The 0^- π' and K' together with the 0^- $_{\eta\pi\pi}$ state at ~1260 MeV[18] provide the best evidence for this multiplet.

6.   THE L = 2 $q\bar{q}$ MULTIPLET

(i) $J^{PC} = 3^{--}$
The states of this multiplet have long been conspicuous. The $\omega(1670)$ can be seen in the CEX reaction partial waves of Fig.4 and the K*** (~1780) has been observed by the MPS experiment in the 3^- K_ρ f wave right at the end of their acceptance. The only missing member of this multiplet is the ϕ object which will presumably appear in analyses of $K\bar{K}\pi$ final states.

(ii) $J^{PC} = 2^{-+}, 2^{--}$
Whereas only the I=1 2^{-+} state can decay into 3_π both the strange 2^{-+} and the 2^{--} can appear in $K_{\pi\pi}$. Bumps have been observed in the 3_π mass spectrum at ~1650-1700 MeV - the A3 region - and in the $K_{\pi\pi}$ mass spectrum at ~1700-1800 MeV - the L region. The total intensity in the A3 region from the ACCMOR experiment is shown in Fig.24 and the enhancement clearly visible. In the past there have been conflicting results on the A3 and no reliable analyses (or data) have existed in the L region. The only new data bearing on these questions comes from the ACCMOR experiment.

(a)   A3 resonance $J^{PC} = 2^{-+}$
The intensity variation of the individual 2^- waves are shown in Fig.25 and the phase variation of the $2^- Sf_\pi$ wave in Fig.26. The same peak occurs in three different channels with the same mass and width, unmistakable evidence for a resonance. Further confirmation is provided by the $2^- Sf_\pi$ phase variation and the approximate constancy of the $2^-\rho\pi$ and $\epsilon\pi$ waves with respect to the f_π wave i.e. they are executing a similar forward resonant phase variation. A simple fit to the data of Fig.25 results in the following parameters

$$M \sim 1670 \text{ MeV} \quad \Gamma_{tot} \sim 220 \text{ MeV} \qquad f\pi{:}\rho\pi{:}\epsilon\pi = .57{:}.32{:}.11$$

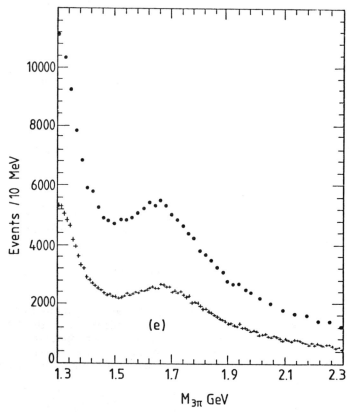

Fig.24 The $\pi^-\pi^-\pi^+$ mass spectrum in the region of 1.6 GeV
showing clearly the A3 enhancement.

However this cannot be the complete story since the 2^-Df_π wave has very different properties peaking at ∼1850 MeV (Fig.25) and a rapid forward phase variation (with respect to the S wave f_π) in this vicinity (Fig.26). These are the hallmarks of a second 2^- resonance and it is clearly necessary to include two 2^- states if an adequate description of these waves is to be obtained. This second resonance will then lead to interference effects in all the 2^- channels and a fit with such a model is shown in Fig.26. The features are clearly reproduced. The parameters from such a fit are shown in Table V, the

Table V   Resonance Parameters

Resonance	M	Γ_{tot}	$\Gamma(f\pi)s$	$\Gamma(f\pi)d$	$\Gamma(\rho\pi)p$	$\Gamma(\varepsilon\pi)d$
1	1710	312	154	35	92	31
	±20	±50	±30	±15	±10	±15
2	2100	650	168	295	123	65
	±150	±80	±40	±30	±30	±35

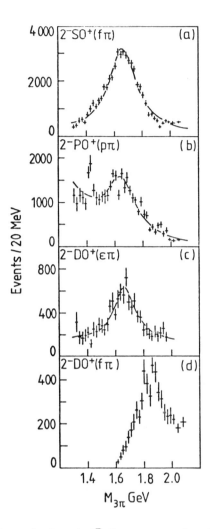

Fig.25 The four dominant $J^P=2^-$ waves, clearly showing
the presence of the A3 in at least three channels.

second resonance being very broad and much more poorly determined
than the first.

Conclusions: The A3 is confirmed as a resonance with

$$M \sim 1710 \pm 20 \text{ MeV} \qquad \Gamma \sim 312 \pm 50 \text{ MeV}$$

A second 2^- resonance, the A3', is almost certainly present with

$$M \sim 2100 \pm 150 \text{ MeV} \qquad \Gamma \sim 650 \pm 80 \text{ MeV}$$

28

This second 2⁻ state can only be ascribed to a radial excitation of the L=2 q̄q multiplet and represents yet further evidence for the existence of radial excitations in light quark systems. Finally it is amusing to note that the A3 and A3' are the first 2⁻ mesons to be identified.

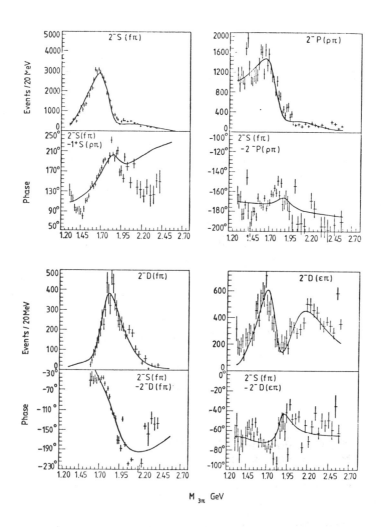

Fig.26 Fits to the intensity and phases of 2⁻ partial waves. The phase of the 2⁻πf wave is measured relative to the 1⁺0⁺πρ wave, while the phases of all other waves are measured relative to the dominant 2⁻πf.

(b) <u>The L resonances</u> $J^{PC} = 2^{-+}, 2^{--}$

In Fig.27 I show the ACCMOR 2^- partial waves. A clear large
enhancement is seen in the $_\pi K^{**}$ s-wave (similar to the $_\pi f$ wave in the
A3), with a corresponding phase variation against the $1^+0^+K_\rho$ wave
indicating the presence of a resonance. A rapid turn on and peak is
also observed in the 2^- Kf wave with some hint of phase variation

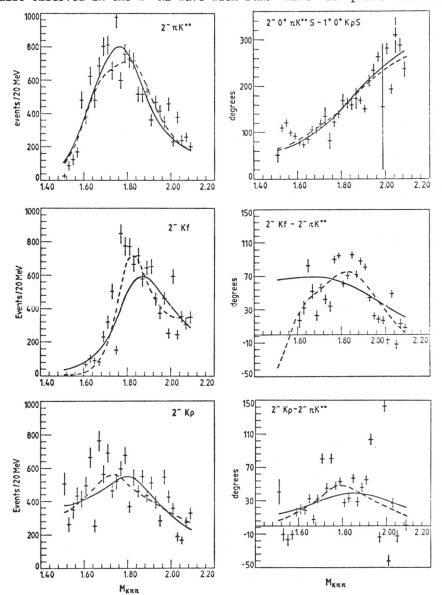

Fig.27 The $J^P=2-$ $K^-_\pi{}^+_\pi{}^-$ waves, with fits containing one
(solid line) and two (dashed line) L resonances.

with respect to 2^- $_\pi K^{**}$ wave. This data requires at least one resonance, the L_1, but there is no clear evidence which forces the need for a second L_2, even with this quality of data. From a one resonance fit the parameters obtained are

$$M \sim 1820 \text{ MeV} \qquad \Gamma \sim 200 \text{ MeV} \qquad _\pi K^{**}:Kf:_\pi K^*=0.6:0.16:0.24$$

Conclusions: There is without doubt one L resonance. However very high statistics experiments, much greater than ACCMOR, will be needed to reliably reveal a second. Even then the situation may be even more complicated because the A3' would imply the existence of a further two 2^- strange mesons, both of which could be present.

(iii) $JPC = 1^{--}$
These states have been alluded to in the discussion of the L=0 $q\bar{q}$ radial excitations. The I=1, ρ like objects do not couple to the $\pi\pi\pi$ system (G parity) and there is no substantial information on the I=0 ω states other than from e^+e^- annihilation where the situation is somewhat confusing.

(iv) Conclusions
The L=2 $q\bar{q}$ multiplet is clearly becoming more populated. The 3^{--} states are well represented, except for the ϕ like object. There is clearly a resonant A3 (2^{-+}) and at least on L meson (2^-). However there is no sign yet of the I=0 2^{--} state and there remain many gaps to be filled.

7. OTHER MULTIPLETS AND STATES

In this section I want to gather together a variety of comments and some effects which point to further resonant states but which do not have the compelling statistical significance of the resonances of the previous sections.

(i) L=3 $q\bar{q}$ multiplet
An enhancement has been claimed[19] in the $_\pi g$ s-wave at ~ 2.2 GeV. No evidence for such a state in this vicinity exists in the ACCMOR data.

(ii) L=1 $q\bar{q}$ radial excitations
Evidence for such states appears in both 3_π and $K_{\pi\pi}$ data.

(a) 1^+ states
In Fig.28 is the $1^+_\pi \rho D$ wave mass spectrum from the 3_π data. A peak is seen in the vicinity of ~ 1650 MeV and is comparatively narrow.
I have already remarked on the apparent bump in the total 1^+ $K_{\pi\pi}$ system around 1750 MeV shown in Fig.17. This peak is associated with the $K\epsilon$ P wave and the $_\pi K^*$ D wave, the latter demonstrated in Fig.29.
Thus there is evidence for further 1^+ states in both S=0 and S=1 systems which could only find positions as radial excitations of the L=1 $q\bar{q}$ system i.e. the A1' and the Q'.

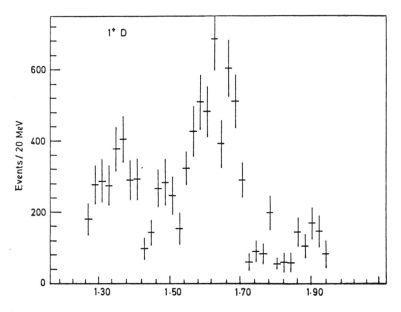

Fig.28   The 1⁺ πρD wave in π⁻π⁻π⁺ data exhibiting an
enhancement at the vicinity of 1.60 GeV.

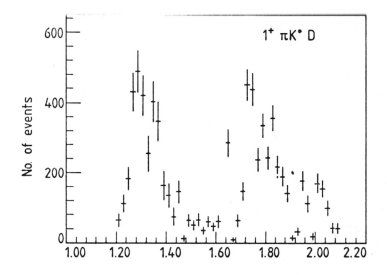

Fig.29   The 1⁺D πK* wave. Peaks in the Q region
and around 1700 MeV are obvious.

(b) <u>2+ state</u>

In Fig.30 the high mass 2+ waves in $\pi\pi\pi$ are shown. There is evidence for activity in the 2+ $_\pi$f P wave around 1700 MeV accompanied by very weak structure in the 2! $_\pi\rho$ D wave. Similar observations have been reported previously(20).

If such a state exists then it could be alternatively assigned to an L=3 $q\bar{q}$ multiplet.

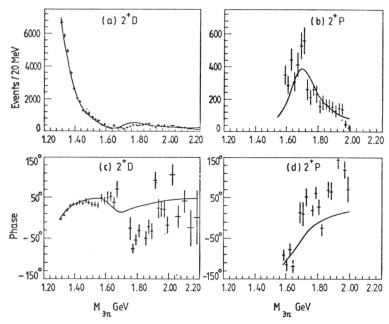

Fig.30 The JP=2+ waves in $\pi^-\pi^-\pi^+$ at high masses.

(iii) <u>L=2 $q\bar{q}$ radial excitation</u>

The presence of an A3', discussed in section 6, is quite convincing. It has a mass of ~2100 MeV and would have to be a $q\bar{q}$ radial excitation. Its existence implies a potentially highly confused L region with as many as four 2⁻ resonances possibly being present.

(iv) <u>Conclusions</u>

There begin to appear a number of possible radial excitations of higher lying $q\bar{q}$ multiplets. Future experiments must focus on this region for any further progress to be made in light quark $q\bar{q}$ spectroscopy.

8. SUMMARY AND CONCLUSIONS

The evidence I have presented clearly demonstrates that the L=0 and L=1 $q\bar{q}$ systems are well 'understood'. The only remaining questions are the identification of the H' and the resolution of the difficulties surrounding the D and E meson widths. This resolution may however reveal some exciting new physics, perhaps in the shape of glue balls.

In the case of the L=0 q$\bar{q}$ radial excitations the π' and K' are now fairly convincing andthe effort must centre on tidying up the vector meson situation.

At the L=2 q$\bar{q}$ level many states are missing even though there is now solid evidence for the A3 and L$_1$ mesons. This means a lot of hard work in the future.

Focussing on the L=2 q$\bar{q}$ region automatically implies that the L=1 and L=2 radial excitations can be investigated. The states I have mentioned need corroboration and there are plenty more to find. The pattern now appears to place these radial excitations at lower masses than suggested by a SHO potential.

Without doubt future experiments must accent the mass region in 3_π and K$_{\pi\pi}$ above ~1600 MeV. If the statistics acquired exceed those of current experiments by a factor of ~10 then we can look forward to a similar improvement in our understanding of this high mass region to that produced at lower masses by the current sets of experiments.

ACKNOWLEDGEMENTS

In preparing this review I would like to acknowledge the help I have received from the members of all three collaborations cited in References 1, 3 and 4. I have particularly benefitted from discussions with M.G.Bowler, J.Dankowych and R.S.Longacre.

REFERENCES

1.   The ACCMOR Collaboration
      C.Daum, L.Hertzberger, W.Hoogland, S.Peters and P.Van Deurzen
      V.Chabaud, A.Gonzalez-Arroyo, B.Hyams, H.Tiecke and P.Weilhammer
      A.Dwurazny, G.Polok, M.Rozanska, M.Turala and J.Turnau
      H.Becker, G.Blanar, M.Cerrada, H.Dietl, J.Gallivan, M.Glaubmann,
      R.Klanner, E.Lorenz, G.Lutjens, G.Lutz, W.Manner and U.Stierlin
      I.Blakey, M.Bowler, R.Cashmore, J.Loken, W.Spalding and
      G.Thompson
      B.Alper, C.Damerell, A.Gillman, C.Hardwick, M.Hotchkiss and
      F.Wickens
      PL 89B, (1980) 27b; PL 89B (1980) 281; PL 89B (1980) 285 and
      articles in preparation.
2.   The ACCMOR Collaboration - article in preparation.
3.   J.A.Dankowych, J.F.Martin, A.J.Pawlicki, J.D.Prentice, T.S.Yoon,
      R.S.Longacre, K.W.Edwards, D.Lesacey, N.R.Stanton, P.Brockman,
      J.Gandsman, P.M.Patel, E.Shabazian and C.Zanfino, Private
      Communication and Articles in Preparation.
4.   A.Etkin, K.J.Foley, J.H.Goldman, R.S.Longacre, W.A.Love,
      T.W.Morris, S.Ozaki, E.D.Platner, A.C.Saulys, C.D.Wheeler,
      E.H.Willen, S.J.Lindenbaum, M.A.Kramer, U.Mallik, Brookhaven
      Preprint BNL-27267.
5.   D.J.Herndon et al PR D11 (1975) 3165.
6.   J.D.Hansen et al, N.P.B81 (1974) 403.
7.   I.Aitchison and M.Bowler, J.Phys. G3 (1977) 1503;
      M.Bowler et al, N.P.B97 (1975), 227;

M.Bowler, J.Phys. G5 (1979) 203.

8.   J.Basdevant and E.Berger, P.R.D16 (1977) 657.

9.   R.Aaron et al, PRL 38 (1977) 1509.

10. P.Gavillet et al, PL 69B (1977) 119.

11. G.Alexander et al, PL 73B (1978) 99;
     J.A.Jaros et al, PRL 40 (1978) 1120;
     J.L.Basdevant et al, PRL 40 (1978) 994.

12. G.W.Brandenberg et al, PRL 36 (1976) 703; PRL 36 (1976) 706;
     NP B127 (1977) 509.

13.   R.Armenteros et al, PL 9 (1964) 207;
     P.Gavillet et al, PL 76B (1978) 517;
     D.J.Crennel et al, PR D6 (1972) 1220.

14. C.Dionsi et al, CERN/EP 80-1.

15. D.Scharre, Review at this conference.

16. R.Longacre, Private Communication.

17. G.W.Brandenburg et al, PRL 36 (1976) 1239.

18. N.P.Stanton et al, PRL 42 (1979) 346.

19. C.Baltay et al, PRL 39 (1977) 591.

20. Yu M.Antipov et al, NP B63 (1973) 153.

DISCUSSION

S.D.PROTOPOPESCU (Brookhaven) - Why should the 1^+ intensity in CEX reaction look similar to the 1^+ diffractive intensity at low t? Particularly since the CEX reaction is not expected to have Deck background?

R.J.CASHMORE - There is no good reason why the $J^{PM\eta}=1^+1^+$ CEX reaction should look like the $J^{PM\eta}=1^+0^+$ diffractive production. There are two points I want to emphasize. First; the partial wave amplitudes, from which the deductions about mass and width are made, are very similar and hence the different Al parameters (which might be concluded) are due solely to the method of analysis. Secondly; in diffractive production we know that it is impossble to explain the $J^P=1^+$ 3π spectrum without including Deck and resonance contributions, even though the Deck contributions are comparatively small (as I showed in our analysis). Hence when studying other reactions one should be alert to the presence of background amplitudes and the effects that they might introduce. In the case of CEX such backgrounds will occur from, for example, multi-Regge diagrams with shapes which are not dissimilar to the Deck diffractive amplitude. Therefore care must be taken and not too much emphasis placed on naive interpretations.

T.FERBEL (Rochester) - What parameterization did you use for the A_2 meson.

R.J.CASHMORE - A straightforward relativistic Breit Wigner with a D-wave barrier à la Blatt and Weisskopf together with a small % of incoherent background. The narrow width is due to a careful job of unfolding the experimental resolution.

P.WEILHAMMER (CERN) - It seems amazing that the CEX data in the A_2 region are almost identical to those of the WA3 experiment in diffraction production at low t. In diffraction production the shape of the 1^+S spectrum in any given t interval is well understood by a superposition of Deck amplitudes, Deck plus rescattering through a A_1 resonance and direct production of the A_1. This Deck amplitude is however absent in CEX reaction and makes it difficult to understand this coincidence.

R.J.CASHMORE - As I mentioned before there are backgrounds in the CEX, particularly as it is at much lower energies. I agree that such close similarity is surprising although perhaps it is not so wild if it is more a property of the $\pi\rho$ phase shift (from for example rescattering) than the production mechanism. In the case of our WA3 data in the absence of rescattering the Deck Amplitude itself is quite small.

G.KNIES (DESY) - The PLUTO data is far superior to the data you quoted and can only be fitted with an Al resonance of mass 1200 MeV and width 400 MeV.

R.J.CASHMORE - I showed the SPEAR data because it was to hand. However I would like to emphasize that in dealing with the τ decays it is absolutely necessary to properly include a weak interaction matrix element as, for example, Basdevant and Berger did. Furthermore even the PLUTO data of 40 events cannot claim to be overwhelming.

R.LONGACRE (BNL) - Could you make a comment about the 1^- amplitudes seen in the K^*_π from the MPS experiment.

R.J.CASHMORE - I didn't have time to cover the MPS 1^- results in the talk. I had hoped to, but the chairman appeared to be rather fidgety. I will include some comments in the proceedings. The MPS observe interesting hints of activity in the 1^- $_\pi K^*$ wave around 1400-1600 MeV. This is, of course important since it could be identified with a vector meson from either the L=0 $Q\bar{Q}$ radial excitation or the L=2 $Q\bar{Q}$ system. However the data is rather sparse, only $\sim$500 events, and I feel that the analysis is being pushed as hard as possible. Improving the statistics by a factor of 10 would lead to a very valuable experiment and perhaps some positive statement about strange vector mesons in this region.

R.HEMINGWAY (Carleton) - I don't agree with the flippancy with which you dismissed the 3π 1^+ data from backward production and τ-decay. They represent important data that are not consistent with your preferred parameters from the ACCMOR analysis.

Would you like to comment on the reliability of resonance parameters in the higher waves, especially the radial excitations. You have shown for the A_1 that mass and width values can move very substantially depending on the background parameterisation - so what hope do we have at higher masses where the backgrounds become almost impossble to handle?

R.J.CASHMORE - I expected somebody not to agree with my evaluation of the backward 1^+ 3π system from K⁻p and the τ decay 3π systems. Let me first make the comment that I don't dispute the measurements but I certainly feel the analyses are, in general, of the naive variety. In the case of backward 3π production the evidence rests on a shoulder in the 3π mass spectrum around 1.04 to 1.08 GeV. The partial wave analysis shows a broad 1^+ structure which jumps up in this mass range. Now at these low energies it seems very naive to believe that there are no background amplitudes and hence correspondingly naive to think that a shoulder gives you a resonance mass. Of course the statistical significance of the data is such that it would not bear a more 'complete' analysis. We can only regret that there is not a factor 10 greater statistics. In the case of τ's the 3_π data is sparse, only $\sim$40-50 events, and many of the same comments apply. Turning to the rest of your comment. I hope I implied that the reliability of the higher mass structures were not as great, except of course for the A3. Even in the $K_{\pi\pi}$ system it becomes difficult to identify more than one L resonance – even with the WA3 data. If attempts are then made to extract resonance parameters the resulting quantities should be treated with caution and I feel one can only give, at the moment, ranges for the parameters. I think a major improvement will only occur when there is a massive increase in statistics in these high mass regions, a factor of 10 over the WA3 experiment. When that sort of improvement is made we always seem to see a corresponding increase in our understanding.

R.AARON (Northeastern) – There might be a fundamental reason why diffractive and CEX $\rho\pi$ data look similar – i.e., they have same background which can be attributed to a broad $q^2\bar{q}^2$ state predicted by Jaffe to be at about $\sim$1250 MeV.
R.J.CASHMORE – I quite agree. The $\pi\rho$ phase shift might have some other contribution besides the Al pole e.g. potential scattering. This adds another possible effect besides background production amplitudes. There are $q^2\bar{q}^2$ states on the books, as you mention, but I think it would be brave to identify such a state as the explanation of the present observations.

W.DUNWOODIE (SLAC) – The (3_π) mass spectrum from $K^-p_{\rightarrow\Sigma^-}$ $(3_\pi)^+$ (4.2 GeV B.C. data CERN) just below 1.1 GeV is very similar to the WA3 1^+ data for $0.3 < t < 0.7$ GeV. Also the WA3 phase shift variation in this region is more rapid in the data than in the fit. Any comment?
R.J.CASHMORE – Remember the $(3_\pi)^+$ spectrum from $K^-p_{\rightarrow\Sigma^-}(3_\pi)^+$ contained all waves i.e. was not partial wave analysed in the figure I showed. The WA3 data for $0.3 < t' < 0.7$ GeV2 was purely the 1^+ component. Hence it is not correct to make a direct comparison. Furthermore the fit to the data is actually very good as we can see from the figure. There is no possibility of smuggling in a further resonant effect in this region.

K⁻p INTERACTIONS AT 11 GeV/c: NEW RESULTS ON STRANGE MESON SYSTEMS*

Blair N. Ratcliff
Stanford Linear Accelerator Center,
Stanford University, California

ABSTRACT

The status of our programmatic study of states containing a strange quark is briefly reviewed. An 11 GeV/c K⁻p experiment run on the LASS spectrometer at SLAC is discussed and preliminary results presented for several inelastic channels based on approximately one half of the available data sample. In addition, new results utilizing the full statistics of the experiment are presented for the elastic $K^-\pi^+$ channel from the $K^-\pi^+n$ final state. The well established leading $K^*(890)$, $K^*(1435)$, and $K^*(1780)$ resonances are observed, and clear evidence is presented for a new $J^P = 4^+$ resonance at ~ 2070 MeV. Preliminary results from an energy independent partial wave analysis of these data are presented which display unambiguous evidence for resonant structure in the non-leading 0^+ and 1^- waves.

---

*Work supported by the Department of Energy, contract DE-AC03-76SF00515.

I.   INTRODUCTION

During the last several years, we have been engaged in a systematic study of strange quark spectroscopy and reaction dynamics for the production of these states, using the rf separated $K^{\pm}$ beams at SLAC.   The physics goals of this program are, in general, to provide a "great leap forward" in our understanding of the K^* (e.g., $s\bar{u}$), and ϕ' ($s\bar{s}$) "strangeonium" mesons, and of the strangeness -2 and -3 hyperons (e.g., ssu).   Since nature has provided us with no stable meson or strange baryon targets, these areas have lagged far behind during the rapid progress made in the last decade toward understanding the hadron spectrum.

The first experiment in this series was a 13 GeV $K^{\pm}$ experiment using a forward dipole spectrometer.[1]   This experiment was very successful in studying the low mass, low multiplicity states.   For example, the two 1^+ Q mesons required by the quark model were observed through a partial wave analysis of the $K\pi\pi$ channel,[2] and a partial wave analysis of the $K\pi$ channel was also carried out from low mass through the $K^*(1780)$ region.[3]   However, forward dipole spectrometers have acceptance limitations which hinder their application to higher masses or to higher multiplicity states.   We have therefore continued these studies in our current experiment, which uses the LASS spectrometer, to study K^-p interactions at 11 GeV/c.

This experiment triggered on essentially the total charged-inelastic cross section and attained a visible sensitivity of ~ 700 events per μb in most channels.   A substantial portion of the events from this experiment has now been passed through the tracking programs and provides the data on which this talk is based.   Starting this summer, we will be extending this data sample to over 5 events per nb in a sequel LASS experiment which uses both K^+ and K^- beams at 11 GeV.[4]

II.   EXPERIMENT

The experiment I will be discussing today was performed by a collaboration of physicists from SLAC, Carleton University, and the National Research Council of Canada.[5]   The experiment uses the LASS

spectrometer facility which is shown in figure 1. It is serviced by
an rf separated beam which can deliver usable kaon fluxes up to
14 GeV/c. LASS contains two large magnets filled with tracking detectors. The first magnet is a superconducting solenoid with a 23 kG
field parallel to the beam direction. This is followed by a 27 kG-m
dipole magnet with a vertical field. The solenoid is effective in
measuring the interaction products which have large production angles
and relatively low momenta. High energy secondaries, tending to stay
close to the beam line, are not well measured in the solenoid, but
will pass through the dipole for measurement there. Particle identification is provided by Čerenkov counters C1 and C2, and a time of
flight (TOF) hodoscope. The trigger for this experiment required two
or more charged particles to exit the target and was essentially
defined by requiring more than two hits in the full-aperture proportional chamber which immediately followed the target.

The total data sample for this experiment contains some 4×10^7
good events. Analyzing this amount of data clearly requires a substantial amount of computer time and, therefore, calendar time as
well. In order to make the full statistics of the experiment available in specific channels relatively early, we split the analysis
into two phases. First, we used multiplicity information from the

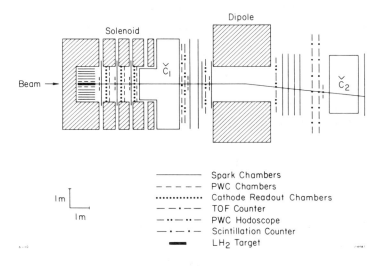

Fig. 1. Plan view of the LASS spectrometer.

proportional chamber system as a software prefilter to select two prong data from the standard data tapes. The production of this data onto data summary tapes was completed in January of this year. Preliminary physics results from the Kπn channel based on the full statistics of the experiment are now available, as I will be discussing in a few minutes. At the same time, we have been continuing with the analysis of the multiprong channels using the central computers of SLAC and the NRC. During the last few months, we have also begun production running with the LASS microprocessor system, the 168/E.[6] The 168/E processors are the result of development efforts to provide the substantial computer analysis power required to handle the data processing load from LASS. They emulate the subset of the instructions of an IBM 370 computer required by a standard FORTRAN-compiled scientific program, so that we are able to run our track finding and fitting programs interchangeably on the central computer or on the emulator. Each 168/E provides about one-half the analysis power of a fully dedicated 370/168 to this experiment. We have been running with a single processor since early January and a two processor system since March. We expect to have six processors in operation by summer with an additional three being added in the fall for processing the data from our sequel experiment. At the present time we have processed over 50 percent of the multiprong data from the present experiment, and expect to finish the production processing by the late summer.

III. MULTIPRONG CHANNELS

Before discussing the Kπn channel in detail, I would like to briefly describe preliminary results on some selected multiprong channels to indicate the breadth of topics that will be available for study in this experiment. These data are based on partial samples from our present analysis and typically contain some 30 percent of the total statistics of the current experiment.

First of all, even though this is a meson spectroscopy conference, it is hard to resist showing a hyperon state. In spite of the rather complete understanding of baryon spectra in general, the Ξ^*

and Ω^* excited states have remained poorly understood. In particular, only three of the 16 Ξ^* and zero of seven Ω^* states are well established in the otherwise clearly defined baryon supermultiplets. The fundamental reason for this lack of information is that these states can be observed only in production experiments. Their decay topologies are also quite complex so that they have generally been studied only in bubble chamber experiments. The largest of these experiments has a sensitivity of about 100 events per μb so that it produces about 100 Ω^- and a few thousand Ξ^-. The sensitivity of our present experiment is several times this level, and the experiment which is now beginning will increase the sensitivity of these studies by a factor of about 50.

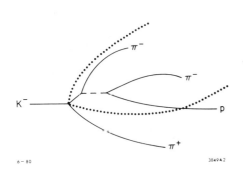

6−80   3849A2

Fig. 2.   Schematic of a $\Xi^-\pi^+$ system.

One reconstructs Ξ^- states as indicated in figure 2. A π^-p V^0 vertex with a Λ invariant mass is combined with a π^- to form a secondary V^- vertex, giving the clear Ξ^- signal shown in figure 3(a). If one then forms combinations of Ξ^- and π^+ at the primary vertex, the invariant mass distribution, figure 3(b), shows strong production of the $\Xi^*(1530)$, as expected.

Let us turn now to the strangeness −1 meson systems. Let me remind you that the goal of meson spectroscopy is not simply to catalogue states in the quark model, but also to confront basic questions about the quark-quark forces, such as understanding the spin and radial excitation dependence of these systems. This effort requires careful study of the classification, level splittings, and mixing of the various states. This, in turn, requires a large amount of data in both elastic and inelastic channels. Indeed, it will often be essential to observe states in several different channels in order to understand them in detail. For example, in the case of the $J^P = 1^-$ $K^*(1650)$, which was observed in two of four solutions through the

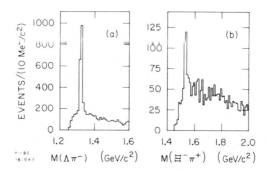

Fig. 3. Invariant mass spectra for inclusive
hyperon production. (a) $\Lambda\pi^-$ mass combination.
(b) $\Xi^-\pi^+$ mass combination.

partial wave analysis of the Kπn channel in our 13 GeV experiment,[3]
it is not clear if it should be classified as a radial excitation or
as a non-leading member of the D-wave $q\bar{q}$ triplet. This question can
be addressed by measuring the interference between the $K^*(1780)$ and
the $K^*(1650)$ in an inelastic final state and determining the relative
sign of the couplings. Using a higher symmetry scheme such as SU(6),
the classification can then be attempted.

With this as an introduction, let me display a few examples of
the present status of our data samples in the inelastic channels. In
figure 4(a), we show the invariant $K^\circ\pi^+\pi^-$ mass from the reaction
$K^-p \rightarrow K^\circ\pi^+\pi^-n$. Clear structures are observed in the 1435 and 1780
mass regions.

There is substantial evidence for "cascade" type decays in this
channel as is displayed in figures 4(b-e). Figures 4(b-c) respec-
tively show the $\pi^+\pi^-$ and $K^\circ\pi^-$ mass combinations to be dominated by ρ
and $K^*(890)$. One also sees a smaller amount of $K^*(1435)$ in the $K^\circ\pi^-$
distribution. If one selects on the ρ and $K^*(890)$ isobars indicated,
the invariant mass distributions of figures 4(d) and 4(e) are obtained. In both distributions, structure in the 1435 and 1780 regions is clearly seen, presumably indicative of the expected leading
K^* states.

Next, let us turn to the reaction $K^-p \rightarrow K^-\pi^+\pi^-p$. Historically,
this reaction has been the most important source of information about
the unnatural spin-parity K^* mesons, as Cashmore reviewed for us in

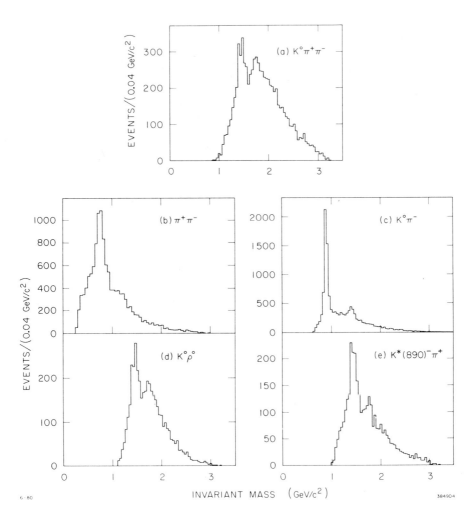

Fig. 4.   Invariant mass spectra for the reaction $K^-p \to K^\circ \pi^+\pi^- n$.
Events selected for $|t| < 0.5$ GeV^2, $M(\pi^+n) > 1.34$ GeV, and
$M(K^\circ n) > 1.50$ GeV.   (a) $K^\circ \pi^+\pi^-$ mass combination.   (b) $\pi^+\pi^-$ mass
combination.   (c) $K^\circ \pi^-$ mass combination.   (d) $K^\circ \rho$ mass combination
where $.65 < M_{\pi\pi} < .89$ GeV.   (e) $K^*(890)\pi$ mass combination where
$.81 < M_{K\pi^-} < .98$ GeV.

the previous talk.[2,7]   In order to extract the detailed information
required, it is necessary to apply a sophisticated partial wave analysis program to the data.   We use the SLAC-LBL approach which directly
fits the scattering amplitudes to the data.   The present data sample
corresponds to about 45 percent of the total data sample and typically

contains 3500 events per 20 MeV bin in the Q region. At the present
time, we have analyzed only the region below 1.6 GeV. In general,

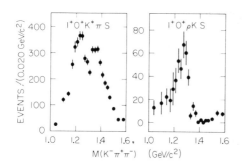

our results are consistent
with those from our earlier
analysis.[2] Figure 5 displays
the amplitudes for the 1^+0^+S
wave $K^*\pi$ and ρK channels which
dominate the Q mass region.
We see clear structure in both
channels, indicating the presence of the two Q mesons previously observed,[2,7] one with
a mass of $\sim$ 1300 MeV coupling
mainly to $K\rho$, and the other at
$\sim$ 1400 MeV coupling mainly to
$K^*\pi$.

Fig. 5. Preliminary results on the
1^+0^+S wave in the Q region taken
from the partial wave analysis of
$K^-\pi^+\pi^-$ scattering in the reaction
$K^-p \rightarrow K^-\pi^+\pi^-p$

IV. THE $K\pi$ ELASTIC CHANNEL

We now turn to an analysis of the natural spin-parity K^* mesons
taken from the full experimental statistics on the reaction
$K^-p \rightarrow K^-\pi^+n$. The raw data sample for this reaction contains 142,000
events in the invariant $K\pi$ mass region $(M(K\pi))$ between .7 and 2.5 GeV.
In figure 6, we see sharp peaks in the invariant mass corresponding

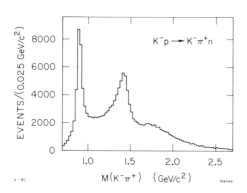

to the $K^*(890)$ and $K^*(1435)$
with perhaps a small hint of
some structure in the region
around 1800 MeV. There is also
some asymmetry on the lower
side of the $K^*(1435)$ bump hint-
ing, perhaps, at other underlying effects, but in general,
little clear evidence for
states other than the two well
known leading K^* states.

A great deal of interest-
ing structure is hidden in this

Fig. 6. Invariant $K\pi$ mass
distribution from the reaction
$K^-p \rightarrow K^-\pi^+n$ with $|t'| < 1.0$ GeV^2.

plot, however, as becomes evident when we increase its dimensionality.
Figure 7 is a computer generated apparent 3-d image where the two

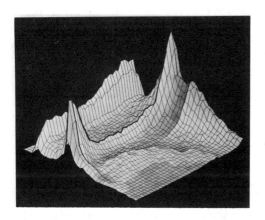

variables in the lower plane
are $M(K\pi)$ and the cosine of
the t-channel kaon decay angle
in the $K\pi$ center of mass
($\cos \theta_J$). The height of the
surface above the plane is the
number of events per bin arbitrarily scaled. The $K^*(890)$
stands out clearly as a ridge
at low mass while two dramatic
spikes, at both forward and

Fig. 7. Apparent 3-d plot of the
invariant $K\pi$ mass ($M_{K\pi}$) versus the
t-channel decay angle of the K in
the $K\pi$ center of mass ($\cos \theta_J$). The
axes in the plane are: $.7 < M_{K\pi} <$
2.5 GeV along the lower left-hand
side; and $-1.0 < \cos \theta_J < 1.0$ along
the lower right-hand side. The
number of events, arbitrarily
scaled, is plotted above the plane.

backward $\cos \theta_J$, signal the
2^+ $K^*(1435)$. As we continue
to higher $M(K\pi)$, the structure
grows more complex. There is
a clear bump in forward $\cos \theta_J$
in the region around 1740 MeV
on top of a ridge extending to
high mass. This ridge becomes
progressively steeper as $M(K\pi)$ increases and is indicative of $K\pi$ diffractive scattering. It falls off as the mass increases due to a
fixed angle cut generated by our two prong trigger. At smaller
$\cos \theta_J$, there is a complex plateau structure which peaks at a $\cos \theta_J$
of $\approx -.5$ at about 1800 MeV. As $|Y_{30}|^2$ has a peak at this same
$\cos \theta_J$, one might wish to interpret this behavior as being due to
the 3^- K*(1780).

Continuing to still higher $M(K\pi)$, the structure grows even more
complex. At a mass of about 2100 MeV, there are small peaks at
$\cos \theta_J \approx 0$ and $\cos \theta_J \approx -1$ which one might speculate as being due to
a high even-spin object such as the 4^+ K^* expected in this mass re-
gion. In any event, the obvious interference structure observed in
this plot is both a promise and a caution.... The promise that there
is a great deal of interesting physics in the data coupled with the

caution that a sophisticated analysis is likely to be necessary to understand the obvious inherent complications. In particular, one must be very cautious about the interpretation of bumps in invariant mass distributions. This is particularly obvious in figure 8, which shows one-dimensional M(Kπ) plots cut in ten different slices of cos θ_J. The change in apparent width and position of the various structures as one changes cos θ_J is very dramatic. For example, the apparent structure in the region around 1800 MeV moves substantially in both mass and width as one moves across the cos θ_J region. Note too, that the "shift" effect can be substantial even in the region of a large leading state. One can clearly see a shift of the structure around 1400 for example. The forward and backward peaks are centered at $\sim$ 1435 but the structure becomes quite asymmetric and substantially lower in mass for cos θ_J near 0. As I will show later, this effect is due to a non-leading 0^+ wave which peaks just below the 2^+. In any event, a fit to the projected distributions in M(Kπ) will clearly give a mass value for the 2^+ object which is substantially too low. It was this effect as observed through the earlier partial wave

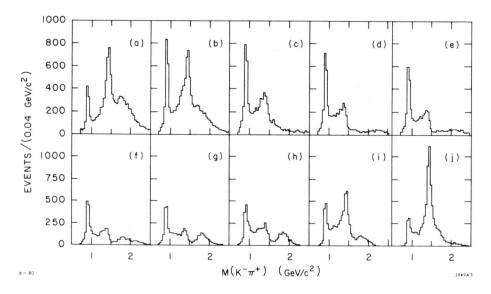

Fig. 8. Invariant Kπ mass spectra cut on 10 different cos θ_J slices $|.2|$ in width. (a) through (j) are ordered from the forward to the backward direction so that, for example, (a) contains the slice .8 < cos θ_J ≤ 1.0 and (j) contains −1.0 ≤ cos θ_J ≤ −.8.

analysis of the 13 GeV data,[3] and confirmed by the present analysis, which shifted the 2^+ mass from its historical 1420 MeV to the presently accepted value of 1434 MeV.[9]

Having hopefully convinced you that there is a great deal of interesting structure in the $\cos \theta_J$ versus $M(K\pi)$ plot which does not appear clearly as bumps in the invariant mass distribution, let us proceed to look at this structure as observed in the spherical harmonic moments of the decay angular distribution. We select the data for this analysis to emphasize the π exchange contribution by cutting on $|t'| < .2$. We also make tight cuts to remove the small competing background reactions and are left with 51,000 events in the small $|t'|$ region. Figure 9 shows the t-channel $M = 0$ moments for this data. In general, at each mass value they are plotted only up to the smallest value of L_{max} required to fit the data. The well known $K^*(890)$ and $K^*(1435)$ bumps are apparent, and their spins can be simply confirmed as 1^- and 2^+ respectively from the behavior of the $\langle Y_{20} \rangle$ and $\langle Y_{40} \rangle$ in these mass regions. Note also the interference patterns in the odd L moments as we move through the resonances. As we go to higher mass we see a clear bump in the $\langle Y_{60} \rangle$ at 1780 MeV indicating the $J^P = 3^-$ $K^*(1780)$ state. Continuing still higher in mass, we see a bump structure in the $\langle Y_{60} \rangle$ and $\langle Y_{80} \rangle$ coupled with a classic interference pattern in the $\langle Y_{70} \rangle$. Since higher moments are consistent with zero in this region, we interpret this structure as the next leading $J^P = 4^+$ K^* required by the quark model.

The probable existence of a large number of underlying states in this region makes a determination of the parameters of this new state rather difficult. At the present time, we have performed a naive fit assuming a relativistic Breit-Wigner in the F and G waves and a simple polynomial background in the D, F, and G waves. After fixing the leading 3^- resonance at values based on our preliminary partial wave analysis results as noted below, we performed a fully correlated fit to the $\langle Y_{60} \rangle$, $\langle Y_{70} \rangle$, and $\langle Y_{80} \rangle$ moments in the region from 1500 to 2300 MeV. We obtain

$$M_F \approx 1790 \text{ (fixed) MeV,}$$
$$\Gamma_F \approx 190 \text{ (fixed) MeV,}$$

48

$$M_G \approx 2070 \text{ MeV,}$$

$$\Gamma_G \approx 220 \text{ MeV,}$$

where the errors are dominated by systematics inherent in the assumptions of the model. The significance of the $J^P = 4^+$ state in this model is greater than 4.5 σ, and as such, constitutes definitive evidence for a new leading state, the $K^*(2070)$.[8]

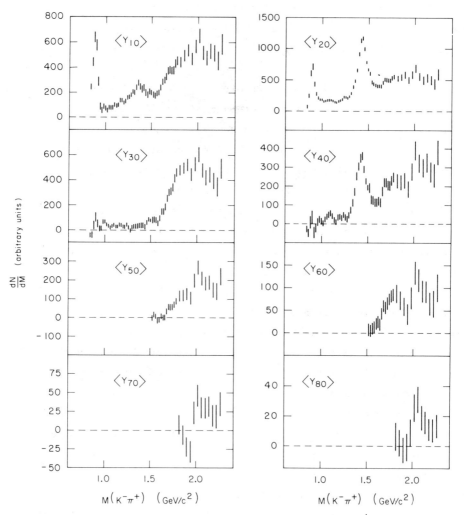

Fig. 9. The unnormalized t-channel moments of the $K^-\pi^+$ angular distribution for $|t'| < .2$ GeV^2. The bins are overlapped by a factor of two to guide the eye.

As we have just illustrated, a great deal can be learned about
the leading states by studying the mass dependence of the angular
moments of the data. In addition, however, the quark model leads us
to expect lower lying states in this same region which will be hidden
in the interferences seen in the moments. In order to understand
these effects, we are in the process of performing an energy independent partial wave analysis. As of today, preliminary results are
available to a Kπ mass of 1.84 GeV. This analysis is based on a
t-extrapolation to the pion pole of the π-exchange contributions to
the t-channel helicity-zero Kπ production amplitudes. It is described in detail in reference 3 so it will not be discussed here.
Let me, however, delineate a few of the important assumptions of this
analysis: (I) we fit to the smallest number of partial waves required by the data; (II) we constrain the S- and P-waves to be
elastic below 1.3 GeV; (III) we assume the I = 3/2 wave as given by
the 13 GeV SLAC experiment;[3] and (IV) we set the overall phase in the
inelastic region by assuming the phase of the leading state is given
by a relativistic Breit-Wigner.

In general, the results obtained are quite consistent with our
earlier 13 GeV analysis. However, the errors are smaller (and the
features therefore clearer) in the high mass region due to the substantially better acceptance. As I will show below, one particularly
important consequence of this is that there is only a single unambiguous solution below ∿ 1700 MeV which, therefore, determines the
existence of the observed structures uniquely.

The S- and P-waves for this solution are shown in figure 10. In
addition to the leading 1⁻ K*(890) resonances, there is interesting
additional structure indicative of non-leading resonance-like effects.
The S-wave exhibits a smooth slow rise in both phase and magnitude
until 1.3 GeV. The magnitude then rises rapidly followed by a precipitous drop in the region just above 1.4 GeV. This is coupled with
a rapid change in the phase motion indicating resonance behavior. As
the mass continues to increase the magnitude becomes very small and
the phase becomes indeterminate. Both the magnitude and phase then
begin to move again but, unfortunately, right at the end point of the

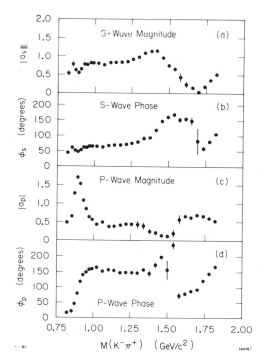

Fig. 10. The S- and P-wave Kπ scattering amplitudes from the reaction K⁻p → K⁻π⁺n. See text for a full explanation.

present analysis. In the P-wave we see the expected elastic K*(890) resonance, with little happening above that until about 1500 MeV. The magnitude is very small here, but rises to a broad peak centered around 1.7 GeV. This rise is accompanied by a fairly rapid phase variation again indicative of resonance-like behavior.

The Kπ partial waves as shown in figure 10 are perhaps easier to interpret when plotted as Argand diagrams. Figure 11 shows the clear circular motion expected for the leading 1⁻, 2⁺, and 3⁻ waves. We can also see motion associated with the non-leading S- and P-wave features described above. In particular, the S-wave wanders up slowly from low mass, passing through 90° at about 1300 MeV, but with very little evidence of resonance behavior there. However, at about 1450 MeV, we see an essentially elastic S-wave loop which is relatively narrow (perhaps 250 MeV) looking very much like a resonance. The P-wave amplitude traces out the elastic K*(890) resonance and then curls around relatively quickly in the region around 1650, again indicating resonance-like behavior.

One of the most exciting features of the present analysis is the indication that it is unambiguous below ∿ 1700 MeV. The ambiguities are most readily apparent as an indeterminacy in the signs of the imaginary parts of the complex zeros of the scattering amplitudes which were emphasized by Barrelet.[10] Since elastic unitarity allows

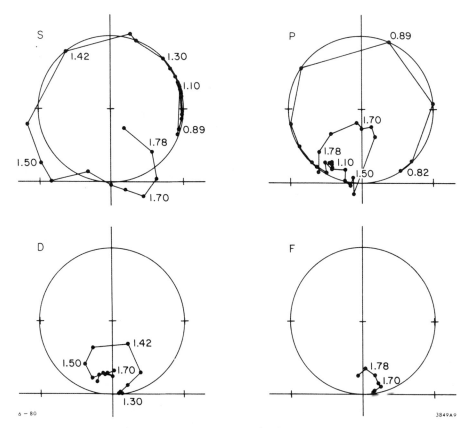

Fig. 11. Argand plots for the I = 1/2 Kπ partial waves. The I = 1/2 S-wave was obtained by subtracting the I = 3/2 S-wave amplitude determined in the 13 GeV K± experiment from the full S-wave amplitude. The I = 3/2 components of the other waves are neglected. See text for a full explanation.

only one solution below 1 GeV, no ambiguities arise until one of the zeros becomes small. Figure 12 shows the real and imaginary parts of the three zeros plotted with error estimates. None of the imaginary parts of the zeros appear close enough to the axis to change signs in the region below 1700 MeV. Hence, we conclude that there is a unique solution in this preliminary analysis.

52

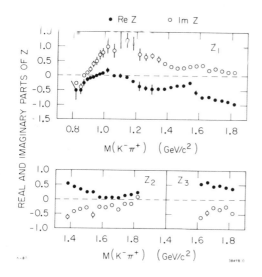

Fig. 12. The real and imaginary parts of the complex zeros of the Kπ scattering amplitudes. Note that the error bars lie within the point diameters unless shown.

V. CONCLUSION

The Kπ moments clearly show the "old" leading K* resonances; the 890 (1^-), 1435 (2^+), and 1780 (3^-). In addition, a new 4^+ resonance is observed at about 2070 MeV, which is naturally interpreted as a SU(3) partner of the h meson. Preliminary results from a Kπ partial wave analysis in the mass region below 1840 MeV clearly confirm the 1^-, 2^+, and 3^- resonances. The most interesting new result is that this analysis has no ambiguities below 1700 MeV so that the evidence for resonance behavior of the nonleading 0^+ wave at about 1450 MeV and the 1^- wave at about 1650 is also unambiguous.

These results and those reviewed earlier today[7] demonstrate the usefulness of having very high statistics data in different channels. We have a great many final states in this experiment most of which are just beginning to be analyzed. We will be extending the data sample by about a factor of seven during the next year, and are looking forward to a productive investigation of the properties of states containing the strange quark during the next few years.

ACKNOWLEDGMENTS

I would like to thank my fellow collaborators and the technical staffs of SLAC Group B, Carleton University, and the LASS Operations group on whose efforts this talk is based. Special thanks are due David Aston, Stan Durkin, Alan Honma, William Johnson, David Leith, and Robert Richter for invaluable help in preparing the specific material of this talk.

REFERENCES

1. See, for example, G. W. Brandenburg, R. K. Carnegie, R. J. Cashmore, M. Davier, T. A. Lasinski, D.W.G.S. Leith, J.A.J. Matthews, P. Walden, and S. H. Williams, Nuclear Physics B104: 413, 1976.

2. G. W. Brandenburg, R. K. Carnegie, R. J. Cashmore, M. Davier, W. M. Dunwoodie, T. A. Lasinski, D.W.G.S. Leith, J.A.J. Matthews, P. Walden, S. H. Williams, F. C. Winkelmann, Physical Review Letters 36:703, 1976.

3. P. Estabrooks, R. K. Carnegie, A. D. Martin, W. M. Dunwoodie, T. A. Lasinski, and D.W.G.S. Leith, Nuclear Physics B133:490, 1978.

4. D. Aston, W. Dunwoodie, S. Durkin, A. Honma, W. B. Johnson, P. Kunz, D.W.G.S. Leith, L. Levinson, B. N. Ratcliff, R. Richter, S. Shapiro, R. Stroynowski, S. Suzuki, G. Tarnopolsky, S. Williams, N. Horikawa, S. Iwata, R. Kajikawa, T. Matsui, A. Miyamoto, T. Nakanishi, Y. Ohashi, C. O. Pak, T. Tauchi, T. Shimomura, K. Ukai, and S. Sugimoto, SLAC-Proposal-E/135, October 1979.

5. D. Aston, W. Dunwoodie, S. Durkin, T. Fieguth, A. Honma, D. Hutchinson, W. B. Johnson, P. Kunz, T. Lasinski, D.W.G.S. Leith, L. Levinson, W. T. Meyer, B. N. Ratcliff, R. Richter, S. Shapiro, R. Stroynowski, S. Suzuki, S. Williams (SLAC), R. Carnegie, P. Estabrooks, R. Hemingway, R. McKee, A. McPherson, G. Oakham, J. Va'Vra (Carleton University and National Research Council, Ottawa).

54

6. P. F. Kunz, R. H. Fall, M. F. Gravina, J. H. Halperin, L. J. Levinson, G. J. Oxoby, Q. H. Trang, SLAC-PUB-2418, October 1979.

7. R. Cashmore, invited talk at this conference; The VI International Conference on Experimental Meson Spectroscopy, Brookhaven, April 1980.

8. Evidence for a charged spin-four K* state at 2060 MeV has been presented by C. Nef, this conference; The VI International Conference on Experimental Meson Spectroscopy, Brookhaven, April 1980.

9. Particle Data Group, LBL-100, April 1978.

10. E. Barrelet, Nuovo Cimento 8A:331, 1972.

NEW RESULTS ON Kπ, KK̄ AND Λ̄p MESONIC STATES

W.E. Cleland*, A. Delfosse, P.A. Dorsaz, M.N. Kienzle-Focacci,
G. Mancarella, M. Martin, P. Muhlemann, C. Nef, T. Pal,
and J. Rutschmann
Université de Genève, Geneva, Switzerland

J.L. Gloor and H. Zeidler
Université de Lausanne, Lausanne, Switzerland

A.D. Martin
University of Durham, Durham, United Kingdom

presented by C. Nef

ABSTRACT

We have studied the diffractive production of mesons with natural and unnatural parity at high energy. Partial wave analyses of the Kπ and KK̄ systems give evidence for $J^P = 4^+$ mesons with masses near 2.06 and 2.00 GeV, respectively. The Breit-Wigner behaviour of the resonant phases is clearly seen in the data. Evidence for unnatural parity states comes from a partial wave analysis of the Λ̄p system. We find three broad states with $J^P = 2^-$, 3^+ and 4^- and masses 2.26, 2.32 and 2.49 GeV, respectively. An interpretation in terms of resonances is consistent with the data.

INTRODUCTION

In this paper we report new results obtained from an analysis of high-statistics, high-energy data samples of the reactions

$$K^+p \rightarrow (K^0\pi^+)p \qquad 19'000 \text{ events}$$
$$K^-p \rightarrow (\bar{K}^0\pi^-)p \qquad 15'000 \qquad (1)$$
$$\pi^+p \rightarrow (\bar{K}^0K^+)p \qquad 19'000$$
$$\pi^-p \rightarrow (K^0K^-)p \qquad 21'000 \qquad (2)$$
$$K^+p \rightarrow (\bar{\Lambda}p)p \qquad 3'400$$
$$K^-p \rightarrow (\Lambda\bar{p})p \qquad 2'700 \qquad (3)$$

The mesonic Kπ, KK̄ or Λ̄p system is produced in the forward direction, in a mass region from threshold up to about 4 GeV, and a

*Permanent address: University of Pittsburgh, Pittsburgh PA 15260

momentum-transfer range $0.05 < -t < 1(GeV/c)^2$. The data have been taken at the CERN SPS at a beam momentum of 50 GeV/c, in experiment WA10. This experiment is a continuation of a similar experiment done at 10 GeV/c[1]; its aim is to study new high-mass states, as well as the energy dependence of resonance production.

Why is it interesting to study production of mesons at high energy? A good reason is the existence of a simple production mechanism for diffractive reactions of the type (1)-(3). It is well known that these reactions are dominated by isoscalar natural parity exchange (P+ f ± ω). Unnatural parity exchange is found to be small already at 10 GeV/c [2] and can be neglected at 50 GeV/c. Furthermore, production at high energy will tend to a single exchange mechanism, P -exchange. As a consequence, we expect to see little difference between positive and negative beam polarity. This is well verified by the data, and for this reason we have combined the data from both beam polarities in the present analysis. The remaining differences are well understood in terms of the interference of even and odd-C exchange, and can be studied in the separate samples. It is this simple production mechanism which makes an amplitude analysis feasible up to relatively high mass and in the presence of high spins.

SPECTROMETER

We have built a non-magnetic spectrometer with the aim of measuring simple event topologies with high statistical accuracy. Good acceptance up to high mass has been achieved due to a large forward solid angle and the absence of a magnetic field. A schematic layout is shown in Fig.1.

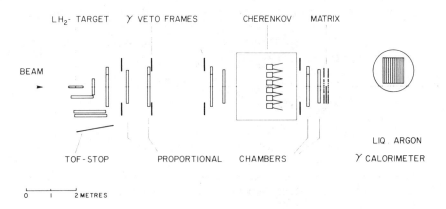

Fig.1 Experimental set-up

An incident beam spectrometer measures the direction, momentum and identity of the incident particle; a recoil detector measures the direction and momentum of slow recoil protons at large laboratory angles; a forward arm gives the directions of the fast forward particles. The forward momenta are not measured. The set-up is similar to the one used in a previous experiment at 10 GeV/c at the CERN PS [1,3]. The Cerenkov hodoscope and the liquid Argon detector in the forward arm are not used in the present analysis.

A high data-taking capability has been essential for this experiment. The data acquisition system included 7 minicomputers, similar to PDP 11/45 processors, built at the University of Geneva [4], which allowed full track reconstruction in the 9 second interval between bursts, at a rate of up to 800 events per burst. The logics of the system permitted event filtering at various levels of complexity: a first level (300 ns after the occurence of an event) used scintillator and Cerenkov information only, a second level (2-6 μs) used MWPC information and a third level (several seconds), by software, used the results of the on-line track reconstruction.

EVIDENCE FOR $K\pi$ AND $K\bar{K}$ MESONS WITH $J^P = 4^+$

Our $K\pi$ and $K\bar{K}$ effective mass spectra, corrected for acceptance, are shown in Fig.2. We observe beautiful K*(890), K*(1430) and A_2 production. Together with similar data taken in our previous experiment at 10 GeV/c [2,5,6], these are the basis of a detailed study of the exchange mechanism, which is in progress at present. In the present analysis we investigate the region above these low-mass resonances up to 2.2 GeV.

First we discuss the results of a moment and amplitude analysis of the high mass $K\pi$ system. Fig.3 shows the acceptance corrected, unnormalized, moments $\langle Y_L^M \rangle$ of the angular distribution of the K, in the t-channel helicity frame. Statistically significant signals are present up to L = 8, implying the presence of spins up to J = 4 (at least). We find that the M = 0 and M = 2 moments, shown in the figure, are dominant. This implies that the $K\pi$ system is dominantly produced with helicity $\lambda = 1$. In other words, the data can be described by $P_+, D_+, F_+ \ldots$, where the subscript + denotes a natural parity exchange, helicity $\lambda = 1$ amplitude.

The peak structure of the $\langle Y_8^0 \rangle$ moment indicates a resonance in G_+ just above 2 GeV. The shape of $\langle Y_8^2 \rangle$ is less clear; we will give a possible reason for this below. The $\langle Y_6^0 \rangle$ moment also shows a very clear peak near the same mass. This cannot be due to $|G_+|^2$, however, since the coefficient coupling $|G_+|^2$ to $\langle Y_6^0 \rangle$ is very small. It can be accounted for by D_+G_+ interference, and therefore determines the G_+ phase, under the assumption that the D_+ phase is given by the tail of the K*(1430). The L = 6 moments clearly show an F_+

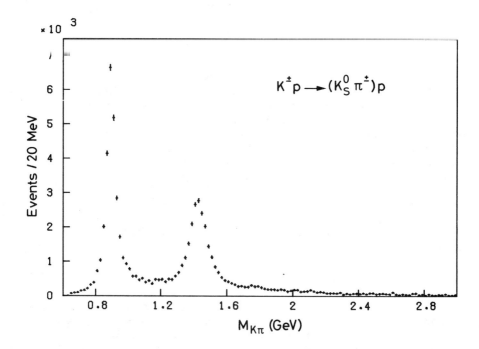

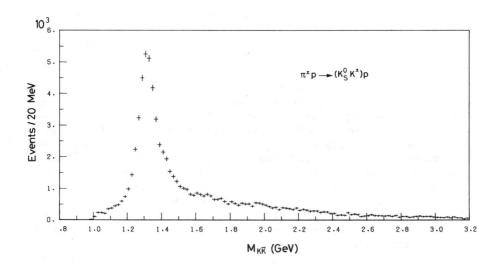

Fig.2   The $K\pi$ and $K\bar{K}$ mass spectra

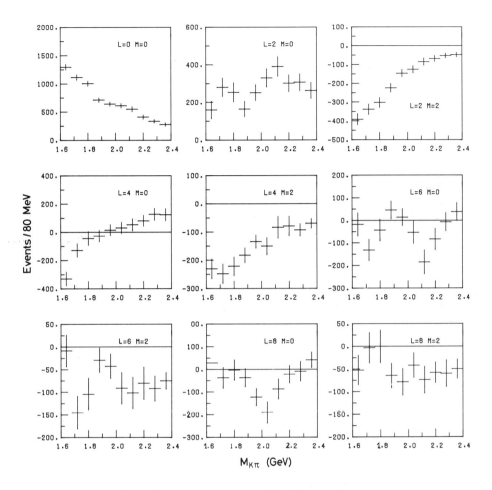

Fig.3 Kπ moments as function of mass

contribution due to the K*(1780). From the lower moments it is
also evident that D_+ gives indeed still a large contribution in
this mass region.

Fig.4 displays the result of a fit of D_+, F_+, G_+ and D_+G_+ to
the even L moments of Fig.3. There is a unique solution. We find
a decreasing $|D_+|$, consistent with the tail of the K*(1430), a peak
in $|F_+|$ at the mass of the K*(1780) and a broad $|G_+|$ wave peaking
near 2 GeV. The relative phase $\phi_{D_+G_+}$ is particularly interesting,
as its mass dependence is in good agreement with that expected for
resonant Breit-Wigner phases (dashed line in Fig.4). This behaviour
is further evidence for a $J^P = 4^+$ resonance. Approximate values
for the mass and width are M = 2.06 GeV and Γ = 0.15 GeV.

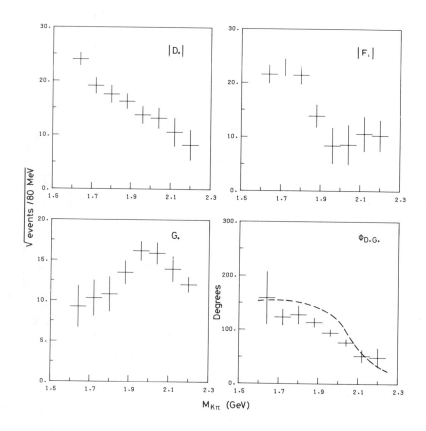

Fig.4 Kπ amplitudes

Next, we discuss $\overline{K}K$ production, which we find in many respects very similar to Kπ production. Fig.5 shows the corrected moments of the $K^{\pm}$ angular distribution. Again, we find statistically significant structure up to L = 8, implying spins up to J = 4. A peak structure, indicating a resonance in G_+, appears cleanest in $\langle Y_8^2 \rangle$. The L = 6 moments show a striking D_+G_+ interference pattern. Notice that there is little F_+ wave in the L = 6 moments; the tail of the g cannot be seen in these moments. This could be expected, as the $\overline{K}K$-production of odd spins is suppressed by G-parity conservation requiring odd-G (ω) exchange.

We have used the same parametrization, incorporating D_+, F_+, G_+ and D_+G_+, to fit the moments of Fig.5. The resulting amplitudes are plotted in Fig.6. We find $|D_+|$ consistent with the tail of the A_2, and $|G_+|$ indicating a broad resonance near 2 GeV. The $|F_+|$ intensity (not shown) is smaller and fluctuates. The relative phase $\phi_{D_+G_+}$ is in excellent agreement with that expected for resonant

61

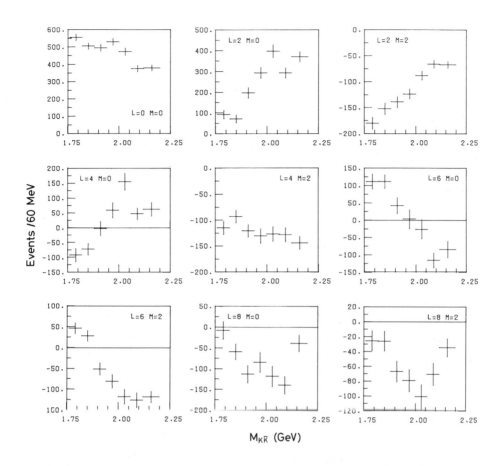

Fig.5 KK̄ moments as function of mass

Breit-Wigner phases. The new 50 GeV/c data therefore confirm the existence of the A_2^* with $J^{PC} = 4^{++}$ first seen in our previous experiment at 10 Gev/c [7,8]. The resonance parameters are approximately M = 2.00 GeV and Γ = 0.25 GeV.

In the present, preliminary analysis we have attempted to extract the dominant amplitudes from the moments. The fit of D_+, F_+ and G_+ amplitudes indicates, however, that further, smaller, contributions are required. This can be seen particularly well in $<Y_O^O>$, the cross section, which is systematically underfitted by about 10%. The following amplitudes, neglected at present, will be included in a more refined fit:

- Helicity λ = 2 amplitudes like D_{2+}. We find significant signals in some M = 1,3 moments which are sensitive to terms like $Re(D_+D_{2+}^*)$.

62

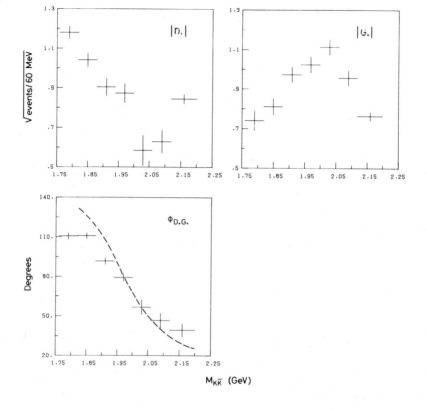

Fig.6 KK̄ amplitudes

- Higher spins. A small contribution from at least spin 5 is
  expected in this mass region. The $|H_+|^2$ intensity will con-
  tribute to the L = 8 moments in the following way:

$$\langle Y_8^0 \rangle = \ldots -0.67 \, |G_+|^2 - 0.10|H_+|^2 \ldots$$

$$\langle Y_8^2 \rangle = \ldots -0.42 \, |G_+|^2 - 0.24|H_+|^2 \ldots \qquad (4)$$

Here the coefficients are such that the $|G_+|^2$ signal is
cleaner in $\langle Y_8^0 \rangle$ than in $\langle Y_8^2 \rangle$. This appears to be the case
in the Kπ data (Fig.3). The situation is different in KK̄,
as there odd spins are suppressed by G-parity.

PARTIAL WAVE ANALYSIS OF THE $\bar{\Lambda}p$ SYSTEM

Fig.7 shows the $\bar{\Lambda}p$ and $\Lambda\bar{p}$ effective mass spectra from reactions (3) before acceptance corrections. We find smooth mass spectra with broad enhancements near threshold, but no statistically significant narrow structure. The two samples are very similar and have been combined for the subsequent analysis. In what follows we will use the notation corresponding to the K^+ reaction when referring to the combined K^+ and K^- data.

A preliminary analysis of these data has been published previously [9]. The present work is based on the formalism discussed in detail in ref.10, and a full presentation of the results is given in ref.11.

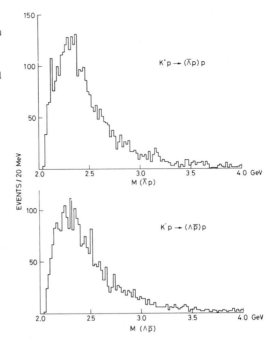

Fig.7 $\bar{\Lambda}p$ and $\Lambda\bar{p}$ mass spectra

We have studied the production of mesonic states R^+ in the reaction $K^+p \rightarrow R^+p$ and their sequential decay $R^+ \rightarrow \bar{\Lambda}p$, $\bar{\Lambda} \rightarrow \bar{p}\pi^+$ in terms of double moments, which we define as follows:

$$H(LM\ell m) \equiv \sum_{i=1}^{N} w_i \, D_{Mm}^{L}(\phi_i, \theta_i, 0) \, D_{mo}^{\ell}(\phi_{1i}, \theta_{1i}, 0) \quad (5)$$

where ϕ_i, θ_i are the Λ-angles of event i in the decay $R^+ \rightarrow \bar{\Lambda}p$, and ϕ_{1i}, θ_{1i} are the $\bar{p}$-angles in the decay $\bar{\Lambda} \rightarrow \bar{p}\pi^+$. As reference frame we choose the t-channel helicity frame for R+. The acceptance correction is done by giving individual events a weight w_i.

Fig.8 shows the corrected moments used in the present analysis. We find that the dominant structures all occur in the M = 0 moments shown in the figure, which implies that R^+ is produced with helicity $\lambda = 0$. The upper part of Fig.8 shows the moments H(L000), which describe the angular distribution of the R-decay, $W(\Omega)$. We see broad, very significant peaks in the L = 4, 6 and 8 moments, which we have reported in a previous publication [9]. The lower part of Fig.8 shows the combination Im(H(L011)–H(L01-1)) which

64

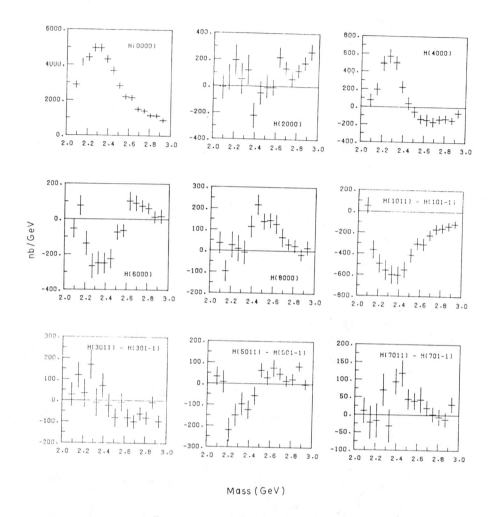

Mass(GeV)

Fig.8 $\bar{\Lambda}$p double moments as function of mass

describes the transverse polarization of the $\bar{\Lambda}$, $Py(\Omega)W(\Omega)$, along an axis normal to the plane containing the momentum vectors of the $\bar{\Lambda}$ and the incident K^+. This is the only non-zero polarization component, as parity conservation in production and decay of R requires $P_x \equiv P_z \equiv 0$ for M = 0 [10]. In Fig.8 we find evidence for large $\bar{\Lambda}$ polarization; it will provide strong constraints for a partial wave analysis.

The double moments can be expanded in terms of bilinear combinations of amplitudes of the form [10]

$$H(LM\ell m) = \sum_{AA'} \cdots \left(H^A_\lambda F^A_{\mu\nu} \right) \left(H^{A'}_{\lambda'} F^{A'}_{\mu'\nu} \right)^* \cdots \qquad (6)$$

where the sum is over interfering states A, A' with A ≡ J^P, as well as over the various helicity indices λ, λ', μ, μ' and ν. The individual terms contain an amplitude H^A_λ describing production of a state R with spin-parity A and helicity λ, and an amplitude $F^A_{\mu\nu}$ describing the decay $R^+ \to \bar\Lambda p$, where μ, ν are the helicities of the decay baryons. The general analysis is much more complicated than in the case of the $K\pi$ and $K\bar{K}$ systems, because R can have both natural or unnatural parity, and furthermore each spin-parity state can decay via two helicity amplitudes $F^A_{\mu\nu}$, which are related to the spin-singlet and spin-triplet configuration of the $\bar\Lambda p$ system.

Fortunately, one can make two reasonable assumptions for diffractive reactions at high energy:

- first, we consider only natural parity exchange in the t-channel, which can be isolated by forming the linear combination

$$N^A_\lambda = c_\lambda \left[H^A_\lambda - \sigma (-1)^\lambda H^A_{-\lambda} \right] \; ; \quad c_\lambda = \begin{cases} \tfrac{1}{2} & \lambda = 0 \\ \tfrac{1}{\sqrt{2}} & \lambda \neq 0 \end{cases} \qquad (7)$$

where $\sigma = P(-1)^J$ is the naturality of the produced state. As noted above, helicity $\lambda = 0$ dominates these data. Eq.(7) then implies that the produced states are of the unnatural parity series $J^P = 0^-$, 1^+, 2^- ... [10].

- second, the generalization of C-parity conservation at the meson vertex within SU(3) [9,10] selects a single decay amplitude for each spin state with unnatural parity. Even J states decay via F_{++}, odd J states via F_{+-}. This can be tested directly from the moments, as it leads to the prediction

$$H(L000) = 0 \quad \text{for L odd}$$
$$H(L011) - H(L01\text{-}1) = 0 \quad \text{for L even} \qquad (8)$$

This is well verified by the data. The Λ polarization arises, in this model, from the interference of even and odd spin states.

66

These two assumptions make a partial wave analysis feasible.
The data determine the overall production and decay amplitudes

$$S_0 \equiv N_0^{0^-} F_{++}^{0^-}, \quad P_0 \equiv N_0^{1^+} F_{+-}^{1^+}, \quad \cdots$$

(9)

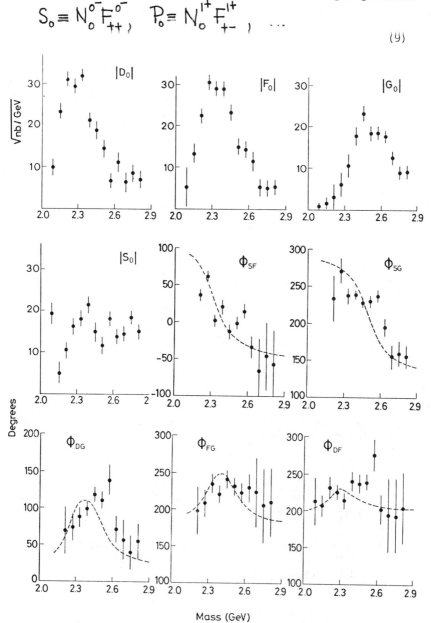

Fig.9 $\bar{\Lambda}p$ amplitudes

Our analysis includes the 5 partial wave magnitudes up to G_O wave, and the 4 relative phases, which we determine from.the 9 moments in Fig.8. The set of equations to be solved does, however, not have a unique solution. There are altogether 8 solutions which we have generated using the technique of Barrelet zeros [11].

In Fig.9 we show the relevant results, illustrated by one particular solution. We find broad, overlapping peaks in $|D_O|$, $|F_O|$ and $|G_O|$ in the mass region 2.3 - 2.5 GeV. This structure is very similar in all solutions. The main differences occur in the less well determined S_O and P_O amplitudes. The relative phases are solution dependent. For the solution shown, they are in good agreement with Breit-Wigner resonant phases for D_O, F_O and G_O, and a constant phase for S_O.(dashed line in Fig.9).

Figure 10 compares the strange mesons with natural and unnatural parity on a Chew-Frautschi plot. The new 2^-, 3^+ and 4^- states lie well below the natural-parity K* resonances, in the neighbourhood of the peripheral curve J = pR with p = $\bar{\Lambda}$-momentum in the c.m. and R = 1 fm. Above this curve, the excitation of resonances is kinematically suppressed in the $\bar{\Lambda}$p channel

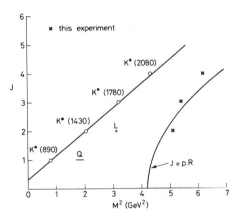

Fig.10 Chew-Frautschi plot of
S = ±1 mesons

CONCLUSIONS

We have studied new high-statistics data samples on the diffractively produced Kπ and K$\bar{\text{K}}$ system. A moment and amplitude analysis gives evidence for a new K* meson with $J^P = 4^+$, M = 2.06 and Γ = 0.15 GeV. The K$\bar{\text{K}}$ data confirm the existence of the A_2^* with $J^{PC} = 4^{++}$, M = 2.00 and Γ = 0.25 GeV, which has first been seen in our previous experiment at 10 GeV/c.

The diffractive $\bar{\Lambda}$p production provides a new possibility for studying unnatural parity high-mass mesonic states. A partial wave analysis gives evidence for three broad states with $J^P = 2^-$, 3^+ and 4^- at masses 2.26, 2.32 and 2.49 GeV, respectively. The interpretation of these states as resonances provides a natural explanation of the data.

REFERENCES

1. R. Baldi et al., Nucl. Phys. B134, 365 (1978)
2. A.D. Martin et al., Nucl. Phys. B134, 392 (1978)
3. W.E. Cleland et al., Phys. Lett. 88B, 409 (1979)
4. V. Hungerbühler et al., Nucl. Instr. and Meth. 137, 189 (1976)
5. R. Baldi et al., Phys. Lett. 70B, 377 (1977)
6. A.D. Martin et al., Nucl. Phys. B140, 158 (1978)
7. R. Baldi et al., Phys. Lett. 74B, 413 (1978)
8. A.D. Martin et al., Phys. Lett. 74B, 417 (1978)
9. W.E. Cleland et al., Phys. Lett. 89B, 290 (1980)
10. A.D. Martin and C. Nef, Spin-parity analysis of the decay dis-
    tributions of unnatural parity mesons, Nucl. Phys., to be pub-
    lished
11. W.E. Cleland et al., A partial wave analysis of diffractively
    produced $\bar{\Lambda}p$ and $\Lambda\bar{p}$ states, Nucl. Phys., to be published.

HEAVY VECTOR MESONS

B. R. Kumar

Rutherford and Appleton Laboratories, Chilton, Didcot, U.K.

ABSTRACT

The current status of heavy vector meson resonances in the ρ', ω' and ϕ' families is discussed in the light of recent results, particularly from the CERN photoproduction experiment WA4.

I INTRODUCTION

On theoretical grounds one expects, in addition to the ground state vector mesons ρ, ω, ϕ, ψ and Υ, higher mass $J^P = 1^-$ objects denoted ρ', ω', ϕ', ψ' and Υ'. In the Veneziano model these resonances correspond to poles on the daughter trajectories under the leading Regge trajectory. Both the presence of these daughter states and their spacing on the trajectory need to be established experimentally; on the ρ-f-g trajectory, for example, one wants to know whether the f has a daughter as well as the g, i.e. whether the first ρ' is to be expected at ~ 1.2 GeV or ~ 1.6 GeV. Another viewpoint is provided by quark models, where heavy vector mesons are seen as radial excitations of the $q\bar{q}$ system, with principal quantum number n>1. With the introduction of orbital as well as radial excitations, a more complex vector meson spectroscopy (denoted $n^{2S+1}L_J$ in the usual spectroscopic notation) is possible. The ψ' and Υ' spectroscopies have been successfully explained in these terms; in the case of ψ', radial excitations up to n = 4 have been established.

Progress in identifying the recurrences of the conventional vector mesons ρ, ω and ϕ has been slower. According to the Vector Dominance Model, vector mesons are coupled directly to the photon, so it is not surprising that most of our knowledge of these resonances has come from observations of the neutral charge states in e^+e^- and photoproduction experiments. A steady growth of results since the early 1970's, generally of low statistics, gave rise to a confusing picture characterised by a multitude of higher vector meson candidates. Within the last few years, however, higher statistics data have become available from the e^+e^- storage rings DCI, VEPP-2M and Adone, and from the photoproduction experiments at Fermilab (broad band beam) and CERN (tagged Υ beam). In this paper recent results on heavy vector mesons are discussed with particular emphasis on results from the CERN photoproduction experiment. The main ρ' candidates ρ' (1600) and ρ' (1250) are discussed in sections 2 and 3 while the ω' and ϕ' candidates are reviewed in section 4.

II $\rho(1600)$

From its quantum numbers I^G I^{PC} = 1^+ 1^{--} the following decay modes are the simplest expected for a ρ'.

$$\rho'^0 \to \pi^+\pi^- \tag{1}$$

$$\left.\begin{array}{l} \rho^0\pi^+\pi^- \\ \rho^0\ \varepsilon^0 \\ A_1{}^{\pm}\pi^{\mp} \end{array}\right\} \quad \pi^+\pi^-\pi^+\pi^- \tag{2}$$

$$\left.\begin{array}{l} \omega\pi^0 \\ \rho^0\pi^0\pi^0 \\ \rho^+\rho^- \end{array}\right\} \quad \pi^+\pi^-\pi^0\pi^0 \tag{3}$$

$$\eta\pi^+\pi^- \qquad\qquad \pi^+\pi^-\pi^+\pi^-\pi^0 \tag{4}$$

If quasi 2 body final states dominate, I spin relates the ρ' branching ratios into the final states (2) and (3). For the ratio $R = \rho' \to \pi^+\pi^-\pi^+\pi^-/\rho' \to \pi^+\pi^-\pi^0\pi^0$ one expects $R = 2$ for $\rho\varepsilon$ intermediate states, $R = 1$ for $A_1\ \pi$ and $R = 0$ for $\omega\pi^0$ and $\rho^+\rho^-$. The decays $\rho' \to \rho\pi$ and $\rho' \to \omega\pi\pi$ are forbidden by G parity.

2.1 Previous results on $\rho'(1600)$

The $\rho'(1600)$ is the only light radial excitation well enough established to appear in the Particle Data Group tables. It has been seen in the mode $\rho' \to \pi^+\pi^-$ in photoproduction [1] and in π^-p phase shift analyses [2]. It has been seen in the mode $\rho' \to 4\pi$ in e^+e^- interactions [3] and in photoproduction [4]. However, these early observations left many unanswered questions about the $\rho'(1600)$:

a)   Which decay modes contribute? Only the decays $\rho'(1600)$ $\to \pi^+\pi^-, \pi^+\pi^-\pi^+\pi^-$ have been identified. The other decay modes listed above have not been seen; in particular there is no evidence for a $\rho'(1600)$ decaying into $\pi^+\pi^-\pi^0\pi^0$. The ratio $\rho'\to\pi^+\pi^-/\rho' \to \pi^+\pi^-\pi^+\pi^-$ is known to be small (< 0.2) but the exact value remains to be established; in fact some dipion phase shift analyses have favoured solutions with no $\rho'(1600) \to \pi\ \pi$ at all. Furthermore the reason for the smallness of this ratio is not well understood theoretically.

b)   Is the observed broad peak in $\gamma p \to 4\pi p$ really a resonance, or is it a threshold enhancement created by a Deck-type effect, or a Ferbel-Slattery mechanism [5]? In the process $e^+e^- \to \rho'(1600) \to 4\pi$, although a Deck-type mechanism is excluded, other processes exist which could simulate a resonance peak [6]. Hence observation of a $\rho'(1600)$ signal in several different modes is required to prove the resonant nature of the enhancement.

c)   What is the structure of the $\pi^+\pi^-\pi^+\pi^-$ final state? While it
is known that this state is dominated by $\rho\pi\pi$ with the dipion pair
most likely to be in a low mass $I = 0$ s wave (often characterised
as the ε), it has not been established whether such a $\rho\varepsilon$ mode
can be distinguished clearly from an $A_1\pi$ decay mode. While it is
generally considered that the ρ and the ε are in a relative s wave,
it is not known whether any d wave component is necessary.

d)   Are there in fact two resonances in this region? Estimates of
the ρ' mass have varied from 1430 to 1620 MeV and of the width from
310 to 710 MeV. The DCI/M3N group have claimed that their e^+e^-
$\rightarrow \pi^+\pi^-\pi^+\pi^-$ and $e^+e^- \rightarrow \pi^+\pi^-\pi^0\pi^0$ data cannot be fitted by one
resonance, and suggested that there are 2 resonances of $\Gamma \sim 200$ MeV
in this region, centred at 1530 and 1690 MeV [3]. The DESY-Frascati
experiment [7] studying the interference pattern between the Bethe-
Heitler amplitude and the real part of the Compton diffractive
photoproduction amplitude in the reaction $\gamma p \rightarrow e^+e^-p$, find that the
mass region above 1.5 GeV cannot be described by the $\rho'(1600)$ alone,
and that the data gives indication for a structure at 1700 MeV.
Note that in the quark model one would expect two ρ' resonances in
the first radially excited state, the 2^3S_1 and the 2^3D_1 states;
these could interfere.

2.2   New results on $\rho'(1600)$

Five experiments have recently reported new results on the
$\rho'(1600)$:

a)   A phase shift analysis [8] by the CERN-Cracow-Munich collaboration of the channel

$$\pi^-p \rightarrow \pi^+\pi^-n,$$

where the sample now includes polarised target data, gives a unique
solution that requires a $\rho'(1600)$ with parameters

$$M = 1598^{+24}_{-22} \text{ MeV} \quad \Gamma = 175^{+98}_{-53} \text{ MeV}.$$

The elasticity $R = 28.7 \pm 4.3\%$

b)   The data from DCI [9] Adone [25] and VEPP-2M [24] now cover
the full mass spectrum of the reaction

$$e^+e^- \rightarrow \pi^+\pi^-\pi^+\pi^-$$

as shown in Fig 1. There is a clear broad resonance centred at
~ 1.6 GeV.

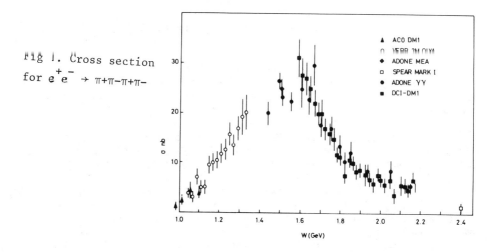

Fig 1. Cross section for $e^+ e^- \to \pi^+\pi^-\pi^+\pi^-$

The Adone group have fitted the data with one relativistic Breit-Wigner with an energy-dependent width and find for the resonance parameters

$$M = 1666 \pm 39 \text{ MeV} \qquad \Gamma = 700 \pm 160 \text{ MeV}.$$

This mass and width differ from the peak and FWHM of the resonance shape, estimated at $\sim$ 1.53 GeV and $\sim$ 0.53 GeV respectively, because of the energy-dependent from assumed for Γ. The VEPP-2M group find that a 2 resonance fit with $\rho(760)$ and $\rho'(1600)$ gives a better fit than a single $\rho'(1600)$. In contrast to the older DC1/M3N data the DC1/DM1 group find no evidence for structure around 1700 MeV. All groups agree that the dynamics is dominated by $\rho\pi\pi$ production. The Adone group's analysis favours the dipions to be in an s wave state. The DC1/DM1 group find no structure in the $\rho\pi$ spectrum.

c)   The Fermilab broad band photoproduction experiment (the Columbia-Illinois - FNAL Callaboration) studied the reactions

$$\gamma C \to \pi^+\pi^-$$

$$\gamma C \to \pi^+\pi^-\pi^+\pi^-$$

in the energy range 50 - 200 GeV [10]. In the first reaction they see a clear signal for $\rho' \to 2\pi$, Fig 2a. The results of a fit with a monotonically decreasing background plus a Breit-Wigner resonance shape gives for the resonance parameters

$$M = 1600 \pm 10 \text{ MeV} \quad \Gamma = 283 \pm 14 \text{ MeV}$$

with an additional 40 MeV systematic uncertainty on both parameters.

In the 4π final state, the mass distribution, Fig 2b, shows a broad resonance centred at ∿ 1.55 GeV which cannot be described by a single Breit-Wigner shape of arbitrary mass and width. The solid curve in Fig 2b shows a Breit-Wigner with the same mass and width as the observed 2π resonance. The discrepancy in shapes is taken as evidence that there are significant additional sources of four pion events besides the ρ' (1600).

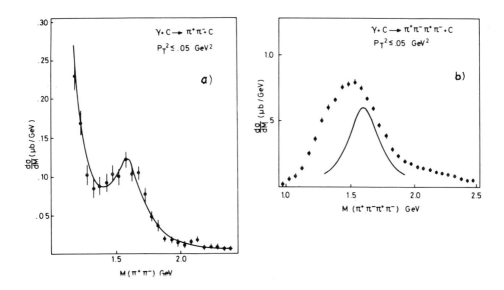

Fig 2. FNAL data

a) M (π+π-) from γC → π+π-
b) M (π+π-π+π-) from γC → π+π-π+π-
Solid curve shows Breit - Wigner with same mass and width as observed 2π resonance

d)   The LAMP2 group at Daresbury have obtained data with a low energy γ beam on the reaction [21]

$$\gamma p \to \pi^+ \pi^- \pi^+ \pi^- p$$

for low 4π masses. They have fitted the then available world data (i.e. not including the CERN WA4 results) on this reaction up to a 4π mass of 1.6 GeV and conclude that an enhancement at ∿ 1.2 GeV is needed as well as a ρ'(1600) to explain the data, and find ρ' (1600) parameters of M = 1.54 ± 0.03 GeV, Γ = 478 ± 135 MeV.

e)   The CERN tagged photon beam experiment WA4 (the Bonn-CERN-Ecole Polytechnique-Glasgow-Lancaster-Manchester-Orsay-Paris VI-Paris VII -Rutherford Sheffield Collaboration) studied the exclusive channel

$$\gamma p \rightarrow \pi^+ \pi^- p$$

with a tagged γ beam in the energy range 20-70 GeV._ Experimental details are given in ref. 37. Fig.3a shows the $\pi^+\pi^-$ mass spectrum from this channel [11]. In addition to the ρ one sees an excess of events at $\sim$ 1.6 GeV. Fitting a relativistic Breit-Wigner with a mass-independent width and a second order polynomial to the mass region 1.1 - 2.4 GeV gives the resonance parameters

$$M = 1590 \pm 20 \text{ MeV}, \quad \Gamma = 230 \pm 80 \text{ MeV}.$$

This result is in good agreement with the mass and width found for $\rho' \rightarrow 2\pi$ by the CERN π-p phase shift and the Fermilab γp experiments quoted above, but it should be noted that a larger width can be accommodated with a different background shape.

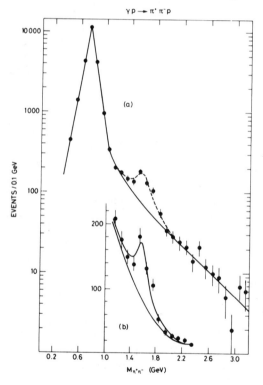

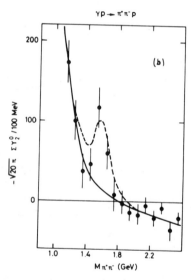

Fig 3. $\gamma p \rightarrow \pi^+\pi^-$p   (WA4)

a) $M(\pi^+\pi^-)$ logarithmic & linear scales. Curves are fits to 2nd order polynomial + Breit-Wigner. Background on log (linear) plot is handrawn (2nd order polynomial)

b)   Moment-$\sqrt{20\pi} \ \Sigma Y^0_2$ as function of 4π mass

The spin of the enhancement was investigated by calculating the spherical harmonic sums $\sum Y_{\ell}^{0}(\theta)$ where θ is the polar angle of the π^{+} in the two pion rest frame. The only moment which gives a significant peak in the 1.6 GeV region is Y_{2}^{0} Fig 3b shows $-\sqrt{20\pi} \sum Y_{2}^{0}$ which, for an s channel helicity conserving (SCHC) 1^{-} state, should be equal to the number of events in the mass plot (acceptance biases are unimportant for this moment). The dotted curve on Fig 3b is the result of the fit to the mass spectrum superimposed on a smooth handrawn background. From the good agreement between the curve and the data points one concludes that the enhancement is consistent with being due to an SCHC 1^{-} state.

From the analysis of the channel $\gamma p \to \pi^{+}\pi^{-}\pi^{+}\pi^{-}p$ in the same experiment, where one finds that 0.7 ± 0.1 of the channel contains a ρ' (1600), one obtains $B\sigma$ (ρ' (1600) $\to \pi^{+}\pi^{-}\pi^{+}\pi^{-}p$) $= 0.6 \pm 0.2 \mu b$. Hence one finds $R = 0.16 \pm 0.09$. Assuming that $\rho' \to \pi^{+}\pi^{-}\pi^{+}\pi^{-}/\rho' \to$ all is ~ 0.5, this corresponds to $B = \rho' \to \pi^{+}\pi^{-}\pi^{+}\pi^{-}/\rho' \to$ all about 0.1. However, larger values of B or R cannot be excluded, in particular from the possibility that the $\rho' \to 2\pi$ mass spectrum could be distorted through interference with the tail of the ρ. The smallness of the ratio R has been explained [12] in the framework of the quark pair creation model for strong decay amplitudes as being due to a suppression of the 2π mode resulting from the presence of nodes in the spatial wave function of the radially excited ρ'. (A similar explanation has been invoked to explain the relative suppression in ψ' (4.414) decays of the $D\bar{D}$ and $D^{*}\bar{D}+D\bar{D}^{*}$ modes with respect to $D^{*}\bar{D}^{*}$ decays). More simply, viewing the ρ' as a radially excited state of the ρ, the predominance of the sequential decay $\rho' \to \rho\pi\pi$ is perhaps not surprising.

The WA4 experiment also studied the reaction

$$\gamma p \to \pi^{+}\pi^{-}\pi^{+}\pi^{-}p$$

in some detail [13]. The 4π mass spectrum, Fig 4a, shows a broad peak at a mass of 1.6 GeV. The $\pi^{+}\pi^{-}$ spectrum, Fig 4b shows a strong ρ (760) peak which, if possible interference effects are neglected, corresponds to almost 1 ρ/event. There is no evidence for $\rho^{0} \rho^{0}$ production.

The spin characteristics of the 4π peak were studied using as an analyser θ_{++}, the angle in the 4π CM system between the resultant of the momenta of the two π^{+} mesons and the momentum of the recoil p in the s channel system. The moments $\langle Y_{0}^{0} \rangle$, $\langle Y_{1}^{1} \rangle$ and $\langle Y_{2}^{0} \rangle$ as obtained by fitting to the $\cos \theta_{++}$ distribution are shown as a function of 4π mass in Fig 4c. $\langle Y_{1}^{0} \rangle$ is consistent with zero throughout. For the production of a $1^{-} \rho'$ by SCHC one expects $\langle Y_{0}^{0} \rangle = -\sqrt{5} \langle Y_{2}^{0} \rangle$ at low masses. Comparison of the relative size of these two moments shows that a major part of the production of this peak is by an SCHC mechanism. The difference in the two moments above 2 GeV indicates the onset of a different production process which can be interpreted as a jet-like mechanism.

76

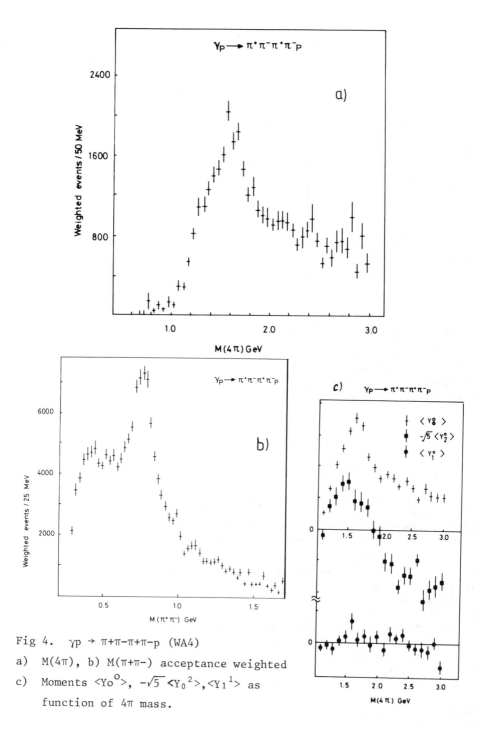

Fig 4.   γp → π+π−π+π−p (WA4)

a)   M(4π), b) M(π+π−) acceptance weighted

c)   Moments $\langle Y_0^0 \rangle$, $-\sqrt{5}\,\langle Y_0^2 \rangle$, $\langle Y_1^1 \rangle$ as

     function of 4π mass.

The substructure of the 4π system in the ρ' decay has been investigated. It is clear from the π⁺π⁻ mass plot that the dominant fianl state is ρππ. However decay schemes such as ρ'→ ρππ with the ρππ in a phase space distribution or ρ'→ ρε with the ρ and ε in a relative s wave give poor fits to the π⁺π⁻ mass spectrum for M(4π) < 1.45 GeV – the experimentally observed dip between threshold and the ρ peak is not reproduced. This can be remedied in two alternative ways:

A)   The ρ' (1600) is assumed to decay through the chain

$$\rho' (1600) \rightarrow A_1^{\pm} \pi^{\mp} \rightarrow \rho^0\pi^+\pi^+$$

Good fits are found with M (A₁) ∿ 1.3 GeV, Γ (A₁) ∿ 0.3 GeV but not with lower masses. The fit is improved by reducing the probability in the lower mass tail of the A₁ by an empirical factor.

B)   In the ρ'(1600) decay, d wave relative motion of the ρ and ε is allowed. The resulting centrifugal repulsion deforms the 4π structure in the required way. It is found that interference between s waves and d waves with relative amplitude d/s ∿ 0.25 gives a good fit to the π⁺π⁻ and 3π mass spectra while also leaving good agreement with the θ₊₊ distribution.

Fig 6. γp→π+π−π+π−p (WA4)
Results of maximum likelihood fits as function of 4π mass

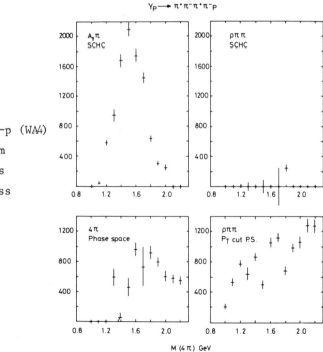

The proportion of the different substates contributing to the 4π final state was estimated by a maximum likelihood fit. In addition to i) a 'deformed' $\rho\pi\pi$ intensity as described above, other intensities, such as ii) 4π phase space, iii) Γ_π truncated phase space, iv) $\rho\pi\pi$ with SCHC alignment, and v) $\rho\pi\pi$ with P_T truncated phase space, were included. Typical results are shown in Fig 6, where the deformed $\rho\pi\pi$ intensity was model A with SCHC alignment. The major feature is the dominance of the $A_1\pi$ type contribution; its intensity reaches a maximum at $\sim$ 1.55 GeV and falls to small values above 2 GeV. A similar result is obtained for model B.

It is interesting that this peak does not have quite the same shape as the comparable e^+e^- cross section, $\sigma(\ e^+e^- \rightarrow \pi^+\pi^-\pi^+\pi^-)\ xW^2$, which can be calculated from Fig 1: the shape of the dominant 1^- component as obtained from photoproduction has a faster fall off beyond the maximum than the e^+e^- shape. Nevertheless if one takes such an $A_1\pi$ or $\rho\epsilon$ contribution to represent the ρ' (1600) peak, the ρ' mass and width are given by

$$M \sim 1.52 \text{ GeV} \qquad \Gamma \sim 0.40 \text{ GeV}$$

This result confirms the discrepancy between the ρ' width as measured in the 4π and 2π final states. The fact that the ρ' width, as deduced from the 4π mass spectrum, is significantly higher than the 2π one has been explained as being due to the large background expected from current algebra calculations [14] in the 4π case. It is not clear whether this idea can satisfactorily explain the present result where the ρ' width is measured by its $A_1\pi$ or $\rho\epsilon$ decay intensity rather than from the raw mass spectrum. It is also possible that the ρ' width as measured in the 2π mode is underestimated because of distortions in the $\pi\pi$ spectrum from the tail of the ρ.

The WA4 experiment have also looked for evidence of other decay modes of the ρ' (1600). In the channel [33]

$$\gamma p \rightarrow \pi^+\pi^-\pi^+\pi^-\pi^0 p$$

the $\pi^+\pi^-\pi^0$ spectrum shows a clear η signal, Fig 10a. The $\eta\pi\pi$ spectrum, Fig 10c shows a broad enhancement centred at $\sim$ 1.6 GeV. There is appreciable ρ production in the $\pi^+\pi^-$ spectrum recoiling against the η. If the $\eta\pi\pi$ enhancement is due to ρ' (1600) this would imply a ρ' (1600) branching ratio of $\eta\pi^+\pi^-$ to $\pi^+\pi^-\pi^+\pi^-$ of $\sim$ 0.1. In the channel [23]

$$\gamma p \rightarrow \pi^+\pi^-\pi^0\pi^0 p$$

discussed in the next section, there is no evidence within the statistics for ρ' (1600) $\rightarrow \omega\pi^0$. In the channel [37]

$$\gamma p \rightarrow \pi^+\pi^-\pi^+\pi^-\pi^+\pi^- p$$

there is no evidence in the 6π mass spectrum for a ρ'(1600) signal.

III ρ' (1250)

3.1 Previous results on ρ' (1250)

The ρ' (1250) has been hinted at by data on $e^+e^- \to \pi^+\pi^-$[15] where the pion form factor has a strong deviation from the shape expected from the ρ tail, and from the rapid rise in the $e^+e^- \to \pi^+\pi^-\pi^0\pi^0$ cross section, dominated by the $\pi^0\omega$ channel [16]. It has been argued [17] however that the enhancement in the pion form factor reflects the inelastic effect coming from the $\omega\pi^0$ channel, whose resonant nature has not itself been established. There has been some evidence for a ρ' (1250) from $\bar{p}p$ annihilations at rest in the reaction $\bar{p}p \to e^+e^-$ + neutrals [18]. Good evidence for an effect at 1250 MeV comes from the SLAC-LBL photoproduction experiment [19] in the reaction

$$\gamma p \to p\pi^+\pi^- + \text{neutrals}$$

There is a strong peak at 1250 MeV of $\Gamma \sim$ 150 MeV in the $M(\pi^+\pi^-\text{MM})$ plot, which is enhanced by selecting 0.32 <MM < 0.60 GeV, a requirement which corresponds effectively to demanding a $\pi^0\omega$ intermediate state. Since the π^0's were not detected, no spin analysis could be carried out and it was not possible to distinguish between $J^P = 1^+$ and 1^- and in particular to rule out the 1^+ B(1235) meson, which is known to be strongly coupled to $\pi\omega$, as the source of the enhancement. The DESY-Frascati experiment claim good evidence for a ρ'(1250) of $M \sim$ 1265, $\Gamma \sim$ 110 MeV in their interference measurements [20] in $\gamma p \to e^+e^-p$. They also see, below the ρ'(1250) signal, a very strong narrow effect at $M \sim$ 1100 MeV of width $\Gamma \sim$ 30 MeV whose interpretation is unclear.

Theoretically, interpretation of an enhancement in $\omega\pi$ as a ρ' is complicated by the fact that some enhancement is expected anyway in this region from the Renard mechanism [22] describing the coupling of the tail of the ρ to $\pi\omega$.

3.2 New results on ρ' (1250)

New results have come from the Daresbury and CERN photo-production experiments and from the low energy e^+e^- storage rings.

a)   The low energy Daresbury photoproduction experiment (LAMP 2) studied the channel [21]

$$\gamma p \to \pi^+\pi^-\pi^0\pi^0 p$$

where only one of the π^0 is detected. After subtracting out the copious $\omega\Delta^{++}$ production, the $\omega\pi^0$ mass spectrum shows a peak at $\sim$ 1.3 GeV with a width of $\sim$ 0.3 GeV whose angular distribution is consistent with a 1^- state. This group also carried out an analysis of the reaction

$$\gamma p \rightarrow \pi^+ \pi^- \pi^+ \pi^- p$$

for all data available then and found that although the 4 pion spectrum in some individual experiments may be acceptably described by $\rho'(1600)$ production, this resonance alone cannot satisfactorily explain all the data taken together, where there is a 1^- signal around 1.2 GeV in excess of that expected from the low mass tail of the $\rho'(1600)$ which can be attributed to coherent production of the $\rho'(1250)$.

b)   The CERN WA4 experiment studied the channel [23]

$$\gamma p \rightarrow \pi + \pi^- \ \pi^0 \pi^0 \ p$$

where both π^0 are detected.   A strong ω is seen in the $\pi^+ \pi^- \pi^0$ spectrum, and in the $\omega \pi^0$ spectrum, Fig 7a, there is evidence for an enhancement of mass $\sim$ 1.25 GeV and width $\sim$ 0.3 GeV.   Observation of an enhancement in the $\omega \pi^0$ channel uniquely identifies its I^{GC} as 1^{+-}.   The spin-parity of the enhancement was investigated by studying the two angles ψ and Θ where ψ is the direction of the ω in the $\omega \pi$ rest frame and Θ is the normal to the ω rest frame with the z axis defined as the $\omega \pi$ direction in the overall γp rest frame, i.e. the s channel axis.   Assuming SCHC one gets the following predictions for ψ and θ depending on the spin state of the resonance.

State	ψ	θ
1^+ s wave	isotropic	$\sin^2\theta$
1^+ B meson (s-d mixture)	$1 + \cos^2\psi$	$\sin^2\theta$
1^- p wave	$1 + \cos^2\psi$	$1 + \cos^2\theta$

The experimental distributions of these angles are shown in Fig.7b.   Also shown are the acceptance-corrected predictions for a 1^+ s wave $\omega \pi$ state and for an isotropic decay of the $\omega \pi$ state. The shape of the acceptance distributions in cos ψ means that this angle is not suitable to distinguish 1^+ from 1^-.   The experimental θ distribution is inconsistent with the acceptance-corrected $\sin^2\theta$ prediction, thus ruling out indentification of the entire enhancement with the 1^+ B meson.   The fact that the θ distribution is consistent with isotropy implies that the enhancement consists of 1^+ and 1^- in the ratio of approximately 1:2.   Even without the assumption of SCHC one could rule out the B as the only source of the peak since the width of the enhancement is inconsistent with that of the B of width 130 MeV.   Within the statistics, however, it is not possible to rule out other interpretations, such as that the enhancement is due to a Deck effect or to the Renard mechanism.

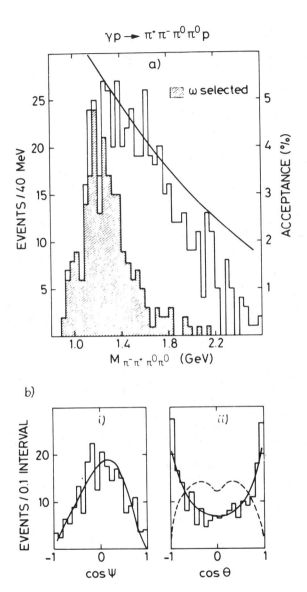

Fig 7. $\gamma p \to \pi^+\pi^-\pi^0\pi^0 p$ (WA4)
a) $M(\pi^+\pi^-\pi^0\pi^0)$ Shaded area shows $M(\omega\pi^0)$ with ω defined as $0.72 < M(\pi^+\pi^-\pi^0) < 0.84$ GeV. Solid line shows acceptance.
b) Polar decay angles ψ and θ as defined in text. Solid curve shows acceptance, dotted curve shows acceptance – biased prediction for 1+ decay

c) New results on the reaction

$$e^+e^- \to \pi^+\pi^-\pi^0\pi^0$$

have come from the VEPP-2M OLYA experiment [24] and the Adone γγ experiment [25], giving the spectrum shown in Fig.8. The expected shape with no ρ'(1250) production, as calculated by the Adone group, is shown too. This is computed as the incoherent sum of the effect from the Renard mechanism plus the expected cross section for the ρ'(1600) decay into ρ⁰π⁰π⁰

$$\sigma\ (\rho'\ (1600) \to \rho^0\pi^0\pi^0) = \tfrac{1}{2}\ \sigma\ (\rho'\ (1600) \to \rho^0\pi^+\pi^-)$$

when the dipion system is in an s wave.

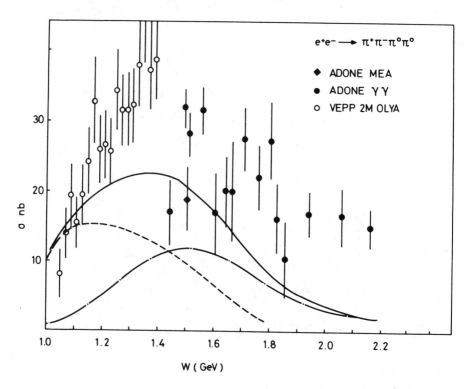

Fig 8. Cross section for $\gamma p \to \pi^+\pi^-\pi^0\pi^0 p$ The dashed and broken curves show the effect expected from the Renard mechanism and the ρ' (1600) → ρ⁰π⁰π⁰ decay respectively The solid curve is the incoherent sum of these two effects.

There is an excess of events around 1.3 GeV above this expectation which might be identified as the ρ' (1250). However this excess appears to be too small to fit in with a simple scaling law of the following form which can be derived from GVMD or local duality arguments [26]

$$\frac{M(\rho') \ \Gamma \ (\rho' \to ee)}{M^2(\rho'') - M^2(\rho')} \quad = \quad \frac{M(\rho) \ \Gamma \ (\rho \to ee)}{M^2(\rho') - M^2(\rho)}$$

where $\rho \equiv \rho(760)$, $\rho' \equiv \rho(1250)$, $\rho'' \equiv \rho'(1600)$.
This gives $\Gamma(\rho' \to ee) \sim 4$ KeV which corresponds to a peak cross section of 150-250 nb. From Fig. 8 it is clear that any observed excess is too small by a large factor. On the other hand one can convert the $\rho'(1250)$ cross section as found in photoproduction to an expectation in e^+e^- via the approximation

$$\sigma^{peak} \ (e^+e^- \to \rho') = \frac{\sigma(\gamma p \to \rho' p)}{\sigma(\gamma p \to \rho p)} \ \frac{M(\rho) \ \Gamma \ (\rho)}{M(\rho')\Gamma(\rho')} \qquad \sigma^{peak} (e^+e^- \to \rho)$$

Assuming that 2/3 of the $\omega\pi^0$ peak seen in the WA4 result comes from $\rho'(1250)$, one obtains a figure of ~ 34 nb for the peak cross section expected in $e^+e^- \to \pi^+\pi^-\pi^0\pi^0$ from $\rho'(1250)$ production, in closer agreement with the data.

The conclusion on the $\rho'(1250)$ is that although there is definite evidence for an enhancement in this region, its interpretation as a ρ' is problematical. If is is a ρ' one would have to explain why its coupling to γ appears to be smaller than that of the $\rho'(1600)$, how exactly it fits into the quark model, whether it is consistent with P wave $\pi\pi$ phase shifts, and why it is not clearly seen in other decay modes such as $\pi^+\pi^-$. It is hoped that the CERN photoproduction experiment WA57, one of whose main aim is the investigation of the $\pi^+\pi^-\pi^0\pi^0 p$ channel, will provide more information on this state.

IV ω' AND ϕ' CANDIDATES

These higher vector mesons must have $I^G = 0^-$. By analogy with the $\rho'(1600) \rightarrow \rho\pi\pi$ decay, one expects the decay modes

$$\omega' \rightarrow \omega\pi\pi$$
$$\phi' \rightarrow K^*K\pi$$

but $\phi' \rightarrow \phi\pi\pi$ is forbidden by the OZI rule.   By analogy with ρ' (1600) $\rightarrow \pi\pi$, one would expect

$$\omega' \rightarrow \pi^+\pi^-\pi^0$$
$$\phi' \rightarrow K^+K^- \quad .$$

Although the decays $\rho' \rightarrow \rho\pi$ and $\omega' \rightarrow \omega\pi$ are forbidden by G parity, the decay

$$\phi' \rightarrow K^*K$$

is allowed.   The ϕ' is expected to be narrower than the ρ' as it is closer to threshold but the ω' is expected to be at least as broad as the ρ' [27]

Previous evidence for these states has been scanty.   The DCI /M3N group found evidence [3] for a state at 1770 MeV with $\Gamma \sim 50$ MeV in $e^+e^- \rightarrow 4\pi^\pm \pi^0$, but the more recent higher statistics DCI/DM1 data [9] do not confirm this object.   Three Adone groups have found a signal [28] at 1.82 GeV with $\Gamma \sim 30$ MeV taking the sum of the two channels $e^+e^- \rightarrow 2\pi^+ 2\pi^- \pi^0$ and $e^+e^- \rightarrow 2\pi 2\pi^-2\pi^0$. (The separation between the two channels is difficult.)   The favoured interpretation of the signal is as a baryonium effect rather than as a ϕ' or ω'.   Preliminary results from two Adone groups [29] indicating a narrow signal, $\Gamma < 4$ MeV, at $M \sim 1500$ MeV in $e^+e^- \rightarrow$ > 2 charged particles have not been confirmed by later data from the same groups [30].   The Adone/MEA group has also reported [31] a resonance at 2130 MeV with $\Gamma \sim 30$ MeV from a study of K^{*0} inclusive final states; however this object does not appear to be confirmed by the preliminary DCI/DM1 results [9] although an identical analysis has not been performed.   An experiment carried out with the Cornell 11.5 GeV bremsstrahlung photon beam [32] found, in the reaction $\gamma p \rightarrow K^+K^- + X$, a dip in the K^+K^- mass spectrum which they interpret as possibly resulting from the interference of a resonance of $M \sim 1.83$ GeV and $\Gamma \sim 120$ MeV with a kaon Söding amplitude background.

A state which seems to be confirmed by subsequent experiments is the enhancement observed in the $\pi^+\pi^-\pi^0$ and $\pi^+\pi^-\pi^+\pi^-\pi^0$ spectra by the DCI/M3N group [3] with $M \sim 1660$ MeV and $\Gamma \sim 40$ MeV.   A similar object was seen by the DCI/DM1 group [9] in $e^+e^- \rightarrow \pi^+\pi^-\pi^+\pi^-\pi^0$ On selecting an ω in the final state a strong peak is observed around 1640 MeV with $\Gamma \sim 90$ MeV, Fig.9.   More data gathered since has increased the significance of the structure.

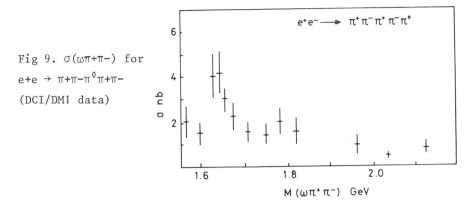

Fig 9. $\sigma(\omega\pi^+\pi^-)$ for

$e^+e^- \to \pi^+\pi^-\pi^0\pi^+\pi^-$

(DCI/DM1 data)

The CERN WA4 experiment searched for ω' in the reaction [33]

$$\gamma p \to \pi^+\pi^-\pi^+\pi^-\pi^0\, p \ .$$

A clean ω signal is obtained in the $\pi^+\pi^-\pi^0$ spectrum. Fig 10a.
The $\omega\pi^+\pi^-$ mass spectrum, Fig 10b, shows a broad threshold
enhancement, in contrast to the behaviour of peripheral 3-body phase
space, at M $\sim$ 1.7 GeV with $\Gamma \sim$ 0.5 GeV.

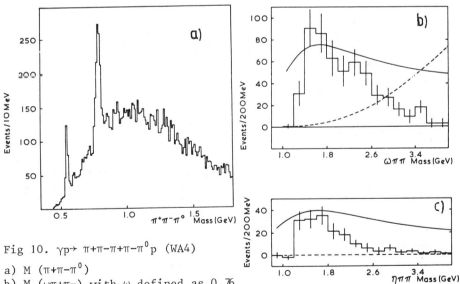

Fig 10. $\gamma p \to \pi^+\pi^-\pi^+\pi^-\pi^0 p$ (WA4)

a) M $(\pi^+\pi^-\pi^0)$
b) M $(\omega\pi^+\pi^-)$ with ω defined as 0.76
$< M(\pi^+\pi^-\pi^0) < 0.815$ GeV
c) M$(\eta\pi^+\pi^-)$ with η defined as 0.53
$< M (\pi^+\pi^-\pi^0) < 0.56$ GeV.
Continuous curves show acceptance;
dashed curve in b) shows
peripheral 3 body phase space.

In the region of the enhancement the distributions of the acceptance-corrected polar angle ψ between the ω and the s channel axis in the $\omega\pi\pi$ CM is consistent with isotropy. The distribution of the angle θ between the normal to the ω decay plane and the s channel axis is consistent with isotropy or a $\sin^2\theta$ distribution. These results are consistent with expectations from the decay of a 1^- object by SCHC. The cross section of the enhancement is ~ 0.1 μb, consistent with $1/9 \times \sigma$ ($\rho'1600$) as expected on vector dominance grounds for an ω' in the same octet as the ρ' (1600).

In the same channel, $\gamma p \to \pi^+\pi^-\pi^+\pi^-\pi^0$ p, there is some preliminary evidence for a narrower object around 1660 MeV with $\Gamma \sim 50$ MeV in the 5π mass spectrum, Fig 11.

Fig 11.

$\gamma p \to \pi+\pi-\pi+\pi-\pi^0 p$ (WA4)

M ($\pi+\pi-\pi+\pi-\pi^0$)

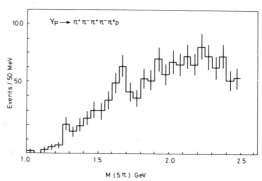

Plotting the $\pi^+\pi^-\pi^0$ spectrum for events in the peak shows that the peak region has about 0.5 ω/event. A similar peak is seen by the WA4 collaboration in the channel.

$$\gamma p \to \pi^+\pi^-\pi^0 \, p$$

A plot of the $\pi^+\pi^-\pi^0$ mass spectrum above the ω peak, Fig 12, shows some indications for a peak at around 1690 MeV with a width of ~ 130 MeV.

Fig 12. $\gamma p \to \pi+\pi-\pi^0 p$ (WA4)

M($\pi+\pi-\pi^0$) for masses

above the ω.

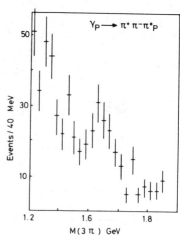

The interest of these two indications in photoproduction of 3π and 5π states of a fairly narrow peak around 1670 MeV is that, as mentioned above, similar peaks have been observed in $e^+e^- \rightarrow 3\pi$ and 5π. However the width of this object, $\Gamma \sim 50 - 100$ MeV, is too narrow to fit into the conventional picture of an ω' resonance. One possible interpretation is that it could in fact be a ϕ' resonance decaying into these pion channels via a slight non-ideal $\omega'-\phi'$ mixing. If that were the case one would still expect to see evidence for a ϕ' of this mass decaying into kaon channels. Preliminary results from the WA4 experiment do in fact show some evidence for an enhancement in the reaction [34]

$$\gamma p \rightarrow K^+ K^- p$$

in the K^+K^- mass spectrum at M $\sim$ 1.69 GeV with $\Gamma \sim$ 120 MeV.

The WA4 experiment also studied the reaction (35)

$$\gamma p \rightarrow K^+K^- \pi^+\pi^- p$$

The channel was extracted with an estimated 20% contamination from the reaction $\gamma p \rightarrow p\bar{p}\pi^+\pi^- p$. The two body mass spectra show that the channel is dominated by K*(890) and ϕ (1020) production, contained in 35% and 10% of events respectively. The KK$\pi\pi$ spectrum shows a broad threshold enhancement peaking at $\sim$ 1.9 GeV. There is no evidence for a narrow ϕ at $\sim$ 1.7 GeV. The K*Kπ component of the enhancement was investigated by plotting the KK$\pi\pi$ mass spectrum selecting on the K* peak (M (Kπ) in 0.83-0.95 GeV) and the K* wings (M(Kπ) in 0.71-0.83, 0.95-1.07 GeV) respectively, as shown in Fig 13a.

The K*Kπ spectrum shows an enhanced peak at M $\sim$ 1.9 GeV of $\Gamma \sim$ 0.4 GeV which disappears when the non-K* events are selected. The $\phi\pi\pi$ spectrum shows no signs of a resonance, Fig 13b. By analogy with the ρ'(1600) decay one would expect the $\phi \rightarrow$ K*Kπ decay to be SCHC with the K* in a relative s wave with respect to the s wave Kπ system. The spin structure of the enhancement was investigated by plotting the 3 polar angles described in Fig 13c and comparing to the acceptance - corrected prediction for a ϕ' decaying in this way. From the figure one concludes that the data are compatible with the enhancement having spin-parity 1^- but also with an isotropic decay. Hence a mixed spin parity state such as might arise from a Deck effect cannot be excluded.

The DCI/DM1 data for this channel, $e^+e^- \rightarrow K^+K^-\pi^+\pi^-$, confirm that the channel is dominated by K*^{0} production. The KK$\pi\pi$ mass spectrum they find is not inconsistent with the CERN results but they claim that in 3 body spectra including the K* there is no evident structure [9]. The Cornell electroproduction experiment studying the channel ep $\rightarrow$ K$^+$K$^- \pi^+\pi^-$p also find a KK$\pi\pi$ spectrum similar to the photoproduction one but the K* Kπ plot is not shown [36].

88

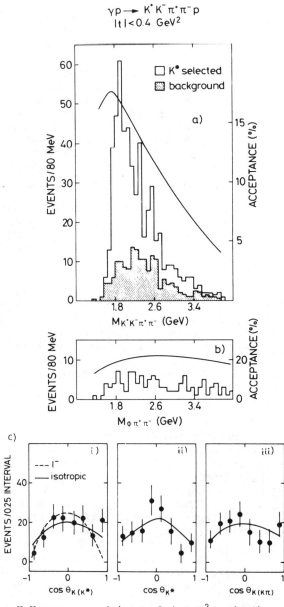

$$\gamma p \rightarrow K^* K^- \pi^+ \pi^- p$$
$$|t| < 0.4 \text{ GeV}^2$$

Fig 13. $\gamma p \rightarrow K+K-\pi+\pi-p$ $|t| < 0.4$ GeV^2 (WA4)
a) M(K*Kπ) for K* peak and background. b) M(φπ+π−) Smooth curves
show acceptance c) Polar decay angles for K* Kπ mass in range
1.8 − 2.1 GeV for i) K from K* in K* rest frame, ii) K* in K*
Kπ rest frame, iii) K not from K* in Kπ rest frame, all with Z
axis as K* Kπ direction in γp CM. Solid lines show acceptance,
dashed line acceptance − biased prediction for 1$^-$ SCHC s−wave
decay.

The WA4 experiment looked for the decay mode $\phi' \rightarrow K^*K$ in the channels

$$\gamma p \rightarrow K^0 K^{\pm} \pi^{\mp} p$$
$$\gamma p \rightarrow K^+ K^- \pi^0 p$$

The combined number of events in these channels is only about 25% of that in the $K^+K^-\pi^+\pi^-p$ channel. The two channels are dominated by K^* production (both K^{*0} and $K^{*\pm}$), Fig 14a, with about 0.5 $K^*/$ event.

Fig 14.

$\gamma p \rightarrow \begin{cases} K^0 K^{\pm} \pi^{\mp} p \\ K^+ K^- \pi^0 p \end{cases}$

a) $M(K\pi)$ all charges
b) $M(K^*K)$ with K^* peak-wings subtraction

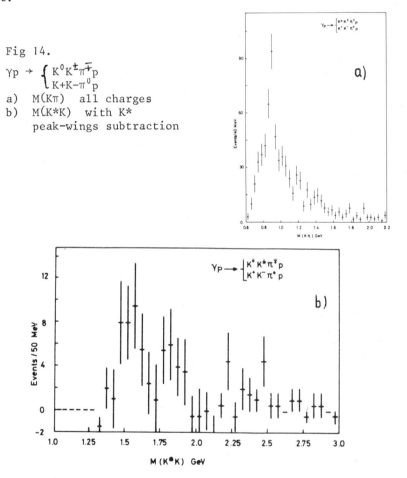

The K^*K spectrum Fig 14b, shows some evidence for 2 enhancements of $\Gamma \sim 100$ MeV at about 1550 and 1820 MeV but the statistics are too low to make any further statement.

V CONCLUSIONS

The recent high statistics γp and $\pi^+ \pi^-$ experiments have clarified the confused state of heavy vector meson spectroscopy somewhat. Many of the states found in earlier experiments have not been confirmed. The ρ' (1600) is now well established with three recent experiments agreeing on its parameters as measured in the 2π decay mode. However more understanding is needed of the different widths found in the 2π and 4π decay modes, and the full details of the substructure of the system are not resolved. The resonant nature of the 1250 MeV enhancement is still to be proved, and there is no evidence for its ω' and ϕ' partners. The evidence for a broad ω' (1700) state and the growing evidence for a narrow I=0, G= - state at $\sim$ 1.7 GeV, whose interpretation is unclear but which could be a ϕ', offers the possibility that the ρ' (1600), ω' (1700) and ϕ' (1700) could be members of an octet of radially excited vector mesons. The broad ϕ' candidate at $\sim$ 1.9 GeV would then be either a threshold enhancement or a member of a higher octet. Several other candidates discussed above remain to be disproved or confirmed. The tentative nature of these conclusions emphasises the somewhat surprising fact that light quark radial excitations are still less well understood than upsilon radial excitations.

Acknowledgements

I would like to thank Prof. A.B.Clegg and Dr.I.O. Skillicorn for many useful discussions.

References

1. F Bulos et al., Phys. Rev. Lett. 26, 149 (1971)
   H Alvensleben et al., Phys. Rev. Lett. 26, 273 (1971)
2. C D Froggatt and J L Petersen, Nucl. Phys. B129, 89 (1977)
   G Grayer et al, Nucl. Phys. B75, 189 (1974)
3. G. Cosme et al., Nucl. Phys. B152, 215 (1979)
4. H H Bingham et al., Phys. Lett. 41B, 635 (1972)
   M Davier et al., Nucl. Phys. B58, 31 (1973)
   P Schacht et al., Nucl. Phys. B81, 205 (1974)
5. T Ferbel and P Slattery, Phys. Rev. D9 824 (1973)
6. A C Hirschfeld, Nucl. Phys. B74, 211 (1974)
7. S Bartalucci et al., DESY 77/60(1977)
8. H Becker et al., Nucl. Phys. B151, 46 (1979)
9. B Delcourt et al., 1979 Int. Symp. on Lepton and Photon
   Interactions, Fermilab, p.499
10. M S Atiya et al., Phys. Rev. Lett. 43, 1691 (1979)
11. D Aston et al., Phys. Lett. 92B, 215 (1980)
12. W B Kaufmann and R J Jacob, Phys. Rev. D10, 1051 (1974)
13. A Kemp, 19th Int. Conf. on High Energy Physics, Tokyo, p.300
    (1978); D Aston et al., 'The reaction $\gamma p \rightarrow p\pi^+ \pi^+ \pi^- \pi^-$ at
    photon energies from 25-70 GeV', in preparation.
14. T N Pham, C Roiesnel and T N Truong, Phys, Lett.80B, 119
    (1978)
15. A D Bukin et al., Phys. Lett. 73B, 226 (1977)
16. M Conversi et al., Phys. Lett. 52B, 493 (1974)
17. B Costa de Beauregard, Phys. Lett. 67B, 213 (1977)
18. G Bassompierre et al., Phys. Lett. 65B, 397 (1976)
19. J Ballam et al., Nucl. Phys. B76, 375 (1974)
20. S Bartalucci et al., Nuovo Cimento 49A, 207 (1978)
21. D P Barber et al., Z.Physik C4, 169 (1980)
22. F M Renard, Nuovo Cimento 66A, 134 (1971)
23. D Aston et al., Phys. Lett. 92B, 211 (1980)
24. V Sidorov, 1979 Int. Symp on Lepton and Photon Interactions,
    Fermilab, p.490.
25. M Spinetti, 1979 Int. Symp. on Lepton and Photon Interactions,
    Fermilab,p.506; C Bacci et al., LNF-80/25(P) (1980)
26. M Greco. Nucl. Phys. B63, 398 (1973)
27. M Greco, Phys. Lett. 70B, 441 (1977)
28. C Bacci et al., Phys. Lett. 68B, 393 (1977)
    G Barbiellini et al., Phys. Lett. 68B, 397 (1977)
    B Esposito et al., Phys. Lett. 68B, 389 (1977)
29. C Bemporad, 1977 Int. Symp. on Lepton and Photon Interactions,
    Hamburg, p. 165
30. G P Murtas, 19th Int. Conf. on High Energy Physics, Tokyo,
    p.269 (1978)
31. B Esposito et al., Lett. Nuovo Cimento, 22, 305 (1978)
32. D Peterson et al., Phys. Rev. D18, 3955 (1978)
33. D Aston et al., CERN-EP/80-31, to appear in Nucl. Phys.B.
34. F Richard, 1979 Int. Symp.on Lepton and Photon Interactions,
    Fermilab, p.469

35. D Aston et al., Phys. Lett. 92B, 219 (1980)
36. Result quoted by E Gabathuler, 19th Int. Conf. on High Energy Physics, Tokyo, p.841.
37. D Aston et al., Nucl. Phys. B166, 1 (1980)

DISCUSSION

J E Wiss (Illinois)   :   Please summarise the evidence that the K* Kπ 'enhancement' is not an artefact of acceptance or phase space.

Kumar   :   The figure of the K* Kπ mass spectrum I showed (Fig 13) also shows the acceptance – it varies smoothly in the region of the enhancement. The phase space distribution for the K* Kπ system is of course the standard 3 out of 4 body phase space which does not have a threshold enhancement.

M Roos (Helsinki)   :   I think one should stress that there exists no evidence for the ρ'(1250) whatsoever. In $e^+e^-{\to}\pi^+\pi^-$ it is not needed; $e^+e^-{\to}4\pi$ one cannot analyse reliably in terms of $A_1\pi$ or ρε decays; ππ phase shifts show no ρ'(1250), and ωπ phase shifts show that the strong $J^P{=}1^-$ amplitude at this energy is not resonant. The evidence against the ρ'(1250) is about as strong as the evidence for the ρ'(1600).

The φ' that Orsay sees at 1640 MeV in the 5π channel is not absent in the $K^+K^-\pi^+\pi^-$ channel; it just cannot be separated from the ρ'(1600).

Kumar   :   Although I would probably not go so far as you in doubting the existence of the ρ'(1250), I think that I stressed in my talk that there are many difficulties concerning this state. One hopes that the CERN experiment WA57 will provide a more definitive answer on this.

As for the narrow φ' candidate at ∿1640 MeV, although there is some evidence in photoproduction for such a state decaying into K^+K^-, 3π and 5π modes, there is no narrow structure at this mass in KKππ; whether it is masked by ρ'(1600) decaying into this channel I don't know.

NEW RESULTS ON RADIATIVE MESON DECAYS

D. Cutter, S. Heppelmann, I. Joyce, Y. Makdisi, M. Marshak,
E. Peterson, K. Ruddick, M. Shupe, University of Minnesota,
Minneapolis, Minnesota 55455; D. Berg, C. Chandlee, S. Cihangir,
T. Ferbel, J. Huston, T. Jensen, F. Lobkowicz, M. McLaughlin,
T. Ohshima, P. Slattery, P. Thompson, University of Rochester,
Rochester, New York 14627; J. Biel, A. Jonckheere, P.F. Koehler,
C.A. Nelson, Fermi National Accelerator Laboratory, Batavia,
Illinois 60510.

Presented by K. Ruddick

ABSTRACT

We have investigated Coulomb dissociation (the Primakoff effect) of high energy charged π and K mesons on heavy nuclei. New values for the electromagnetic transition rates $\Gamma(\rho^- \to \pi^- \gamma)$ and $\Gamma(K^{*-} \to K^- \gamma)$ have been extracted from the data. Some preliminary data for the Primakoff production of higher meson excitations will also be presented.

INTRODUCTION

A measurement of meson radiative transition rates can provide direct tests of SU(3) symmetry and quark model ideas. In particular, transitions between the vector and pseudoscalar mesons (V → Pγ and P → Vγ) have been considered by many authors.[1] These are M1 transitions with matrix elements proportional to quark spin-flip magnetic dipole operators.

Three distinct methods have been used to obtain the electromagnetic couplings or partial widths $\Gamma(X\pi\gamma)$ where X represents a meson[2]:

1. The direct determination of the branching ratio X → $\pi\gamma$, yielding $\Gamma(X\pi\gamma) = $ B.R. x Γ_{total}. This has been successful where the branching ratio is relatively large, as for example $\omega \to \pi\gamma$, $\eta' \to \rho\gamma$, $\omega\gamma$, $\phi \to \eta\gamma$.

2. Isolation of the one pion exchange amplitude in the photoproduction of X, as for example A2$\pi\gamma$. This requires a sufficiently large coupling so that the one pion exchange amplitude may be cleanly separated.

3. Isolation of the one photon exchange amplitude as illustrated in Figure 1. This one photon exchange or Primakoff effect, is best known for its application to the determination of the $\pi^0\gamma\gamma$ and $\eta\gamma\gamma$ couplings in π^0 and η photoproduction. Even though this is an electromagnetic process, its isolation from other processes is facilitated by the very sharp forward peak in the differential cross section. Since the process occurs coherently in the nuclear coulomb field, it may be enhanced further by a factor Z^2 by using

ISSN:0094-243X/81/670094-10$1.50 Copyright 1981 American Institute of Physics

high Z targets. The technique has been exploited to isolate the K^{*0} $K^0\gamma^3$ and $\rho^-\pi^-\gamma$ [4] couplings in experiments performed in the energy range 10-20 GeV, where, however, the competing processes of ω^0 exchange and A2 exchange interfere with comparable amplitudes.

The cross-section for Primakoff production rises approximately logarithmically with incident momentum, while that for ω exchange falls inversely as the square of the incident momentum. A much more direct isolation of the photon exchange amplitude therefore will result from an increase in the incident momentum by an order of magnitude over the earlier measurements. This paper will report some new measurements of meson production via the Primakoff effect at Fermilab momenta.

THE MINNESOTA-FERMILAB-ROCHESTER EXPERIMENT

Our collaboration has constructed a high resolution spectrometer which has operated in a charged secondary beam at Fermilab. The main features of the spectrometer are shown in Figure 2.

Individual particles in the beam are tagged by a system of three Cerenkov counters and their directions determined in multiwire proportional chambers J1 and J2 before striking a target consisting of typically one tenth radiation length of lead, copper, aluminum or carbon. Magnetic analysis of secondary particles is achieved by drift chambers D1-D4 placed both upstream and downstream of a large aperture magnet. Photons are detected in a liquid argon ionization calorimeter at the downstream end of the spectrometer. Track counting in multiwire proportional counters P1 and P2, coupled with pulse height information from scintillators downstream of the target and from the liquid argon calorimeter, enable us to selectively trigger the spectrometer for those final state configurations which are of interest. The target is surrounded by anticoincidence counters to reduce triggers from events with large momentum transfer to the target nuclei.

An important feature of the experiment is the detection of K^- decays in flight, to final states of $\pi^-\pi^0$ and $\pi^-\pi^+\pi^-$, concurrent with data acquisition. This provides checks of the overall normalization, spectrometer acceptance, and the efficiencies of the various components of the system.

The spectrometer has been operated in two separate experimental runs. The first run was with a negative beam at incident momenta of 156 and 260 GeV/c with particle fractions π^-:K^-:p of .95:.04:.01 typically. The second run used a positive beam at 200 GeV/c with slightly improved triggering logic and data acqusition rates. The principal change in the latter run was the incorporation of a beryllium absorber, typically 2m in length, placed just downstream of the beam production target. The beryllium absorbed protons and pions at a greater rate than kaons, yielding particle fractions π^+:K^+:p of .45:.10:.45, thus significantly enhancing the K^+ fraction in the beam. Triggers were recorded for only one sixteenth of the incident protons.

SOME RESULTS FROM RUN 1

An estimate of the resolution of the spectrometer can be gauged from the following typical resolutions (one standard deviation) obtained from a measurement of 156 GeV/c K$^-$ decays in flight:

$$\Delta m(\pi^0 \to 2\gamma) = 7 \text{ MeV/c}^2$$
$$\Delta m(K^- \to \pi^-\pi^0) = 10 \text{ MeV/c}^2$$
$$\Delta m(K^- \to \pi^-\pi^+\pi^-) = 5 \text{ MeV/c}^2$$
$$\Delta p_\perp = 10\text{-}15 \text{ MeV/c (target dependent)}$$

1. $\pi^-\gamma \to \rho^-$

Data from the reaction $\pi^-A \to \pi^-\pi^0 A$ have been presented elsewhere.[5] A selection of these data are shown in figures 3 and 4. The differential cross-sections show the strong forward peaks corresponding to the Primakoff process with a background due to ω exchange at the one percent level, approximately, for the lead target. The lower mass region of the ρ^- is strongly enhanced and the decay angular distribution behaves as $\sin^2\theta$, as expected for photon exchange. Analysis is somewhat complicated by the relatively large total width of the ρ, requiring a convolution of a relativistic p-wave Breit-Wigner form with the differential cross-section. The ω exchange amplitude has been treated according to the prescription of Fäldt.[6] We have obtained a value $\Gamma(\rho\pi\gamma) = 67 \pm 7$ keV from these data.

2. $K^-\gamma \to K^*(890)$

Figure 5 shows preliminary data for $K^*(890)$ production in final states of both $K^-\pi^0$ and $\pi^-K^0_S$. As a preliminary estimate we quote $\Gamma(K^{*-} \to K^-\gamma) = 50 \pm 15$ keV where the uncertainty is primarily statistical.

3. $\pi^-\gamma \to A2^-$

We have observed A2 production in final states of $\eta^0\pi^-$ and $K^-K^0_S$ as shown in Figure 6. A Primakoff peak is observed but the analysis is complicated by the possible presence of a competing production process: diffraction dissociation. Full analysis is not yet complete but it appears that $\Gamma(A2^- \to \pi^-\gamma)$ is in the range 200-500 keV depending on the background subtraction.

4. $\pi^-\gamma \to A1^-$

Figure 7 shows data for the differential cross-section for $\pi^-Pb \to \pi^-\pi^-\pi^+Pb$. An excess is seen at very low t, above the diffraction dissociation, corresponding to a total Primakoff cross-section ~ 0.5 - 1.0 mb and $\Gamma(A1^- \to \pi^-\gamma)$ ~ several hundred keV. The complete analysis requires a phase shift analysis, which is at present under way.

SOME RESULTS FROM RUN 2

Calibration data for the liquid argon detector (using

incident electrons) and for the drift chambers are not yet fully analysed. However, using nominal calibrations resulting in correspondingly poor resolution, we have inspected approximately 20 percent of the data. Part of these data are presented to illustrate some of the new results to be obtained from this run.

Figure 8 shows effective masses for $(\pi K)^+$ final states. The improvement due to an enhanced incident K flux is immediately apparent. In addition to $K^*(890)$ production we now observe a pronounced $K^*(1420)$ signal. Assuming a branching ratio for the latter of 0.53 and an acceptance of approximately 0.5 relative to $K^*(890)$, it appears that their production cross sections are of comparable magnitude. If we assume equal Primakoff production for these states, this would imply $\Gamma(K^*(1420)) \approx 10 \times \Gamma(K^*(890))$.

Figure 9 shows the effective mass of $\pi^+\pi^-\pi^0$. The η peak corresponds to A2 production primarily, but a selection on the ω peak yields a strong $B(\omega\pi)$ signal in the 4π effective mass.

SUMMARY AND SOME OBSERVATIONS

To date, we have obtained a new value for $\Gamma(\rho^- \to \pi^-\gamma)$ which is a factor two higher than the previous measurement made at low energies,[4] and a first measurement of $\Gamma(K^{*-} \to K\gamma)$. These results, along with the recent measurement of the η' total width[7] which has resulted in the determination of the $\eta \to \rho\gamma$ and $\eta \to \omega\gamma$ partial widths, make worthwhile a new assessment of the PVγ and VPγ couplings in the framework of SU(3). This has been carried out by T. Ohshima of this collaboration.[8] The new value for $\Gamma(\rho\pi\gamma)$ in particular has helped ameliorate previous problems.

These new data make possible some interesting comparisons with quark model predictions for the electromagnetic couplings, which depend on the quark spin-flip magnetic dipole operators. For example

$$\frac{\Gamma(\omega \to \pi\gamma)}{\Gamma(\rho \to \pi\gamma)} = \left(\frac{\mu_u - \mu_d}{\mu_u + \mu_d}\right)^2 \times \text{P.S. factor}$$

$$= 9 \times 1.04 = 9.4 \text{ for } \frac{\mu_u}{\mu_d} = -2$$

This is to be compared with

$$\frac{889 \pm 60 \text{ keV}}{67 \pm 7 \text{ keV}} = 13.3 \pm 1.6$$

If we permit the magnetic moments to vary from the simple quark model, we find $\mu_u/\mu_d = -1.75 \pm .09$.

It is interesting to note that a similar deviation from the simple quark model occurs in the static baryon magnetic moments. For example

$$\frac{\mu_p}{\mu_n} = \frac{4\mu_u - \mu_d}{4\mu_d - \mu_u} = -1.46$$

where the quark model predicts a ratio - 1.5. The data then imply
μ_u/μ_d = -1.91.

A possible explanation for this effect has been presented
recently by Geffen and Wilson.[9] They interpret the data as providing evidence for an anomalous quark magnetic moment which should
increase rapidly with quark mass. An important test of this idea
should occur in the decay $T \to n_b\gamma$ where the effect of the anomalous
moment should be exaggerated in the charge -1/3 system.

Another interesting ratio involves the strange quarks

$$\frac{\Gamma(K^{*0} \to K^0\gamma)}{\Gamma(K^{*-} \to K^-\gamma)} = \left(\frac{\mu_s + \mu_d}{\mu_s + \mu_u}\right)^2 = 4 \text{ for } \mu_u : \mu_d : \mu_s$$
$$= 2 : -1 : -1$$

If we assume $m_u = m_d$, m_s/m_u = 1.25, this ratio becomes 2.25, while
Geffen and Wilson predict 1.4. The data show

$$\frac{75 \pm 35}{50 \pm 15}^3 = 1.5 \pm 0.9$$

While ultimately we shall be able to measure the K^{*-} transition rate
to an accuracy of better than 10 percent, the error in this ratio
will still be large, unfortunately. A new, high precision measurement of the K^{*0} rate would be useful.

Predictions for the electromagnetic decay rates of higher $q\bar{q}$,
$L \geq 1$ states have been made by Babcock and Rosner.[10] Our present
rough estimates for A2 $\to \pi\gamma$ and $K^*(1420) \to K\gamma$ appear to be in
reasonable agreement with these predictions although a quantitative test must await completion of our analysis.

REFERENCES

1. For example, C. Becchi and G. Morpurgo, Phys. Rev. 104B, 687 (1965).
2. For example, E. H. Thorndike, EMS 77 Proceedings, (1977).
3. W. C. Carithers et al., Phys. Rev. Lett. 35, 349 (1975).
4. B. Gobbi et al., Phys. Rev. Lett. 33, 1450 (1974) and 37, 1439 (1976).
5. D. Berg et al., Phys. Rev. Lett. 44, 706 (1980).
6. G. Faldt, Nuc. Phys. B43, 591 (1972).
7. D. M. Binnie et al., Phys. Lett. 83B, 141 (1979), and G. S. Abrams et al., Phys. Rev. Lett. 43, 477 (1979).
8. T. Ohshima, University of Rochester Preprint UR-733 (1980).
9. D. Geffen and W. Wilson, Phys. Rev. Lett. 44, 370 (1980).
10. J. Babcock and J. Rosner, Phys. Rev. D14, 1286 (1976).

Figure 1.

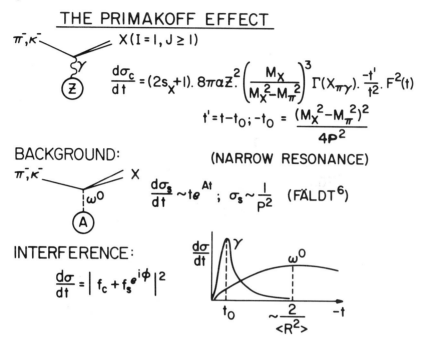

THE PRIMAKOFF EFFECT

$\pi^-, \kappa^- \quad\quad X(I=1, J \geq 1)$

$$\frac{d\sigma_c}{dt} = (2s_X+1) \cdot 8\pi a \bar{z}^2 \cdot \left(\frac{M_X}{M_X^2 - M_\pi^2}\right)^3 \Gamma(X_{\pi\gamma}) \cdot \frac{-t'}{t^2} \cdot F^2(t)$$

$$t' = t - t_0; \quad -t_0 = \frac{(M_X^2 - M_\pi^2)^2}{4P^2}$$

BACKGROUND: (NARROW RESONANCE)

$\pi^-, \kappa^- \quad\quad X$

$$\frac{d\sigma_s}{dt} \sim t e^{At}; \quad \sigma_s \sim \frac{1}{P^2} \quad \text{(FÄLDT}^6\text{)}$$

INTERFERENCE:

$$\frac{d\sigma}{dt} = \left| f_c + f_s e^{i\phi} \right|^2$$

Figure 2. The spectrometer.

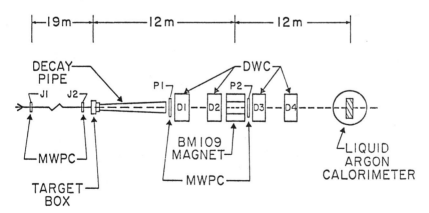

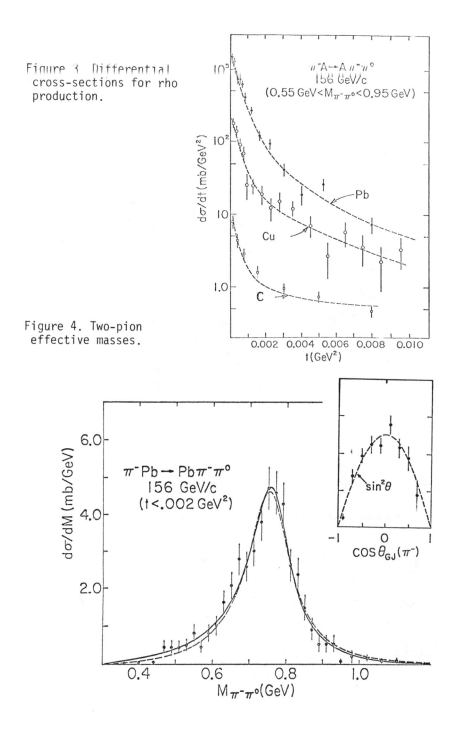

Figure 3. Differential cross-sections for rho production.

Figure 4. Two-pion effective masses.

101

Figure 5.
K⁻ → K*(890) production (all elements combined).

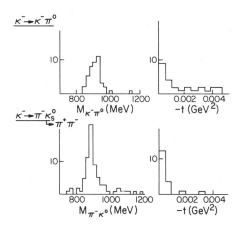

Figure 6.
π⁻ → A2⁻ production (all elements combined).

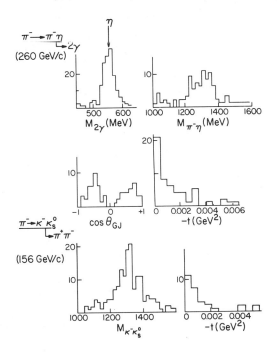

Figure 7.
$\pi^- \to \pi^- \pi^- \pi^+$ (Pb target)

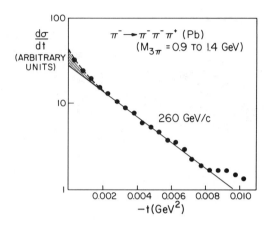

Figure 8.
$K^+ \to (K\pi)^+$ at 200 GeV/c.

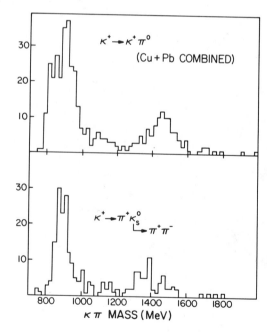

Figure 9.
$\pi^+ \rightarrow (4\pi)^+$ production. The $\pi^+\pi^-\pi^0$ effective
mass spectrum contains two entries per event.
The $(\pi^+\eta)$ effective mass is mainly A2 (not shown).

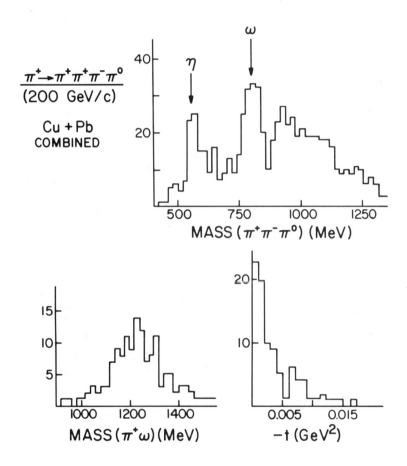

DYNAMICS OF LIGHT HADRONS - MOSTLY GLUEBALLS

John F. Donoghue
Massachusetts Institute of Technology, Cambridge, Ma. 02139

ABSTRACT

Current ideas on the spectra, production and properties of gluon bound states are reviewed.

INTRODUCTION

On the program, the title of this talk is "Theory of Light Quark Dynamics". However, I will deceive you and talk predominantly about mesons without quarks - glueonium or glueballs.[1-13] Throughout I will assume that Quantum Chromodynamics (QCD) is in fact correct. In QCD the important extra degree of freedom - the gluons - also carry color and are confined. Our models of quark confinement tell us that gluons by themselves also form bound states, and these states are potentially the most interesting aspect of future meson spectroscopy. Perhaps the most important message of this talk is that, if present understanding is correct, glueballs are sitting right there in the resonance region, accessible to experiments. I will describe the quantum numbers of such states, and review suggestions for producing them.

I. PRELUDE -- QUARK STATES

First let us review the properties of quark states in order to present the framework for discussing glueballs. The essential features of light quark states are rather general: 1) Quarks are confined to a region of space with a radius of about 1fm. The confinement scale generates a characteristic energy $E \sim \sqrt{3}/R \sim 300\text{--}350 \text{MeV}$ (the $\sqrt{3}$ is for 3 dimensions). This is just a reflection of the uncertainty principle for light particles. 2) Despite this confinement, quarks are fairly free within hadrons. By this I mean that interactions do not destroy the naive quark model predictions for the spectrum, magnetic moments, coupling constants etc., which are calculated by neglecting interactions. In practice it has not been hard to maintain properties 1 and 2 simultaneously. 3) The third general feature is the existence of a rich spectroscopy of ground and excited states, with characteristic energies of excitation again of order 300 MeV.

Various models build these features in different ways. Perhaps the model with the greatest phenomenological success is the MIT bag model.[14] It is built on a guess about the vacuum structure of the theory, namely that there are two types of "vacuum". "Outside" of hadrons is the true vacuum which is presumably quite complex. The guess is that whenever strong quark or gluon fields are present (i.e. inside hadrons) the vacuum undergoes a change of phase and is close to the "perturbative vacuum", i.e. one where properties can be calculated in a power series in the coupling constant about zero field strength. The complexity of the true vacuum is circumvented as quarks do not penetrate there. The "inside" vacuum must have an energy per unit volume (called B) larger than the true vacuum. From the phenomenological application it appears that

$$B^{1/4} \approx 130 \text{MeV}$$
$$B \approx 7 \times 10^{34} \text{ ergs/cc} \qquad (1)$$

This vacuum anzatz satisfies properties 1 & 2 quite easily. The boundary between the two phases occurs where $\bar{q}q=0$ (for quarks) or

$$\frac{1}{4} F^A_{\mu\nu} F^{A\mu\nu} = B \text{ (for gluons)}. \qquad (15)$$

Another class of hadron theories are string models. Motivated by considerations of local gauge invariance, these theories postualte that two or three quarks are connected with colored strings. While many of the studies of strings have been rather formal, there have been some applications to date. One of the features of these models is the ability to handle light or massless quarks in a relativistically correct way. The dimensional parameter of string theories is the string tension

$$T_o = 1/2\pi\alpha' = .19 \text{ GeV}^2 = 1.5 \times 10^5 \text{nt}. \quad (\alpha' \text{ is the Regge Slope}).$$

Finally there are potential models[16]. These have proven extremely useful for classification of states and for heavy quarks. Their drawback as dynamical theories is that they require nonrelativistic quarks, which is clearly not the case for light hadrons. (One indication of this last statement is the observed excitation energies. The hallmark of a nonrelativistic system is excited states whose energy above the ground state is small compared to the constituent's mass. However, in the hadron spectrum the excitation energies are of the order of or larger than the quark mass).

The various models appear the most similar for states of high angular momentum. In all pictures the quark and antiquark are imagined to be rotating around a common center with a string of flux between them. This generates the Regge trajectories. For example in the bag model one finds the mass of these states to be[17]

$$M = 2 \left(\frac{8\pi^3 \alpha_s}{3} \right)^{1/4} B^{1/4} J^{1/2} \qquad (2)$$

which implies a Regge slope of

$$\alpha' = \frac{1}{8\pi^{3/2}} \left(\frac{3}{2} \right)^{1/2} \frac{1}{\alpha_s^{1/2}} \frac{1}{B^{1/2}} = 0.88 \text{ GeV}^{-2} \qquad (3)$$

consistent with experiment

For the ground states of the spectrum, the quarks are assumed to be in spatially symmetric (S-wave) configurations, with the confining force being spin independent. The degeneracies of such a situation are lifted by allowing the exchange of a transverse gluon of QCD[19] (Fig. 1). This generates a spin-spin interaction which naturally makes $M_\pi < M_\rho$ and $M_n < M_\Delta$. With some fitting the observed spectrum can be understood. Even the standard downfall of quark models, the pion, can be accomodated, at least within the bag model. Recent work[19] has shown that the standard quark model techniques are not applicable for very light states (i.e. those whose radius is smaller than their Compton wavenumber $1/m$). The appropriate new methods easily generate a light pion and obtaining $M_\pi = 135 \text{MeV}$ is not a problem. Some contact can be made with the pion of PCAC, although in this regard there is much left to do.

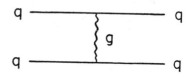

Fig. 1. One gluon exchange

In between these two extremes - ground states versus high angular momentum - lie a wealth of radial and orbital excitations as well as four quark states. Perhaps the greatest unsolved problem of quark spectroscopy is why this host of states have not all been seen. In the mesons only one nonet (the 2^{++}) above the ground state is known entirely. It would be advisable to keep this in mind when discussing the spectrum of glueballs.

II. GLUEBALLS

Consider first a pure Yang-Mills theory, i.e. QCD without quarks. Such a theory is assumed to be confining. Indeed it is for just this case that we have the strongest evidence of confinements in the form of Creutz's Monte Carlo calculations on a lattice.[20] Such a theory must have a spectrum and that spectrum must be composed of bound states of glue. In fact the lightest glueball (or perhaps several) will be stable.

If we now add quarks, there will be more states in the spectrum but the glue states should remain. It is conceptually easiest to use this procedure in the limit of a large number of colors.[21] In such a limit, all states are narrow, and any mixing between quark and glue states vanishes. It is then clear that gluon bound states do exist in addition to quark states. If present understanding is correct, this holds true also for N=3 (Note that throughout much of the remainder of this talk the phrase, "if present understanding is correct", is implied but not stated.) Besides rendering all glueballs unstable, the addition of quark states may change the properties of glueballs in more subtle ways, as will be disdussed below. First, let us look at the pure glue spectrum.

a) Spectroscopy: Since the gluon carries no flavor quantum numbers the most obvious feature of the glueball spectrum is that all states will be isospin and SU(3) singlets.

At high angular momentum the structure of glueballs would be similar to the corresponding quark states. The gluon fields would be well separated by the centrifical barrier and a flux tube or string would connect the two ends. The main difference is that the gluon's color charge is larger than the quarks charge by a factor of 3/2. This increases the mass slightly for a fixed angular momentum. In the bag model

$$M = \left(\frac{3}{2}\right)^{1/2} 2 \left(\frac{8\pi^3 \alpha_s}{3}\right)^{1/4} B^{1/4} J^{1/2} = 1.3 \text{GeV } J^{1/2} \tag{4}$$

and the Regge slope is smaller

$$\alpha'\text{glue} = \frac{2}{3} \alpha'\text{quark} \tag{5}$$

I am not mentioning these states expecting them to be observed. Indeed they should decay quite easily by pulling a pair of gluons out of the flux tube. However they do provide an upper bound on the guleball spectrum, using a dynamical framework quite different from that of the ground state. For example if this picture should work for J=5, the bag model would predict a mass of 3 GeV. Suura has considered states in a string-like model, and he is able to bound the ground state mass to be less than 2 or 3 GeV. [22] From these high angular momentum states we have the clear message that the Regge trajectories point back to low mass states for both quarks and gluons.

	Two Gluons	Three Gluons
Ground States	$0^{++}, 2^{++}$	$0^{-+}, 1^{--}, 3^{--}$
First Excited States	$0^{-+}, 1^{-+}, 2^{-+}$	$0^{++}, 1^{++}, 2^{++}, 1^{+-}, 2^{+-}, 3^{+-}$
Others possible	$1^{++}, 1^{--}?, 3^{++}, 3^{-+}$	all J^{PC}

TABLE I

Glueball Spectrum in an Atomic Classification

For the low lying states, we can duplicate the quark techniques and assume that the gluons are confined by a spin independent force. This leads to the spectrum [23] listed in Table I. The ground states of two bound gluons is $J^{PC} = 0^{++}$ and 2^{++}. Among the first existed states the 0^{-+} is interesting because the known pseudoscalar states have received so much attention that an extra state would stand out. In addition the 1^{-+} state is exotic. There is a question mark next to the 1^{--} state because although you can form it from 2 massive gluons, Furry's theorem appears to prevent its formation for massless gluons. [24] Three gluons can also form a color singlet and the ground state there contains a 1^{--} state, which could be useful to look for. If our experience with quark states is any indication, the ground states should be the most visable, with some of the first excited states detectable. In general we can probably forget about the rest, at least at the start.

One feature which is disturbing about the above classification is the role of Non Abelian gauge invariance. A Non Abelian gauge transformation can change two gluons into three gluons, linking the two segments of the table. Nevertheless, the simple classification scheme is probably a valid guide (For quark states an analogous problem arises. The SU(6) symmetry which is used to construct the quark analogue of Table I is not Lorentz invariant. A Lorentz transformation mixes the various states. However the symmetry remains extremely useful for classification.) One can show how to obtain the above quantum numbers in a gauge invariant way. Quark states can be generated by acting on the vacuum by quark bilinears

$$|\pi> \sim \bar{q}\gamma_5 \vec{\tau}q|0>$$

$$|\rho> \sim \bar{q}\gamma_\mu \vec{\tau}q|0> \tag{6}$$

$$\text{etc.}$$

Likewise gauge invariant combinations of gluon field strength tensors can serve as interpolating fields for glueballs.

$$|G(0^{++})> \sim F^A_{\mu\nu} F^{A\mu\nu}|0>$$

$$|G(2^{++})> \sim F^A_{\mu\lambda} F^{A\lambda}{}_\nu|0>$$

$$|G(0^{-+})> \sim F^A_{\mu\nu} \tilde{F}^{A\mu\nu}|0>$$

$$= \epsilon^{\mu\nu\alpha\beta}F^A_{\mu\nu} F^A_{\alpha\beta}|0> \tag{7}$$

$$|G(1^{--})> \sim d_{ABC} F^A_{\mu\nu} F^{B\nu\lambda}F^C_{\lambda\sigma}|0>$$

A completely different analysis of the glueball spectrum is given by the work of Kogut, Sinclair, and Susskind, using a spatial lattice and no quarks in the theory. They identify 3 glueball states 0^+, 2^+ and 1^+, and working in a strong coupling limit find

$$\frac{m_T}{m_S} = 1.003 \qquad ; \qquad \frac{m_A}{m_S} = 1.575 \tag{8}$$

The appearance of the axial vector is somewhat unexpected from the point of view of the atomic classification, although the mass ratio $\frac{m_A}{m_S} \approx 1.6$ cound be quite reasonable in this framework.

It is important to try to estimate the scale for the start of the spectrum. The simplest estimate just relies on the uncertainty principle. If we confine massless gluons to a region of size 1fm, they have an energy

$$E > \sqrt{3}/R \tag{9}$$

so that the spectrum would start at roughly twice this

$$M_o \gtrsim \frac{2\sqrt{3}}{1fm} \approx 700 \text{ MeV} \tag{10}$$

Most confinement techniques would have the ground state energy only slightly higher

$$M_o \sim 1 \text{ GeV} \tag{11}$$

This can be made more precise by studying a specific model. Jaffe and Johnson have studied the glueball masses in the naive bag model; naive in neglecting spin-spin interactions.[3, 25] They find

$$M(0^{++}, 2^{++}) = 960 \text{ MeV}$$
$$M(0^{-+}, 1^{-+}, 2^{-+}) = 1290 \text{ MeV} \tag{12}$$
$$M(3 \text{ gluons}) = 1460 \text{ MeV}$$

The spin-spin interaction could be very important, in general moving state around by $O(300 \text{ MeV})$. (see also the question and answer period for this session). However not all states will be pushed up in mass by this, so that 1 GeV is still a reasonable start of the spectrum.

A final estimate can be made from a potential model framework if applicable. The only dimensional factor in the potential is the coefficient (b) of the linear term $V \sim br$ which has dimensions of $(mass)^2$. At maximum this would be increased by a factor of $(charge)^2 = 9/4$ for gluon states. The mass scale of the problem then goes up by 50% over quark states leading to

$$m = \frac{3}{2} m_\rho \approx 1.1 \text{ GeV} \tag{13}$$

There is of course some room for error in these estimates, but I feel not by a factor of two. We expect glueballs to appear somewhere in the resonance region of 1 GeV→2GeV, with most likely the lower end favored.

b) Mixing: At this point a caution should be given: in fact There are no pure glueballs. In general glueballs will mix with quark states of the same spin, parity, etc. Even exotics can mix with 4 quark states. The lowest order diagrams are given in Fig. 2 a,b. At present no reliable estimate of this mixing has been produced. However, the naive expectation is that it will be sizeable. For example, compare diagram 2c,d with 2a. Fig. 2c gives the lowest order diagram responsible for the splitting of SU(6) multiplets. Experimentally these effects are about 200 - 300 MeV. Likewise 2d gives the similar splitting in glueballs. I know of one calculation for bagged glue where the result was a mass shift of 300 MeV. These diagrams are similar to 2a and are suggestive that it also would have a matrix element of order (100-300 MeV). Some special cases may be less. Mixing angles would be

$$\tan 2\theta = 2\left(\frac{\text{\image_ref_inline}}{M_Q - M_G} \right)$$

(14)

or

$$\theta \sim 0 \rightarrow 45°$$

(15)

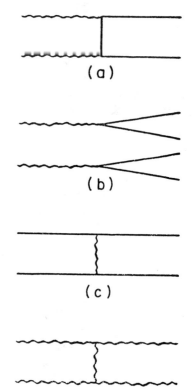

Fig. 2. Quark-glue mixing and spin splittings.

depending on the location of the glueball. If this estimate is true, glueballs will not look much different from ordinary hadrons, because they in part are composed of ordinary hadrons. However the gluon degree of freedom is still manifested in the existence of an extra state. In general there will be 3 isoscalar states, i.e. those which are predominently

$u\bar{u} + d\bar{d}$, $s\bar{s}$, and gg. It is amus-
ing that we expect most of the ground state glueballs to fall close in mass to quark states of the same quantum numbers. Thus the octets of SU(3), which grew into nonets via the quark model, now become approximate decuplets in QCD.

c) Production: Far and away the best mechanism for producing glueballs is in the radiative decay of charmonium. In QCD perturbation theory this occurs through the diagram of Fig. (3), i.e. through

$$\psi \rightarrow \gamma\, g\, g \quad : \tag{16}$$

The two gluons give us an opportunity to look at

gluon-gluon scattering.
The energy can even be
tuned by observing the
photon energy. One can
imagine a spectrum like
Fig. (4), with the bumps
being glueballs. In fact
any single particle state
seen in $\psi \rightarrow \gamma M$ is a gluon
gluon resonance. This can
constitute a definition of
a state which is part glueball, even if this encom-
passes several known states.
As Scharre has (subsequently)
discusses at this conference,
$\eta', \eta, f(1270)$ and $E(1420)$ have
been seen so far in this reaction.

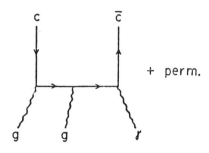

+ perm.

Fig. 3. Diagram for
ψ radiative decay.

Given infinite data,
it would be of interest to
compare $\psi \rightarrow \gamma M$ with $\psi \rightarrow \omega M$ or ϕM.
In the hadronic decay $\psi \rightarrow 3g$,
any pair of gluons is always
in a color octet and cannot
easily form a color singlet
glueball. The ωM final state
would then presumably emerge
through a rearrangement of
quarks. If M were dominently
a quark state it would favor
production in $\psi \rightarrow \omega M, \phi M$, while
glueballs would be signaled
by states that show up rela-
tively more strongly in $\psi \rightarrow \gamma M$
than $\psi \rightarrow \omega M$ or ϕM.

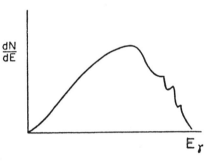

Fig. 4. Hypothetical spectrum
in $\psi \rightarrow \gamma x$

Of course this mechanism also works for upsilon
radiative decay. However the inclusive decay rates,
due to $T \rightarrow 3g$, $gg\gamma$ scale up with mass, while any single
channel rises less fast. The relative branching ratios
will therefore fall

$$\frac{\Gamma(V \rightarrow \gamma M)}{\Gamma(V \rightarrow \gamma \, \text{all})} \leq K \left(\frac{M_H}{M_V} \right)^2 \qquad (17)$$

where M_H is some typical fixed hadronic mass and M_V is
the mass of the vector meson. Individual branching
ratios at the upsilon will be at least an order of magnitude smaller than with charm (excepting the existence

of states close to or above the J/ψ in mass).

For the upsilon, another mechanism, suggested by Roy and Walsh[7], is more appropriate. The dominent upsilon decay is into three gluons, and these gluons fragment into hadrons. A likely fragment for a gluon jet is a glueball, and these authors estimate that 80% of gluon jets do indeed contain a glueball. Since in upsilon decay one knows one has a gluon jet, one can compare final states on and off resonance to look for new states.

It has been suggested that the fabled Pomeron is none other than two gluon exchange.[26]. If this is true glueballs could be easily produced whenever the pomeron plays a large role. Robson has suggested double Pomeron exchange[6] (Fig. 5) as possible source of glueballs.

A 1^{--} glueball, composed of three gluons, can be photoproduced. The coupling to the photon is induced through a quark loop Fig. 6a. Either diffractive scattering, Fig. 6b or e^+e^- production, Fig. 6c can be used to search for it. The coupling is not difficult to estimate, although the calculation is not yet complete.[27]

Fig. 5. Double Pomeron production.

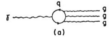

(a)

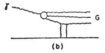

(b)

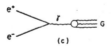

(c)

Fig. 6. Photon-glueball mixing.

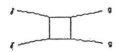

Fig. 7. 2 photon-2 gluon coupling.

Two photon physics which promises to be a growth industry at PEP and PETRA (especially since the top quark has not been found), may be quite useful. Two photons couple to the same quantum numbers on two gluons, and they can mix as in Fig. 7. The two photon experiments will provide a clean way to scan through the scalar, tensor and pseudoscalar spectra.[28]

Finally one should not forget the standard processes, i.e. hadron probes. It is quite likely that glueballs can be produced here -- after all, hadronic interactions are via glue. This is especially true if the mixing with quark states is large.

d) Decay: It is here that theory is found to be embarrassingly weak. We must admit that we have no quantitative understanding of the decays of quark states, much less glueballs. At present there is not even a way to reliably relate quark and gluon states. Some people have suggested that glueball decay may be partially Zweig suppressed, as the gluons must convert into quarks.(2,6) However, this arguement is not very convincing especially for light states, and I feel that the width is an open question.

Naively glueball decays should be flavor independent, as the form of the gluon coupling is the same for

$s\bar{s}$, $u\bar{u}$, and $d\bar{d}$. However, mass effects will be important and will suppress strange particle production. The resulting flavor distribution will be somewhere between

that of $s\bar{s}$ states and $u\bar{u}$ and $d\bar{d}$ states, somewhat favoring the decay into non-strange particles.

If glueballs are heavy, with masses near 2GeV the final states may be similar to those in $\bar{p}p$ annihilation or D decay. Roy and Walsh have suggested a statistical treatment,(7) which yields a predominence of four pion final states, with

$$B(G \rightarrow 4\pi) \approx 50\%$$
$$B(G \rightarrow \pi^+ \pi^+ \pi^- \pi^-) \approx 20\% \qquad (18)$$

However for lighter states we have much less guidance.

III. OPTIONS AND A LOOK AT THE DATA TABLES

Despite the prejudices which I have embraced above, we have three options for glueballs.

a) They could have mass greater than 2 GeV. Most experiments have a hard time reconstructing particles in this region, and it would be difficult to find glueballs. However, all our estimates go against this possibility.

b) The mass may be less than 2 GeV, with narrow widths and small mixing with quark states. Such states could easily be missed by standard experiments. To uncover these glueballs one needs reactions which couple to glue strongly -- such as ψ radiative decay. When found these glueballs would present a very clear signal. However, the SLAC experiments on ψ radiative decays (see Scharre's talk at this conference) probably already rule out this option.

c) The third and favored (by me, at least) option is that the mass is less than 2 GeV and they are typical

in character. Typical in character means that at least
some of their widths are within a factor of a 2 or 3
either way of hadronic states which have been observed.
Unfortunately this is still a wide range (Glueball
widths cannot be excessively large, or they would make
the widths of quark states large also through mixing.
However they may be wide enough to present difficulty
in searching for them). For such glueball mixing with
quark states will most likely be appreciable. If this
option is correct the data tables may already contain
states with a high glue content.

There is a good deal of theoretical work which
could be useful in glueball physics. The most important,
for the experimenter, is to decide among these three options by understanding better the mass, width and mixing
of glueballs. It is also crucial to understand the role
of the non-Abelian character of confined gluons.

It is worthwhile to take a stroll through the data
cards looking at the isoscalar states for glueball candidates.[29] It is clear from a cursory glance that the
quark model has already claimed the most interesting
states as its own and that one can squeeze glueballs in
only with difficulty. If this situation persists indef-
initely, option (c) above may be in trouble.

The η' is an example of the dilemma we face. It
has long been thought of as a quark state, but recent
work on QCD has made it clear that the η' has a high
glue content. Indeed, most of its mass comes from its'
glueball component. The theory of this goes under the
name of the U(1) problem.[30]

One can define an isoscalar axial current

$$A_\mu^o = \sum_{i=u,d,s} \bar{q}_i \gamma_\mu \gamma_5 q_i \qquad (19)$$

This current naively appears to be conserved in the
limit that the quark masses are zero.

$$\partial_\mu A_\mu^o \Big|_{naive} = 0 \qquad (20)$$

In the massless quark limit the octet of pseudoscalar
mesons must be massless and one can show using the naive
divergence Eq.(20) that the η' is also massless. Turning on the quark masses cannot accomodate the large η'
mass. However Eq. (20) is not in fact correct, as this
U(1) axial current has an anomaly in its divergence.[31]
The correct result is (again for massless quarks)

$$\partial_\mu A^{\circ\mu} = \frac{3 \; g_s^2}{16\pi^2} \; \varepsilon^{\mu\nu\alpha\beta} \; F^A_{\mu\nu} \; F^A_{\alpha\beta} \qquad (21)$$

$$= \frac{3\alpha_s}{4\pi} \; F^A_{\mu\nu} \; \tilde{F}{}^{A\mu\nu}$$

Taking matrix elements of these currents

$$\langle\eta'|A^\circ_\mu|0\rangle \; F_{\eta'} p_\mu \qquad (22)$$

$$\langle\eta'|\partial^\mu A^\circ_\mu|0\rangle = F_{\eta'} m^2_{\eta'} = \frac{3\alpha_s}{4\pi} \langle\eta'|F^A_{\mu\nu} \tilde{F}{}^{A\mu\nu}|0\rangle$$

Here the mass of the η' explicitly arises from the matrix element of a gluonic operator (Remember that we earlier described $F\tilde{F}$ as the interpolating field for a pseudoscalar glueball). The η' in QCD must contain a mixture of quark and glue. In fact the matrix element of $F\tilde{F}$ is quite large. Various authors have included quark masses and estimated $F_{\eta'}$, to extract

$$\langle\eta'|\alpha F^A_{\mu\nu} \tilde{F}{}^{\alpha\mu\nu}|0\rangle = \begin{array}{ll} (.89\text{GeV})^3 & \text{Ref}(32) \\ (.75\text{GeV})^3 & \text{Ref}(33) \end{array} \qquad (23)$$

(The η also picks up a similar but smaller matrix element through SU(3) breaking). To see how large this is on a hadronic scale, one can compare it to the predicted matrix element for a <u>pure</u> bag model pseudoscalar glueball[27]

$$\langle G0^{-+}|\alpha F^A_{\mu\nu} F^{A\mu\nu}|0\rangle = (.71\text{GeV})^3 \qquad (24)$$

The results are comparable within the uncertainties in the various calculations.

A quite strong signal has been seen in $\psi \to \gamma\eta'$. From the viewpoint of QCD this is quite reassuring, and is consistent with our present understanding of the η'.

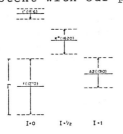

Fig. 8. The 2^{++} nonet

The 2^{++} spectrum is quite well-known (see the nonet in Fig. (8), and, since we expect the glueball ground state to also be 2^{++}, it is worthwhile to look around to see where the glueball might be found. Let me present three minor problems that looked at together begin to appear puzzling. 1) The f(1270) has been seen strongly in $\psi \rightarrow \gamma f$, however, the f'(1516) has not.

$$\text{B.R.}(\psi \rightarrow \gamma f) - (2.0 \overset{+}{-} 0.7) \times 10^{-3} \text{ PLUTO}$$

$$(1.1 \pm 0.3) \times 10^{-3} \text{ DASP} \qquad (25)$$

$$\text{B.R.}(\psi \rightarrow \gamma f') < 0.35 \times 10^{-3} \qquad (26)$$

This is disturbing as we expect the ratio to be $f\gamma : f'\gamma = 2:1$, from the equality of the two glue coupling to up, down, and strange quarks. 2) In the bag model, including spin corrections, the 2^{++} glueball falls right at the f (1270) mass.[34] There is of course some uncertainty in this calculation, but it is expected to be a good first approximation. 3) The nonet appears reasonable in structure; however the f mass is lower than one would naively expect. It is expected to be composed of $u\bar{u} + d\bar{d}$ and is 40MeV below the A2 ($u\bar{u} - d\bar{d}$). In other nonets the isoscalar is slightly heavier. In addition the f($u\bar{u} + d\bar{d}$):K*(us):f'(ss) splittings are not equal but are 150 MeV and 96 MeV respectively.

In conjunction these make it difficult to imagine where to put the glueball. Putting it above the f' would do violence to the bag estimate, and would also mix the glueball with the f' to generate a larger glue component than the f, making the radiative transition difficult to understand. It is probably impossible to put it below the f, as there the phase shift information is very good here. Between the f and f' would be marginly possible if it didn't couple too strongly to $\pi\pi$, but the radiative transition experiments probably would have seen it. It is possible to imagine it hidden under the f peak. This would explain the strong $\psi \rightarrow \gamma f$ signal and perhaps the mass-shift of the f. It also could be very narrow and decouple from $\pi\pi$, but this should be answerable by the $\psi \rightarrow \gamma M$ data. A sharpening of the limits for $\psi \rightarrow \gamma f'$ and a search for other 2^{++} states would be most useful.

In the 0^{++} system, we have two isoscalar states known, S*(993) and ε(1200), in addition to the broad shoulder near 700 MeV. However the status of the 0^{++} nonet is shaky. If an extra state were found the supposition would be that one is a glueball. (But which one?)

Finally, data presented at this conference indicates a strong $\psi \rightarrow \gamma E$(1420) signal. Could this be a glueball? The hadronic data seems to be settling on 1^+ for this state, which would suggest a quark interpretation,

in the same nonet as the Al and D(1285), with the E
being the $s\bar{s}$ partner, and D(1285) being $u\bar{u} + d\bar{d}$. Again
a problem arises as the universality of the gluon coup-
lings would predict $\psi \to D\gamma$ to be twice as strong as $\psi \to E\gamma$,
while the D is not seen at all. Scharre has also
claimed that the E that they see differs in some respects from the E of hadronic reactions. Perhaps they
are different, in which case the object seen in radiative decays is almost certainly a glueball. Even if
they are the same, and the 1^+ assignment holds up, a
glueball interpretation may be warrented in light of
the special position of the axialvector glueballs in
the lattice calculations.

IV. SUMMARY

The gluons of QCD should lead to additional isoscalar states in the meson spectrum. Estimates of their
mass put them in the resonance region, and they are then
clearly accessible to experiment. They may be produced
in a rich variety of ways, and the "best" channels to
look for them are 0^{++}, 2^{++}, 0^{-+}, 1^{--}, and perhaps 3^{--},
although several others are possible. Although theory
falters in being unable to detail their structure or decay, glueballs should still be recognizable by their
relationship to known quark states.

The observation of glueballs would be especially
valuable as a very direct confirmation of existence of
the gluons of QCD. To conclude, I'd like to briefly
recall the early days of meson spectroscopy. It was the
proliferation and character of hadrons which gave the
first clue that quarks were needed, and which started
us down the road to where we are today. It would be
fitting if meson spectroscopists could top this off by
finding the crucial other ingredient of the theory to
which they gave original impetus.

118

REFERENCES

1. H. Fritzsch and P. Minkowski, Nuove Cimento 30A, 393 (1975).
2. P.G.O. Freund and Y. Nambu, Phys. Rev. Lett. 34, 1645 (1975).
3. R.L. Jaffe and K. Johnson, Phys. Lett. 34, 1645 (1976).
4. J. Willemsen, Phys. Rev. D13, 1327 (1976).
5. J. Kogut, D.K. Sinclair and L. Susskind, Nucl. Phys. B114, 199 (1976).
6. D. Robson, Nucl. Phys. B130, 328 (1977).
7. P. Roy and T. Walsh, Phys. Lett. 78B, 62 (1978).
8. K. Koller and T. Walsh, Nucl. Phys. B140, 449 (1978).
9. K. Ishikawa, Phys. Rev. D20, 731 (1979).
10. J.D. Bjorken, SLAC Summer Institute on Particle Physics (1979).
11. J.J. Coyne, P.M. Fishbane, and S. Meshkov, Phys. Lett. 91B, 259 (1980).
12. V. Novikov, M. Shifman, A. Vainshtein and V. Zakharov, Nucl. Phys. B165, 55 (1980); ibid B165, 67 (1980).
13. R.H. Capps, Purdue Preprint (1980).
14. A. Chodos, R.L. Jaffe, K. Johnson, C. Thorn and V. Weisskopf, Phys. Rev. D9, 3471 (1974);
    A. Chodos, R.L. Jaffe, K. Johnson and C. Thorn, Phys. Rev. D10, 2599 (1974);
    J.F. Donoghue, E. Golowich, and B.R. Holstein, Phys. Rev. D12, 2875 (1975);
    T. De Grand, R.L. Jaffe, K. Johnson, and J. Kiskis, Phys. Rev. D12, 2060 (1975);
    J.F. Donoghue and K. Johnson, Phys. Rev. D21, 1975 (1980).

15. S. Mandelstam, Phys. Rep. 13C, 260 (1974);
    H.B. Nielsen and P. Olesen, Nucl. Phys. B61, 43 (1973);
    K. Wilson, Phys. Rev. D10, 2445 (1974);
    W. Bardeen, I. Bars, A. Hanson and R. Peccei, Phys. Rev. D13, 2364 (1976);
    A. Chodos and C. Thorn, Nucl. Phys. B72, 509 (1974);
    S.-H.H. Tye, Phys. Rev. D13, 3416 (1976);
    Y.J. Ng and S.-H.H. Tye, Phys. Rev. D16, 2468 (1977);
    R.E. Cutkosky and R.E. Hendrick, Phys. Rev. D16, 786 793 (1977).

16. R.P. Feynman, M. Kislinger and F. Ravndal, Phys. Rev. D3, 2706 (1971);
    Y.S. Kim and M.E. Noz, Phys. Rev. D8, 3521 (1973);
    A. De Rújula, H. Georgi and S. Glashow, Phys. Rev. D12, 147 (1975);
    J.F. Gunion and R.S. Willey, Phys. Rev. D12, 174 (1975);
    W. Celmaster, Phys. Rev. D15, 1391 (1977);
    C.Y. Hu, S. Moszkowski and D. Shannon, Phys. Rev. D17, 874 (1978).
    N. Isgur and G. Karl, Phys. Rev. D20, 1191 (1979).
17. K. Johnson and C. Thorn, Phys. Rev. D13, 1937 (1976).
18. De Rújula, Georgi, and Glashow, Ref. 16.
19. Donoghue and Johnson, Ref. 14.
20. M. Creutz (to be published).
21. A review and references can be found in S. Coleman Erice Lecture

Notes, SLAC-PUB-2484, March (1980).

22. H. Suura, University of Minnesota Preprint - to be published in Phys. Rev. Lett.

23. The spectrum is discussed more thoroughly in Refs. 6 and 11.

24. W. Furry, Phys. Rev. 51, 125 (1937).

25. R. Giles (private communication) has shown that the 0^{++} spherical bag state is unstable against flattening into a pancake. This instability and its effects need to be better understood.

26. F.E. Low, Phys. Rev. D12, 163 (1975);
    S. Nussinov, Phys. Rev. Lett. 34, 1286 (1975).

27. J.F. Donoghue, to be published.

28. F.J. Gilman, SLAC-PUB-2461 (1980).

29. Particle Properties, Rev. Mod. Phys. 52, S1 (1980).

30. Recent work includes:
    E. Witten, Nucl. Phys. B156, 269 (1979);
    G. Veneziano, CERN preprint TH-2651 (1979);
    P. Di Vecchia, CERN preprint TH-2680 (1979);
    C. Rosensweig, J. Schecter and G. Trahern, Syracuse preprint SU-4217-148 (1979);
    J. Schecter, Syracuse preprint SU-4217-155 (1979).

31. S. Adler, Phys. Rev. 177, 2426 (1969);
    J. Bell and R. Jackiw, Nuovo Cimento 60A, 47 (1969).

32. H. Goldberg, Northeastern preprints NUB-2377, 2411 (1979).

33. V. Novikov, M. Shifman, A. Vainshtein and V. Zakharov, Phys. Lett. 86B, 347 (1979).

34. K. Johnson, private communication.

120

DISCUSSION PERIOD

1. N. Samios (BNL): I was a little puzzled by your coupling
to the 1^{--}. Earlier you said the Furry theorem or something like
that. Could you reverse it and say that the 1^{--} meson nonet works
so well in perfect mixing - in fact it's the only one that is really
perfect - you might say that the fact that the others don't work so
well - especially the pseudoscalars - might be evidence for the
glueballs?

Donoghue: Yes.

Samios: Second comment that I would like to make is that it is
difficult to say if they are glueballs or quarks if they all have
the same spin parity. However, as you pointed out, there are some
states such as, I believe, the 1^{-+}, that don't exist under the
quark system. If one found such a state it would be interpreted as
a gluon state. The question I have is, if you don't find one will
the theorists come back and say that's not the lowest state and
therefore you shouldn't be producing it well? So my real question
is - it would be nice to search for states which don't exist in the
quark model, and can you give us any estimation of the width? If it
were very broad then we have extra difficulties.

Donoghue: One of the puzzles that we don't understand about
the quark spectroscopy is why some of the states that are predicted
are not found. Presumably the same thing could be said about the
glue excited states. If some of these glue excited states are not
found maybe it's for the same unknown reason. The ground states,
however, are much better - that's why I pointed them out in particular. The ground states are the ones that you'd expect to see. As
far as the widths go - I wish I could give you a better estimate of
that. If you go through the references you'll find that approximately half the papers assume that there is some Zweig rule and the
others assume that they have typical hadronic widths. We also don't
have much handle on the calculation of the absolute rate of decay
of quark states - we use SU(3) but don't calculate their absolute
rates. If we could do that we could also do this, but unfortunately
we are stuck.

2. D. Scharre (SLAC): I have two questions: One is in regard
to a 1^{--} state of two gluons. I thought two gluons need to have a
negative C parity.

Donoghue: 1^{--} is the one I put a question mark after in the
two gluon sector. 1^{--} is three gluons. The reason it was on the
table at all is that some other people in the literature include it
in their tables.

Scaharre: The other question is more important. Can you set
an upper limit on the width? You said that the widths may be narrow or they may be like normal hadronic resonances. Can you say
that they are not 500 MeV? For instance if you looked in some experiment like ψ radiative decays, and say nothing with widths less
than 200 MeV, could you rule out gluonium states of a particular
spin parity?

Donoghue: If you caught me on even days I would say yes, and on odd days I would come up with a reason to increase the widths. In fact, if you are not careful when attempting to develop the Zweig rule arguements you can get turned around and argue that the states are wide. I really don't know anything solid to say about a width of a state less than 1.5 GeV.

Scharre: In other words, one cannot rule out gluonium states, one would have to find one to show that it is there.

L. Heller (Los Alamos): I'd like to ask - in the bag model of the gluons - are you able to satisfy the nonlinear boundary condi- at the surface?

Donoghue: In fact, you do not. The point is that in the bag model there is a pressure balance equation. The inward pressure of the vacuum must be balanced by a field pressure of the quarks or gluons. For the glueball states of Jaffe and Johnson the pressure is balanced on the average but not point by point.

R. Jaffe (MIT): As chairman, let me exercise a prerogative, and say something about that. The non-relativistic quark model violates all kinds of symmetries that we believe nature has, like chiral symmetry when we give the quarks masses of 300 or 400 MeV. So no one takes it very seriously. But if you want to know the spectra of quark excitations, you just take quarks of 300 MeV and add them up, and it works beautifully. So I think that one shouldn't make too much of a demand on half-assed gluon models which violate all kinds of symmetries, like gauge invariance, by giving gluons a mass and putting them in a potential or bag models which put them all in 1 fermi and add up the quantum numbers. We are in an early stage in the spectroscopy and it would be a very nice idea just to have a rough idea of where and what to look for.

A. Hey (Southhampton): I don't wish to dump too heavily on the MIT bag model, because I need it for my talk, but I seem to remember doing a calculation of glueball spin-spin mass splittings on Charles Thorn's blackboard at MIT and in fact they are much big- ger than quark spin-spin splittings because the Casimir is bigger. In fact, we did a naive calculation and came up with negative masses when you put this in. So I was slightly suspicious of your esti- mates.

Donoghue: That was the scalar state that was pushed far down in mass. I didn't mention it because it hasn't ever been published. Would you like to publish it? With the scalar state you have to worry because there is another scalar state with a lot of glue in the theory. That is the vacuum. The vacuum is known to be complex and contain a lot of glue. If you push this scalar state down to low mass, you have to wonder if you are just finding a component of the vacuum. You have to, in some sense that I don't understand, orthoganalize the two states. That will push them apart. I don't know what the final answer there is.

Hey: It would be nice if someone published the calculation.

J. Ballam (SLAC): Do you have any prediction for the mass of the 1^{-+} state, because there is a way you could look for it with polarized photons? You could look for a 1^{-+} state by the decay $\eta\eta'$.

Donoghue: 1^{-+} was a first excited state of 2 gluons, so without spin-spin splittings that comes out at about 1300 MeV. Pretend that it is within 300 MeV of that.

Ballam: That would be a way of looking for it then.

S. Lindenbaum (BNL/CCNY):   I'll be very brief. I think that if you are going to look for glueballs in light quark dynamics you really have to look for something special that is going to knock down all the backgrounds. And one special thing that has occurred to me is that if you get a disconnected quark diagram where you are dealing with 2 or 3 gluons going accross you essentially have a glueball. If you have just one particle at the end, that doesn't do you much good, but if you have two, like $\phi\phi$, then you are sweeping a spectrum of masses and you may in that way find a glueball.

In the BNL/CCNY $\pi^- p \rightarrow \phi\phi n$ experiment of a couple of years ago, you have just such a situation. In that experiment we found the OZI suppression for this OZI forbidden reaction is effectively absent. I believe the most logical explanation for this is the intervention of glueball resonances. We may have already seen glueball resonances in this experiment. We are planning to do it over with MPS II and obtain twenty times more data. I am hopeful that we will discover glueballs in that experiment.

REVIEW OF p̄p PRODUCTION EXPERIMENTS AT CERN

A. Ferrer

CERN, Geneva, Switzerland

(On leave of absence from IAL, Orsay, France)

INTRODUCTION

A plentiful spectroscopy of states coupled to baryon-antibaryon has been reported by many experiments, using different experimental techniques. An illustration of the abundancy of peaks claimed in 1977 can be seen in Figure 1, taken from D. Treille's report at the Campione d'Italia meeting [1].

Maybe the most exciting states among those reported in Fig. 1 are the narrow states found below the $\bar{N}N$ threshold, in $\bar{p}p \rightarrow \gamma X$ annihilation and $\bar{p}d$ capture experiments. However, very unexpected results were found by two Ω experiments:

1) The "fast proton" experiment of P. Benkheiri et al., [2] reported two narrow $\bar{p}p$ bumps at 2020 and 2204 MeV in the nucleon-exchange processes

$$\pi^- p \rightarrow \Delta^0_f \ (\bar{p}p)$$
$$N^0_{1520} \ (\bar{p}p)$$

   at 9 and 12 GeV/c.

2) The experiment of Evangelista et al., [3], gave evidence of a narrow bump in the $(\bar{p}p\pi^-)$ mass spectrum at 2950 MeV, in the diffractive reactions

$$\pi^- p \rightarrow (\bar{p}p\pi^-)_F + X$$

   at 16 GeV/c.

All these observations gave the first serious evidence for the long time searched exotic 4-quarks $(qq\bar{q}\bar{q})$ mesons predicted many years ago by Rosner [4], according to the duality arguments.

A clear interest has grown at CERN these last years to study the baryon-antibaryon system. And this is particularly true for the experimental teams that have been lately using the Ω facility at CERN.

In this review, we will concentrate on the various experiments that have recently been done at CERN, or are presently in progress, and which are studying the production of $\bar{p}p$ states.

In Table I we show the list of recent and current high sensitivity CERN experiments which deal with $\bar{p}p$ final states. These experiments have all been installed in the experimental West Area. Of these, one was carried

out in the ACCMOR (WA3) and one in the Geneva-Lausanne (WA10) spectrometers
- which have been described in other talks given in this Conference -; all
the others have used the Ω spectrometer (equipped with optical spark
chambers) or more recently, since the beginning of 1979 in its new version,
quoted in Table I as Ω', where the spark chambers inside the magnet have
been replaced by fast multi-wire proportional chambers of wide acceptance.

Figs. 2 and 3 show respective views of the old and new equipment for two
particular experiments: The first one corresponds to the experiment of
P. Benkheiri et al.,[2], already mentioned. There one can see the 80 forward
spark chamber gaps (SC1) and the 20 side chambers (SC2) located on both
sides of the hydrogen target.

Figure 3 shows an incomplete version of the detector layout as it was
available early 1979 for the first Phase of WA56 experiment[5].

One can see that the forward spark chamber modules have been replaced by
two sets of MWPC's: the so-called type B (densely packed 2 plane chambers)
and type A (widely spread 3 plane chambers). Since the end of 1979, two
more type A modules have been added, and 8 type C modules (2 planes each)
have been installed on each side of the target that play the role of the
old SC2 chambers of Figure 2, giving therefore a good detection efficiency
for "slow" charged particles around the target.

The new Ω' equipment has allowed the installation of a new generation of
experiments, that can use very high incident fluxes (up to 10^7 particles/
burst), within reasonable dead time limitations and keeping its good
acceptance for detecting complex multi-particle final states.

Let us finally recall that the Ω spectrometer facility at CERN can be
served by 3 secondary beams:

- the tagged photon beam El (20 - 70 GeV)
- the unseparated hadron beam S1 ($\leq$ 40 GeV/c)
- the separated hadron beam H1 ($\leq$ 80 GeV/c)

The tagged photon beam has been used by WA4 collaboration and obtained
results on photoproduction of $\bar{p}p$ states (the first experiment listed in
Table I).

The remaining experiments of Table I, use hadron beams. They are classified
according to three main $\bar{p}p$ production mechanisms: Inclusive, Diffractive
and Backward (baryon-exchange).

In Table I, also are reported the CERN code numbers of each experiment (see
the grey CERN Report "Experiments at CERN in 1979" published in August 1979
and finally the status of each experiment.

Experiment WA63, a CERN-Saclay collaboration, is presently on the floor
at Ω. The trigger is based on inclusive $\bar{p}$ and $\bar{\Lambda}$ produced centrally by
± 40 GeV/c hadrons.

I. PHOTOPRODUCTION OF $\bar{p}p$ STATES

The British-French-German collaboration which has carried out experiment
WA4, has reported results on photoproduction of $\bar{p}p$ states[6]. These events
were obtained from two different triggers using the Ω spectrometer:

a. The 2 forward particles trigger
b. The multiplicity (4, 9)-charged particles trigger.

a) The exclusive reaction

$$\gamma p \to (\bar{p}p)p \qquad 44 < E\gamma < 70 \text{ GeV.}$$

was obtained from the 2 forward particle trigger events. An unambiguous
identification of forward p and $\bar{p}$ was achieved with the forward Cerenkov
information in the momentum range (20-35 GeV/c). Throughout this range
the protons give no light, while both π's and K's should give light, the
the measured efficiency being 98.7% for π's. No structure appears in
the $\bar{p}p$ mass spectrum (Fig. 4). The total cross section for $M_{\bar{p}p} < 2.7$ GeV
is $\sigma = 20 \pm 8$nb averaged over the quoted Eγ range. The differential
cross section peaks, as expected for diffraction, following an e^{at}
distribution with $a = 4.7 \pm 0.7$ GeV^{-2}.

b) The inclusive reaction

$$\gamma p \to (\bar{p}p) + X \qquad 50 < E\gamma < 70$$

was analysed using the higher multiplicity (4 to 9)-charged particles
trigger. X in the reaction above means therefore ≥ 2 charged particles.
To improve the reliability of the channel identification, an energy of
the $\bar{p}p$ system greater than 44 GeV is required, and the p and $\bar{p}$ are then
selected and identified as in a).

A bump is observed in the $\bar{p}p$ mass distribution (Fig. 5a) which cannot be
explained by misidentification of protons (or $\bar{p}$) as shown in Fig. 5b. Here
those events for which one of the forward fast particles gives a Cerenkov
signal are analysed as if they were $\bar{p}p$ events, and show the reflections of
ϕ and ρ^o mesons at about 1900 and 2000 MeV respectively, which are due to
events where the Cerenkov has been inefficient for one K or π.

The properties of the bump found in WA4 experiment are the following

i) $M = 1930 \pm 2$ MeV
    $\Gamma = 12 \pm 7$ MeV   (M resolution $\sim$ 3.5 MeV)

ii) $d\sigma/dp_T2 \sim e^{-a \ p_T^2}$ $m(\bar{p}p < 1.96 \to a = 1.25 \pm 0.29$ GeV^{-2}
     $m(\bar{p}p > 1.96 \to a = 2.48 \pm 0.20$ GeV^{-2}

     (see Fig. 6)

iii) Central rather than peripheral production, as can be observed in
     the Feynman x_F distribution (Fig. 7).

iv) $\sigma = 5 \pm 2nb$ for $0.6 < x_F < 0.95$

e.q., $\sim 1/30$ of the total inclusive $\bar{p}p$ production cross section

The absence of signal in the diffractive channel

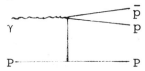

and the properties iii) above would imply that the spin parity of the bump is not 1^-, if interpreted as a resonance.

II. HADROPRODUCTION OF $\bar{p}p$ STATES

II.1. Inclusive Production

The ACCMOR Collaboration has recently reported results[7] on the inclusive production of $\bar{p}p$ pairs. The data came from their 2 prong trigger, which was designed to trigger on ϕ mesons. They used an incident beam of π and p of 93 GeV/c and are able to identify unambiguously secondary p and $\bar{p}$'s in the range (13-24) GeV/c.

No structure is found in the $\bar{p}p$ mass spectrum from reactions

$$\pi^{\pm}p \rightarrow (\bar{p}p) + X$$

as can be seen in Fig. 8 b) and 8 c).

However a 4-standard deviation from background appears in the $\bar{p}p$ mass spectrum for reactions induced by incoming protons:

$$pp \rightarrow (\bar{p}p) + X$$

(Fig. 8a).

The properties of the bump found are the following:

i) $M = 1940 \pm 1$ MeV with $\Gamma \sim 10$ MeV

ii) $\sigma \sim 1\mu b$ in the $0.2 < x_F < 0.48$ interval.

The cross section in the $-0.1 < x_F < 0.2$ interval is similar after subtraction of the ϕ contamination which is strong in this central x region. So the $\bar{p}p$ (1940) cross section shows rather little dependence on x_F.

iii) Strongly asymmetric $\bar{p}p$ decay distribution. Events in the 1940 bump seem to be associated with a proton which is faster than the $\bar{p}$.

II.2. Diffractive p̄p Production

Two experiments have reported negative results in the search of the narrow
2.95 GeV p̄pπ⁻ bump[3]:

II.2.1 Ω Experiment WA40

This experiment - Aachen, Bari, Bonn, CERN, Glasgow, Liverpool, Milan
collaboration - used an incident π⁻ beam of 16 GeV/c and triggered on a
fast forward p̄ (or K⁻).

Different channels were analysed following the mechanism:

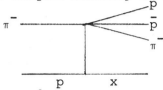

This experiment has reported[8] negative evidence for the narrow 2.95 GeV
bump, with an order of magnitude higher statistics than the earlier
experiment carried out largely by the same team and described in
C. Evangelista et al.[3].

The upper limit of the cross section x branching ratio for production of
a p̄pπ⁻ state with a width Γ < 30 MeV as derived from their data is

$$\sigma \times BR = 50nb$$

which is a factor 6 lower than the value expected on the basis of ref[3].

II.2.2 The Geneva-Lausanne Experiment WA10

Results from the Geneva-Lausanne experiment WA10 have recently been
published[9]. Events from the diffractive process were selected among the

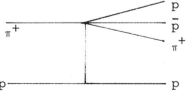

4-prong candidates recorded using a slow proton trigger device. The
incident beam energy was 50 GeV/c. Unlike Ω, the Geneva-Lausanne Spectrometer has no magnet; therefore they are unable to measure the momenta of

the three forward particles. However they can identify the particle masses with their multicell Cerenkov counter, measure their directions, and identify the events of reaction $\pi^+ p \rightarrow (\bar{p}\bar{p}\pi^+) P_{recoil}$ through a 1-C fit, with a final background of $\sim 20\%$.

Figures 9 and 10 show the $\bar{p}\bar{p}\pi^+$ and $p\bar{p}$ mass spectra obtained from the sample of 80000 events (corresponding to a sensitivity of 20 events/nb), where no structures are visible.

The upper limits obtained from this experiment are

$$\sigma \times BR = 15\text{nb for a } \bar{p}\bar{p}\pi^+ \text{ state with } \Gamma \leq 40 \text{ MeV}$$
$$\text{or a } p\bar{p} \quad \text{state with } \Gamma \leq 26 \text{ MeV}$$

II.2.3 The Ω Experiment WA48

WA48 is an experiment done by a Glasgow-Birmingham-CERN collaboration. Here the hadronic separated beam available in Ω was used to trigger on events $K^+ p \rightarrow$ fast $\bar{p}$ or K^- at 13 GeV/c. The obtained sample of events corresponds to a sensitivity of 12 events/nb.

The main purpose of this experiment was to check the $\bar{\Lambda}p\pi^+$ signal at 2460 MeV found earlier by a related collaboration[10]. Preliminary results show no evidence for this peak, again with a strongly increased statistic[11].

From nearly 1500 events identified as

$$K^+ p \rightarrow (p\bar{p}K^+)_F \, p$$

preliminary analysis[11] shows that this four prong reaction is dominated by $\bar{\Lambda}(1520)$ forward production, as can be seen in the $\bar{p}K^+$ mass spectrum of Fig. 11. No indication of a significant narrow bump is visible in the $p\bar{p}$ mass spectrum (Fig. 12).

II.3. Baryon Exchange

Three different experiments have been done this last year in the Ω' Spectro-meter, and have obtained final $p\bar{p}$ states produced via a baryon exchange mechanism. The layout for these experiments is similar to that shown in Figure 3.

II.3.1. Experiment WA49

A CERN-Liverpool Collaboration has collected events on the reaction

$$\bar{p}p \rightarrow p_{fast} \ (or \ K^+_{fast}) + X$$

at an incident $\bar{p}$ momentum of 12 GeV/c.

This experiment has produced preliminary results[12] on a first low statistics exposure (2 ev/nb) done in Ω late in 1978. Different quasi-two body reactions dominated by the mechanism

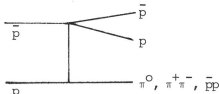

have been isolated. No evidence of any significative narrow bump in their $\bar{p}p$ mass spectra has emerged. An example is shown in Fig. 13 for the channel $\bar{p}p \rightarrow \bar{p}p\pi^+\pi^-$.

Data from the new exposure in Ω' (about 10 times higher statistics) is currently being analysed.

II.3.2. Experiment WA60

This is an experiment that was conceived as a "Strangeonium and Baryonium search" in K^-p interactions. It has been carried out by a Bari-Birmingham-CERN-Milan-Paris VI-Pavia Collaboration.

The trigger selected events with a fast forward p(or K^+) in

$$K^-p \rightarrow (p \ or \ K^+) + X$$

at 18.5 GeV/c. With a total incident flux of 2.10^5 K^-/burst they have recorded 7.6×10^6 triggers in a 12 days data taking period (the corresponding nominal luminosity is $\sim$ 15 events/nb).

From the $\bar{p}p$ background production point of view, preliminary results[13] show that there will be about 2000 events that could be associated to the reaction mechanism:

130

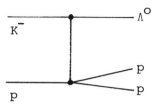

Higher statistics (about 20.000 events in total) will be obtained for meson exchange channels with ($\bar{p}p$) forward or ($\Lambda^0\bar{p}$) forward.

II.3.3 Experiment WA56

This experiment has been conceived[5] as a genuine exotic mesons X^{++} search, in π^+p interactions at 20 GeV/c via baryon exchange, e.g. according to

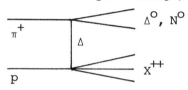

with $X^{++} \to p\bar{n}\pi^+$ or $p\bar{p}\pi^+\pi^+$.

Simultaneously, an improved trigger has been developed to search for narrow $\bar{p}p$ states via the process:

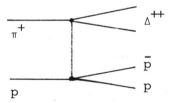

for which more favourable isospin coefficients are expected than for the corresponding graph of the π^-p experiment reported in ref. 2.

The layout of the first phase of WA56 experiment, where only the search for narrow $\bar{p}p$ states was considered, is shown in Fig. 3. In the second phase, the detector layout was completed with the new side MWPC's and two more forward MWPC's.

A very selective fast proton trigger, with extra multiplicity and γ veto conditions, was built with the new MBNIM electronics developed by the Ω-staff[14]. Of an average incident total flux of $\sim 5 \times 10^6$/burst, about 50% were useful π^+'s and produced about 100 triggers per burst

The trigger cross section was ∿ 10μb, and the total luminosity of the sample obtained is 700 events/nb for 100% acceptance.

We will present here some preliminary results obtained in the first phase (Summer 1979) of this experiment, e.g. ∿ 10^6 triggers concerning the search of narrow $\bar{p}p$ states. Roughly 10 times higher statistics will be available from phase 2 (Feb/March 1982). A clear signal, with very low contamination (∿ 1%) of events from reaction

$$\pi^+ p \to p_f \pi^+ \, p_s \bar{p}$$

is obtained. The momentum imbalance is better than 100 MeV along the beam axis and about 30 MeV for the transverse components.

The main characteristics of the reaction above is the strong production of $\Delta^{++} \to p\pi^+$. This is visible in Fig. 14 where both mass combinations with the fast and the slow proton are plotted. Fig. 15 shows the projection for Δ^{++}_f combination.

If we retain the events where $m(p_f\pi^+)$ is inside the Δ^{++} bounds (1.1,1.3 GeV) and plot the $\bar{p}$ decay angle in the $\bar{p}p$ rest system, we observe a strong asymmetric distribution (Fig. 16) where the forward peak could be interpreted as coming from diffractive $(p_f\bar{p})$ or $(\Delta^{++}_f \bar{p})$ production.

In analogy with the work of Benkheiri et al.[2] we try to suppress this diffractive background by selecting $\bar{p}p$ events with backward Jackson angles of the $\bar{p}$ ($\cos\theta_J < 0.7$), Fig. 17 shows the resulting $\bar{p}p$ mass spectrum, compared with that of Ref. 2.

The whole trigger + geometrical acceptance for this backward channel has been estimated to be 6%, therefore the total sensitivity of the sample plotted in Figure 17 is 5 ev/nb to be compared to ∿ 2 ev/nb reported by the old $\pi^- p$ experiment.

These partial results do not show much evidence of the 2020 MeV bump. However a ∿ 3 standard deviation effect is seen in the exact 2204 MeV location.

If the events in our 2204 MeV peak are compared to the old $\pi^- p$ data we see that the cross section is reduced by a factor ~ 5. This result is somewhat in disagreement with the expectations from a pure nucleon-exchange mechanism for the two body reaction $\pi p \to \Delta M(2204)$, estimated by Pennington[1] to lead to a ratio of 1.2 between the cross section for 12 GeV/c π^- and 20 GeV/c π^+, in the observed charge states.

III. CONCLUSIONS

The available results on $\bar{p}p$ production experiments recently done at CERN show:

1) A significative evidence for the inclusive production of the now very controversial[16] S(1936) meson, in photon and proton produced reactions, at small Feynman x values.

2) Negative evidence for the narrow $(\bar{p}p\pi)$ state of 2.95 GeV/c^2, from two high statistics experiments.

3) A preliminary indication (3 standard deviations effect) for the narrow 2204 MeV state produced backwards in reaction $\pi^+ p \to \Delta^{++} (\bar{p}p)$ at 20 GeV/c, but with a cross section ~ 5 times smaller than naively expected.

4) More data on $\bar{p}p$-spectra from several experiments will become available in the forthcoming months.

REFERENCES

1. D. Treille. Talk given in Triangular Meeting at Campione d'Italia 3-7 October 1977.

2. P. Benkheiri et al., Phys. Lett. 68B (1977) 483.
   P. Benkheiri et al., Phys. Lett. 81B (1979) 380.

3. C. Evangelista et al., Phys. Lett. 72B (1977) 139.

4. J. Rosner, Phys. Rev. Lett. 21 (1968) 950; 22 (1969) 689.

5. A. Ferrer et al., Proposal to Study $N\bar{N}$ states produced via Baryon Exchange in $\pi^{+}p$ Interactions using the Ω Prime Spectrometer. CERN/SPSC/78-119; SPSC/P.117 (12 October 1978).

6. D. Aston et al., CERN/EP/80-40 (1 April 1980).

7. C. Daum et al., Phys. Lett. 90B (1980) 475.

8. T.A. Armstrong et al., Phys. Lett. 85B (1979) 304.

9. W.E. Cleland et al., Phys. Lett. 86B (1979) 409.

10. T.A. Armstrong et al., Phys. Lett. 77B (1978) 447.

11. R.M. Turnbull. Private Communication.

12. Status Report on WA49. CERN/SPSC/79-82. SPSC/M.190 (24 August 1979).

13. L. Mandelli. Private Communication.

14. A. Beer et al., Nucl. Inst. and Meth. 160 (1979) 217.

15. M.R. Pennington. CERN/EP/PHYS 78-44 (21 November 1978).

16. G. Smith. $\bar{p}p$ formation experiments. This Conference.

TABLE I

List of $\bar{p}p$ Production Experiments at CERN

	Beam energy	Spectrometer	Code	Status
I. Photoproduction of $\bar{p}p$				
$\gamma p \rightarrow (\bar{p}p)\, X$	40 – 70 GeV	Ω	(WA4)	Published (Ref. 6)
II. Hadroproduction of $\bar{p}p$				
II.1 INCLUSIVE				
$\binom{\pi}{p} p \rightarrow (\bar{p}p)\, X$	93 GeV/c	ACCMOR	(WA3)	Published (Ref. 7)
$\binom{K}{\pi} p \rightarrow \bar{p}_{x\simeq 0}\, X$	40 GeV/c	Ω'	(WA63)	Data taking
II.2 DIFFRACTIVE				
$\pi^- p \rightarrow (\bar{p}p\pi^-)\, X$	16 GeV/c	Ω	(WA40)	Published (Ref. 8)
$\pi^+ p \rightarrow (\bar{p}p\pi^+)\, p$	50 GeV/c	GVA-LAUSANNE	(WA10)	Published (Ref. 9)
$K^+ p \rightarrow (\bar{p}p)\, pK^+$	13 GeV/c	Ω	(WA48)	Prel. results
II.3 BARYON EXCHANGE				
$\bar{p}p \rightarrow (\bar{p}p)\, X$	12 GeV/c	Ω Ω'	(WA49)	Prel. results
$K^- p \rightarrow \Lambda^0\, (\bar{p}p)$	18.5 GeV/c	Ω'	(WA60)	Analysis
$\pi^+ p \rightarrow \Delta_F^+\, (\bar{p}p)$	20 GeV/c	Ω'	(WA56)	Prel. results

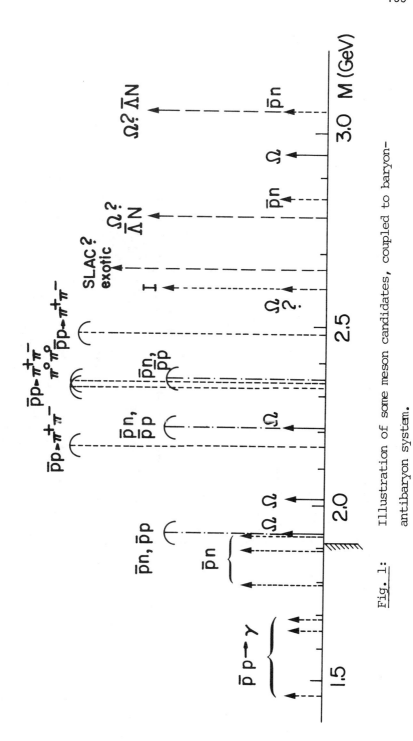

Fig. 1: Illustration of some meson candidates, coupled to baryon-antibaryon system.

136

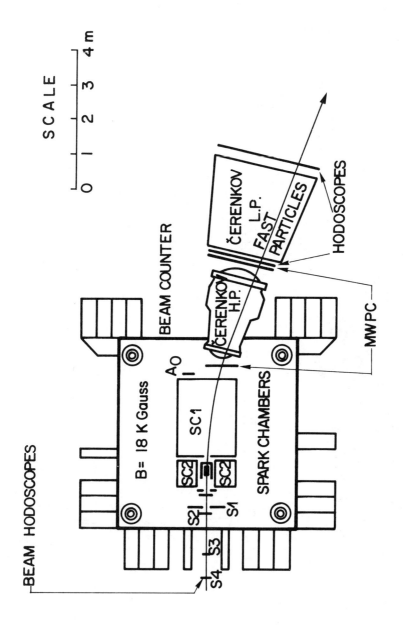

Fig. 2:   The Ω layout as used in the fast proton experiment (ref. 2).

137

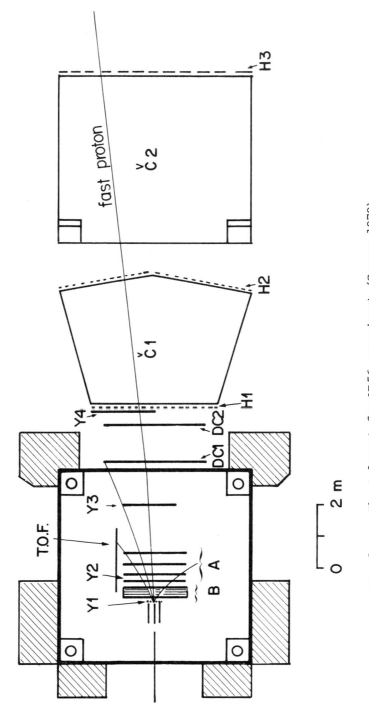

Fig. 3: The Ω layout for WA56 experiment (Summer 1979).

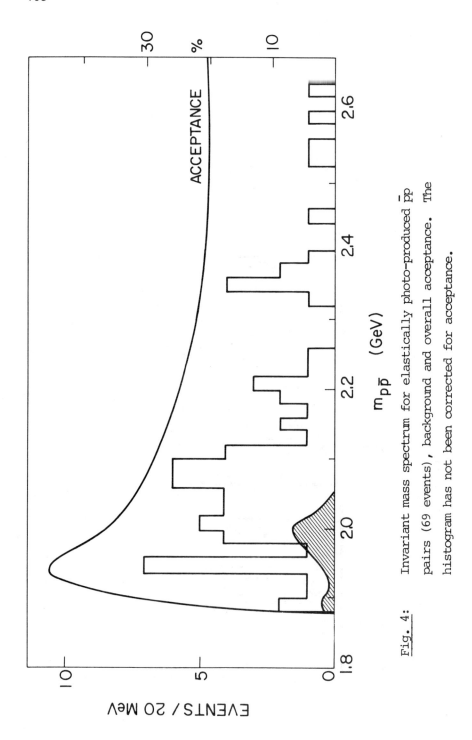

Fig. 4: Invariant mass spectrum for elastically photo-produced p̄p pairs (69 events), background and overall acceptance. The histogram has not been corrected for acceptance.

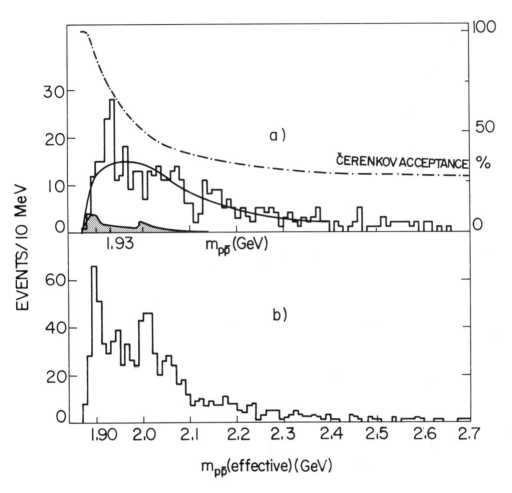

Fig. 5:   (a) Inclusive p̄p mass spectrum (462 events), background,
               acceptance for Cerenkov identification of both
               particles, and fitted curve.
          (b) Effective p̄p mass spectrum for "light/no-light" pairs
               (897 events). Both histograms are uncorrected for
               acceptance.

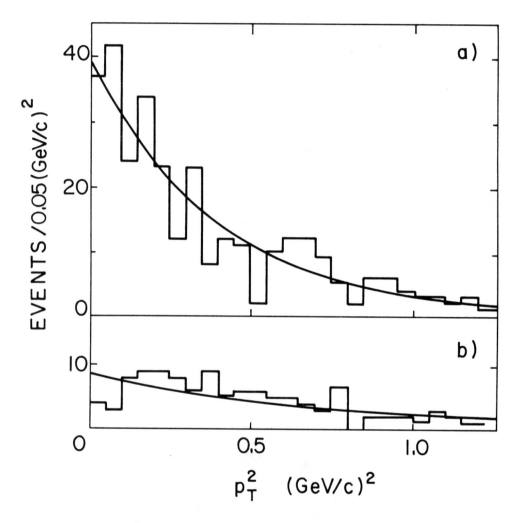

Fig. 6: p_T^2 distributions, with fitted curves:
(a) $m(\bar{p}p) > 1.96$ GeV (b) $m(\bar{p}p) < 1.96$ GeV.

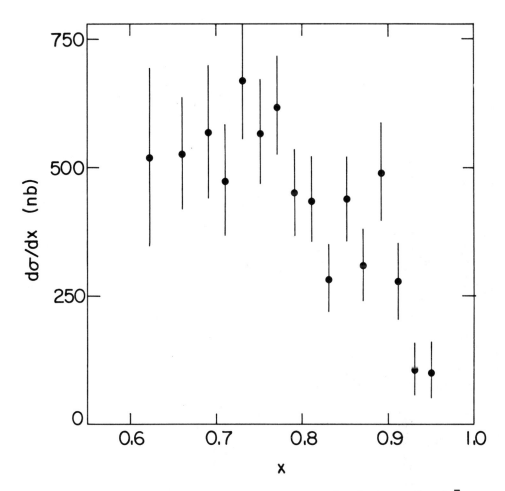

Fig. 7: Feynman x distribution for inclusively photo-produced p̄p pairs (total charged track multiplicity $\geq$ 4). All p̄p masses are included, and full acceptance corrections have been applied; there is a systematic uncertainty of ± 40% on the vertical scale.

142

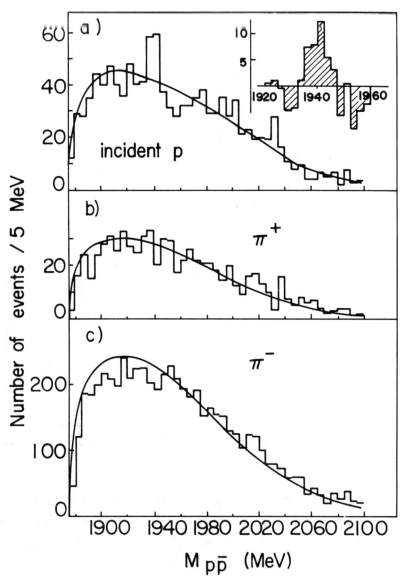

Fig. 8: $\bar{p}p$ invariant mass distribution for a) incident p,
b) incident π^+ and c) incident π^- at 93 GeV. The (anti)
protons have momenta between 13 and 24 GeV. The background
curves are obtained by pairing a proton and an antiproton
from two different events and are normalised to the total
number of events in the plot. The inset in Fig. 8a is an
enlarged view of the background subtracted mass distribution
around 1940 MeV.

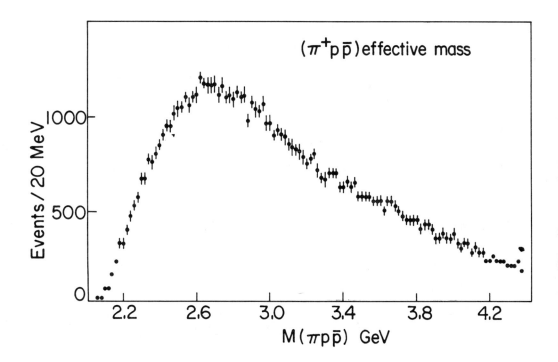

Fig. 9: π⁺pp̄ effective mass spectrum from the diffractive reaction
π⁺p → (π⁺pp̄)p at 50 GeV/c.

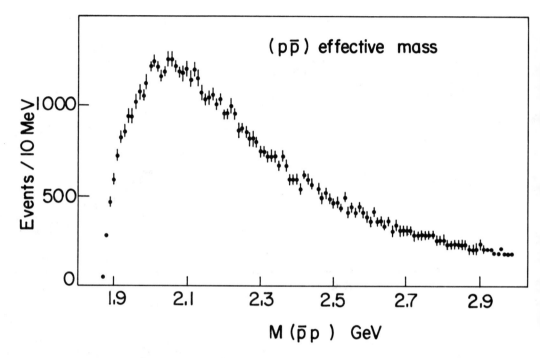

p̄p effective mass spectrum ($\pi^{+}p \rightarrow (\pi^{+}p\bar{p})p$ at 50 CeV/c).

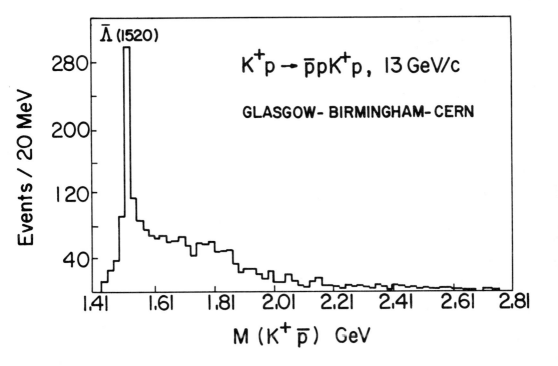

Fig. 11:   K$^{+-}_{}$p effective mass spectrum (K$^+$p → K$^{+-}_{}$ppp at 13 GeV/c).

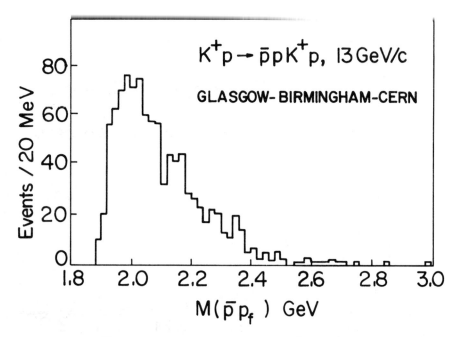

Fig. 12: $\bar{p}p$ effective mass spectrum ($K^+p \rightarrow K^+\bar{p}pp$ at 13 GeV/c).

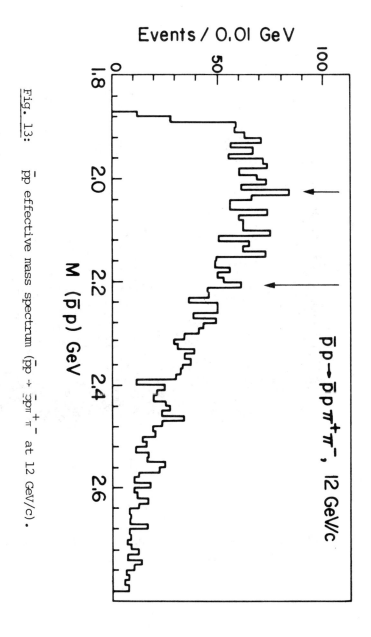

Fig. 13: $\bar{p}p$ effective mass spectrum ($\bar{p}p \rightarrow \bar{p}p\pi^+\pi^-$ at 12 GeV/c).

148

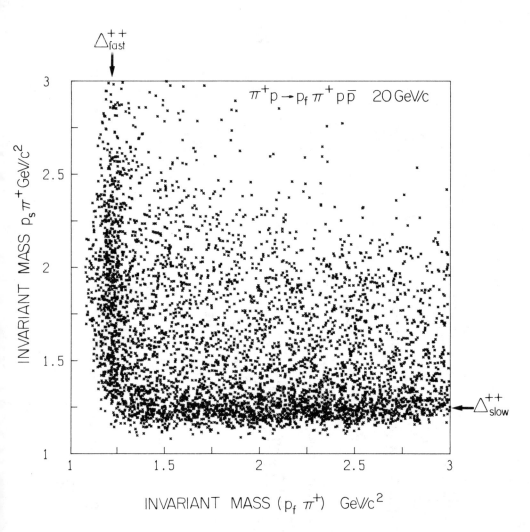

Fig. 14: Effective mass $(p_f \pi^+)$ versus $(p_s \pi^+)$ from reaction $\pi^+ p \to p_f \pi^+ p_s \bar{p}$ at 20 GeV/c.

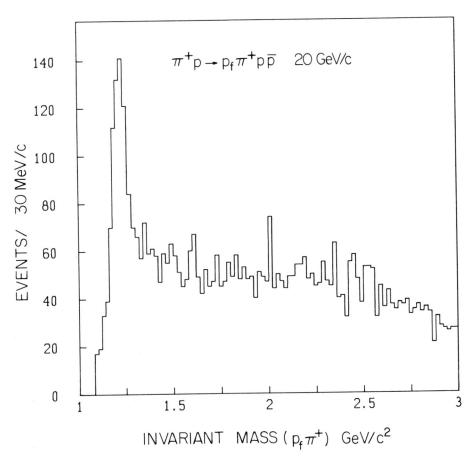

<u>Fig. 15:</u>   Effective mass $(p_f \pi^+)$ as projected from previous plot.

150

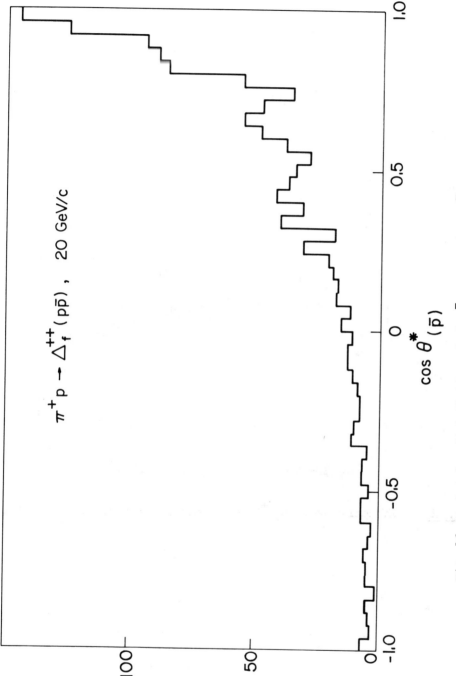

Fig. 16: Angular distribution of the $\bar{p}$ decay in the $(\bar{p}p)$ rest mass system for Δ_f^{++} $\bar{p}p$ events.

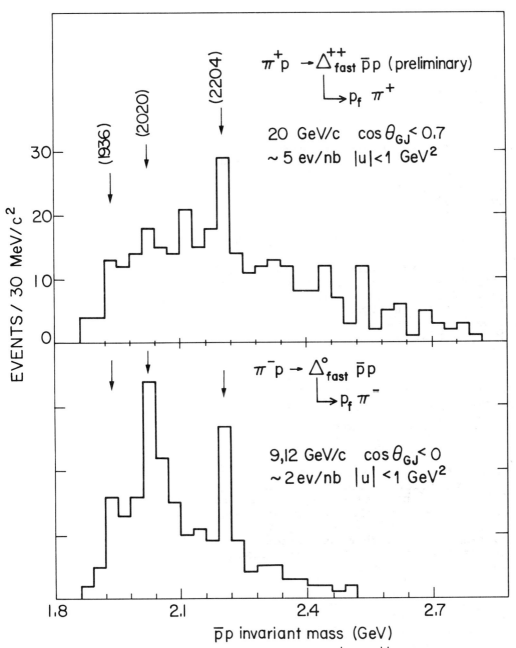

Fig. 17: $\bar{p}p$ effective mass spectrum from $\pi^+ p \to \Delta^{++}_f \bar{p}p$ reaction at 20 GeV/c with a selection in $\cos\theta$ Jackson of $\bar{p}$ decay < 0.7 and with a cut in the momentum transfer $|u(\pi^+ \to \Delta^{++}_f)| < 1$ GeV^2.

Search for Narrow $\bar{p}p$ States in 10 GeV/c
π^+p Interactions

D. R. Green
Fermilab, Batavia, Illinois 60510

ABSTRACT

Narrow $\bar{p}p$ states were searched for in the reaction $\pi^+p \rightarrow \Delta^{++} (\bar{p}p)$ at 10 GeV/c. No states were observed with masses less than 2.3 GeV/c^2 and cross sections greater than 40 nb at the 95% c.l.. In particular, the two previously reported states[1] at 2020 and 2200 MeV/c^2 were not observed at the 11 s.d. level.

INTRODUCTION

This paper describes results from E716 which was a collaboration between people from BNL, CMU, FNAL, and SMU. The experiment was run at the Brookhaven National Laboratory multiparticle spectrometer (MPS). The experiment used a fast forward proton trigger with 10 GeV/c incident π^+.

$$\pi^+p \rightarrow p_F X^+ \qquad (1)$$

Incident π^+ were chosen to allow one to study exotic meson[2] systems, for example:

$$\pi^+p \rightarrow \Delta^0 X^{++} \qquad (2)$$
$$\Lambda^0 S^{++} \qquad (3)$$

In addition, the Clebsch-Gordon coefficients for forward Δ^{++} production are more favorable than those for forward Δ^0 production by π^-. An incident momentum of 10 GeV/c was chosen as a compromise between the s^{-2} p dependence expected for baryon exchange, and the desire to search for high mass states. At 10 GeV/c, with our trigger, we are sensitive to masses ≈ 2.5 GeV/c^2 which is well above the baryon-antibaryon threshold.

ISSN:0094-243X/81/670152-18$1.50 Copyright 1981 American Institute of Physics

APPARATUS AND TRIGGER

The apparatus has been described elsewhere.[3]
Basically it consists of a set of recoil scintillators
and spark chambers, and a set of forward spark chambers
interspersed with PWC planes used in the trigger. All
these detectors completely surround a liquid hydrogen
target and are immersed in a magnetic field. Downstream
of the magnet there are lever arm spark chambers, a
multicell Cerenkov counter, and a scintillation counter
hodoscope.

The trigger for the experiment consisted of requiring a) an incident π^+ b) No light in the Cerenkov counter
(C) c) one and only 1 hit in the hodoscope d) a multiplicity for the event of $\geq$ 3 particles e) and a 3 point
logic[4] which selected a positive track with momentum $\geq$
5.5 GeV/c. The trigger selected fast forward leading
baryons and 4 prong or higher final multiplicities.

Simultaneous with the main data taking a small
sample of data was taken of the reaction;

$$\pi^+ p \rightarrow \pi^+_F X^+ \qquad (4)$$

This trigger had all the trigger requirements of the main
trigger (1) save that $C \rightarrow C$. Finally, elastic scattering
data with incident $\pi^\pm$ at 6, 8, and 10 GeV/c were taken
during special runs.

ACCEPTANCE

The first input to our acceptance calculation came
from measured elastic data discussed previously. Seven
different elastic triggers were taken which allowed us to
measure the efficiency of the C counter, the hodoscope,
and the multiplicity counters and PWC. These measurements, and a Monte Carlo program for the geometric
acceptance allowed a calculation of $d\sigma_{el}/dt$. In fact,
this was done two ways; once by missing mass alone, and
once by 2 prong reconstruction. This procedure allowed a
check on the recoil track pattern recognition efficiency.
As seen in Fig. 1 a our agreement with published data[5],
both in magnitude and shape, is good.

In order to study a known 4 prong reaction, we used
the minority trigger data from (4) to reconstruct.

$$\pi^+ p \rightarrow \rho^0 \Delta^{++} \qquad (5)$$

154

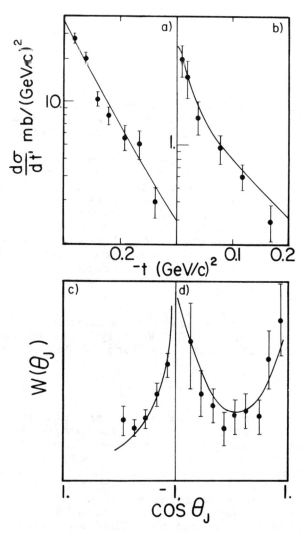

Fig. 1. a) Elastic scattering dσ/dt. The line represents the data of Ref. 5.
   b) dσ/dt' for reaction (5). The line represents the data of Ref. 6.
c),d) Jackson angles for $\rho^0 \rightarrow \pi^+\pi^-$, $\Delta^{++} \rightarrow p\pi^+$. The lines represent the data of Ref. 6.

Note that this reaction is topologically very similar to the reaction we wish to study.

$$\pi^+ p \rightarrow \Delta^{++} (\bar{p}p) \tag{6}$$

As with the elastic scattering data, we have a cross check on the recoil chamber pattern recognition efficiency in that we found $d\sigma/dt'$ both by missing mass and by complete 4 prong reconstruction. In Fig. 1 b is shown $d\sigma/dt'$ for this experiment and for bubble chamber data[6]. In addition, the Jackson angles were reconstructed for ρ^0 and Δ^{++} decays. As seen in Fig. 1c and Fig. 1d the good agreement with known data[6] gives us confidence that the acceptance is understood over all of phase space.

RESOLUTION

Several checks were made on the mass resolution used in this experiment. First the elastic scattering data at 6, 8, and 10 GeV/c was used in order to study the trigger track momentum resolution. By varying the momentum, and by dropping sparks artificially, we could verify that $dp \sim p^2/L^2$ as expected. Thus, we feel that the fast track momentum resolution is understood.

Next we studied the effective mass resolution of the forward system. The effective mass resolution for reaction (3) is shown in Fig. 2a. The resolution is ± 4 MeV/c^2 and the Λ^0 mass is 1114.6 MeV/c^2. These data check that we can propagate the forward system through the inhomogeneous magnetic field to a common vertex.

Finally, since we wish to examine the slow $\bar{p}p$ system, we studied a topologically similar reaction which has a recoil δ function in mass,

$$\pi^+ p \rightarrow K^+_F \ Y^{*+}$$
$$\quad \quad \quad \llcorner K^0_S \ p$$
$$\quad \quad \quad \quad \quad \llcorner \pi^+ \pi^- \tag{7}$$

The agreement with the Monte Carlo results implies that the recoil effective mass resolution is well understood. Both data and Monte Carlo predictions are shown in Fig. 2b.

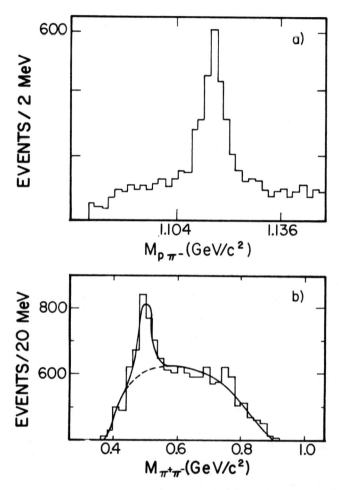

Fig. 2. a) Effective mass distribution for reaction (3)
        b) Effective mass distribution for reaction (7). The line is a smooth background plus the Monte Carlo resolution shape.

MASS ASSIGNMENTS

All triggers with 4 prongs from a common vertex were examined for momentum balance. For the reaction $1+p \rightarrow (2 + 3)_F + (4 + 5)_B$ one removes missing neutrals by requiring

$$\delta p \equiv | \vec{P}_1 - \vec{P}_2 - \vec{P}_3 - \vec{P}_4 - \vec{P}_5 | \tag{8}$$

to be small. For example δp is shown for reaction (5) candidates in Fig. 3a. The curve is a Monte Carlo sum of shapes for reaction (5) and

$$\pi^+ p \rightarrow \rho^0 \Delta(1670)$$
$$ \llcorner p\pi^+\pi^0 \tag{9}$$

Clearly, the main features of the δp spectrum are well understood. Events with $\delta p < 150$ MeV/c were retained for subsequent analysis.

To assign the masses one picks that set which gives the minimum energy imbalance, $\delta E \equiv E_1 + M_p - E_2 - E_3 - E_4 - E_5$. Incorrect mass assignments will have $\delta E \neq 0$; $d(\delta E) = -dM_2/\gamma_2 - dM_3/\gamma_3 - dM_4/\gamma_4 - dM_5/\gamma_5$. Since 2 and 3 are fast and 4 and 5 are slow, $d(\delta E) \sim -(dM_4 + dM_5)$. In Fig. 3b is shown the data for reaction (5) candidates with $\delta p < 150$ MeV/c. Again the agreement with the Monte Carlo is excellent.

We have reconstructed events from the reactions

$$\pi^+ p \rightarrow \quad \Delta^{++}(\pi^+\pi^-) \tag{10}$$

$$\Delta^{++}(K^+K^-) \tag{11}$$

$$(K^+_F \pi^+)(pK^-) \tag{12}$$

in order to check backgrounds. Since $d(\delta E)$ is expected to be -1.6 GeV, -0.9 GeV, and -0.5 GeV respectively, and since we cut on $|\delta E| < 100$ MeV we find no ambiguity in mass assignments for events with no missing neutral energy.

158

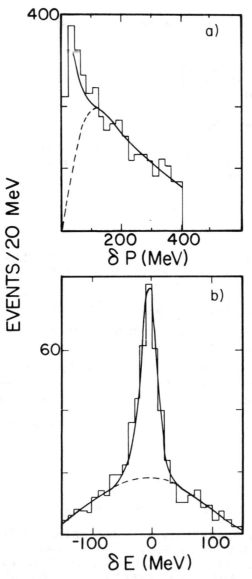

Fig. 3. a) Momentum imbalance for reaction (5)
              candidates. The curve is the Monte Carlo
              sum of δp for reaction (9) (dashed line)
              and reaction (5)
         b) Energy imbalance for reaction (5)
              candidates. The curve is the sum of a
              smooth background and the Monte Carlo
              shape for δE.

In Fig. 4 is shown the energy imbalance for reaction (6) candidates; $\delta p < 150$ MeV/c, missing mass to $(p_F\pi^+) > 1.8$ GeV/c^2 and forward effective mass of $(p_F\pi^+)$ < 1.35 GeV/c^2. The curve is a smooth background due to reactions (10) through (12) with added neutrals, plus data for reaction (12), plus Monte Carlo results for reaction (6). The agreement in shape gives confidence that backgrounds are understood. In addition, since δE is related to the $(\bar{p}p)$ mass resolution, another check is provided. We estimate that our sample of reaction (6) candidates contains 30% background of events from reactions (10) through (12) with added neutrals.

BARYON EXCHANGE

In Fig. 5 is shown the $(\bar{p}p)$ effective mass distribution for all events of the type

$$\pi^+ p \rightarrow p_F \pi^+ \bar{p}p \tag{13}$$

which satisfy $\delta p < 150$ MeV/c, $|\delta E| < 100$ MeV. Clearly, the spectrum is smooth and featureless. However, it is important to note that merely requiring a fast forward baryon does not select a baryon exchange process. A longitudinal phase space analysis of reaction (13) with a cross section of ~ 18 μb leads to cross sections.[7]

$$
\begin{array}{llll}
& (\bar{p}p)_F \ (p\pi^+)_B & \sim 9.7 \ \mu b & (14) \\
\pi^+ p \rightarrow & (\bar{p}p\pi^+)_F \ (p)_B & \sim 3.6 \ \mu b & (15) \\
& (p)_F \ (\bar{p}p\pi^+)_B & \sim 2.9 \ \mu b & (16) \\
& (p\pi^+)_F \ (\bar{p}p)_B & \sim 1.8 \ \mu b & (17)
\end{array}
$$

Thus, 90% of reaction (13) consists of meson exchange (74%) or baryon exchange with an isolated leading proton (16%).

The diagrams for reactions (14) through (17) are shown in Fig. 6. Clearly, reactions (14) and (16) will be reduced by requiring the $(p_F\pi^+)$ mass to be small.

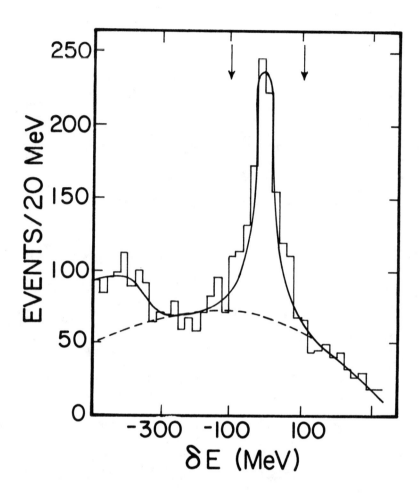

Fig. 4.   Energy imbalance for reaction (6) candidates.
The curve is the sum of a smooth background,
Monte Carlo shape for reaction (6), and data
for reaction (12).

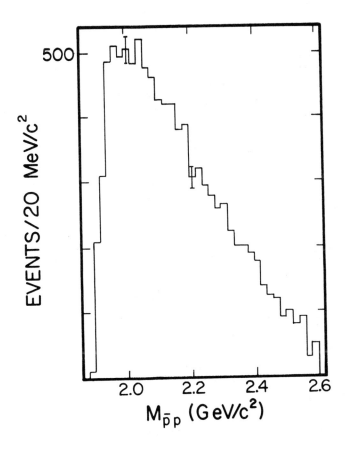

Fig. 5.   All reaction (13) candidate events satisfying
δE and δp cuts.

162

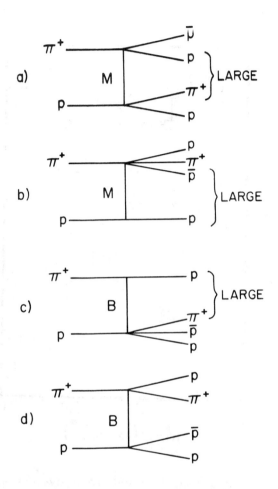

Fig. 6.   Diagrams contributing to reaction (13)
    a),b)   Meson exchange
    c),d)   Baryon exchange

Similar backgrounds exist for the data of Ref. 1. In Fig. 7a is shown the ($p_F\pi^-$) mass distribution from that experiment.[8] Contrary to the statements of Ref. 8, there is no evidence for forward Δ^0 production. By comparison, our data is shown in Fig. 7b. As expected, reactions (14) and (16) contribute to large ($p_F\pi^+$) masses. A cut was made at ($p_F\pi^+$) mass $\leq$ 1.35 GeV/c^2. This cut largely eliminates reactions (14) and (16) and some fraction of reaction (15).

Additional studies were made to attempt to further isolate reaction (17). Since $d\sigma/dt'$ $\backsim$ $e^{-2t'}$, where t' = -momentum transfer from π^+ to Δ^{++}, a peripheral cut was not very useful. Examination of the distribution of ($\bar{p}p$) Jackson angle after the Δ^{++} cut indicated substantial residual meson exchange as expected. Requiring $Xp_F>0$, $X_{\pi^+}>0$, $X_{\bar{p}}<0$, $X_p<0$ resulted in an essentially pure Δ^{++} signal and a much more isotropic Jackson angle distribution. However, the $\bar{p}p$ mass distribution in both size and shape was essentially unchanged for $\bar{p}p$ masses $\leq$ 2.3 GeV/c^2 whether one made a simple Δ^{++} cut, or the full set of X cuts. Since the residual meson exchange background surviving the Δ^{++} cut was largely confined to ($\bar{p}p$) masses $\geq$ 2.3 GeV/c^2 as expected from Fig. 6b, we considered the sample surviving a simple Δ^{++} cut, an essentially pure baryon exchange sample for ($\bar{p}p$) masses $\leq$ 2.3 GeV/c^2.

NARROW RESONANCE SEARCH

The final data sample for reaction (17) is shown in Fig. 8. The $\bar{p}p$ mass distribution is featureless. This is twice the data set of our recent publication.[9] There are roughly 2000 events surviving the Δ^{++} cut. The sensitivity is > 1 event/nb for ($\bar{p}p$) mass < 2.3 GeV/c^2. The cross section is $\sigma(\pi^+p \to \Delta^{++}\bar{p}p)$ = 2.3 ± .04 μb.

As mentioned above, we have indirect checks on our acceptance calculations using elastic scattering and reaction (5). We use these results, and our δE background uncertainty to quote a 20% absolute normalization uncertainty. In Fig. 9a is shown our cross section compared to bubble chamber results.[10,11] Clearly, we agree with the other data within the quoted errors.

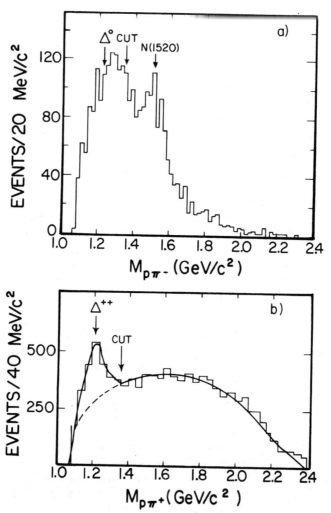

Fig. 7.   a)   ($p_F \pi^=$) mass distribution at 9 GeV/c from Ref. 8.

   b)   ($P_F \pi^+$) effective mass distribution for reaction (13). The curve is the sum of a smooth background plus Δ^{++} resonance.

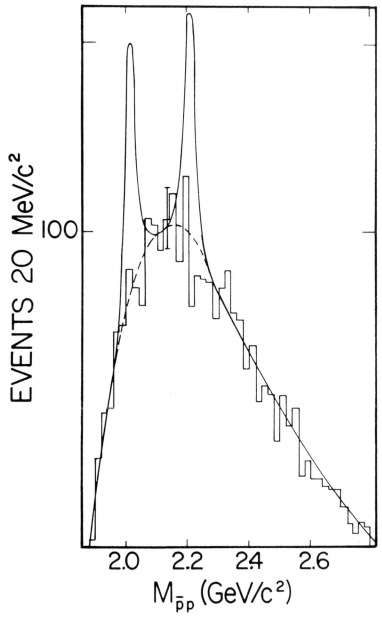

Fig. 8.   Final data set for reaction (17). The smooth curve is a polynomial, $\chi^2/\text{DOF} = 47/49$, plus the signals expected for nucleon exchange production of the states of Ref. 1.

Our calculated ($\bar{p}p$) mass resolution is shown in Fig. 9b. As discussed above, this result must be consistent with our data on elastic scattering missing mass at 6, 8, and 10 GeV/c, with a forward system δ function (Λ^0) and with a recoil system δ function (K^0).

Given these results for cross section and resolution one can confidently set a model independent upper limit on the production cross section of possible narrow resonances. Of course, narrow means with respect to the resolution shown in Fig. 9b. The 3 standard deviation (95% c. l.) limit curve is shown in Fig. 9c. We see no narrow resonances at the 95% c.l. with cross sections of $\geq$ 40 nb and masses $\leq$ 2.3 GeV/c^2. To set the scale, 4 quark exotic mesons are expected to be produced as copiously as normal mesons[2], i.e. σ $\sim$ 300 nb for reaction (10) for the $\pi^+\pi^-$ in the ρ^0 region.

SEARCH FOR THE 2020 and 2204 MeV/c^2 STATES

Two states were reported in Ref. 1 at 2020 and 2204 MeV/c^2 respectively at incident momenta of 9 and 12 GeV/c. As mentioned above, no evidence of forward Δ^0 production was seen. In any case, applying a cut on the ($p_F\pi^-$) mass of < 1.35 GeV/c^2, the authors of Ref. 1 quote a cross section times branching ratio of $\sim$ 14 nb and $\sim$ 19 nb and a width $\sim$ 24 MeV/c^2 and $\sim$ 10 MeV/c^2 for the 2020 and 2204 states respectively.

In order to directly compare the two experiments one needs to understand the production mechanism. Most baryon exchange processes proceed via nucleon rather than Δ exchange.[12] In particular, the authors of Ref. 1 have searched for the two states in[13]

$$\pi^- p \rightarrow p\,(\bar{p}n) \qquad\qquad (18)$$

The absence of signals in reaction (18) may be interpreted as due to nucleon exchange dominance.

The mass spectrum of Ref. 1 is very similar to that shown in Fig. 8 outside the resonance regions. If we assume nucleon exchange dominance, then we should see $\sim$ 140 events above backgrounds of $\sim$ 160 events in a single 40 MeV/c^2 bin. This is illustrated by the smooth curves in Fig. 8 and the • symbols in Fig. 9c. Under these assumptions the two states would appear as 11 s.d. fluctuations.

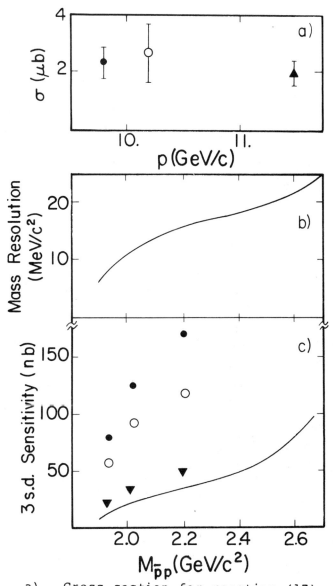

Fig. 9.   a)   Cross section for reaction (17), ● this
                experiment, o Ref. 10, ▲ Ref. 11.
           b)   Mass resolution of the (p̄p) system
           c)   3 s.d. upper limit for narrow resonances
           ●   states from Ref. 1 scaled by
                     nucleon exchange
           o   states from Ref. 1 scaled by
                     cross section ratio
           ▼   states from Ref. 1 scaled by Δ
                     exchange.

It instead one merely scales to the cross section ratio of 6.5 the two states would appear as 8 s.d. fluctuations. This assumption appears as the o symbols in Fig. 9c. Finally, against all evidence, if one assumed Δ exchange dominance, the states would still appear as 3 s.d. fluctuations. This assumption appears as the ▼ symbols in Fig. 9c.

In conclusion we find no evidence for the states reported in Ref. 1 at the 95% c.l. (Δ exchange), 99.97% c.l. (σ scaling), or 99.998% c.l. (N exchange).

CONCLUSIONS

We have searched for narrow ($\bar{p}p$) states produced by baryon exchange reactions in π^+p interactions at 10 GeV/c. Numerous checks of the apparatus efficiency and geometric efficiency have been performed. The quoted cross section agrees with other data to ± 15%. Our mass resolution is checked using a slow K^o_s decay.

The desired baryon exchange fraction of the reaction (13) is only 10%. For low ($\bar{p}p$) masses it is cleanly isolated by cutting on the forward ($p_F\pi^+$) mass. The Δ^{++} signal is rather cleaner than the Δ^o signal of Ref. 1 indicating that nucleon exchange dominates.

In general we do not observe any narrow resonances at the 40 nb level for ($\bar{p}p$) masses $\leq$ 2.3 GeV/c^2 (95% c.l.) In particular we do not observe the 2020 and 2200 MeV/c^2 states reported in Ref. 1 at the 3 s.d. level assuming Δ exchange, at the 8 s.d. level scaling to the same cross section, or at the 11 s.d. level assuming N exchange (as advocated by the authors of Ref. 1 and Ref. 13).

REFERENCES

1) P. Benkheiri et. al., Phys. Lett. <u>68B</u>, 483 (1977).

2) R. Jaffe, Phys. Rev. <u>D17</u>, 1444 (1978).

3) N.A. Stein et. al., Phys. Rev. Lett. <u>39</u>, 378 (1977).

4) E.D. Platner, IEEE Trans. Nucl. Sci. <u>24</u>, 225 (1977).

5) K.J. Foley et. al., Phys. Rev. Lett. <u>11</u>, 503 (1963).

6) O. Maddock et. al., Nuovo Cimento <u>A5</u>, 433 (1971).

7) A.W. Key et. al., in Proceedings of the Fourth European Antiproton Symposium, ed. A. Fridman (editions du CNRS, Strasbourg, 1978), Vol. I, p. 611.

8) L. Montanet, in Experimental Meson Spectroscopy 1977, ed. E. VonGoeler, R. Weinstein (Northeastern University Press, Boston, 1977), p. 281.

9) R.M. Bionta et. al., Phys. Rev. Lett. <u>44</u>, 909 (1980).

10) C.N. Kennedy et. al., Phys. Rev. <u>D16</u>, 2083 (1977).

11) A. W. Key et al., submitted to the XIX International Conference on High Energy Physics, Tokyo, Japan 23-31 August, 1978.

12) N. Sharfman, Ph.D. Thesis, Carnegie-Mellon University, 1979 (unpublished).

13) P. Benkheiri et. al., Phys. Lett. <u>81B</u>, 380 (1979).

SEARCH FOR NARROW p̄p STATES IN THE REACTION π⁻p › pπ⁻ p̄p at 16 GeV/c

S.U. Chung, A. Etkin, R. Fernow, K. Foley, J.H. Goldman,[*]
H. Kirk, J. Kopp, A. Lesnik,[†] W. Love, T. Morris,
S. Ozaki, E. Platner, S.D. Protopopescu,
A. Saulys, D. Weygand, C.D. Wheeler, E. Willen
Brookhaven National Laboratory, Upton, New York   11973

J. Bensinger, W. Morris
Brandeis University, Waltham, Massachusetts   02154

S.J. Lindenbaum
Brookhaven National Laboratory and City College of New York

M.A. Kramer, U. Mallik
City College of New York, New York, New York   10031

Z. Bar Yam, J. Dowd, W. Kern, M. Winik
Southeastern Massachusetts University, N. Dartmouth, Massachusetts

J. Button-Shafer, S. Dhar,[†] R. Lichti[§]
University of Massachusetts, Amherst, Massachusetts   02154

(Presented by S.D. Protopopescu)

ABSTRACT

We have carried out a sensitive ($\gtrsim$5 events/nb) search for
narrow p̄p states at the BNL Multiparticle Spectrometer. We found
no evidence of narrow p̄p states at 2020 and 2200 MeV in the reaction
π⁻p → pπ⁻p̄p at 16 GeV/c. We quote 2 σ upper limits of $\lesssim$ 3 nb for
these states in our data. Based on the cross sections of the CERN
Ω experiment at 12 GeV/c and assuming baryon-exchange processes for
the production, we should have seen $\gtrsim$5 σ signals at 2020 and 2200
MeV.

*Present address: Florida State University, Tallahassee, FL 32306
†Present address: Fairchild Republic Company, Farmingdale, NY 11735
‡Present address: Raytheon Corporation, Wayland, MA 01420
§Present address: Texas Tech University, Lubbock, TX 79409

ISSN:0094-243X/81/670170-16$1.50 Copyright 1981 American Institute of Physics

In recent years narrow $N\bar{N}$ states have attracted much interest from both theorists and experimentalists. They represented prime candidates for being four-quark exotic states, often referred to as baryonium. Several $\bar{p}p$ states with narrow widths ($\lesssim$ 24 MeV), have been reported. Of these, the two states with mass 2020 and 2200 MeV have been seen with good statistical significance in only one experiment. This was a production experiment with π^-p interactions at 9 and 12 GeV/C carried out in 1977 at the CERN Ω by Benkheiri et al.[1] A number of experiments have since looked for these states in both formation as well as other production processes, all with negative results.[2,3,4]

Our experiment is the first to search for these states in the same reaction as the Ω -spectrometer experiment using similar trigger techniques and acceptance.[1] The reaction studied is

$$\pi^-p \rightarrow (p_f\pi^-) \, (\bar{p}p)_s \tag{1}$$

where the subscript f(s) refers to a fast (slow) system in the laboratory. Our experiment was performed at the BNL Multiparticle Spectrometer (MPS) with a π^- beam at 16 GeV/c impinging on a 60 cm long LH$_2$ target (see Fig. 1). The trigger required a fast forward proton with good acceptance for a baryon-exchanged, fast $p\pi^-$ system going downstream of the target. The slow-recoil $\bar{p}p$ system was then kinematically identified with the aid of spark-chamber modules on both sides as well as downstream of the target. The trigger elements included two planar PWC's, T_1 and T_2, and two scintillation counter hodoscopes, H_5 and H_7, which were used to select on-line positive tracks with momenta between 8 and 12 GeV/c, and two Čerenkov counters, C_6 and C_7, with γ thresholds of 20 and 13, respectively. For proton identification the trigger utilized two sets of three dimensional coincidence-matrix logic systems implemented via two random access memories, RAM1 and RAM2. The elements in the logic system were (T1, T2, H5.$\bar{C}6$) in RAM1 and (T1, T2, H7.$\bar{C}7$) in RAM2. With these systems the efficiency for rejecting fast forward π^+ and K$^+$ was better than 99%. In addition, a multiplicity trigger around the target was required to select events with charged tracks $\geq$ 3. A total of 3.4 x 10^6 proton triggers were recorded, and $\sim$80% of the sample have been analyzed to date, corresponding to a total path length of 62 nb^{-1}.

Events have been processed in two stages. The first stage consisted of a pattern-recognition and a crude vertex-fitting program. One can already glean at this stage much of the physics information contained in our data, as is demonstrated by a plot of $M(\pi^-p)$ for V^0 events (see Fig. 2). Note that a clean Λ^0 peak is seen with mass (1115.1 ± 0.2) MeV and σ = 3.5 MeV. From a total 450K 4-prong events collected at this stage, we have selected 40K 4C candidates by requiring missing momenta to be small ($|\Delta P_x|$ < 300 MeV, $|\Delta P_y|$ < 200 MeV and $|\Delta P_z|$ < 1 GeV), and processed them through a more elaborate fitting program.

The second stage of our data reduction chain consists of a fitting program designed to perform iterative fits to spark-chamber measurements and beam parameters simultaneously, where the parameters in the fit are the vertex position and the momentum of each track at the vertex plus kinematic constraints (if any). The 40K sample has been processed through this program, first without the kinematic constraints (OC-fits), and then with the 4C kinematic

constraints for the hypothesis corresponding to Reaction (1). Figs.
3a-c show the distributions in missing momenta ΔP_x, ΔP_y and ΔP_z after
the OC-fit. Fig. 3d shows the difference in CM energy, $\Delta\sqrt{s} = \sqrt{s}$
(initial) $\sqrt{s}$ (final), after making cuts on the OC missing momenta
($|\Delta P_x|$ ‹ 75 MeV/c, $|\Delta P_y|$ ‹ 60 MeV/c and ΔP_z ‹ 200 MeV). It is seen
that a clean peak in $\Delta\sqrt{s}$ emerges with practically no non-4C background. From this we estimate that non-4C background in our final
sample is at most a few percent of the events in the sample. A
total of ~7K events survive the 4C-fit for Reaction (1) with
acceptable χ^2. The surviving events are shown as shaded histograms
in Figs. 3a-d for comparison with the OC events.

In Fig. 4 we present $M(p_f\pi^-)$ from our final ~7K sample, where
p_f is the fast-forward triggered proton. Although the background is
substantial, $\Delta^0(1238)$, $N^0(1520)$ are clearly produced in our data.
Fig. 5 shows the effective mass of the recoil system, $M(\bar{p}p_s)$, where
p_s is the slow proton not associated with the triggered particle.
There is no evidence for the production of 2020 and 2200 MeV states
in our data. We have attempted to enhance the baryon-exchanged
N^0 or Δ^0 production by making cuts on $M(p_f\pi^-)$, on the corresponding
t', and on the Jackson angle for the $\bar{p}p_s$ system. None of the cuts
significantly improved the signal of the two claimed $\bar{p}p$ states.

Our resolution for the $\bar{p}p_s$ system has been estimated from
Monte-Carlo (MC) events generated according to the observed resolution and efficiency of the MPS spark chambers, PWC's and hodoscopes.
By examining the spread in mass after the MC events generated at a
given $M(\bar{p}p_s)$ have been processed through our data-reduction programs,
we conclude that the mass resolution is less than that shown in
Fig. 6a. Thus, our resolution at 2020 (2200) MeV is less than 7(11)
MeV, sufficient for us to have seen narrow states, had they been
produced in our data. As a check of our mass resolution calculation
we have calculated the four known masses of the final state $p_f\pi^-\bar{p}p_s$
for each particle from the remaining three plus the beam and the
target for the OC fits from the data and from the MC events, and
found very good agreement.

Our acceptance from finite geometry and program inefficiency
as a function of $M(\bar{p}p_s)$ has been estimated using again the same
MC events. The results are shown in Fig. 6b at two different values
of $M(p_f\pi^-)$ corresponding to $\Delta^0(1238)$ and $N^0(1520)$, respectively.
It is seen that our acceptance for $M(\bar{p}p_s)$ at 2.02 (2.20) GeV is 23%
(16%) with $M(p_f\pi^-)$ at $\Delta^0(1238)$.[5] Our estimate for the additional
loss due to inefficiencies in the trigger components, χ^2 cut, etc.,
is ~44%. Thus, the overall visible sensitivity for our present data
is

$M(\bar{p}p_s)$	$\Delta^0(1238)$	$N^0(1520)$
2.02 GeV	8.0 evts/nb	7.0 evts/nb
2.20 GeV	6.0 evts/nb	5.0 evts/nb

to be compared with the original CERN data with sensitivities in
the 1-2 evts/nb range.

Since the CERN experiment saw their $\bar{p}p$ states most clearly with $M(p_f\pi^-)$ and Jackson angle cuts, we display in Figs. 7a-c the $M(\bar{p}p_s)$ spectra selecting for $\Delta^0(1238)$, $N^0(1520)$ and with $\cos\theta_J < 0$. Again we see no evidence for the 2020 and 2200 MeV states. The dotted histograms show our estimate of the peaks we should have seen, had they been produced with the cross sections quoted in reference (1) but reduced via $\sim P_{lab}^{-2.5}$, a typical behavior of baryon-exchange processes. The absence of the $\bar{p}p$ states in our data corresponds to $\gtrsim 7\sigma$ and $\gtrsim 5\sigma$ discrepancies at 2020 and 2200 MeV, respectively. We show in Fig. 8, 2σ upper limit cross sections of 3.0 nb for 2.02 state (obtained by combining Δ^0 and N^0 events) and 2.0 nb for 2.20 state (for Δ^0 events alone), along with the quoted cross sections of the CERN data.

We must point out, however, that we do see a marginal signal at 2.02 GeV, if our $\sim$7K sample is enlarged by relaxing the χ^2 cut. This effect is demonstrated in Fig. 9a, where the $M(\bar{p}p_s)$ spectrum is shown with $\cos\theta_J < 0$ and with two different χ^2 cuts. This signal is closely associated with ρ^0 events from the reaction

$$\pi^- p \to (p_f\pi^-)\ (\pi^+\pi^-)X^0 \qquad (2)$$

Note that, if X^0 is slow in the laboratory and has mass $\sim$500 MeV, the events resulting from this reaction would be impossible to distinguish from the events of Reaction (1). If we take the $\bar{p}p_s$ system to be a $\pi^+\pi^-$ system and plot the resulting effective mass (see Fig. 9b), we see a broad enhancement in the ρ^0 vicinity. If we take those events within the 2020 region (dotted lines in Fig. 9a), we find that they are indeed associated with the ρ^0 peak (shaded area in Fig. 9b). The effect is even more noticeable if we take events which fail the 4C fit for Reaction (1) and satisfy Reaction (2) with X^0 mass < 800 MeV.

The effect of a ρ^0 contamination can be further illustrated with MC events, shown in Fig. 10. Potential sources for difficulties with the ρ^0 events are twofold. First, the apparent width of the ρ^0 is reduced by a factor of $\sim$2 if the $\pi^+\pi^-$ system is taken to be a $\bar{p}p$ system. Second, if the ρ^0 happens to be on a steeply rising background (dashed curve in Fig. 10a), one is tempted to overestimate the background with the resultant apparent decrease in width of the resonance. Again, the 2020 MeV region is associated with the ρ^0 events: the shaded area in Fig. 10b, peaked at the ρ^0 region, corresponds to the 2020 MeV region (dotted lines in Fig. 10a).

We wish to emphasize, therefore, that it is very important to demonstrate that, if one observes a signal in $\bar{p}p$ at 2020 MeV, it is not due to ρ^0 contamination.

In summary, we do not observe the 2020 and 2200 MeV $\bar{p}p$ states in our π^-p data at 16 GeV/c. We find that the 2σ upper limits for these states are less than 3 nb. From the cross sections of the 12 GeV/c CERN data and the assumption of nucleon exchange, we should have seen $\gtrsim 5\sigma$ signals at 2020 and 2200 MeV in our data. We conclude, therefore, that our experiment contradicts the results of the CERN data. In order for our data and the CERN data to agree, one will have to invent a precipitous energy dependence, unlike that encountered so far in the studies of exclusive exchange processes.

174

This research was supported by the U.S. Department of Energy under contracts DE-ACO2-76CH00016, EY-76-S-02-3230, DE-ACO2-76-ERO-3330 and by the National Science Foundation under contract PHY-7924579. The City College of New York was supported by the National Science Foundation and the City University of New York PSC-BHE Research Award Program.

REFERENCES AND FOOTNOTES

1. P. Benkheiri et al., Phys. Lett. 68B, 483 (1977); Phys. Lett. 81B, 380 (1979). A marginal signal was also seen in e⁻p interactions by Gibbard et al., Phys. Rev. Lett. 42, 1593 (1979).

2. Several p̄N formation experiments do cover the 2020 MeV region, but find little evidence for the peak: W. Bruckner et al., Phys. Lett. 67B, 222 (1977); M. Alston-Garnjost, Phys. Rev. Lett. 43, 1901 (1979); R.P. Hamilton et al., Phys. Rev. Lett. 44, 1179 (1980); Phys. Rev. Lett. 44, 1182 (1980).

3. Several p̄p backward production experiments report no evidence of either 2020 or 2200 MeV states: C. Cline et al., Phys. Rev. Lett. 43, 1771 (1979); R.M. Bionta et al., Phys. Rev. Lett. 44, 909 (1980); A.S. Carroll et al., BNL preprint BNL 27492.

4. Peripherally produced p̄p systems have also been examined, with equally negative results: T.A. Armstrong et al., Phys. Lett. 85B, 304 (1979); W.E. Cleland, Phys. Lett. 86B, 409 (1979): C. Daum et al., Phys. Lett. 90B, 475 (1980).

5. Quoted acceptances include absorption effects in the LH₂ target. All other losses due, for example, to equipment inefficiency, etc., have been lumped together separately and commented on elsewhere in the text.

175

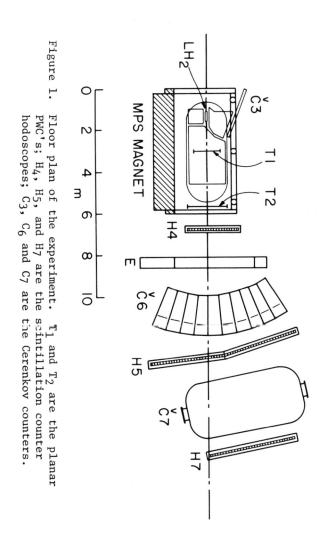

Figure 1. Floor plan of the experiment. T₁ and T₂ are the planar PWC's; H₄, H₅, and H₇ are the scintillation counter hodoscopes; C₃, C₆ and C₇ are the Cerenkov counters.

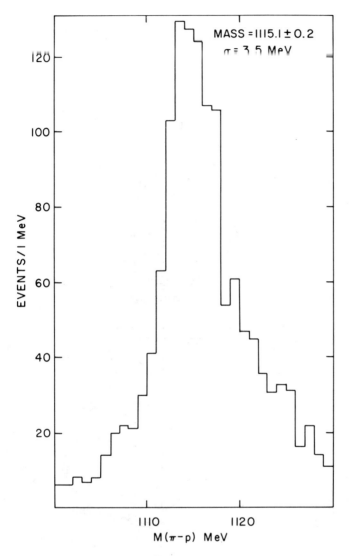

Figure 2. M(π^-p) in the Λ region.

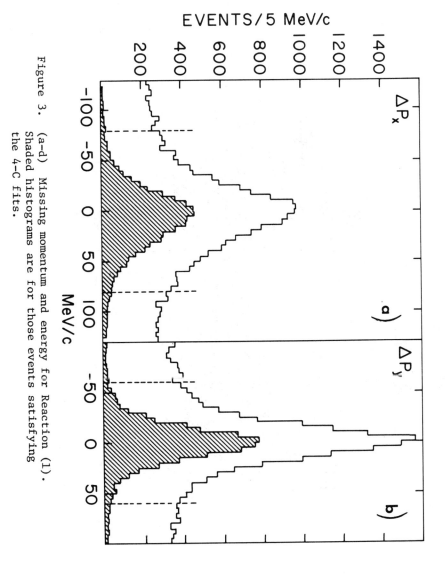

Figure 3.   (a–d) Missing momentum and energy for Reaction (1).
Shaded histograms are for those events satisfying
the 4-C fits.

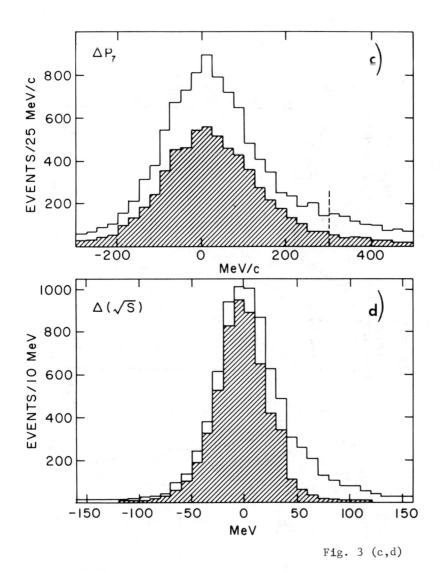

Fig. 3 (c,d)

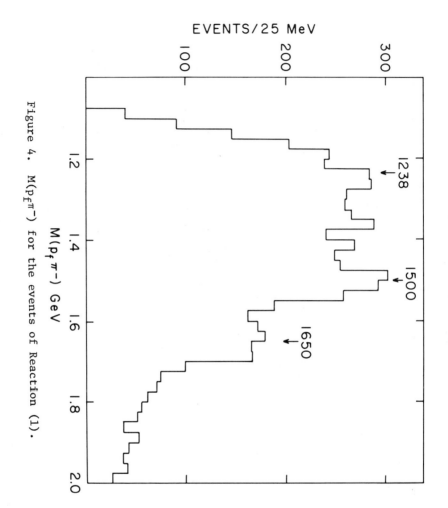

Figure 4. M($p_f\pi^-$) for the events of Reaction (1).

180

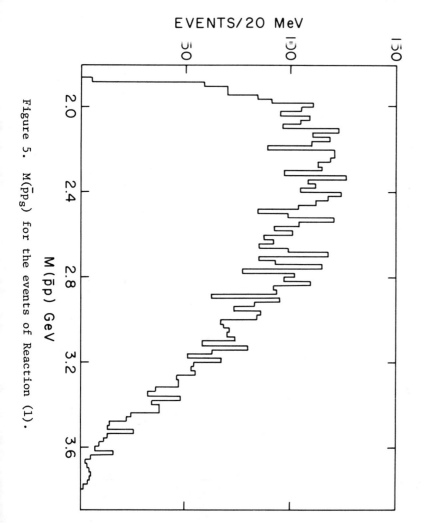

Figure 5. M($\bar{p}p_s$) for the events of Reaction (1).

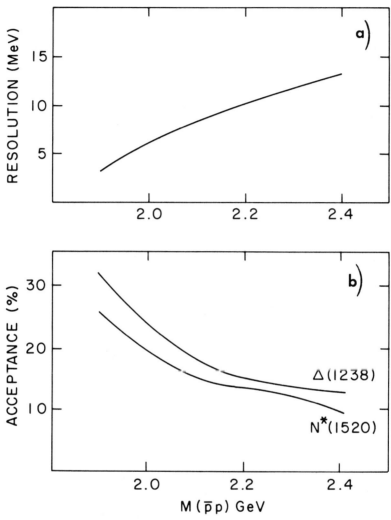

Figure 6. (a) Mass resolution for the $\bar{p}p_s$ system.
(b) Acceptance as a function of $M(\bar{p}p_s)$ with the recoil system in the region of $\Delta^0(1238)$ or $N^0(1520)$.

182

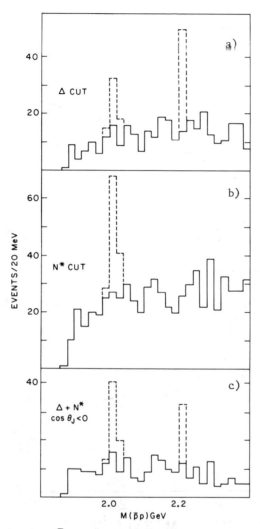

Figure 7. (a–c) M($\bar{p}p_S$) for the events in the region of Δ^0(1238),
N^0(1520), and Δ^0 plus N^0 with $\cos\theta_J < 0$. Dotted bins
delineate the 2020 and 2200 peaks expected from the CERN
data.

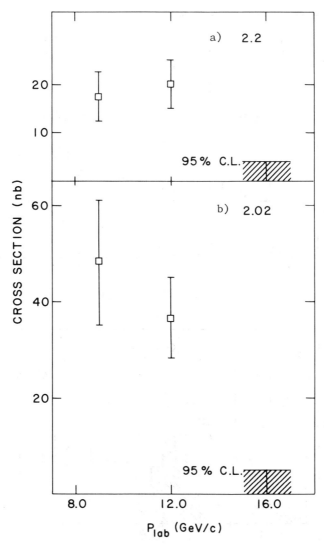

Figure 8. 2 σ upper limits for the 2200 and 2020 states in our data, along with the cross sections quoted in the CERN paper.

184

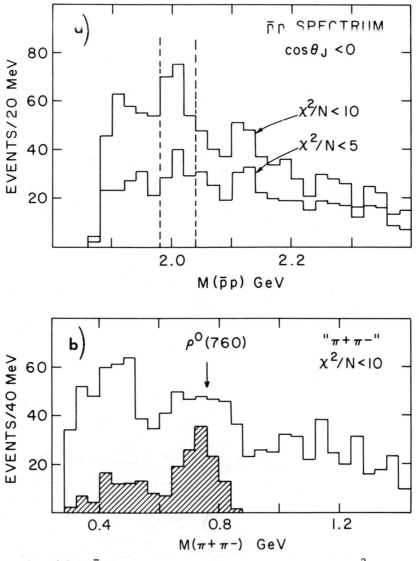

Figure 9. (a) $M(\bar{p}p_s)$ for the events with $\cos\theta_J < 0$ and χ^2 (per degree-of-freedom) < 5 [this is the cut adopted for Reaction (1) throughout this paper], and also for those with $\cos\theta_J < 0$ and χ^2 (per degree-of-freedom) < 10.

(b) Mass spectrum for the events with the larger χ^2 cut in (a), under the hypothesis that the $\bar{p}p$ system is a $\pi^-\pi^+$ system. The shaded histogram corresponds to those events in the 2020-MeV region [dotted lines in a)].

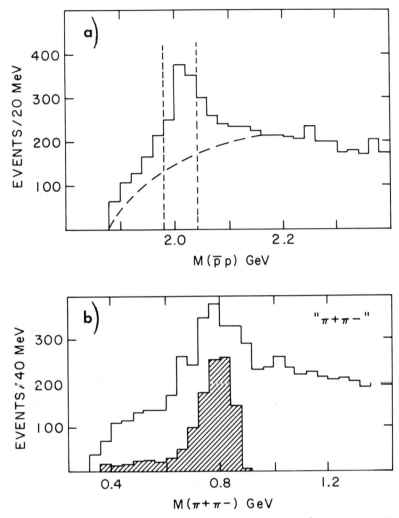

Figure 10. Monte Carlo events generated with a ρ^0 on a smooth, rising background. (a) and (b) are similar to those of Figure 9.

REVIEW OF $\overline{N}N$ FORMATION EXPERIMENTS

Gerald A. Smith*
Department of Physics, Michigan State University,
East Lansing, Michigan   48824

ABSTRACT

Experiments in search of narrow states ($\lesssim 10$ MeV) produced in $\bar{p}p$ formation, sometimes called baryonium, are reviewed. First generation experiments which cite evidence for the S-meson are compared with more recent experiments, which do not generally support the claim for its existence. The situation appears to be one in which either the S-meson does not exist, or perhaps does exist, but in a much less conspicuous form than previously suspected.

INTRODUCTION

The subject of experimental $\overline{N}N$ formation physics may be currently thought of in terms of three general categories: (1) narrow states ($\Gamma \lesssim 10$ MeV), sometimes called baryonium; (2) broad states ($\Gamma \simeq$ several hundred MeV), involving complex phase shift analyses and (3) states (usually narrow and hence candidates for baryonium) produced in the impulse approximation from low energy interactions with deuterium. Because of limits of time in this talk, as well as my own current interests, I will discuss only the first topic, the <u>direct</u> search for <u>narrow</u> states in $\bar{p}p$ (and $\bar{p}d$) formation experiments.

Since the first observation of the S-meson in 1974, there has developed a strong interest in the subject due to the possibility that the S-meson might represent the first of many states made up of four quarks (di-quark, di-antiquark). In such states, the color degree of freedom manifests itself through the separated physical structure of di-quark and di-antiquark clusters, similar to atoms bound together to form a molecule. The narrow width of such states is a distinguishing feature of the model. This analogy sometimes suggests the name "color chemistry" for this subject. Of course, if true, color chemistry would be extremely valuable in shedding light on the true quark dynamics inside hadrons.

For the remainder of this talk, I will first discuss several experiments completed since 1974 and review the evidence (pro and con) for the S-meson. I will then report on results from two recent BNL experiments studying $\bar{p}p$ annihilation. Finally, I will summarize the evidence for the S-meson and risk a conclusion on its existence.

*Research supported in part by the U.S. National Science Foundation

REVIEW OF PREVIOUS EXPERIMENTS

As I stated previously, the S-meson appeared on the scene in 1974, with the work of Carroll et al.[1] In a total cross section measurement, an enhancement of 18^{+3}_{-6} mb was observed in the $\bar{p}p$ (and $\bar{p}d$) total cross section. The mass and width were reported as 1932 ± 2 and 9^{+4}_{-3} MeV respectively. In 1976 Chaloupka et al.[2] "confirmed" the effect by observing a 10.6 ± 2.4 mb enhancement in the $\bar{p}p$ total cross section at a mass of 1936 ± 1 MeV. This result was confined mainly to the elastic channel, with a cross section of 7.0 ± 1.4 mb and a width $8.8^{+4.3}_{-3.2}$ MeV. The annihilation showed a small (∿2 mb) enhancement, but in its own right was not particularly signficant. A third total cross section measurement was reported in 1978 by Sakamato et al.[3], also seeing an enhancement at 1936 ± 1 MeV equal to 16 ± 4 mb with a width of 2.8 ± 1 MeV.

Another important piece of evidence for the S-meson was reported in 1977 by Brückner et al.[4] This experiment, in contrast to the previously discussed experiments, reported evidence for the S-meson in $\bar{p}p \rightarrow$ charged mesons (annihilation) at a mass of 1939 ± 3 MeV and with a width <4 MeV, consistent with their resolution. These four experiments, with independent observations in the elastic, annihilation and total cross section, provided extremely strong evidence for the existence of the S-meson. The lack of consistency among the measured masses, widths, and cross sections is disturbing, however.

At about this time (1978), a number of second generation experiments were conceived to explore the properties of the S-meson and look for additional states of baryonium. Among these were a series of experiments carried out by R. Tripp (Berkeley) and collaborators at BNL. These experiments are completed and just now appearing in the literature. First, as reported by Hamilton et al.[5], they have been unable to reproduce the 18 mb narrow enhancement of Carroll et al. in the total $\bar{p}p$ (and $\bar{p}d$) cross section. Instead, they see a much smaller (3 ± 0.7 mb) and broader (22 ± 6 MeV) effect, which, in fact, is almost entirely associated with annihilation. Second, Alston-Garnjost et al.[6] see no signal at the S-meson in 180° elastic scattering. A model based on a single, non-interfering state would predict an enhancement of ∿0.5 mb/sr, assuming the 7 mb elastic cross section of Chaloupka et al. The upper limit set by this experiment is ∿0.1 mb/sr. Third, Hamilton et al.[7] have searched for an enhancement at the S-meson in $\bar{p}p \rightarrow \bar{n}n$ (charge exchange). They see no effect, although again an effect is predicted based on a model of a single, non-interfering state. Previous results[1,2] suggest the elasticity of the S-meson should be ∿40%. Hamilton et al. measure ≲4%, an order of magnitude smaller.

At this conference we have heard from Dr. Nakamura of the recent results from KEK[8]. In a total cross section measurement, Nakamura and his colleagues have been <u>unable</u> to confirm the existence of the S-meson. Their data are shown in Figure 1. At the 90% confidence level, as a function of an assumed Γ, their upper limit cross sections at the S-meson are 10 mb (3 MeV), 7.5 mb (5 MeV) and 5 mb (10 MeV).

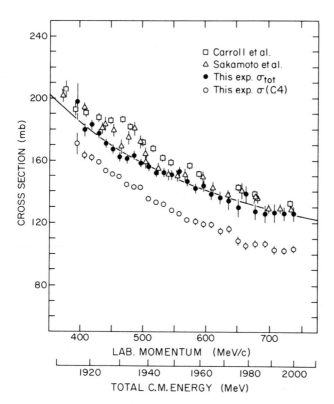

Closing this portion of my talk, we see that the first of the second generation experiments searching for baryonium are essentially unable to confirm the basic results of the first generation experiments, let alone find new states. <u>At this point, it is probably fair to say that baryonium, if it exists, is not manifest with the very large cross sections suggested by the first generation experiments.</u>

Fig. 1.   Total cross section measurements from Kamae <u>et al</u>. (Ref. 8)

RECENT ANNIHILATION RESULTS

In the past two years, the new LESBII beam channel <u>at</u> BNL has been used by two experiments (E-701 and E-708) to study $\bar{p}p$ annihilation in the region of the S-meson. Both experiments have submitted their first results to this conference, and should have an important bearing on the question of the existence of the S-meson, as they constitute a check on the earlier data of Brückner <u>et al</u>.

Jastrzembski <u>et al</u>.[9] (E-701) have utilized the apparatus shown in Figure 2. Annihilation fragments produced in a one meter long hydrogen target are detected by means of detectors PT, WT,

PB and WB placed above and below the target. A large span in C.M. energy (∼40 MeV) is realized by two beam momentum settings (604

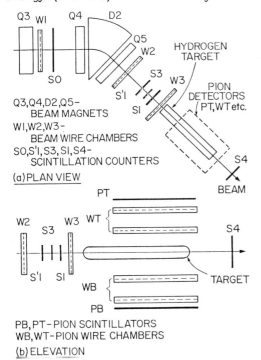

Q3,Q4,D2,Q5-
BEAM MAGNETS
W1,W2,W3-
BEAM WIRE CHAMBERS
S0,S'1,S3,SI,S4-
SCINTILLATION COUNTERS

(a) PLAN VIEW

PB,PT-PION SCINTILLATORS
WB,WT-PION WIRE CHAMBERS

(b) ELEVATION

Fig. 2. Beam and apparatus from the experiment of Jastrzembski et al. (Ref. 9)

and 586 MeV/c) and energy loss across the long target. The event interaction vertex is reconstructed with an accuracy of ±1 cm. The interacting antiproton is tagged in the beam channel with detectors S0-S4 and W1-3, which give an incident beam momentum resolution of ±0.7%. These effects combine to give a mass resolution of ±1.5 MeV at 1935 MeV.

Their results are shown in Figure 3. A suggestive structure is apparent near 1937 MeV. However, using a Breit-Wigner resonance of fixed integrated cross-section 26 (mb-MeV) and fixed width Γ = 4 MeV, as would be suggested by Brückner et al., but allowing the mass to vary, they get a very poor fit (dashed line on Fig. 3), while a simple quadratic in 1/p gives a good fit

(full line on Fig. 3). If all the parameters of the resonance are allowed to vary, the best fit, for masses between 1930 and 1942, always has a very small integrated cross-section ($\Gamma\sigma_R$ < 10 (mb-MeV)), where Γ is the full width and σ_R the cross-section at the peak). If they assume that the non-resonant background is more complex than a quadratic in its dependence on 1/p, they can find a fit to an enhancement with mass 1937 MeV, width ∼3 MeV, and integrated cross-section 12 ± 6 (mb-MeV). However, they do not feel that the required momentum dependence of the cross-section is a reasonable one, especially in view of the fact that a good fit is found with a simpler dependence and no resonance. In all their attempts at fitting they have assumed narrow resonances, consistent with the values reported for the S-meson; values greater than 8 MeV cannot be excluded by their data. They conclude that they do not confirm the structure reported by Brückner et al.

The second annihilation experiment I want to discuss is that of BNL-DOE-Michigan State-Syracuse Collaboration (BDMS), or

E-708. The BDMS apparatus is shown in Figure 4. The device, as a whole, serves a combined function of a gamma-ray and annihilation spectrometer. However, as applied to the current problem, the apparatus is similar to that of E-701. A combination of planar and cylindrical drift chambers measure annihilation vertices to an accuracy of ±2cm. in the target. The beam is momentum tagged to a precision of ±1.5% at 500 MeV/c with drift chambers, resulting in an overall C.M. energy resolution of ±2 MeV at 1939 MeV. The liquid hydrogen target is 50 cm. long, necessitating the use of seven carbon degrader

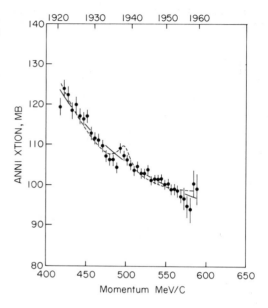

Fig. 3. Annihilation cross section measurement of Jastrzembski et al. (Ref. 9)

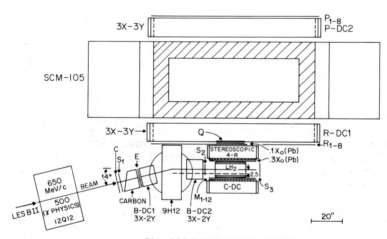

BNL AGS EXPERIMENT #708
ANNIHILATION and GAMMA RAY SPECTROMETER
BNL–DOE–MICHIGAN STATE – SYRACUSE

Fig. 4. Apparatus of BNL-DOE-Michigan State-Syracuse (BDMS) Collaboration

settings in the beam to give good overlap of data in the desired mass region (∿1920-1960 MeV).

Figure 5 shows the distribution of annihilation events along the axis of the target for the seven settings, as well as three lower momentum settings where the beam stops in the target.

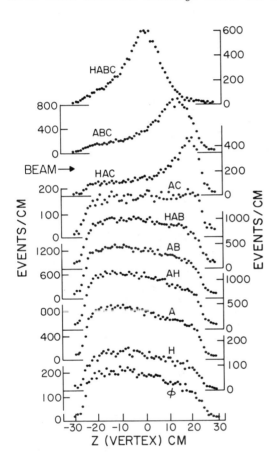

The data for each setting are compared with a detailed Monte Carlo calculation as a function of C.M. energy. The ratio of data to Monte Carlo events plotted in one MeV bins for the separate settings are then merged into one set of ratios covering the entire mass range. A cross section is calculated, assuming an overall smooth representation of the annihilation cross section based on a composite of other published data, and normalized to the composite value at 500 MeV/c. The results are shown in Figure 6. The data show no apparent enhancement in the 1939 MeV region, and are fitted overall quite satisfactorily with an a+b/p dependence. The difference between the cross sections and the fit is plotted as $\Delta\sigma_A$. The result of Brückner et al., namely an enhancement of integrated area 26 ± 6 (mb-MeV) and width Γ = 4 MeV centered at 1939 MeV, is not in agreement with these data. Using the same width (Γ = 4 MeV), the

Fig. 5. Axial profile of annihilation events from the BDMS experiment for seven degrader settings, plus three additional settings which stop the beam in the target

BDMS group has fit their data to a background of the form a+b/p plus a Breit-Wigner resonance. For resonance masses between 1934

192

and 1944 MeV, they find a 95% confidence level upper limit of 10 (mb-MeV) for the integrated area of the resonance. In summary, the BDMS collaboration finds no evidence for the S-meson as reported by Brückner et al.

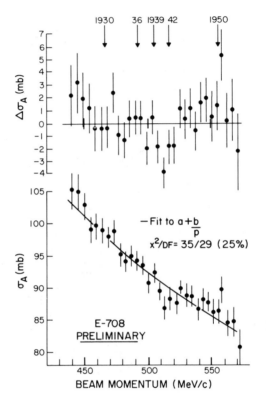

Fig. 6. Annihilation cross section from the BDMS experiment. The quantity $\Delta\sigma_A$ refers to the difference of the data and the fit as shown in the figure.

CONCLUSIONS

The second generation of experiments described herein seriously challenge the very existence of the S-meson. Certainly the large effects seen earlier in the total cross section by Carroll et al., Chaloupka et al., and Sakamoto et al. are unsupportable. Furthermore, the results of Brückner et al. in the annihilation reaction seem to be on very unfirm ground, based on the results of the two recent BNL experiments. No supporting evidence for the S-meson is provided in backward elastic or charge exchange scattering experiments, although these results are model dependent. It would be quite useful at this time to check the results of Chaloupka et al. with yet another elastic scattering experiment. In conclusion, the S-meson (and probably baryonium) is in deep trouble!

REFERENCES

1. A. Carroll et al., Phys. Rev. Lett. 32, 247 (1974).
2. V. Chaloupka et al., Phys. Lett. 61B, 487 (1976).
3. S. Sakamoto et al., see review talk of S. Ozaki, Proceedings of the XIX International Conference on High Energy Physics, Tokyo, 1978, p. 101; Nucl. Phys. B158, 410 (1979).
4. W. Brückner et al., Phys. Lett. 67B, 222 (1977).
5. R. P. Hamilton et al., Phys. Rev. Lett. 44, 1182 (1980).
6. M. Alston-Garnjost et al., Phys. Rev. Lett. 43, 1901 (1979).
7. R. P. Hamilton et al., Phys. Rev. Lett. 44, 1179 (1980).
8. "Evidence Against the Narrow S(1936) in a Measurement of $\bar{p}p$ Total Cross-Section", T. Kamae et al., preprint from the Toyko-Hiroshima collaboration, January 1980.
9. "Search for Structure in the Low-Energy $\bar{p}p$ Annihilation Cross-Section", E. Jastrzembski et al., preprint from the Temple-UC Irvine-New Mexico collaboration, April, 1980.

THEORIES OF BARYONIUM, EXOTICS AND MULTIQUARK SYSTEMS?

Anthony J. G. Hey
Physics Department, Southampton University, SO9 5NH, U.K.

EPITAPH

The great tragedy of science - the slaying of a beautiful hypothesis by an ugly fact.

T.H.Huxley

0. INTRODUCTION AND OUTLINE

It is my brief at this conference to review theories of multiquark states. In the past few years, theorists have been much excited by the appearance of narrow states in baryon-antibaryon (BB) channels and many theoretical papers have appeared on the subject of "Baryonium".1) I will therefore begin by reviewing the early theoretical ideas underlying the description of both high and low mass multiquark systems. These 'first thoughts' undoubtedly had the merits both of simplicity and also intuitive appeal. Alas, I must then summarize the current experimental situation on both broad and narrow BB states. In essence, no claims for theoretically-beloved narrow states have been confirmed. The situation for broad states is also more complex than first thought. What should we now think of all those brave theoretical words? After a short pause to take stock, I therefore sketch some 'second thoughts' about multiquark states, before closing with some remarks which can hardly be dignified by the title of conclusions.

I. THEORY: SOME FIRST THOUGHTS

The candidate theory of strong interactions is Quantum Chromo-Dynamics - an SU(3) gauge theory of colored quarks interacting via colored massless vector gluons. The quark and antiquark fields transform as a triplet and anti-triplet representation, respectively, under the color SU(3) group, and the gluon fields as an octet:

$$q \sim 3_c \; ; \; \bar{q} \sim \bar{3}_c \; ; \; g \sim 8_c$$

The color confinement hypothesis corresponds to the statement that only color singlet hadrons are observable.

To understand the implications of color for multiquark states we begin by recalling some group theory results for products of SU(3) representations. We shall need the following results

$$3 \times 3 = 6 + \bar{3}$$

$$3 \times 6 = 10 + 8$$

$$3 \times \bar{3} = 8 + 1$$

$$6 \times \bar{6} = 27 + 8 + 1$$

Notice that color singlets only appear in the products $3 \times \bar{3}$ and $6 \times \bar{6}$ listed above. Assigning quarks one third baryon number, we immediately arrive at the simplest quark content for color singlet mesons and baryons:

Mesons : $B = 0$; $\underline{1}_c$; $(q\bar{q})$ states

Baryons : $B = 1$; $\underline{1}_c$; (qqq) states

In a baryon, any two quarks can be coupled to either a 6 or a $\bar{3}$ representation of color: Only the $\bar{3}$ possibility can combine with the third quark to produce a color singlet (Fig. 1)

Fig. 1 Color couplings for quarks in baryons

This observation will be needed in our discussion of M- and T-Baryonium.

All the impressive spectroscopic success of the quark model is based on these two configurations - qq and qqq. In the days before color and QCD, the quark model could be regarded as merely a mnemonic or recipe and the absence of other quark configurations did not particularly alarm us. Nowadays, we are more ambitious and believe in an underlying field theory of quarks and gluons: In such a theory there is no reason, a priori, why other, more complex color singlet configurations of quarks, such as $q^2\bar{q}^2$, $q^4\bar{q}$ and so on, might not be observable as resonant states. Moreover, it is widely speculated that quarkless glue states will exist - the glueballs considered by Donoghue[2] at this conference; gg, ggg and so on. Once we allow the concept of constituent gluons, or alternatively admit the possibility of more than just non-relativistic degrees of freedom in baryons and mesons, there are yet more 'extra' states - which may be characterized as $q\bar{q}g$, qqqg,etc. It is the outstanding problem of hadron spectroscopy to try to identify any of these extra, QCD-inspired states. But, on the other side of the coin, it must be remembered that the

196

mere fact that we can generate such color singlet combinations is
no guarantee that all, or even any, of these configurations will
be observable as resonant states. Ultimately, there is no escape:
theorists will eventually have to calculate the actual spectrum
of mesons in QCD and estimate widths. Only then will we know
what configurations are stable enough to be seen. Until that day,
theorists have to resort to more tractable models and to intuition.
With all this as background, I will now describe two such attempts
to describe "Baryonium" - or more precisely "Diquonium" - $q^2\bar{q}^2$
configurations.

LOW MASS $Q^2\bar{Q}^2$ STATES: ROUND HADRONS

It was in 1977 that Jaffe[3] first opened Pandora's box and
examined $q^2\bar{q}^2$ states in the MIT Bag. For the low mass states, a
spherical bag is assumed and this is roughly equivalent to confining quarks and antiquarks by an infinite square-well
"potential".[4] By applying the appropriate boundary conditions
a series of allowed energy levels emerges, just as in the non-
relativistic problem,even though, in this case, the MIT Bag con-
fines light (or even massless) relativistic Dirac particles.[5]
The familiar ground state pseudoscalar and vector meson multiplets are then obtained from the configuration of one quark and
one antiquark in the lowest eigenstate: The lowest lying multiquark mesons are obtained from configurations with two quarks and
two antiquarks in the lowest level. In this Bag picture, the
mass of a state is characterized by three types of contribution

$$M = \sum_{q,\bar{q}} E_q + U + H_{gluon}$$

The first term corresponds to the quark and antiquark kinetic
energies, U represents the 'confining energy' - a Bag pressure term
and so on - and H_{gluon} is a perturbative correction due to one
gluon exchange. (Fig. 2)

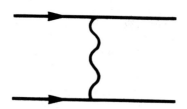

Fig. 2 One-gluon exchange graph between quarks.

Detailed calculations[6] show that the dominant contribution from one-gluon exchange between quarks in their ground state is a short-range 'magnetic' color-spin interaction of the form

$$H_{gluon} \sim \alpha_s \, \underline{\lambda}_1 \cdot \underline{\lambda}_2 \, \underline{S}_1 \cdot \underline{S}_2$$

where α_s is the QCD fine structure constant, and the factors $\underline{\lambda}_1 \cdot \underline{\lambda}_2$ and $\underline{S}_1 \cdot \underline{S}_2$ describe the color and spin couplings of the quarks. The masses of the $q^2 \bar{q}^2$ states may now be calculated: This involves a detailed enumeration of the possible flavor-color-spin wavefunctions and then diagonalization of the gluon exchange perturbation to find the physical states. The results may be summarized as follows:[3]

(a) Masses

(i) The predicted masses span a rather broad range

650 MeV - 2350 MeV

(ii) The lowest mass $q^2 \bar{q}^2$ states have non-exotic SU(3) flavor quantum numbers - so-called "cryptoexotic" configurations.

Both these features stem from the form of the gluon exchange color-spin interaction. For ground state hadrons, this interaction is believed to be responsible for the Δ-N and ρ-π mass differences: In multiquark states, the same color-spin factor can have much larger matrix elements and therefore produce much larger splittings.

(b) Widths

In general, $q^2 \bar{q}^2$ states can decay by a "Zweig-super-allowed" quark line diagram - the so-called "celery-stick" diagram (Fig. 3)

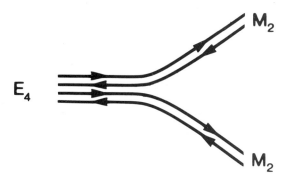

Fig. 3 Celery-stick diagram.

198

Given the possibility of a large amplitude for such "fall-apart" decays into ordinary mesons, one expects $q^2\bar{q}^2$ states to be very broad. There is an important exception: Multiquark mesons which contain an $s\bar{s}$ pair, and which are light, may have very little phase-space available for their allowed fall-apart decay to $K\bar{K}$. Such states could be narrow, and it was for this reason that Jaffe boldly suggested that the rather narrow 0^{++} δ and S^* mesons at around 1 GeV should be identified with $q^2\bar{q}^2$ configurations instead of their usual assignment as $(q\bar{q})$ L = 1 mesons. We defer further examination of this suggestion till later.

HIGH MASS $Q^2\bar{Q}^2$ STATES: THE "MIT BONE"

Johnson and Thorn[7] have considered the question of Regge trajectories in the context of the MIT Bag. For high angular momentum ℓ, a round hadron approximation is inappropriate and a long thin, almost string-like configuration is preferred. For mesons, the quark and antiquark are localized at opposite ends: For Baryons, one expects a quark-diquark configuration. This suggests a simple picture for high ℓ $q^2\bar{q}^2$ states (Fig. 4)

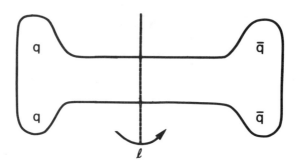

Fig. 4   MIT Bone.

For such "bone-like" bags, the fall-apart decays into ordinary $q\bar{q}$ mesons are suppressed by the angular momentum barrier separating quarks from antiquarks.

Consider the color properties of the diquark and antidiquark systems:

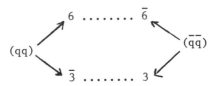

To produce a color singlet, the 6 must be matched with the $\bar{6}$, and the $\bar{3}$ with the 3. In low mass, spherical hadrons the 6-$\bar{6}$ and $\bar{3}$-3 configurations are mixed by the short-range magnetic color-spin interaction. However, for high ℓ states the color-spin mixing between the ends of the 'bone' is expected to be negligible and pure 6-$\bar{6}$ and $\bar{3}$-3 combinations will separately correspond to physical states. Now, the Regge slope in this model depends on the color flux connecting the two ends[7]: Since this is different for 6-$\bar{6}$ and $\bar{3}$-3, two types of Baryonium are expected. What are the distinguishing characteristics of these two families of mesons? Since the fall-apart decays are suppressed at high ℓ, these $q^2\bar{q}^2$ states presumably decay via some quark-pair creation mechanism — in the same way as an excited baryon resonance decays to a ground state baryon plus a meson — and one would naively expect typical hadronic widths. It is at this point that a simple color argument suggests a striking difference between the two varieties of Baryonium.[8,9]

Consider $\bar{3}$-3 Baryonium first: The decay mechanism is indicated schematically in Fig. 5(a)

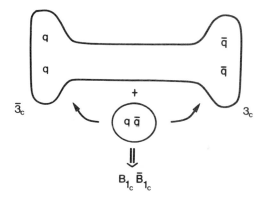

Fig. 5(a)   Decays of $q^2\bar{q}^2$ states: T-Baryonium.

A quark from the extra $q\bar{q}$ pair can combine with the $\bar{3}$ diquark to form a color singlet baryon: Similarly, the antiquark can form an antibaryon with the antidiquark. Hence one expects these states to couple preferentially to B$\bar{\text{B}}$ and Chan and Hogaasen[8] denoted the $\bar{3}$-3 species as T-Baryonium — T for true. This is in contrast with the 6-$\bar{6}$ variety: Fig. 5(b) shows the same mechanism for these states. The simple color argument suggests that they do not couple strongly to B$\bar{\text{B}}$. For this reason the 6-$\bar{6}$ $q^2\bar{q}^2$ mesons are called[8] M-Baryonium — M for mock.

200

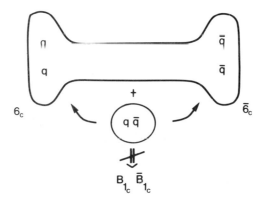

Fig. 5(b)   Decays of $q^2\bar{q}^2$ states: M-Baryonium

Our expectations may be summarized as follows:

T-Baryonium: $(q^2)_{\bar{3}} - (\bar{q}^2)_3$

- Unusually large couplings to $B\bar{B}$ channels

- Typical hadronic widths

  $\Gamma \sim 100$ MeV "Broad"

M-Baryonium: $(q^2)_6 - (\bar{q}^2)_{\bar{6}}$

- Decays to mesons suppressed by high angular momentum barrier

- Couplings to $B\bar{B}$ channels suppressed by color

- Small widths

  $\Gamma \sim 10$ MeV "Narrow"

The broad T-Baryonium states are expected to be seen in direct, $B\bar{B}$ formation experiments: The narrow M-Baryonium states are more likely to be visible in production experiments. For the narrow M-Baryonium family, M_4 ,one expects a whole sequence of such states with 'cascade' decays to lower M_4 states a preferred decay mode

$$M_4^* \to M_4 + \pi$$

There was consequently great excitement generated by reports of just such narrow states in $B\bar{B}$ production experiments, with widths compatible with the experimental resolution of 10 to 20 MeV. In particular, one such state, a 2.95 GeV (ppπ) state,[10] appeared to decay preferentially to (independently observed[11]) 2.20 and 2.02 (pp) states. Many theoretical papers[12] subsequently appeared normalizing the whole spectrum of M-Baryonium to one or other of these states - making some arbitrary spin assignment - and predicting the whole decay spectrum. The T-Baryonium states were usually normalized to the T(2.15) meson, assuming a spin 3 assignment as suggested by the amplitude analysis of Carter et al.[13]

II.A LOOK AT THE DATA

Narrow States?

We have heard the sorry tale of the search for narrow $B\bar{B}$ states from the preceding speakers[14] in this session. Let me briefly summarize their conclusions:

  (1) The narrow 2.95 (ppπ) state has not been confirmed.
  (2) Production experiments also fail to confirm the narrow (pp) states at 2.02 and 2.20.
  (3) The narrow exotic state at 2.46 GeV in $\bar{\Lambda}\Delta^{++}$ and $\bar{\Sigma}^{*+}p$ has not been confirmed.
  (4) Other experiments also report negative results for narrow resonance searches: for details see the preceding talks.

This leaves the well-established, "PDG-approved", S(1936) meson as the only reliable narrow state. Before this conference, despite some doubts about the precise resonance parameters, I was reassured to see that two new experiments claim to see the S-meson in production.[15,16] However, the new data[14] on the pp formation channel seem to case new doubts on even the existence of a narrow S(1936) state. You must draw your own conclusions from the conflicting experimental reports.

Broad States: The T, U and V regions

The T and U are seen as bumps in the $\bar{p}p$ total cross section and in pp → ππ. Let me first recall the results of the first serious attempt at an amplitude analysis of pp → ππ. Carter et al.[13] used a Barrelet-zero type technique to analyse $\bar{p}p \to \pi^+\pi^-$ differential cross section and polarization data in the mass range 2.08 to 2.50 GeV. With some (in retrospect rather too drastic) simplifying assumptions, they obtained the following pattern for the dominant waves:

$$T : J^P = 3^- \quad I = 1 : M = 2.15 \text{ GeV} ; \Gamma = 200 \text{ MeV}$$

$$U : J^P = 4^+ \quad I = 0 : M = 2.31 \text{ GeV} ; \Gamma = 210 \text{ MeV}$$

$$V : J^P = 5^- \quad I = 1 : M = 2.48 \text{ GeV} ; \Gamma = 280 \text{ MeV}$$

However, with the advent of good $p\bar{p} \to \pi^0\pi^0$ differential cross section data,[17] this analysis was seen to be inadequate and the simple, single resonance dominance assumption of Carter et al. is no longer tenable.

There have recently been two new attempts[18,19] at model-independent Barrelet-zero analyses of $p\bar{p} \to \pi\pi$: Both use both the $\pi^+\pi^-$ and the $\pi^0\pi^0$ data. Nevertheless, despite the obvious similarities in technique between the two analyses, the analyses of A.D.Martin and M.Pennington[18] (MP) and of B.R.Martin and D.Morgan[19] (MM) differ quite significantly in details, notably in the choice of partial-wave cut-off – L_{max} versus J_{max} – and in their implementation of threshold constraints. The MP analysis finds only a two-fold ambiguity while MM find more solutions and moreover, do not agree in detail with the solutions of MP. We draw the following tentative conclusions

(1) A simple single resonance picture for the T, U and V regions is invalid

(2) There are probably resonances in all waves from $J = 1$ to $J = 5$ and effects tend to enter with ascending spin

(3) No firm conclusions concerning masses, widths, spin couplings and relative production cross sections can yet be drawn

(4) The combination of the $\pi^0\pi^0$ and $\pi^+\pi^-$ data have reduced the ambiguities to a relatively small number compared with an analysis of the $\pi^+\pi^-$ data alone. To improve on this situation, more data, particularly in the region from threshold to 2 GeV is needed in both the $\pi^0\pi^0$ and $\pi^+\pi^-$ channels. The planned LEAR facility would seem ideally suited to do this.

Other $B\bar{B}$ States

(1) A new experiment reported at this conference[20] claims evidence for three broad ($\bar{\Lambda}p$) states seen in production. In contrast to the natural parity T, U, V sequence, a moments analysis identifies unnatural parity states with $J^P = 2^-$, 3^+ and 4^- in the region 2.3 to 2.5 GeV, all with widths of 200 MeV or more.

(2) There have been several claims for narrow $N\bar{N}$ states near threshold. An analysis[21] of radiative transitions from $p\bar{p}$ atoms suggested the existence of states at 1684, 1646 and 1395 MeV. A new experiment is currently in progress to confirm these states but no results have yet been reported.

Another way of looking for narrow, near threshold states is via the reaction

$$\bar{p} + d \to a + X$$

where a can be π, p or n. Several states have been claimed but all are in need of confirmation. One recent experiment[22] finds no evidence for narrow states.

INTERMISSION

In his review at the Geneva conference,[1] Chan concluded with the remark:

"If the existence of such narrow states is confirmed by future experiment, it may be considered a triumph for the current theory of color confinement."

What then about the converse of this statement? Given the disappearance of essentially all candidates for narrow $B\bar{B}$ states, is this a disaster for QCD? The answer is obviously no. The 'first thoughts' of theory were just that: They were a simple, appealing framework but one not in any sense derived from QCD. Even accepting the framework, the comparisons of the theory with the data, as it then appeared, seemed to lack some credibility. For example, the lowest M-Baryonium candidates were assigned to $\ell = 1$ $q^2\bar{q}^2$ configurations: $\ell = 1$ is certainly bigger than zero, but hardly high enough to give the required, huge suppression of fall apart decays. Moreover, there was confusion about the production mechanism of such states - some were produced in the backward direction, others forward. It is time, therefore, to turn to some second thoughts on the theory of $q^2\bar{q}^2$ states.

III. THEORY : SOME SECOND THOUGHTS

1. Pure theory: The 1/N color expansion

It was pointed out by 't Hooft[23] that the generalization of SU(3) QCD to an SU(N) theory has some attractive features. In the large N limit, the structure of the theory simplifies dramatically, and it is possible to obtain results on the spectrum and decays of states - which are normally intractable, non-perturbative problems. There is no time to discuss the theory in detail, nor indeed is it appropriate. I would, however, like to give some inkling of why the theory simplifies in this limit, and how one can obtain results for strong coupling problems.

In an SU(N) theory, the quark and antiquark color labels run over N indices: quarks and antiquarks are denoted by directed lines

$$q^i \equiv \longrightarrow \qquad \bar{q}_j \equiv \longleftarrow$$

where i, j run from 1 to N. The gluon field, on the other hand, has the color indices of a quark-antiquark system $q_i \bar{q}^j$, apart from subtraction of the trace (singlet). Thus, for large N, the gluon field $A^i_{\ j}$ has $(N^2 - 1) \sim N^2$ components and the color indices can be denoted by the equivalent $q\bar{q}$ diagram

$$A^i_{\ j} = \quad \sim\!\!\sim\!\!\sim \quad \equiv \quad \xrightarrow{\qquad} \atop \xleftarrow{\qquad}$$

The simplification at large N occurs because many graphs of the perturbation series give negligibly small contributions in this limit. An example will make this clear. Consider one loop contributions to the gluon propagator: a quark loop amplitude, A_q, as in Fig. 6(a) and a gluon loop, A_g, as in Fig. 7(a).

Fig. 6a Quark loop correction to gluon propagator: Feynman diagram.

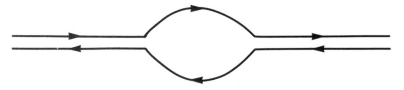

Fig. 6b Quark loop correction to gluon propagator: Color coupling.

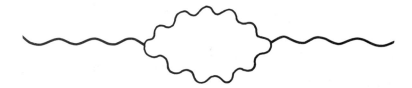

Fig. 7a Gluon loop correction to gluon propagator: Feynman diagram.

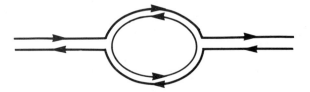

Fig. 7b Gluon loop correction to gluon propagator: Color coupling.

Figures 6(b) and 7(b) show the corresponding color couplings. In the case of the gluon loop, there is a free color loop and consequently N such contributions. Thus, relative to the gluon loop, the quark loop contribution is down by a factor of 1/N

$$\frac{A_q}{A_g} \sim \frac{1}{N}$$

This is why the theory simplifies: By an extension of these arguments one arrives at a set of rules for the dominant contributions in the large N limit. How does one obtain results on the meson spectrum? The trick is to consider Green's functions of currents of quark bilinears, $J \sim \bar{q}q$, and examine the intermediate states that contribute in the large N limit

$$\sum_M < 0 \mid J \mid M >< M \mid J \mid 0 >$$

In this way, one can derive many interesting results concerning the meson spectrum, decays and scattering processes. For my purposes, I wish only to note the following result:[24]
    At large N, the coupling of $\bar{q}q$ mesons to $q^2\bar{q}^2$, $q\bar{q}g$, gg exotic states is suppressed by factors of 1/N.
    In other words, if you start with $\bar{q}q$ mesons it will be hard to produce these other types of mesons. Some comments are in order:
    (1) Is N = 3 large N? Witten[24] makes the (not-too-serious) observation that

$$\frac{1}{4\pi} \frac{1}{N^2} \sim \frac{1}{113} \qquad \text{for N = 3}$$

(It is not clear, however, that this is the relevant expansion parameter.)
    (2) Experimentally, exotics are suppressed – in the sense that no-one has found any unambiguous candidates. Witten[24] turns this defect into a virtue and takes this as evidence that the 1/N expansion is relevant.
    (3) A simple graphical discussion of baryons in the 1/N limit is not possible. The problem is that a color singlet baryon involves the generalized N-index, antisymmetric tensor

$$B_{1_c} \sim \varepsilon_{ij\ell..} q^i q^j q^\ell \cdots$$

so baryons contain N quarks! A discussion of baryons, and of $B\bar{B}$ states is possible for heavy quarks, but is problematic for light quarks.[24]

2. Production Mechanisms : Flavor or Color?

    Old-fashioned duality diagrams based on the flavor properties of hadrons suggest that $q^2\bar{q}^2$ exotics may be best produced in the backward direction via non-exotic baryon exchange[25] (Fig. 8)

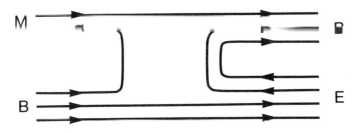

Fig. 8 Duality diagram for backward production of $q^2\bar{q}^2$ states

Unfortunately, a simple color argument would suggest that it is impossible to produce M-Baryonium states via this mechanism.[26] This is because the two 'spectator' quarks in the initial baryon must be coupled to a $\bar{3}$ of color, and so also, are the exchanged pair of quarks. Hence, one can only produce the T-Baryonium, 3-3, series in the backward direction. From a color point of view, the most plausible production mechanism for M-Baryonium would seem to be forward production via exotic exchange.[26] (Fig. 9)

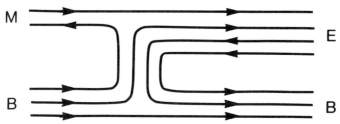

Fig. 9 Color diagram for forward production of $q^2\bar{q}^2$ states

The exchanged exotic object starts life as $3-\bar{3}$ but, for t near zero, one expects mixing with $6-\bar{6}$ configurations as with the low mass states and both types of q^2q^2 state can therefore be produced.

To make these considerations at all quantitative clearly requires a theory of quark line diagrams. Approaches via topological expansions, duel unitarization and so on can in principle provide such a theory,[27] but, in my mind at least, the status of these schemes is unclear.

Another issue concerns the formation of T-Baryonium states and whether or not they are produced peripherally. Peripherality is an idea[28] deriving from two-component duality, in which s-channel resonances are associated with the 'non-Pomeron' part of the amplitude, and are dominated by peripheral partial waves satisfying

$$L \sim kR$$

Here, L is the resonant partial wave, k the CM momentum and R a fixed interaction radius, of the order of $1/M_\pi$. The original analysis of Carter et al.,[13] for the T, U, V regions, was in agreement with this idea applied to $p\bar{p} \to \pi^+\pi^-$. Thus theorists[8,9] have liked to use this as a principle to predict which of their welter of $q^2\bar{q}^2$ states will be dominant. Some words of warning:

   (i) Duality arguments are always somewhat imprecise: They are clearly more so in their extension to $B\bar{B}$ channels.

   (ii) The simple peripheral picture of Carter et al. for $p\bar{p} \to \pi\pi$ is inadequate, and while it is true that the dominant waves do seem to enter with increasing spin, the final words on the T, U and V regions have yet to be written.

3. Quark Potential Models

   The models for Baryonium have invoked a simple intuitive picture of diquark 'molecules' as their starting point. The aim of calculations in quark potential models is to establish whether or not such a picture is at all plausible, using 'realistic' interquark confining forces. 'Realistic' is defined in terms of non-relativistic potential models developed for the baryon spectrum. In such models[29] the Hamiltonian takes the form

$$H = \sum_i \frac{P_i^2}{2M_i} + V_{conf} + H_{gluon}$$

where various forms of confining potential have been used, for for example, $V_{conf}^{ij} \sim (\underline{r}_i - \underline{r}_j)^2$ in the harmonic oscillator model, or some generalization of a linear potential $V_{conf}^{ij} \sim |(\underline{r}_i - \underline{r}_j)|$. The form for H_{gluon} is usually adapted from the Fermi-Breit, non-relativistic reduction of one gluon exchange between quarks.[30]

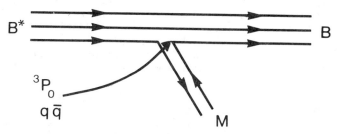

Fig. 10 ${}^3P_0$ model for $B^* \to BM$.

Good agreement with the strange and non-strange baryon spectrum is obtained up to about 2.1 GeV in mass.[29,31] Both models and data show no evidence for diquark clustering within baryons up to $\ell = 2$ internal angular momentum.

For decay widths, various models have been used, among them the so-called $^3P_0$ model of quark pair creation (Fig. 10) initiated by Micu,[32] and by Colglazier and Rosner,[33] and developed in recent years by a group at Orsay.[34] Such models obtain reasonable agreement for decay widths.

I wish to discuss two applications of such models to Baryonium $q^2\bar{q}^2$ systems. One is by the Orsay group[35] who use a harmonic oscillator confining potential: Barbour and Ponting[36] use instead a generalized linear potential model that they have developed for the baryon spectrum.[37] Both groups agree that M-baryonium is un-stable to meson decay up to at least $\ell = 4$! The detailed calculations show that there is no marked clustering for 6-6 states and hence considerably overlap of the quarks and antiquarks allowing fall-apart decays to mesons.

For T-Baryonium, there is some disagreement - even though both groups use the same $^3P_0$ model for decays. Ader et al.[38] claim that T-Baryonium have surprisingly narrow widths, $\Gamma \sim 10$ MeV, in contrast to the naive expectations.

On the other hand, a recent calculation by Barbour and Gilchrist[39] obtains broad widths, $\Gamma \sim 100$ MeV, as expected. They also roughly reproduce the pp cross-sections in the T, U and V regions, although in this model, the 'bumps' contain many resonances.

It is difficult to know quite what to conclude from these calculations, but they do cast doubt, albeit in the context of non-relativistic potential models, on the heuristic picture of M-Baryonium.

4. NN̄ Potential Models

Various groups of theorists[40] have explored the idea that there may be "deuteron-like" states in the NN̄ system, close to threshold. A model for the N-N̄ potential is constructed by analogy with models for the N-N system, usually consisting of one boson exchange graphs, and perhaps two pion exchange contributions. In the N-N̄ system, however, there is clearly an additional, short-range annihilation potential which has no analogue in the N-N case. The hope is that the main features of the NN̄ spectrum will be insensitive to details of this unknown contribution. The results[40] are disappointing in that the absolute binding energies are very model dependent and are sensitive to both the choice of intermediate range potential and to details of the short-range cut-off. Moreover, it is not clear that any of these states would be narrow. A naive estimate of the annihilation potential from the NN̄ cross-section in the form of an imaginary contribution to the potential

$$V + iW$$

leads to huge widths for the $\bar{N}N$ bound states.[41] Shapiro[42] has claimed that this is not so, but there is some debate about the validity of his conclusions.[43]

It seems fair to say that $\bar{N}N$ potential models have little or no genuine predictive power.

5. The P-Matrix: Low mass $q^2\bar{q}^2$ states revisited

We now return to the 'round' $q^2\bar{q}^2$ states discussed first by Jaffe in his Bag calculations.[3] He identified the following states

$$\left. \begin{array}{c} \dfrac{1}{\sqrt{2}}\,(u\bar{u} + d\bar{d})s\bar{s} \sim S^*(993) \\[2ex] u\bar{d}s\bar{s}, \text{ etc. } \sim \delta(976) \end{array} \right\} \quad \text{narrow,} \quad K\bar{K} \text{ decay}$$

$$u\bar{u}d\bar{d} \quad \sim \varepsilon(700) \qquad \text{Broad, } \pi\pi \text{ decay}$$

as discussed in section I. However, the same calculations predict an $I = \frac{1}{2}$, 0^{++} κ state around 900 MeV and a genuine $I = 2$, 0^{++} flavor exotic at about 1100 MeV. The relevant phase-shifts are certainly non-resonant at these energies. Moreover, the identification of the $\varepsilon(700)$ as a genuine resonance is not without some danger. Morgan[44] has discussed a method for subtracting the effect of the S^* resonance from the $I = 0$, 0^{++} phase shift, leaving a residual phase shift which then 'resonates' at a much higher energy, namely around 1200 MeV. It is therefore unclear whether the identification of the $\varepsilon(700)$ as a resonance in its own right involves some double-counting.

On April 1st, 1979, Jaffe and Low[45] published some interesting second thoughts on Bag calculations of $q^2\bar{q}^2$ states and introduced a quantity they called the P-matrix. The message of their paper is that multiquark states calculated in the Bag do not in general correspond to resonant phase shifts. There is no space here to attempt a detailed discussion but I will attempt to sketch the motivation for their analysis.

Their starting point is the observation that $q^2\bar{q}^2$ color singlet states may be resolved into two types of $(q\bar{q})(q\bar{q})$ components

$$[q^2\bar{q}^2]_{1_c} \begin{array}{c} \nearrow \ [(q\bar{q})_{8_c} - (q\bar{q})_{8_c}]_{1_c} \\[2ex] \searrow \ [(q\bar{q})_{1_c} - (q\bar{q})_{1_c}]_{1_c} \end{array}$$

The component with $(q\bar{q})$ coupled to an octet of color is expected
to be confined by the Bag in the usual way, but the color singlet
qq component is not expected to experience strong confining
forces. This component has been 'artificially' confined in the
Bag model calculation   instead of being treated as an open
decay channel. It is clear that the time-honored 'narrow
resonance approximation' of quark model calculations in neglecting decay channels when calculating masses is liable to be mis-
leading for these states. The P-matrix represents a first,
admittedly crude attempt to take into account the presence of
these open, fall-apart decay channels. The basic idea is best
illustrated by considering the $(q\bar{q})_1$ - $(q\bar{q})_1$ component alone.

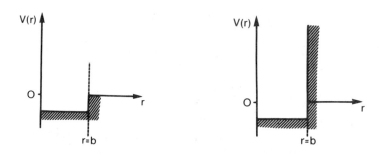

Fig. 11 $(q\bar{q})_1$ - $(q\bar{q})_1$ potentials:

      (a) Realistic potential   (b) Bag 'approximation'.

Fig. 11(a) and 11(b) attempt to give an impression of the real
potential and the Bag "approximation" potential between the $(q\bar{q})_1$
sub-systems. Experiment yields some phase-shift, $\delta(k)$, which
reflects the real potential of Fig. 11(a):The Bag calculation
yields the energy eigenstate of the infinite square well of Fig.
11(b). Jaffe and Low take the experimental phase-shift and construct from it a quantity they call the P-matrix, whose poles
correspond to the eigen-energies or "primitives" of the Bag
problem. For a non-relativistic S-wave potential the connection
is explicit,namely

$$P(k) = k \cot(kb + \delta(k))$$

A realistic calculation involves generalizing this idea to a multichannel situation, including both open and closed (confined) channels.

The result of all this is that multiquark 'primitives', as calculated in the Bag, can be associated with non-resonant phase-shifts. In particular, this resolves the problems concerning the $\varepsilon(700)$ and the $\kappa(900)$ $q^2\bar{q}^2$ Bag states. Moreover, the 0^{++} I = 2 <u>repulsive</u> phase-shift has a <u>P</u>-matrix pole near 1100 MeV which corresponds to the exotic $q^2\bar{q}^2$ primitive. (Fig. 12)

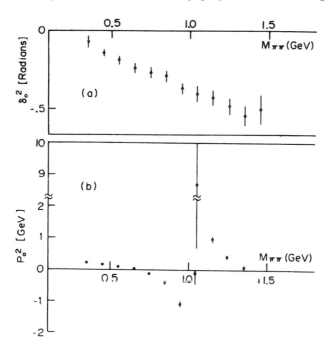

Fig. 12 $\pi\pi$ I = 2 s-wave: (a) Phase shift   (b) P-matrix - the arrow marks the position of the pole.

Jaffe and Low also make the claim that the relative positions of these P-matrix poles are evidence for the color magnetic interaction from one-gluon exchange!

There are many questions raised by this analysis but the claimed conclusions are so striking - namely the first evidence for $q^2\bar{q}^2$ exotics - that the theoretical basis for these calculations deserves further study. It is also desirable to try to construct other tests of a $q^2\bar{q}^2$ vs $q\bar{q}$ L=1 assignment for the 0^{++} mesons. In this connection, I would like to mention two recent calculations. Barnes[46] has taken a non-relativistic potential model for charmonium and extrapolated downwards to

212

strangeonium, which still has some claim to be considered a "heavy quark" system. Under 'reasonable' variations of the parameters, he finds it impossible to accomodate a 0^{++} $s\bar{s}$, L = 1 state below about 1200 MeV. The second paper is by Aaron and Goldberg[47] who construct a pole model for $\eta \rightarrow 3\pi$. They claim that comparison of their calculated decay rate with experiment implies a substantial $q^2\bar{q}^2$ component for the $\varepsilon(700)$. Both calculations therefore lend more grist to Jaffe's mill.

IV SOME FINAL REMARKS

(1) There are many apparently different descriptions of low mass meson states, namely, $q\bar{q}$, $q^2\bar{q}^2$, meson-meson, nucleon-anti-nucleon, gg, and qqg. A consistent theoretical framework which avoids the problem of double-counting and so on, is still lacking.

(2) Although all the current candidates for narrow Baryonium have disappeared, it is still conceivable that narrow M-Baryonia do exist at sufficiently high masses, and genuinely large ℓ. How best one could produce such states is an open question.

(3) Although strictly outside the scope of this conference I would like to conclude with a mention of the latest 6 standard deviation bump to excite theorists interest. In the reaction

$$K^- + p \rightarrow \pi^- + R^+$$

two bubble chamber experiments[48] (at 8.25 GeV/c and 6.5 GeV/c) have looked for events corresponding to the decays

$$R^+ \rightarrow \Sigma K\bar{K} + \pi\text{'s}$$

$$\rightarrow \Lambda K\bar{K} + \pi\text{'s}$$

$$\rightarrow \Xi \quad + \pi\text{'s}$$

With only a cut for backward production, the combined spectrum is shown in Fig. 13. A peak is observed at 3.17 GeV with a width of less than 20 MeV. Their search was motivated by the possibility of Y^*'s containing an extra hidden $s\bar{s}$ pair, and one interpretation of the peak _might_ be in terms of a

$$(qqq)_{8c} - (q\bar{q})_{8c}$$

molecule. However, the production mechanism of such a state seems rather obscure in terms of quark-line diagrams.

(4) One cannot build a whole spectroscope of multiquark states on one experimental candidate. The establishment, or otherwise, of all or some of the new forms of quark and gluon spectroscopy is a vital parameter in our understanding of QCD.

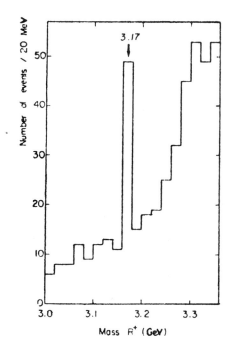

Fig. 13 Mass spectrum of R^+ for $K^- + p \rightarrow \pi^- + R^+$

ACKNOWLEDGEMENTS

Many people were helpful in my preparation of this talk. I
particularly wish to thank, Ian Barbour, Peter Litchfield,
Alan Martin, David Morgan, Mike Pennington, Robert Sekulin,
Robert Turnbull and Jacques Weyers: any errors or misapprehensions
are of course entirely mine.

REFERENCES AND FOOTNOTES

1. For a review and a guide to the early references see the reviews of Povh and Chan at the E.P.S. International Conference on High Energy Physics, Geneva 1979.

2. J.Donoghue, these proceedings.

3. R.L.Jaffe, Phys.Rev. D15, 267 (1977).

4. P.N.Bogolioubov, Ann.Inst.H.Poincaré, 8 163 (1967).

5. An elementary account of the Bag boundary conditions (essentially a translation of Ref.4) may be found in lectures by this author in "Topics in Quantum Field Theory and Gauge Theories", Salamanca 1977, edited by J.A.de Azcárraga, published in the series Lecture Notes in Physics, 77 (Springer-Verlag, 1978).

6. T.A.DeGrand, R.L.Jaffe, K.Johnson and J.Kiskis, Phys.Rev. D12, 2060 (1975).

7. K.Johnson and C.Thorn, Phys.Rev. D13, 1934 (1976).

8. Chan Hong-Mo and H.Høgaasen, Phys.Lett. 72B 121 (1977) and Nucl.Phys. B136, 401 (1978).

9. R.L.Jaffe, Phys.Rev. D17, 1444 (1978).

10. C.Evangelista et al., Phys.Lett. 72B, 139 (1977).

11. P.Benkheiri et al., Phys.Lett. 68B, 483 (1977).

12. It would be uncharitable to list all such papers.

13. A.A.Carter et al., Phys.Lett. 67B, 117 (1977).

14. See conference proceedings.

15. C.Daum et al., CERN-preprint EP/79-157 (1979).

16. D.Aston et al., CERN-preprint EP/80-40 (1980).

17. R.S.Dulude et al., Phys.Lett. 79B, 329, 335 (1978).

18. A.D.Martin and M.Pennington, Durham preprint (1980), (to be published in Nucl.Phys.B).

19. B.R.Martin and D.Morgan, Rutherford preprint RL-80-009, (1980).

20. C.Nef, these proceedings.

21. R.Bertini et al., proposal, CERN/PSC/78-15 (1978).

22. C.Amsler et al., Phys.Rev.Lett. 44, 853 (1980).

23. G.'t Hooft, Nucl.Phys. B72, 461 (1974); B75, 461 (1974).

24. For a recent summary of the results of the 1/N expansion and references to earlier papers, see the paper by E.Witten, Nucl.Phys. B160, 57 (1979).

25. M.Jacob and J.Weyers, Nuovo Cimento 69A, 521 (1970); 70A, 285(E) (1970).

26. R.L.Jaffe, contribution to the proceedings of XIV Recontre de Moriond, edited by J.Tranh Than Van (Editors, Frontières, Dreux, 1979).

27. See, for example, M.Imachi et al., Progr.Theor.Phys.57, 517 (1977); G.C.Rossi and G.Veneziano, Nucl.Phys. B123, 507 (1977); C.Rosenzweig and G.F.Chew, Phys.Lett.58B 93 (1975); Chan H-M. J.E.Paton, Tsou S.T. and Ng S.W., Nucl.Phys. B92 13 (1975).

28. For an up-to-date review of the phenomenological status, see A.C.Irving and R.P.Worden, Physics Reports 34C, 117 (1977).

29. For a review of these models, see the talk by this author at the E.P.S.International Conference on High Energy Physics, Geneva 1979.

30. A. de Rujula, H.Georgi and S.Glashow, Phys.Rev. D12, 147 (1975).

31. R.Koniuk and N.Isgur, Phys.Rev.Lett. 44, 845 (1980); and Toronto preprint (1980).

32. L.Micu, Nucl.Phys. B10, 521 (1969).

33. E.W.Colglazier and J.L.Rosner, Nucl.Phys. B27 349 (1971).

34. A.Le Yaouanc et al. Phys.Rev. D8, 2223 (1973); Phys.Rev. D9, 1415 (1974).

35. M.B.Gavela et al., Phys.Lett. 79B, 459 (1978).

36. I.Barbour and D.Ponting, Glasgow preprint (1980), to be published in Z.Physik C.

37. I.Barbour and D.Ponting, Z.Physik C4, 119 (1980).

216

38. J.P.Ader, B.Bonnier and S.Sood, Phys.Lett. 84B, 488 (1979).

39. I.Barbour and J.P.Gilchrist, Glasgow preprint in preparation.

40. For a review of this field, see for example, W.W.Buck, C.B.Dover and J.M.Richard, Ann.Phys. (N.Y.) 121, 47 (1979).

41. C.M.Dover and J.M.Richard, Ann.Phys. (N.Y.) 121, 70 (1979).

42. I.S.Shapiro, Phys.Rep. C35, 129 (1978).

43. C.M.Dover, private communication .

44. D.Morgan, Phys.Lett. 51B, 71 (1974).

45. R.L.Jaffe and F.E.Low, Phys.Rev.D19, 2105 (1979).

46. F.E.Barnes, Southampton University preprint, SHEP 79/80-4 (1980).

47. R.Aaron and H.Goldberg, North Eastern preprint NUB-2435 (1980).

48. J.Amirzadeh, Phys.Lett. 89B, 125 (1979).

SEARCH FOR THE NARROW S-RESONANCE IN A MEASUREMENT
OF p̄p TOTAL CROSS SECTION USING WIRE CHAMBERS

Presented by K. Nakamura for:

H. Aihara, J. Chiba, H. Fujii, T. Fujii, H. Iwasaki, T. Kamae,
K. Nakamura, T. Sumiyoshi, Y. Takada, T. Takeda and M. Yamauchi
Department of Physics, University of Tokyo, Tokyo 113, Japan

H. Fukuma and T. Takeshita
Department of Physics, Hiroshima University, Hiroshima 730, Japan

ABSTRACT

We have measured p̄p total cross section at beam momenta between
396 and 737 MeV/c by the transmission method using multiwire proportional chambers and a beam monitoring spectrometer. The spectrometer was the key instrument in adjusting and monitoring the beam and
the detectors, and in obtaining various correction factors. No
evidence was found on the existance of the narrow resonance S(1936)
reported by other experiments.

The S(1936) resonance has been attracting special attention
because of the reported narrow width and strong coupling to the p̄p
channel. Theoretically, it has been considered the best candidate
for a baryonium státe. Whereas the observation of a narrow enhancement around 500 MeV/c beam momentum has been reported in the
measurements of total p̄p[1,2,4] and p̄d[1], p̄p elastic[2,3] and p̄p charged-
prong annihilation[2,3] cross sections, no such structure has been
found in a recent series of experiments by Tripp and coworkers on
p̄p backward elastic[5], charge exchange[6], and total[7] cross sections.
Thus the S-resonance has come into a confused situation.

The present experiment was motivated to examine this problem
by measuring the total p̄p cross section with an improved transmission method. The experiment was carried out in the low-momentum
separated beam (K3) of the National Laboratory for High Energy
Physics (KEK) in Japan. The momentum range covered was 395.9 –
737.4 MeV/c, corresponding to the p̄p mass range of 1910 – 2000 MeV.
The mass resolution was determined by the length of the liquid
hydrogen target, 8.6 cm, and was 1.5 MeV (rms) at ∿500 MeV/c. Fig. 1
shows a plan view of the experimental arrangement. The trigger
requirement was the three-fold coincidence C1·C2·C3. A unique feature which distinguishes the present experiment from conventional
transmission counter experiments is the use of multiwire proportional chambers for the tracking of both incoming and outgoing anti-
protons, and of a beam monitoring spectrometer (see Fig. 1). Thus
the beam properties were completely known unlike most of other total
cross section measurements in which possible biases due to unknown
quality of the beam could be rather large and difficult to estimate.
The following points were essential to achieve small systematic
errors.

218

(i) The momenta and trajectories of the incoming and outgoing anti-protons were measured event by event. Specifically, the same beam phase space was always used by making appropriate cuts on the incoming trajectory.

(ii) Particle identification was done with good precision (fractional π^- contamination $\sim 1/5000$) using TOF and dE/dx measurements of the scintillation counters.

(iii) The beam momentum was continuously calibrated using precisely measured magnetic fields of the magnet D3 to an absolute accuracy of ±0.5 % and relative (point-to-point) accuracy of better than ±0.1 %.

(iv) The temperature of the liquid hydrogen target was controlled within ±0.1°K throughout the experiment.

(v) To avoid spurious narrow structure due to time-dependent systematic errors, the full momentum range was repeatedly measured, and in each spectrometer setting three to four full-target-to-empty-target cycles were repeated.

(vi) Forward elastic scattering was concurrently measured and the real-to-imaginary ratio of the forward elastic amplitude, ρ, was determined. Thus, the correction for the Coulomb-nuclear interference effect can be made in a consistent way. (Our preliminary results for ρ are: $\rho = 0.2$ for the antiproton momentum $p \leq 500$ MeV/c and $\rho = 0.0015 - 0.55$ for $p > 500$ MeV/c.)

Fig. 2 shows the results of the experiment (filled circles) plotted with the bin size of 2 % in the beam momentum. We have checked that there is no obvious binning effect by examining the data plotted with the bin size of 1 % in the beam momentum. The error bars represent the statistical errors only. Besides the standard corrections for the single Coulomb, Coulomb-nuclear interference, and forward elastic scattering effects, we have made the correction

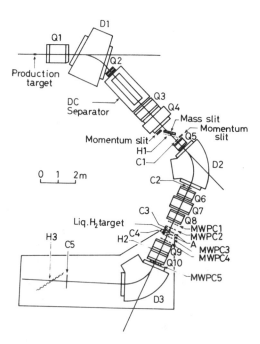

Fig. 1. Plan view of the experimental apparatus. D1-D3 are bending magnets, Q1-Q10 are quadrupole magnets, C1-C5 are scintillation counters, H1-H3 are scintillation counter hodoscopes, and MWPC1-MWPC5 are multiwire proportional chambers of bi-dimensional readout. Scintillation counters A are used to detect charged annihilation events.

to account for the difference in the average momenta of incoming
antiprotons in the full and empty target runs due to the energy loss
in the liquid hydrogen target (typically 0.3 % in σ_{tot}), and the
correction for the inefficiency in the reconstruction of the incoming tracks due to the backsplashing of the charged pions. The last
correction was necessary because we required the uniqueness of the
hits in proportional chambers MWPC1 and 2 for unambiguous selection
of the incoming tracks. The correction factors (smoothly varying
from 5.6 ± 1.7 mb at 400 MeV/c to 2.6 ± 0.8 mb at 700 MeV/c) were
obtained by carefully examining those events having multiple hits in

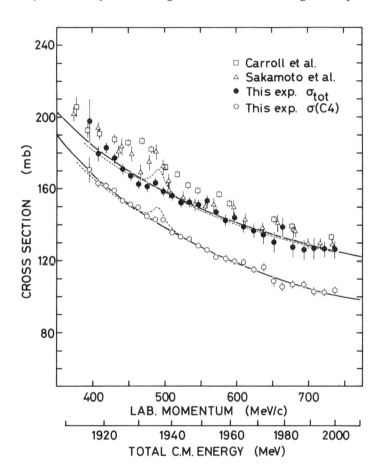

Fig. 2. The total cross section (filled circles) are plotted
together with previous data (empty squares, Carroll et al.[1]; empty
triangles, Sakamoto et al.[4]). The empty circles are σ(C4). The
solid curves are the results of the fits with the form a+b/p to
σ_{tot} and σ(C4) obtained in the present experiment. The dashed
curves correspond to the 90 % confidence-level upper limits for a
narrow (Γ = 5 MeV) resonance at 1936 MeV.

either MWPC1 or 2. They were also estimated by a Monte Carlo calculation which gave consistent results. The total systematic errors
are estimated to be less than 1 %. There is an additional overall
normalization uncertainty of 0.8 % arising from the uncertainty in
the absolute length of the liquid hydrogen target. In Fig. 2 are
also plotted some of the previous data (Carroll et al[1] and Sakamoto
et al[4].) which indicate a narrow (width Γ < 10 MeV) and large (resonant cross section σ_R > 10 mb) enhancement at about 1935 MeV.
Clearly no such structure is observed in our data.

We have also measured the cross section $\sigma(C4)$ defined as

$$\sigma(C4) = \sigma_{tot} - \int_0^{\Omega(C4)} \frac{d\sigma_{el}}{d\Omega} \, d\Omega \, ,$$

where $\Omega(C4) = 160$ msr is the solid angle subtended by the counter C4
placed behind the target and $d\sigma_{el}/d\Omega$ is the elastic differential
cross section in the lab. frame. The dominant contribution to the
second term on the right-hand side of the above equation comes from
diffraction scattering, and the contribution into this small solid
angle from possible (low-spin) resonance is negligible. Therefore
any resonant total cross section should also show up in $\sigma(C4)$
plotted by empty circles in Fig. 2. The cross section $\sigma(C4)$ is
measured with better statistical accuracy than σ_{tot}. No significant
structure is seen around 1935 MeV in $\sigma(C4)$, either.

Both σ_{tot} and $\sigma(C4)$ can be fitted well by a smooth functional
form $a + b/p$. For σ_{tot}, we obtain $a = (54.3 \pm 4.6)$ mb and $b =$
(52.2 ± 2.4) mb·GeV/c with χ^2/DF (degree of freedom) $= 14.4/26$. For
$\sigma(C4)$, we obtain $a = (19.5 \pm 2.4)$ mb and $b = (59.8 \pm 1.3)$ mb·GeV/c
with $\chi^2/DF = 16.4/26$. These fitted curves are subtracted from the
data. The results are shown in Fig. 3 and compared to the Breit-
Wigner curves representing the S-resonance contribution claimed by
the previous experiments[1,2,4,7] We also calculate the 90 %
confidence-level upper limits for a Breit-Wigner resonance at 1936
MeV based on both σ_{tot} and $\sigma(C4)$ and their statistical errors. The
results are given in Table I. Our results are clearly inconsistent
with the results of Carroll et al[1] ($\Gamma = 9^{+4}_{-3}$ MeV, $\sigma_R = 18^{+6}_{-3}$ mb).
Our upper limit based on $\sigma(C4)$ is inconsistent with the results of
Chaloupka et al[2] ($\Gamma = 8.8^{+4.3}_{-3.2}$ MeV, $\sigma_R = 10.6 \pm 2.4$ mb). The results
of Sakamoto et al[4] ($\Gamma = 2.8 \pm 1.4$ MeV, $\sigma_R = 14.5 \pm 3.9$ mb) are not
inconsistent with our upper limit if one considers their statistical
uncertainty. However their results may be regarded as consistent
with absence of the S-resonance, since their fit without the reso-

Table I   90 % confidence-level upper limit
for a possible resonance at 1936 MeV.

	$\Gamma = 5$ MeV	$\Gamma = 10$ MeV
σ_R (based on σ_{tot})	12.6 mb	9.8 mb
σ_R (based on $\sigma(C4)$)	10.2 mb	8.5 mb

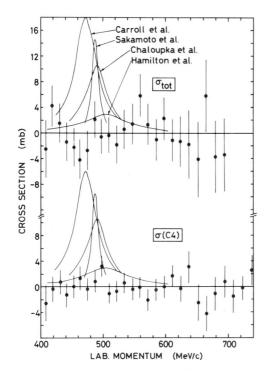

Fig. 3. Data are plotted after subtraction of the fitted curves a+b/p, and compared to the Breit-Wigner curves for the S-resonance contribution claimed by the previous experiments.[1,2,4,7]

nance gave a high confidence level of 49.3 %. As for a "broad resonance ($\Gamma \sim 20$ MeV, $\sigma_R \sim 3$ mb) suggested by Hamilton et al.[7], no statistically significant argument can be made from our data.

In conclusion, our results do not support the existence of the narrow S-resonance. The 90 % confidence-level upper limit of a possible resonance cross section is about 10 mb for the width $\Gamma \lesssim 10$ MeV.

We would like to thank all the staff of the National Laboratory for High Energy Physics (KEK) for excellent machine operation and various help given to us throughout this experiment.

REFERENCES

1. A. S. Carroll et al., Phys. Rev. Lett. 32, 247 (1974).
2. V. Chaloupka et al., Phys. Lett. 61B, 487 (1976).
3. W. Bruckner et al., Phys. Lett. 67B, 222 (1977).
4. S. Sakamoto et al., Nucl. Phys. B158, 410 (1979).
5. M. Alston-Garnjost et al., Phys. Rev. Lett. 43, 1901 (1979).
6. R. P. Hamilton et al., Phys. Rev. Lett. 44, 1179 (1980).
7. R. P. Hamilton et al., Phys. Rev. Lett. 44, 1182 (1980).

CHARMED MESONS FROM e^+e^- ANNIHILATION

Gerson Goldhaber
Department of Physics and Lawrence Berkeley Laboratory
University of California, Berkeley, California 94720

I would like to discuss some of the recent, and a few of the not-so-recent, results on charmed mesons as seen in the e^+e^- annihilation process.

Table I gives a list of the experiments which have contributed to this subject so far: the SLAC-LBL Mark I detector, the Lead-Glass Wall (LGW), both at SPEAR, DASP and PLUTO (at DESY working at DORIS), the SLAC-LBL Mark II detector, DELCO, and the Crystal Ball (all at SLAC working at SPEAR). Of these, the first six detectors are already out of their respective beams; only the Crystal Ball is still taking data in the $E_{cm} = 3 - 7$ GeV region at this time. In the future there will be another detector, the Mark III, coming up at SPEAR to study psions and charmed particles. There are several detectors at DORIS (DASP 2, LENA, DESY-Heidelberg, etc.), which has moved into the Υ energy region.

The topics I will discuss are the following:
(1) The ψ'' has been observed in three experiments: originally in the LGW experiment and DELCO and now there is more recent data from the Mark II.
(2) The branching ratios of D^0 and D^+ mesons into hadronic final states.
(3) The Cabibbo-suppressed decay modes.
(4) Recent evidence from Mark II and from DELCO that the D^+ has a longer lifetime than the D^0 which is obtained from measuring the semileptonic branching ratios.
(5) The semileptonic decay modes of the charmed mesons.
(6) A new result from the Crystal Ball; namely, the direct observation of the π^0 from the decay of the $D^* \to D\pi^0$. Previously in the Mark I detector this reaction was inferred from a detailed analysis of the recoil spectra only.
(7) New results on the production of a vector plus a pseudoscalar; viz., ρK and $K^*\pi$, from the Mark II detector.
(8) New values for the masses of the D^0 and D^+ from the Mark II detector.
(9) The ratio R_D of the cross section for D^0, D^+ production to the μ pair cross section.
(10) The evidence for the F^+. There is the original data from DASP. At present there is a study going on with the Crystal Ball on the behavior of the η inclusive cross section. So far I have no results from this search.

(1) The ψ'' -- a D Factory

The first observation of the charmed mesons[1] D^0 and D^+ in the SLAC-LBL Mark I detector was made in the $E_{cm} = 3.9$ to 4.6 GeV region. In this energy region, and particularly at the cross-section peak at 4.028 GeV, the principal D production processes are

ISSN:0094-243X/81/670223-34$1.50 Copyright 1981 American Institute of Physics

Table I. Charmed mesons from e^+e^- annihilation.

	Mark I	LGW	DASP	PLUTO	Mark II	DELCO	Crystal Ball
(1) $\psi(3770) \equiv \psi''$		X			X	X	
(2) Br D^0, D^+ hadronic		X			X		
(3) Cabibbo-suppressed					X		
(4) $\tau(D^+)/\tau(D^0)$					X	X	
(5) Semileptonic Br		X	X	X	X	X	
(6) $D^{0*} \to \pi^0 D^0, \gamma D^0$	X						X
(7) $D \to \rho K, K^*\pi$					X		
(8) Mass D^0, D^+, D^{*0}, D^{*+}	X	X			X		
(9) $R_C, R_D = \dfrac{\sigma(D^0, D^+)}{\sigma(\mu\mu)}$		X			X	X	
(10) Evidence for F^+			X				S

$X \equiv$ observations reported.

$S \equiv$ search in progress.

$$e^+e^- \to D^* \bar{D} \text{ or } D\bar{D}^*$$

and
$$e^+e^- \to D^* \bar{D}^* .$$

It was only at a later stage that the $\psi(3770)$ or ψ'' resonance was discovered in the LGW and DELCO experiments.[2,3] The mass value of the ψ'' lies below the threshold for the above processes. Thus only the production of $D\bar{D}$ occurs. Namely: $e^+e^- \to \psi'' \to D^0\bar{D}^0$ and D^+D^-. Thus the ψ'' which lies ~ 40 MeV above $D^0\bar{D}^0$ and ~ 30 MeV above D^+D^- threshold is ideally suited for the study of D meson properties. If we compare the width of the $\psi(3684)$ or ψ', $\Gamma = 0.228$ MeV, with that of the ψ'', $\Gamma = 25$ MeV, we note the width has increased by a factor of ~ 100 and thus that the effect of the OZI suppression at the ψ' is no longer present at the ψ''. This is ascribed to the fact that the D production threshold opens up at 3726 MeV (3736 MeV for D^+D^-), and hence channels with c and $\bar{c}$ quarks in the final state can occur at the ψ''.

• Properties of the $\psi(3770)$ Resonance. The ψ'' has been studied extensively in the LGW and DELCO experiments and recently again in the Mark II experiments as I will report here. Figure 1 shows the R distribution observed in the Mark II experiment. Here R is the ratio of the hadronic cross section to the theoretical QED μ pair cross section $\sigma_{\mu\mu}$. The latter is obtained from calibration against observed Bhabha pairs. The $\tau^+\tau^-$ cross section has been subtracted. Figure 1a gives this corrected value. Figure 1b gives the R distri-

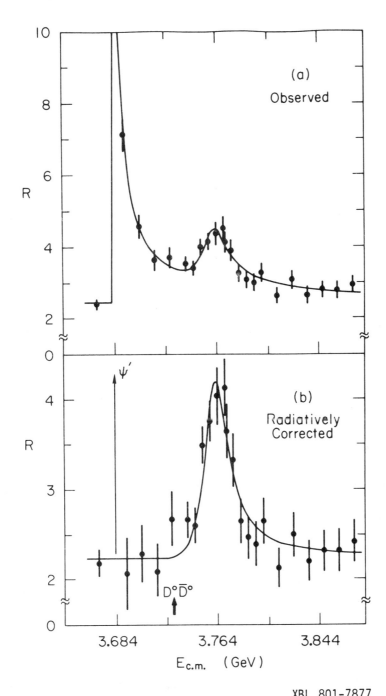

XBL 801-7877

Fig. 1. The value of $R \equiv \sigma_{had}/\sigma_{\mu^+\mu^-}$ in the vicinity of the ψ'' obtained by the Mark II collaboration,[7] (a) before and (b) after radiative correction using the technique of Jackson & Scharre. The curve is a fit of the data to the Breit-Wigner expression.

bution after the radiative tails from the J/ψ and ψ' have been subtracted. The errors shown are statistical. The resonance is fitted to a p-wave Breit-Wigner expression[4] with an energy-dependent total width $\Gamma_{tot}(E_{cm})$ which takes account of the vicinity of the $D^0\bar{D}^0$ and D^+D^- thresholds. Each charmed meson pair is assumed to rise from threshold in a manner characteristic of p-wave production. Here

$$R(E_{cm}) = \frac{1}{\sigma_{\mu\mu}} \frac{3\pi}{M^2} \frac{\Gamma_{ee}\,\Gamma_{tot}(E_{cm})}{(E_{cm} - M)^2 + \Gamma_{tot}^2(E_{cm})/4}$$

and

$$\Gamma_{tot}(E_{cm}) \propto \frac{p_+^3}{1 + (rp_+)^2} + \frac{p_o^3}{1 + (rp_o)^2}$$

where p_+ (p_o) is the momentum of the pair produced D^+ (D^0) and r is the interaction length. The quantities M (the resonance mass) and Γ_{ee} (the partial width to electrons) were determined in a fit to the data points. The fit is not sensitive to r which was taken as 2.5 Fermi.

The new results are consistent with the earlier data from the LGW and DELCO, except for a shift in the central mass value which is now found to be 3764 ± 5 MeV, that is $6 - 8$ MeV lower than the previous values. A more precise measurement is ΔM, the $\psi'' - \psi'$ mass difference which does not include the systematic error of the absolute beam energy calibration. This is found to be $\Delta M = 80 \pm 2$ MeV. Furthermore the Mark II value for the width of decay into e^+e^-, $\Gamma_{ee} = 276 \pm 50$ eV lies in between the earlier two values. The comparison between the new measurements and the earlier results is given in Table II. From theoretical arguments[5] the ψ'' is believed to a $^3D_1$ state of charmonium which is however mixed with the ψ', the 2^3S_1 state. The relatively large Γ_{ee} value gives an estimate for this mixing angle of $20.3° \pm 2.8°$.

Table II. Measurements of the $\psi(3770)$ resonance parameters.

Experiment	Mass MeV/c^2	Γ_{tot} MeV	Γ_{ee} eV	ΔM^* MeV/c^2
DELCO[3]	3770 ± 6	24 ± 5	180 ± 60	86 ± 2
LGW[2]	3772 ± 6	28 ± 5	345 ± 85	88 ± 3
Mark II[7]	3764 ± 5	24 ± 5	276 ± 50	80 ± 2

$^*\Delta M$ is the mass difference between the $\psi(3684)$ and $\psi(3770)$.

(2) Charmed Meson Hadronic Branching Ratios

A measurement of the number of events in a given D decay channel, together with a Monte Carlo calculation of the corresponding detection efficiency for the given detector and the luminosity calibration gives σB. To obtain B, the branching ratio, the following properties of the ψ'' are assumed:

•The ψ'' is a state of definite isospin (0 or 1); this allows a prediction of the D^0/D^+ production ratio, namely $\sigma(D^0)/\sigma(D^+)$ $\simeq p_0^3/p_+^3$ as expected for p-wave production. This reflects the difference between the D^0 and D^+ masses (1863.3 MeV and 1868 MeV respectively).

· The ψ'' decays nearly entirely into $D\bar{D}$ ($\sim 99\%$). This is based on the $\Gamma_{tot}(\psi')$ to $\Gamma_{tot}(\psi'')$ ratio ($\sim 1/100$); i.e., that the OZI-suppressed portion of the ψ'' decay width is of the same magnitude as the $\Gamma(\psi')$.

These assumptions were checked for those events decaying into $D\bar{D}$ mesons which were both identified.

Another feature of the fact that the ψ'' decays into a pair of $D\bar{D}$ mesons is that in addition to the invariant mass of a given state one can also use the beam-energy-constrained mass

$$M_b = \sqrt{E_b^2 - p^2} \quad .$$

This assumes that each particle combination which corresponds to a D decay mode has a total energy equal to the beam energy E_b. In practice particle combinations with energies within 50 MeV of E_b are accepted. Such a procedure results in marked reductions in background as well as much better mass resolutions (~ 3 MeV). Figure 2 shows the M_b distributions obtained for a number of D^0 and D^+ decay modes in the LGW experiment.[6] In the data from the Mark II detector[7] extensive running was carried out at $E_{cm} = 3.771$ GeV, an energy which lies slightly above the peak of the resonance mass, giving a total of 49,000 hadronic events. From the fit to the Breit-Wigner expression above, the $D\bar{D}$ pair cross sections at this energy is 6.85 ± 1.2 nb. When this value is apportioned between the D^0 and D^+, this gives

$$\sigma_{D^0}(3.771) = 7.8 \pm 1.2 \text{ nb} \quad \text{and} \quad \sigma_{D^+}(3.771) = 5.9 \pm 1.0 \text{ nb}$$

for the two inclusive (single D) cross sections. Figures 3-6 show the D^0 and D^+ decay modes obtained with the Mark II detector. Table III gives the hadronic branching ratios obtained in these two experiments.

(3) Cabibbo-Suppressed Decay Modes

An intrinsic feature of the GIM mechanism for charm is the prediction that aside from the principal (Cabibbo-favored) D decay modes, which lead to K^- or $\bar{K}^0$ in the final states, there also be the Cabibbo-suppressed modes leading to zero strangeness final states. The Cabibbo-favored and -suppressed modes for D^0 two-particle final states are illustrated by the quark diagrams in Fig. 7. Here the angle θ_A is familiar Cabibbo angle θ_C while θ_B is the new angle which on the four-quark model is associated with the flavor mixing of charmed quarks. The GIM assumption is that $\theta_A = \theta_B$. Experimentally the two angles can be independently measured as:

$$\tan^2 \theta_A = \frac{\Gamma(D^0 \to K^- K^+)}{\Gamma(D^0 \to K^- \pi^+)} \quad \text{and} \quad \tan^2 \theta_B = \frac{\Gamma(D^0 \to \pi^- \pi^+)}{\Gamma(D^0 \to K^- \pi^+)} \quad .$$

· Studies of the ψ''. Figures 8, 9 and 10 show the experimental result from the Mark II experiment. In Figs. 8 and 9 I show the

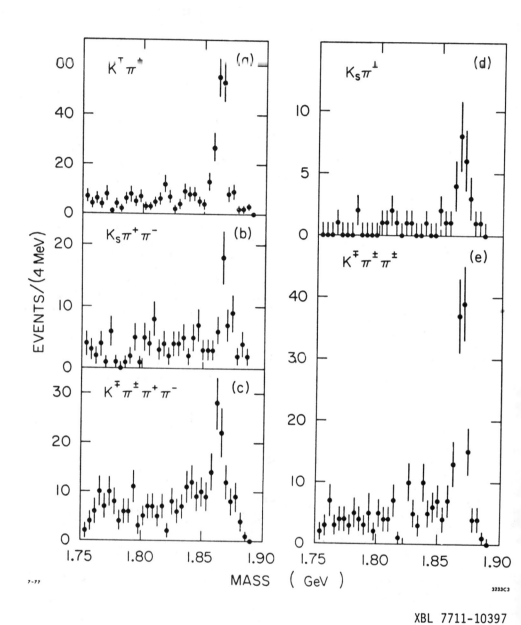

7-77

MASS (GeV)

XBL 7711-10397

Fig. 2. The beam-energy-constrained mass distributions for the indicated D^0, D^+ decay modes obtained by the LGW collaboration at the ψ''.

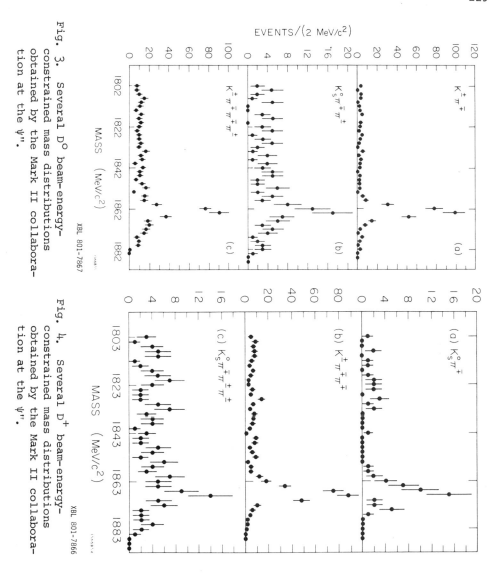

Fig. 3. Several D⁰ beam-energy-constrained mass distributions obtained by the Mark II collaboration at the ψ".

XBL 801-7867

Fig. 4. Several D⁺ beam-energy-constrained mass distributions obtained by the Mark II collaboration at the ψ".

XBL 801-7866

230

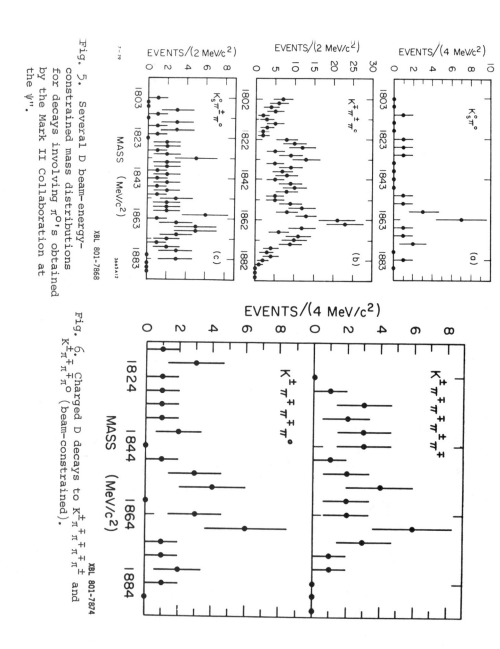

Fig. 5. Several D beam-energy-constrained mass distributions for decays involving π^0's obtained by the Mark II Collaboration at the ψ''.

XBL 801-7868

Fig. 6. Charged D decays to $K^{\pm}\pi^{\mp}\pi^{\mp}\pi^{\pm}$ and $K^{\pm}\pi^{\mp}\pi^{\pm}\pi^0$ (beam-constrained).

XBL 801-7874

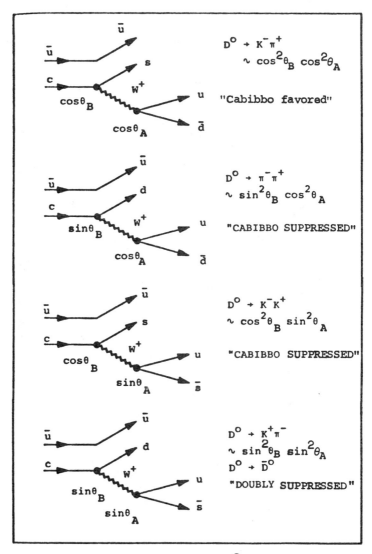

Fig. 7. Quark diagrams for D^0 decays.

232

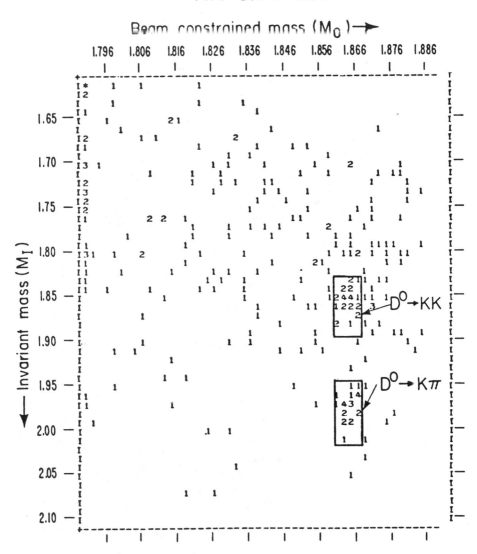

Fig. 8. Scatter plot of the beam-constrained mass versus the invariant mass for K^+K^- pairs as identified from TOF and with $|E_{K^+} + E_{K^-} - E_{beam}| < 50$ MeV/c^2.

XBL 796-1810

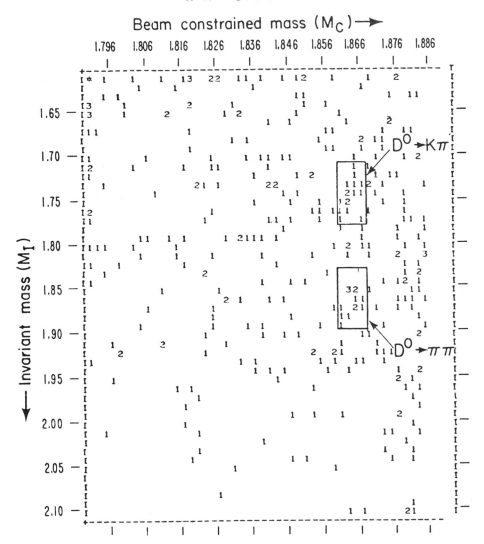

XBL 796-1805

Fig. 9. Scatter plot of the beam-constrained mass versus the invariant mass for $\pi^+\pi^-$ pairs as identified from TOF and with $|E_{\pi^+} + E_{\pi^-} - E_{beam}| < 50$ MeV/c^2.

234

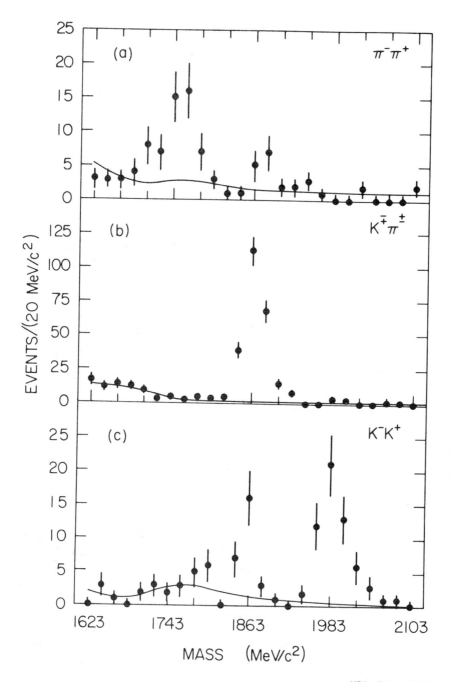

XBL 801-7873

Fig. 10. Evidence for two-body Cabibbo-suppressed decay modes
obtained by the Mark II Collaboration at the ψ". Candidates
are required to have momenta within 30 MeV of that expected
for D-pair production. Both figures (a) and (c) show prominent reflection peaks not centered at the D^0 mass due to π/K
misidentification by their TOF system. The solid curves are
background estimates.

Table III

$\sigma \cdot$ Br and Br for Cabibbo-Favored D Decays

Mode	Events in signal	ε	$\sigma \cdot$ Br (nb)	Mark II Br (%)	LGW Br (%)
$K^-\pi^+$	263.0+17.0	0.386	0.24+0.02	3.0+0.6	2.2+0.6
$\overline{K}^0\pi^0$	8.5+3.7	0.017	0.18+0.08	2.2+1.1	
$\overline{K}^0\pi^+\pi^-$	32.0+7.7	0.037	0.30+0.08	3.8+1.2	4.0+1.3
$K^-\pi^0\pi^+$	37.2+10.0	0.019	0.68+0.23	8.5+3.2	12.0+6.0
$K^-\pi^+\pi^-\pi^+$	185.0+18.0	0.095	0.68+0.11	8.5+2.1	3.2+1.1
$\overline{K}^0\pi^+$	35.7+6.7	0.090	0.14+0.03	2.3+0.7	1.5+0.6
$K^-\pi^+\pi^+$	239.0+17.0	0.221	0.38+0.05	6.3+1.5	3.9+1.0
$\overline{K}^0\pi^0\pi^+$	9.5+5.5	0.004	0.78+0.48	12.9+8.4	
$\overline{K}^0\pi^+\pi^-\pi^+$	21.0+7.0	0.015	0.51+0.18	8.4+3.5	
$K^-\pi^+\pi^-\pi^+\pi^+$	<11.5	0.021	<0.23	<4.1 (at 90% C.L.)	

scatter plots where the beam-constrained mass (M_C) is plotted against the invariant mass (M_I). In Fig. 8 K^+K^- mass combinations (as identified by TOF) are given. We note the enhancement for $M_C \simeq 1863$ and $M_I \approx M_C$. This corresponds to $D^0 \to K^+K^-$; a second enhancement at $M_I \simeq M_C + 120$ MeV/c^2 corresponds to $D^0 \to K^+\pi^-$ where the π^- is misidentified as a K^-. Figure 9 shows the case where $\pi^+\pi^-$ mass combinations are identified by TOF. Here again we note a (small) enhancement at $M_I \simeq M_C$ corresponding to $D^0 \to \pi^+\pi^-$ as well as the misidentified (K^- as π^-) cases at $M_I \simeq M_C - 120$ MeV/c^2. In Fig. 10 the $\pi^-\pi^+$, $K^-\pi^+$ and K^-K^+ invariant mass projections are shown for two-particle combinations with momenta within 30 MeV/c of the expected D pair momentum of 288 MeV/c (for $E_{cm} = 3.771$ GeV). Aside from the signals in the three channels at the D mass one notes here again the kinematic reflections shifted by about ± 120 MeV/c^2 from the D mass due to $\pi \longleftrightarrow K$ misidentifications. A fit to the data yields 235 ± 16 $K^+\pi^\pm$ events, 22 ± 5 K^+K^- events and 9 ± 3.9 $\pi^+\pi^-$ events.[8] Accounting for the relative K and π detection efficiencies give

$$\frac{\Gamma(D^0 \to K^-K^+)}{\Gamma(D^0 \to K^-\pi^+)} = 0.113 \pm 0.03$$

and

$$\frac{\Gamma(D^0 \to \pi^-\pi^+)}{\Gamma(D^0 \to K^-\pi^+)} = 0.033 \pm 0.015 \ .$$

Here the quoted errors include systematic effects. The results clearly demonstrate the existence of the Cabibbo-suppressed decay modes of roughly the expected magnitude: $\tan^2\theta_C \simeq 0.05$.

•Studies at $E_{cm} = 4 - 4.5$ GeV. We can obtain consistency check on the above results by using data from a different region, $E_{cm} = 4 - 4.5$ GeV. Here we have more data but the background is much worse. I want to show that we see the $K\bar{K}$ signal here as well and that it is consistent with the above results.

In Fig. 11 we show the invariant $K\pi$, $K\bar{K}$ and $\pi\pi$ masses on the left-hand side and the recoil mass distribution for $1840 < M_I < 1900$ MeV/c^2 on the right-hand side. For the reaction $D^o \rightarrow K^-\pi^+$ we note the well-known recoil pattern which corresponds to $D\bar{D}$, $D\bar{D}^*$, and $D^*\bar{D}^*$. If we now examine the recoil spectrum against the $K\bar{K}$ signal we see the same characteristic recoil pattern; thus we were sure that we have identified $K\bar{K}$ decays of the D. When you carry out the same procedure for the $\pi\pi$ system, the result is consistent, although the recoil pattern is not obvious in this case. Numerically the consistency check is as follows: there are 497 $K\pi$ events and from the above ratios we then predict 47 ± 12 $K\bar{K}$ events and we observe 50 ± 17 in this data. For the $\pi\pi$ case we predict 20 ± 9 events which is consistent with our observation, although no specific peaks can be identified from the recoil distribution.

(4) D Meson Lifetime Ratios

Work at the ψ'' also allows the study of "tagged" events. This method has been used in LGW experiment and more recently by the Mark II experiment. Here I will quote the recent Mark II results for nearly 300 D^+ and 500 D^o tagged events. In "tagging" we assume that at the ψ'' if a D^+ (D^o) is observed the remaining tracks in the event correspond to D^- ($\bar{D}^o$) decay.

Figure 12 gives an example of a D "tagged" event in the Mark II detector. This is an event at the ψ'' where we see both a D^o and a $\bar{D}^o$ completely reconstructed. The $\bar{D}^o$ decays into $K^+\pi^-$. We identify the K^+ by time of flight. The momentum is measured in the drift chambers and one can see the drift chamber points (×) in this Figure. The time-of-flight counters which fired are also indicated. The outer octagon represents the liquid argon shower counters. The second D, the D^o, decays into a K_S which is reconstructed from the observed $\pi^+\pi^-$ pair. There is an additional $\pi^+\pi^-$ pair and finally there is a π^o which is obtained by reconstructing two γ rays which are observed in the shower counters. The energies of the γ-ray showers, 0.48 and 0.80 GeV are shown in the Figure. The $\gamma\gamma$ angle is fairly well determined, the energies not as well, but the information is sufficient to reconstruct a π^o. If we take the $\bar{D}^o$ one can use it as a "tag" and look for what else occurs in the event. In this particular event there is a complete D^o there as well.

Hence for electron identification we can distinguish "right sign e" and "wrong sign e." The results are shown in Table IV. After background subtraction and allowance for "wrong sign e" contributions we find a larger $B(D^+ \rightarrow e^+ + ...)$ than $B(D^o \rightarrow e^+ + ...)$. On the theoretical assumption that $\Gamma(D^+ \rightarrow e^+ + ...) = \Gamma(D^o \rightarrow e^+ + ...)$ we thus obtain:

$$\frac{\Gamma(D^o \rightarrow all)}{\Gamma(D^+ \rightarrow all)} = \frac{B(D^+ \rightarrow e^+ + ...)}{B(D^o \rightarrow e^+ + ...)} = \frac{\tau(D^+)}{\tau(D^o)} .$$

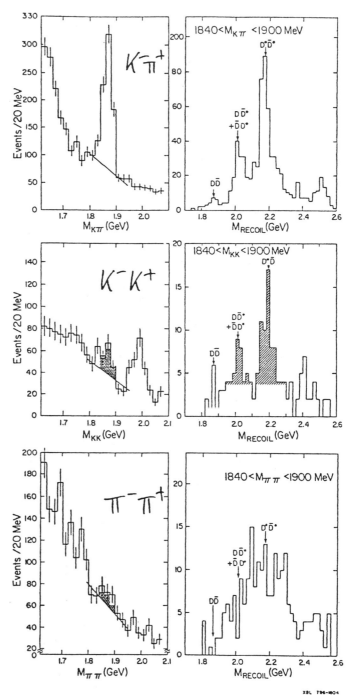

Fig. 11. D^O mass and recoil pattern for 4-4.5 GeV data (Mark II).

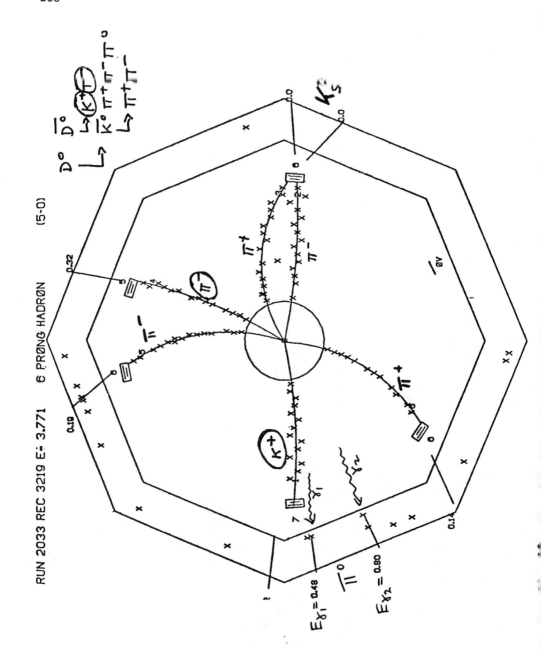

Table IV. Semileptonic decays of D^+ and D^0 (Mark II data).

Decay mode	# tags	# electrons	Background	Br (%)
$D^+ \to e^+$	295 ± 18	38	15 ± 1	16.8 ± 6.4
$\to e^-$		4	3.9 ± 0.5	
$D^0 \to e^+$	480 ± 23	36	19 ± 1	5.5 ± 3.7
$\to e^-$		19	12 ± 1	

Hence from a maximum likelihood fit we obtain

$$\tau(D^+)/\tau(D^0) = 3.1^{+4.1}_{-1.3} \quad ,$$

which is evidence for a larger D^+ than D^0 lifetime.

A similar independent conclusion has been reached by the DELCO experiment.[9] This work consists of an analysis of the two-electron versus one-electron rate at the ψ''. In the limit of perfect accept-ance, the single (N_1) and double (N_2) electron event rates due to D decays are related to the neutral (b_0) and charged (b_+) semileptonic branching ratios by

$$N_1 = 2N_o b_o(1 - b_o) + 2N_+ b_+(1 - b_+)$$
$$N_2 = N_o b_o^2 + N_+ b_+^2$$

where N_o and N_+ are the number of $D^0\bar{D}^0$ and D^+D^- events produced. One sees from the above that in the limit of small branching ratios a measurement of N_1 determines essentially a line in the b_o vs b_+ plane whereas a measurement of N_2 determines an elliptical arc. One thus expects that a simultaneous measurement of N_2 and N_1 will lead to two ambiguous solutions for b_o and b_+. Figure 13 indicates the experi-mental regions in the b_+ vs b_o plane which are consistent to within one standard deviation with the data presented in Table V. The data has been corrected for electron detection efficiency using the two extreme models that $D \to Ke\nu$ (Fig. 13a) or $D \to K^*e\nu$ (Fig. 13b).

Table V. The DELCO multi-prong electron data sample at the ψ''.

Event description	Event topology		
	1 electron	2 electrons	2 electrons + "V" (K_S^0)
Observed	1416	21	8
Background	692	4.6	1.8
Charm signal	724	16.4	6.2

Under either assumption it appears that $b_o \gg b_+$ or $b_+ \gg b_o$.

240

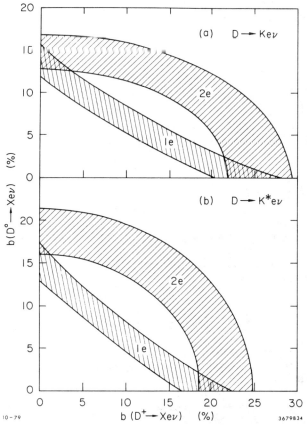

Fig. 13. The allowed solutions for the D⁰ and
D⁺ semileptonic branching ratios in the
DELCO[9] 1e and 2e multiprong data at the ψ".
The shaded regions, which correspond to ±1σ
limits, are plotted for two extreme assump-
tions of the detection efficiencies: (a) all
D → Keν and (b) all D → K*eν.

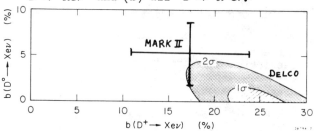

Fig. 14. Neutral versus charged branching ratios
from the Mark II and DELCO experiments.

DELCO uses the K_S content in the two-electron events to distinguish between these two possibilities. Although both D^o's and D^+'s can decay into K_S, the D^o must do so via the decay sequence $D^o \to (K\pi)e^+\nu$ where the isodoublet $K\pi$ system decays into a charged kaon 2/3 of the time ($I = 1/2$) and a neutral kaon only 1/3 of the time. The D^+, on the other hand, can produce K_S vis both $D^+ \to K_S e^+\nu$ and $K_S\pi^o e^+\nu$. In fact a rather large fraction of the two-electron events (8 out of 16.4) have a K_S which suggests the solution $b_+ \gg b_o$. Combining the information on the single-electron and two-electron rate at the ψ'' with the K_S content of the two-electron events, DELCO finds the D^+ and D^o semileptonic branching ratios to be $(22^{+4.4}_{-2.2})\%$ and $< 4\%$ (95% C.L.) respectively; this yields $\tau(D^+)/\tau(D^o) > 4.3$ (95% C.L.). Figure 14 compares the branching ratios from these two experiments which are in reasonable agreement.

(5) Semileptonic Decay Modes

One of the early observations of charmed mesons was via the semileptonic decay mode in PLUTO and DASP at DORIS, and in particular the observation of eK^o coincidences. A later study was carried out in the LGW experiment at SPEAR.

Table VI reviews the average D^o, D^+ semileptonic Br. Here I will just show two of the very recent results on the electron spectra.

Table V. The branching ratio for $D \to e\nu X$.

Experiment	E_{cm} (GeV)	Branching ratio (%)
DASP	$3.99 \to 4.08$	8.0 ± 2.0
LGW	ψ''	7.2 ± 2.8
DELCO	ψ''	8.0 ± 1.5
Mark II	ψ''	9.8 ± 3.0
	Average	8.0 ± 1.1

Figure 15 gives the spectra as obtained in DELCO[9] and shows that both $Ke\nu$ and $K^* e\nu$ decay modes are needed. Figure 16 shows the spectra from the Mark II. Here the two decays D^o and D^+ are separated as obtained from the tagged events.

(6) $D^* \to \pi D$ and $D^* \to \gamma D$ Decay Modes

The
$$D^{*\pm} \to \pi^o D^\pm \quad \text{and} \quad D^{*\pm} \to \gamma D^\pm$$
decay modes had only been measured indirectly from the D^+ recoil spectra in the SLAC-LBL (Mark I) experiment at $E_{cm} = 4.028$ GeV.[10] What is new now is that the Crystal Ball experiment[11] has a (qualitative) result showing the direct observation of low energy π^o's ($E_{\pi^o} < 150$ MeV) at $E_{cm} = 4.028$ GeV where strong D^+ signals have been inferred in the Mark I experiment. On the other hand no such π^o signal is observed at $E_{cm} = 3.77$ GeV (the ψ''). Figure 17 shows the $\gamma\gamma$ mass

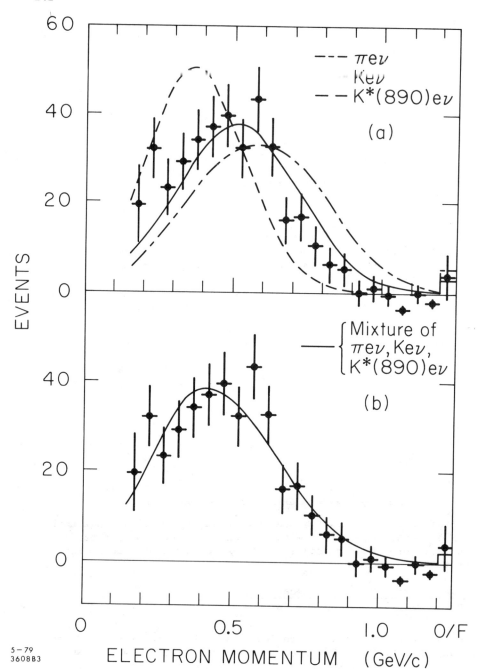

Figure 15. The electron momentum spectrum from D decays at the ψ'', measured by DELCO.[9] The curves have been fitted to the data below 1 GeV/c and correspond to the following hypotheses: (a) D → $\pi e\nu$ (dot-dashed curve, $\chi^2/\text{dof} = 80.9/16$), D → Ke$\nu$ (solid curve, $\chi^2/\text{dof} = 23.4/16$), D → K*(890)e$\nu$ (dashed curve, $\chi^2/\text{dof} = 53.8/16$). (b) Contributions from D → Keν (55%), D → K*eν (39%) and D → $\pi e\nu$ (6%) with $\chi^2/\text{dof} = 11.2/15$.

5–79
360883

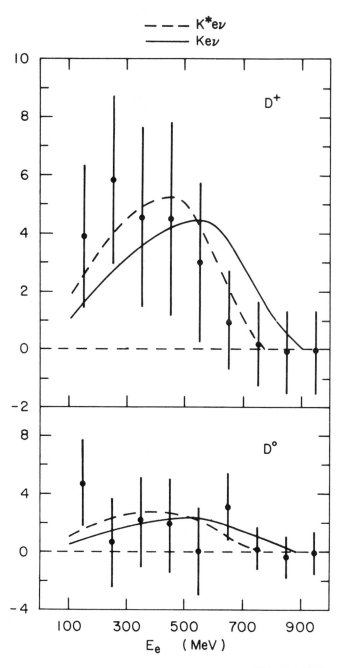

XBL 801-7872

Fig. 16. Electron spectra from tagged events
(Mark II).

CRYSTAL BALL - PRELIMINARY RESULTS

GAMMA GAMMA MASS

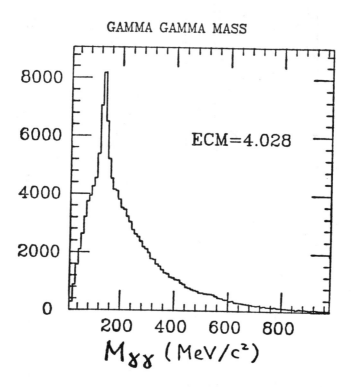

Fig. 17. $\gamma\gamma$ mass from the Crystal Ball experiment.

with a clear π^0 signal from the Crystal Ball experiment. The $\gamma\gamma$ system is fit to the π^0 mass. The π^0 events are selected by a cut at $\chi^2 < 10$. The π^0 energy is then defined as $E_{\pi^0} = E_{\gamma_1} + E_{\gamma_2}$ as obtained from the accepted fits. Figure 18 shows the resulting π^0 energy spectrum at 3.77 and 4.028 GeV respectively. While at ψ'' the π^0 energy spectrum falls off at 150 MeV, at 4.028 GeV, on the other hand, a clear peak is observed below $E_{\pi^0} = 155$ MeV as expected for D^* decay to $D + \pi^0$. Absolute branching ratios are not available at this time.

(7) <u>D Decays Into Vector Plus Pseudoscalar Final States</u>
Out of the five possible D decay modes to K$\pi\pi$:

$$D^0 \to \bar{K}^0 \pi^+ \pi^- \qquad\qquad \text{52 events, } 33 \pm 9\% \text{ background}$$
$$D^0 \to K^- \pi^+ \pi^0 \qquad\qquad \text{56 events, } 43 \pm 9\% \text{ background}$$
$$D^0 \to \bar{K}^0 \pi^0 \pi^0 \qquad\qquad \text{---}$$
$$D^+ \to K^- \pi^+ \pi^+ \qquad\qquad \text{292 events, } 12 \pm 2.4\% \text{ background}$$
$$D^+ \to \bar{K}^0 \pi^+ \pi^0 \qquad\qquad \sim \text{10 events}$$

there are only three decay modes for which the statistics are adequate to study the Dalitz plot in this Mark II data.[7]

$\cdot$D^0 Decays. The $K^0\pi^+\pi^-$ Dalitz plot is a superposition of D^0 and $\bar{D}^0$ decay and shows clear K^* bands. There is very little $K^0\rho^0$ production. The Dalitz plot and $M^2(K\pi)$ projection are shown in Figs. 19a and d respectively. The $K^\pm\pi^\mp\pi^0$ Dalitz plot shows a strong enhancement at one end of the $\rho^\pm$ band, Fig. 19b. The appearance of this plot is influenced by the π^0 detection efficiency which falls off strongly for low π^0 momenta. The reaction gives large $\rho^\pm K^0$ production with little $K^*\pi$ production. The $M^2(\pi^\pm\pi^0)$ projection is shown in Fig. 19e.

These reactions were fitted to a Breit-Wigner amplitude for the ρ and each of the appropriate K^*'s as well as a nonresonant amplitude. In all the fits the variation of the detection efficiency and acceptance across the Dalitz plot was taken into account.

The different amplitudes were added with relative phases, taken as parameters in the fits. In general the fits were not sensitive to the interference phases within the available statistics. Table VI gives the resulting branching fractions for each channel in the absence of interference between channels.

D^0 decay to the $K^*\pi$ and ρK channels can proceed via the $I = 1/2$ and $I = 3/2$ states, while D^+ decay is restricted to the $I = 3/2$ state. For $I = 1/2$,

$$\frac{\Gamma(K^{*-}\pi^+)}{\Gamma(\bar{K}^{*0}\pi^0)} = \frac{\Gamma(K^-\rho^+)}{\Gamma(\bar{K}^0\rho^0)} = 2 \quad .$$

For the K^* channels it thus appears that $I = 1/2$ is dominant (within the experimental errors). For the ρ ratios however an appreciable $I = 3/2$ contribution appears necessary in view of the small mass of the $\bar{K}^0\rho^0$ channel. This would predict a large $D^+ \to \bar{K}^0\rho^+$ branching ratio, a channel for which there is not enough data in the Mark II experiment.

$\cdot$D$^+$ Decays. For the decay $D^+ \to K^-\pi^+\pi^+$ the detection efficiency is quite uniform over the Dalitz plot, except for some drop-off in the regions for low K and π momenta. We note a very clear and distinct

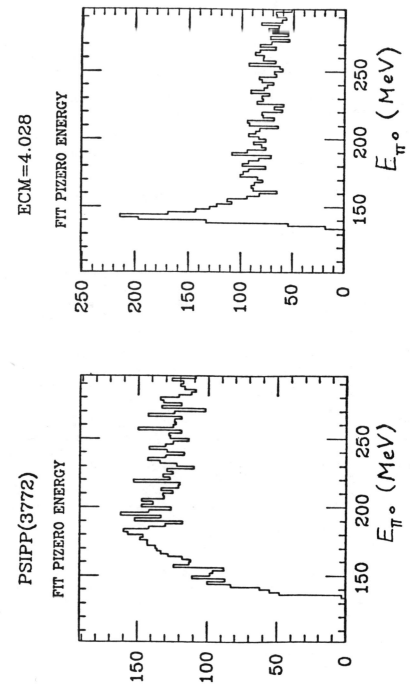

Fig. 18. Fitted π^0 energies at the ψ'' and at 4.028 GeV.

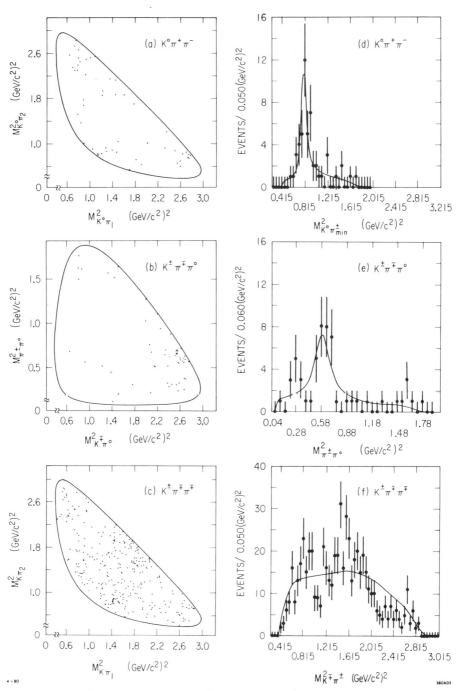

Fig. 19. D → Kππ (Mark II data).

248

Table VI. Summary of pseudoscalar vector branching fractions (%).

Three-body state	$\bar{K}^0 \pi^+ \pi^-$	$K^- \pi^+ \pi^0$	$\bar{K}^0 \pi^0 \pi^0$
Nonresonant $K\pi\pi$	$1.1\ ^{+0.7\ +0.2}_{-0.6\ -0.2}$	$< 1.3^a$	No data
$K^{*-} \pi^+$	$4.0\ ^{+1.1\ +0.3}_{-1.2\ -0.3}$	$1.8\ ^{+1.6\ +1.3}_{-1.2\ -0.5}$	$\times$
$\bar{K}^0 \rho^0$	$0.1\ ^{+0.3\ +0.1}_{-0.1\ -0.1}$	$\times$	
$\bar{K}^{*0} \pi^0$	$\times$	$1.4\ ^{+1.6\ +1.3}_{-0.9\ -1.3}$	No data
$K^- \rho^+$	$\times$	$7.2\ ^{+1.9\ +0.8}_{-2.1\ -0.7}$	$\times$

[a] At 90% C.L.
$\times$ Forbidden.

structure across the Dalitz plot in Fig. 19c. Figure 19f gives the $M^2(K^-\pi^+)$ projection where the curve represents phase space. We note very little K^{*0} production although there is a distinct apparently nonresonant structure to the Dalitz plot. Here the $\pi^+\pi^+$ must lie in an $I = 2$ state and hence either in $L = 0$ or $L = 2$. Furthermore, Bose symmetrization must be applied to the state. The details of this structure are not fully understood at present.

We can set an upper limit to the K^* production of 0.39 at a 90% C.L. This limit is a considerable overestimate as all events in the mass region of the K^* band were considered as K^*'s.

(8) New D^0 and D^+ Mass Values

The D^0 and D^+ mass measurements came from the beam energy constrained mass distributions at the ψ'' as obtained in the LGW experiment. The D^{*+} mass comes from the study of the $D^{*+} \to \pi^+ D^0$ decay, the D^{*0} mass from the study of the recoil spectra. The last two results come from the SLAC-LBL Mark I experiment. Figure 20 shows the well-known decay scheme and gives mass values from Mark I and the LGW experiments.

The Mark II detector has now repeated the mass measurements at the ψ'' with higher statistics and better resolution (drift chamber rather than spark chambers). Figure 21 shows the beam energy constrained mass spectra for the three reactions:

$$D^0 \to K^- \pi^+$$
$$D^0 \to K^- \pi^+ \pi^- \pi^+$$

and
$$D^+ \to K^- \pi^+ \pi^+$$

in the 1 MeV/c^2 mass bins.[7] The results are:

$$M_{D^0} = 1863.8 \pm 0.5 \text{ MeV/}c^2$$
$$M_{D^+} = 1868.4 \pm 0.5 \text{ MeV/}c^2$$

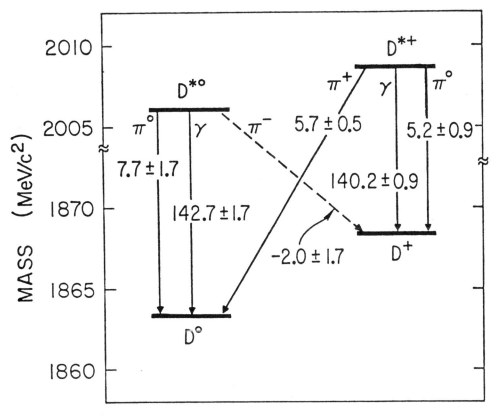

XBL 7812-13692

Fig. 20. Mass level diagram for D* and D states. The arrows represent different decay **modes** of the D*; the numbers across the lines represent the Q for each decay expressed in MeV. The decay $D^{*0} \rightarrow D^+\pi^-$ is kinematically forbidden. The masses as obtained from the Mark I and LGW experiments are (D^0) 1863.3 ± 0.9, (D^+) 1868.3 ± 0.9, (D^{*0}) 2006.0 ± 1.5 and (D^{*+}) 2008.6 ± 1.0 MeV/c², ΔM_D = 5.0 ± 0.8 MeV/c², ΔM_{D^*} = 2.6 ± 1.8 MeV/c².

250

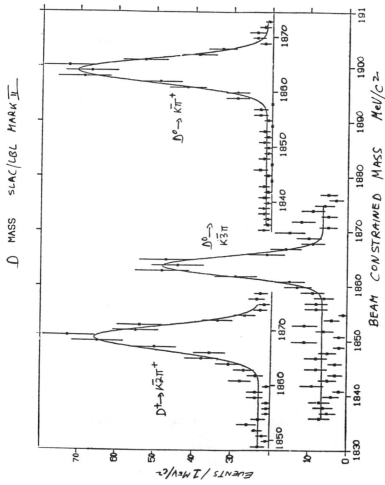

Fig. 21. Beam-energy-constrained mass used for D^0, D^+ mass determination in the Mark II experiment.

where both statistical and systematic errors are taken into account. Some of the systematic errors cancel in the mass difference:

$$\Delta M_D = M_{D^+} - M_{D^0} = 4.7 \pm 0.3 \text{ MeV/c}^2 .$$

It is interesting to note that the precision in ΔM_D is beginning to approach the accuracy of the $K^0 - K^+$ mass difference:

$$\Delta M_K = M_{K^0} - M_{K^+} = 4.01 \pm 0.13 \text{ MeV/c}^2 \text{ (note sign convention!)} ,$$

although not yet that of the $\pi^+ - \pi^0$ mass difference:

$$\Delta M_\pi = M_{\pi^+} - M_{\pi^0} = 4.6043 \pm 0.0037 \text{ MeV/c}^2 .$$

The remarkable coincidence between the magnitudes of the meson mass differences is just that -- a coincidence. The π^+ has the same quark content $u\bar{d}$ as the π^0 (a mixture of $u\bar{u} + d\bar{d}$). Thus the mass difference for the pion is ascribed purely to electromagnetic effects. The K^0 ($d\bar{s}$) and K^+ ($u\bar{s}$) differ in quark content giving a non-electromagnetic $M_d - M_u$ contribution to the $K^0 - K^+$ mass difference to which the electromagnetic effect for the kaon must be added. Here M_d and M_u are quark masses. This electromagnetic part or "photon exchange" part of the K mass splitting can be calculated from the π mass splitting.[12] From the π and K mass differences one can deduce the $M_d - M_u$ mass difference ($\sim 5 \text{ MeV/c}^2$).

The D^+ ($c\bar{d}$) and D^0 ($c\bar{u}$) again have the $M_d - M_u$ mass difference term to which a positive electromagnetic effect for the D must be added. The new improved experimental precision in ΔM_D suggests that the $M_d - M_u$ mass difference should be less than 5 MeV. A later estimate by Weinberg[13] indeed gives $M_d - M_u \approx 3.3$ MeV. With the new higher precision experimental value for ΔM_D the entire set of the above mass differences should now be re-examined.

(9) Determination of R_D (Preliminary Result)

The scan data taken with the Mark II detector allow a cross-section determination for D mesons. Figure 22 shows the distribution of R_C, the charmed cross section to μ pair cross-section ratio as determined from inclusive studies in the DELCO experiment. Superimposed is (the heavy bars) the result for the Mark II data obtained from D^0 and D^+ cross-section measurements.[14]

Figure 23 shows $R_{hadronic}$ from the Mark II experiment (i.e., the τ cross section has been subtracted). Here again we superimpose R_D but measured from a fixed (old physics) value of $R = 2.5$.

The conclusion from these results is that within the errors the preliminary R_D values appear to account for most of the charm cross section.

(10) Evidence for the F Meson

The DASP collaboration at DORIS have used their γ detecting facility, namely, the central detector to look for the F meson and have found evidence for it.[15] The evidence consists of two steps: first, the observation of a signal for η mesons -- their method of detecting η's is to look for the $\eta \rightarrow \gamma\gamma$ decay. They did this as a function of incident energy and find an inclusive η signal at 4.42 and a smaller one at 4.16 GeV. Their event selection criteria were: more than two photons of energy > 140 MeV, two charged prongs

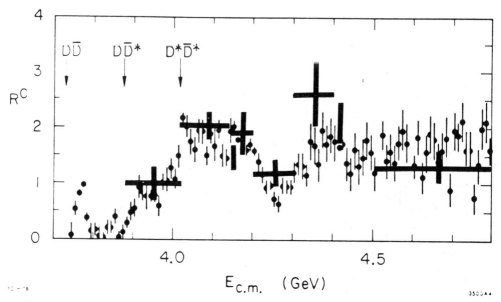

Fig. 22. DELCO data for R_C. Mark II data for R_D superimposed (heavy bars).

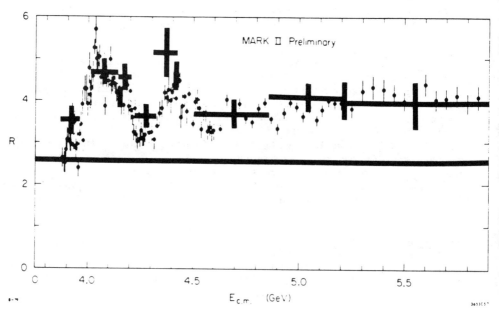

Fig. 23. Mark II data for $R_{hadronic}$ with $R_D + 2.5$ superimposed (heavy bars).

with energies greater than 100 MeV each, and no more than six photons.
For these events they reconstructed the $m_{\gamma\gamma}$ mass for all γ pairs.
This distribution is shown in Fig. 24 as a function of incident energy.
The η signal becomes clearer when they demand either a coincidence
with an electron (Fig. 26) or (at 4.42) an additional low energy gamma
ray, γ_{low} (Fig. 25). Here "low energy" stands for $E_\gamma < 140$ MeV.
    Among the $\eta\gamma_{low}$ events, they now selected a subset with an identi-
fied charged pion of momentum greater than 600 MeV/c. They found a
total of 43 events which satisfied this requirement, as well as additional quality requirements, at 4.42 GeV and 79 events at all other
energies. These were fitted to the two reactions below:

$$
\begin{array}{ll}
e^+e^- \to F^+ & F^{*-} \\
\quad\ \ \lfloor \to \eta\pi^+ & \quad \lfloor \to \gamma_{low}F^- \\
\qquad\qquad \lfloor \to \gamma\gamma &
\end{array}
\tag{a}
$$

and

$$
\begin{array}{ll}
e^+e^- \to F^{*+} & F^{*-} \\
\quad\ \ \lfloor \to \gamma_{low}F^+ & \quad \lfloor \to \gamma_{low}F^- \\
\qquad\qquad\ \ \lfloor \to \eta\pi^+ & \\
\qquad\qquad\qquad \lfloor \to \gamma\gamma &
\end{array}
\tag{b}
$$

The fit is a 2C fit as $m_{\gamma\gamma}$ is constrained to the η mass and the F^+F^-
masses are constrained to be equal. The found 15 events at 4.42 GeV
and 11 events at the other energies, which fitted hypothesis (a) with
a $\chi^2 < 8$. These were reduced to 12 and 10 events respectively by a
further quality cut. At 4.42 GeV they obtain a tight cluster of 6
events at $M(\eta\pi) = 2.04$ GeV and $M_{recoil} = 2.15$ GeV for hypothesis
(a). No such cluster is observed at the other energies. The same 6
events also fitted hypothesis (b) with slightly lower mass values of
2.00 and 2.10 GeV respectively. They thus take the average values and
quote $M(F) = 2.03 \pm 0.06$ GeV and $M(F^*) = 2.14 \pm 0.06$ GeV. The mass
difference between F^* and F is directly determined from the energy of
γ_{low}. This gives $M(F^*) - M(F) = 0.110 \pm 0.046$ GeV. The cluster in
Fig. 27 looks impressive, but the background observed at neighboring
energies is not negligible. Clearly more data on the F is needed.

ACKNOWLEDGMENT

    I want to thank Mrs. C. Frank-Dieterle for her help and meticulous
care in preparing and compiling this manuscript and Ms. Lora Ludwig for
her help in preparing the illustrations.
    This work was supported primarily by the U. S. Department of Energy
under Contract No. W-7405-ENG-48.

REFERENCES

1. G. Goldhaber et al., Phys. Rev. Lett. 37, 255 (1976); I. Peruzzi
   et al., Phys. Rev. Lett. 37, 569 (1976).
2. P. A. Rapidis et al., Phys. Rev. Lett. 39, 526 (1977).
3. W. Bacino et al., Phys. Rev. Lett. 40, 671 (1978).
4. A. Barbaro-Galtieri in Advances in Particle Physics, Vol. 2, R.
   Cool and R. Marshak, Editors, p. 193 (1968).
5. E. Eichten et al., Phys. Rev. D17, 3090 (1978); T. Appelquist et
   al., Ann. Rev. Nucl. Sci. 29, 387 (1978).

6. I. Peruzzi et al., Phys. Rev. Lett. 39, 1301 (1977).

7. R. H. Schindler et al., Phys. Rev. D, in preparation (1980).

8. G. S. Abrams et al., Phys. Rev. Lett. 43, 481 (1979).

9. J. Kirkby, 1979 International Symposium on Lepton and Photon ⟨illegible⟩ ⟨illegible⟩, 1979; SLAC-PUB-2419.

10. G. Goldhaber et al., Phys. Lett. 69B, 503 (1977).

11. H. Sadrozinski, private communication (1980).

12. See for example K. Lane and S. Weinberg, Phys. Rev. Lett. 37, 717 (1976).

13. S. Weinberg, "Rabi Festschrift," Transactions of the New York Academy of Sciences, Series 2, 38, 185 (1977).

14. M. W. Coles, Ph.D. thesis, University of California, Berkeley, CA, in preparation (1980).

15. R. Brandelik, Zeitschrift für Physik C, Particles and Fields 1, 233 (1979).

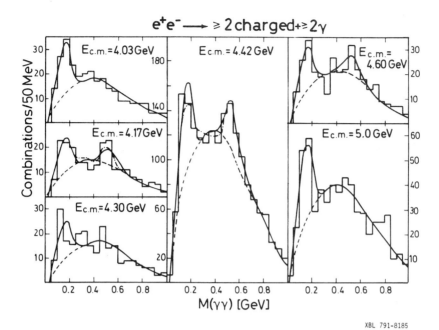

Fig. 24. DASP data on inclusive η production.

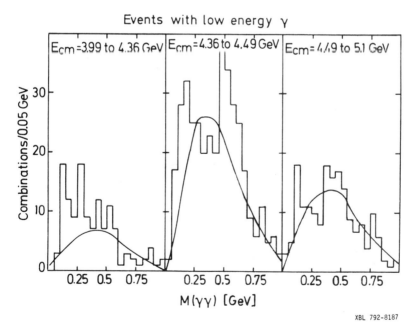

Fig. 25. DASP data on η production for events with low
energy γ-rays.

256

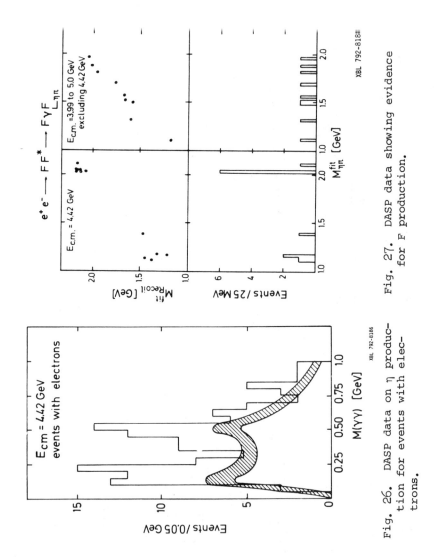

Fig. 27. DASP data showing evidence
for F production.

Fig. 26. DASP data on η production for events with electrons.

D MESON PRODUCTION BY PHOTONS AND NEUTRINOS*

James E. Wiss
Department of Physics
University of Illinois at Urbana-Champaign, IL 61801

ABSTRACT

We discuss the explicit production of D and D^* mesons by neutrinos and photons. Neutrino production of charmed mesons appears to occur at the level of $\simeq$ 10% of the total charged current rate in agreement with indications from neutrino induced dilepton studies. Two photoproduction experiments observe D meson production at the level of a few hundred nanobarns per nucleon but disagree substantially on the production mechanism.

INTRODUCTION

The last two years have produced considerable evidence for the observation of D^0 and D^{*+} mesons in high energy photon and neutrino interactions. I will limit my discussion to explicit observations of D^0's and D^{*+}'s decaying into exclusive hadronic final states through enhancements in mass distributions -- thus leaving the burgeoning field of charm observation by finite lifetime methods to Professor Prentice. We begin by discussing neutrino experiments.

NEUTRINO PRODUCTION

1. Results from the Columbia-Brookhaven Collaboration

Figure la shows the $K_s \pi^+ \pi^-$ invariant mass distribution obtained by the Columbia-Brookhaven Collaboration[1] in approximately 1/2 of their exposure of 150,000 charged current events observed in the 15' helium filled bubble chamber at FNAL.

This data required the presence of a muon with momentum exceeding 2 GeV/c in order to reduce the background due to non-interacting pions. An $\simeq 4\sigma$, 64 event excess is observed with a mass of 1850 ± 15 MeV -- consistent with the known mass of the D^0. The σ = 20 MeV/c^2 width of the enhancement is consistent with their experimental resolution. No such peak is observed in Fig. lb which shows the $K_s \pi^+ \pi^-$ mass distribution for "muonless" neutral current events. This absence reflects both the relative scarcity of charm quarks in the nucleon as well as the lack of charm changing neutral currents.

Correcting for efficiency, and the $\bar{K}_0 \rightarrow \pi^+ \pi^-$ branching fraction, they obtain the result:

$$\frac{\nu Ne \rightarrow \mu^- (D^0 \rightarrow \bar{K}_0 \pi^+ \pi^-) X}{\nu Ne \rightarrow \mu^- all} = (.7 \pm .2)\% .$$

*This work was supported in part by the U. S. Department of Energy under contract DE-ACO2-76ERO1195

258

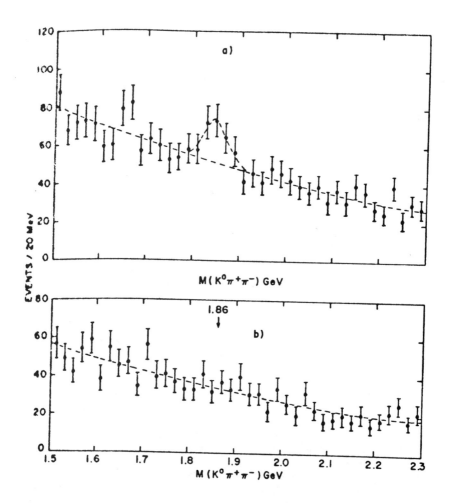

Figure 1

The $K^0\pi^+\pi^-$ invariant mass distribution obtained in the Columbia-Brookhaven Collaboration's bubble chamber ν exposure for events (a) with an additional μ^- and (b) without an additional μ^-. The enhancement present in Figure 1a consists of 64 events centered at 1850 ± 15 MeV/c^2.

Using the $D^o \to \bar{K}^o \pi^+ \pi^-$ branching fraction of $(3 \pm .8)\%$[2] one obtains the result that $(23 \pm 10)\%$ of charged current events in their data contain a D^o. Since the publication of Reference 1, the Columbia-Brookhaven Collaboration has analyzed the remaining half of their data, and indications[3] are that the fraction of charged current events with a $D^o \to \bar{K}^o \pi^+ \pi^-$ decay will drop to $(.4 \pm .15)\%$. This new value is in good agreement with the $\simeq 10\%$ level of total charm production suggested by the rate for ν induced dilepton events.

2. Results from the ABCMO Collaboration

Information on the neutrino production of the D^{*+} comes from the Aachen-Bonn-CERN-Munich-Oxford Collaboration[4] among ~ 6000 charged current events obtained in the hydrogen filled BEBC bubble chamber at CERN. This data required at least 5 GeV of visible energy and a negative muon with momentum exceeding 3 GeV/c.

Two charged current events were observed containing a D^{*+} candidate where all final state particles were identified. The two interactions were of the form $\nu P \to \mu^- P K^- \pi^+ \pi^+$ and $\nu P \to \mu^- P K^- 3\pi^+ 2\pi^-$ with the K^- identified via the secondary interaction $K^- P \to \Sigma^- \pi^+$. Both processes explicitly violate the $\Delta S = \Delta Q$ rule, thus suggesting charm production from a proton valence quark. In addition both processes have a $K^- \pi^+$ combination with a mass consistent with the mass of the D^o, as well as $(K^- \pi^+)\pi^+$ combination with the known mass of the D^{*+}. The well resolved $(K^- \pi^+)\pi^+$, $K^- \pi^+$ mass difference is within a fraction of an MeV of the known $D^{*+} - D^o$ mass difference.

A statistically enhanced sample of D^{*+} candidates is obtained by forming neutral multitrack mass combinations where every negatively charged track is treated in turn as a K^-, and all other tracks are treated as pions. Those neutral mass combinations falling in the range from 1840-1890 MeV/c^2 were considered as possible D^o candidates and an additional positive track in the event was combined with the D^o candidate to form a possible D^{*+} candidate. The resulting D^{*+}, D^o mass difference distributions for the indicated topologies are shown in Figs. 2a and 2b. Figure 2c shows the mass difference distribution obtained for $(K_s \pi^+ \pi^-)\pi^+$ D^{*+} candidates, while Fig. 2d shows the sum of the mass distributions of Fig. 2a through Fig. 2c. The total D^{*+} signal present in Fig. 2d consists of 7 ± 4 events. After correcting for efficiencies and the individual branching fractions, the ABCMO Collaboration finds that $(4.1 \pm 2.4)\%$ of charged current interactions above D^{*+} threshold (65% of all charged current interactions) have a D^{*+} in the final state.

260

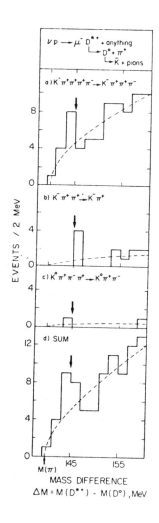

Figure 2

Distribution of the D^{*+}, D^0 mass difference for possible D^0, D^{*+} candidates obtained by the ABCMO Collaboration in their bubble chamber ν exposure. For Fig. 2a) and 2b) each negative track is used in turn as a K^-.

PHOTOPRODUCTION

1. Results from the CERN WA4 Collaboration

I shall review the photoproduced D meson results from this British, French, and German collaboration[5] allowing Dr. Kummar to discuss their recent results on the F^+ meson. The experiment was performed using tagged 20-70 GeV photons impinging on a 67 cm liquid hydrogen target. The photoproduced final state was analyzed by the CERN Ω spectrometer with kaon identification provided from 5.7 to 19 GeV/c by a multicell Cerenkov counter, and neutral particle detection provided by a large lead glass shower counter array. The data discussed here consists of $\simeq$ 300,000 events with charged kaon and an incident photon energy exceeding 40 GeV.

Figure 3 shows the $K^{\pm}\pi^{\mp}$ and $K^{\mp}\pi^{\pm}\pi^0$ mass distributions obtained by the WA4 collaboration. Evidence is found for the photoproduction of $\bar{D}^0 \to K^+\pi^-$, $K^+\pi^-\pi^0$ but there is no evidence for $D^0 \to K^-\pi^+$, $K^-\pi^+\pi^0$ production. These results differ from those of the Columbia-Illinois-Fermilab collaboration which we shall discuss shortly. This group, operating at larger average photon energies, finds nearly equal D and $\bar{D}$ signals.

As a possible explanation for the charm asymmetry observed by the WA4 group, they theorize that anticharmed mesons are produced in association with charmed baryons. In order to further test this hypothesis, in Fig. 4 the WA4 group histograms the invariant mass for various $\bar{D}^0$ topologies subject to the requirement of an additional, heavy, positively charged particle produced against the $\bar{D}^0$. This heavy particle is Cerenkov identified as either a K^+ or P; only the P interpretation makes sense in their model, however. An improvement in signal to noise is seen in Fig. 4 compared to Fig. 3 because of the recoil proton requirement. Again these results disagree with those of the Columbia-Illinois-Fermilab Collaboration to be discussed shortly.

The WA4 Collaboration extracts a 525 ± 140 nb cross section for $\bar{D}^0$ production over their energy range from 40 to 70 GeV using the $\bar{D}^0 \to K^+\pi^-$ signal of Fig. 3. Their acceptance is nearly flat for positive $\bar{D}$ Feynman x_F values; hence this value should be reasonably model independent for forward hemisphere production. Forward hemisphere production is preferred by the WA4 data because it gives consistency between the level of the $K^-\pi^+\pi^0$ signal and the $K^-\pi^+$ signal.

Finally the WA4 group has extracted the cross section for $\gamma P \to \bar{D} \Lambda_c x$, using the $\bar{D}$ subsample with a recoil proton. They obtain a cross section of $\sigma_{\gamma P \to \bar{D} \Lambda_c x}$ = 525 ± 160 nb which is in excellent agreement with their completely inclusive cross section. This again indicates that the WA4 data is primarily associated production (with a charmed baryon) rather than D meson pair production.

262

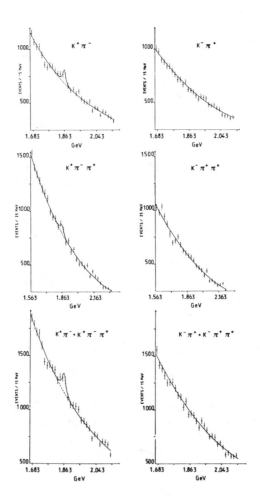

Figure 3

$\bar{D}^0$ signals obtained in the CERN WA4 Collaboration's photoproduction experiment. The data requires the presence of a charged kaon identified by their multicell Cerenkov counter and an incident photon energy exceeding 40 GeV. Although signals appear for $\bar{D}^0$ topologies (left hand side), no signals appear for D^0 topologies (right hand side).

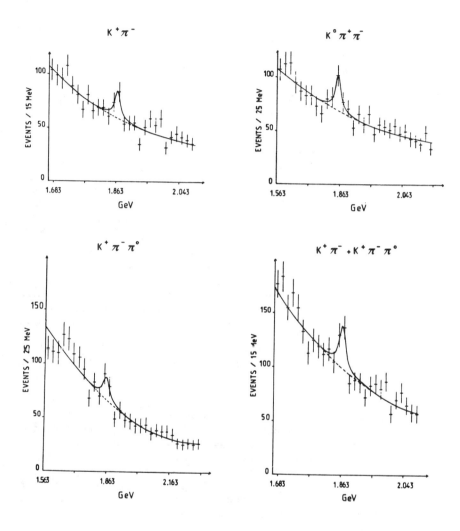

Figure 4

$\overline{D}^o$ signals obtained in the WA4 Collaboration's photoproduction experiment. These signals require the presence
of an additional recoil K^+ or $\overline{P}$ candidate in the event.
The data suggest photoproduced $\overline{D}^o$'s are accompanied by
charmed baryons.

2. Results from the Columbia-Illinois-Fermilab Collaboration

I turn next to the observation of photoproduced D^{*+}'s by the Columbia-Illinois-Fermilab Collaboration in their exposure of 6×10^{11} high energy (> 50 GeV) photons at the Fermilab wide band photon beam facility.[6] Because I am a member of this collaboration, I will frequently employ the first person plural in the discussion to follow.

Figure 5 shows the roughly exponential photon spectrum obtained in the wide band beam. The beam has an $\simeq 1\%$ contamination from K_ℓ's and neutrons which forms a major background to the photoproduced multihadronic data sample discussed here. The beam impinges on an $\simeq 2"$ segmented active scintillator target. The resulting interactions are analyzed by the two magnet, 5 proportional chamber, spectrometer shown in Fig. 6. Two multicell Cerenkov counters separate kaons and protons from pions within a momentum range from 6 to 40 GeV/c.

In order to avoid unwieldy notation in the discussion which follows, reference to a given state will imply a reference to its charge conjugate state.

Figure 7 shows the $\Delta \equiv M_{(K^-\pi^+)\pi^+} - M_{K^-\pi^+}$ mass distribution in one MeV/c^2 bins for all $(K^-\pi^+)\pi^+$ combinations with a $K^-\pi^+$ submass satisfying: (7a) $1.800 < M_{K\pi} < 1.825$ (below the D^0 mass), (7b) $1.850 < M_{K\pi} < 1.875$ (straddling the D^0 mass), and (7c) $1.900 < M_{K\pi} < 1.925$ (above the D^0 mass). Other than the Cerenkov requirement on the kaon, the data is completely uncut. A clear enhancement is observed at a mass difference of 145.5 MeV/c^2 for combination with a $K^-\pi^+$ mass near the D^0. This value is in excellent agreement with the known D^{*+}, D^0 mass difference of 145.3 MeV/c^2 obtained at the Mark I magnetic detector at SPEAR.[7] We thus attribute the peak of Fig. 7b to the process:

$$D^{*+} \to \pi^+ D^0$$
$$\quad\quad\quad \hookrightarrow K^-\pi^+ .$$

The shaded portion of Fig. 7 shows the appropriately normalized Δ distribution obtained in special background runs where the high energy photons in the beam are attenuated with six radiation lengths of lead leaving only the hadronic contamination. We see that the shaded region of Fig. 7b forms a nearly perfect background to the signal in the unshaded region. This demonstrates that the D^{*+} is indeed photoproduced and would appear very clean in the absence of hadronic beam contamination.

In Fig. 8 we show the $K^\mp\pi^\pm$ mass distribution (8a) and the $K_s\pi^+\pi^-$ mass distribution (8b) for combinations within the D^*, D mass difference peak. Because of the larger combinatoric background for the $(K_s\pi^+\pi^-)\pi^\pm$ signal we only plot combinations from events with less than 8 visible tracks.

A fit to the data of Fig. 8 reveals a signal of 143 ± 20 events at a mass of 1860 ± 2 MeV/c^2 for the $K^-\pi^+$, and a signal of 35 ± 13 events at a mass of 1869 ± 4 MeV/c^2 for the $K_s\pi^+\pi^-$ signal. The

265

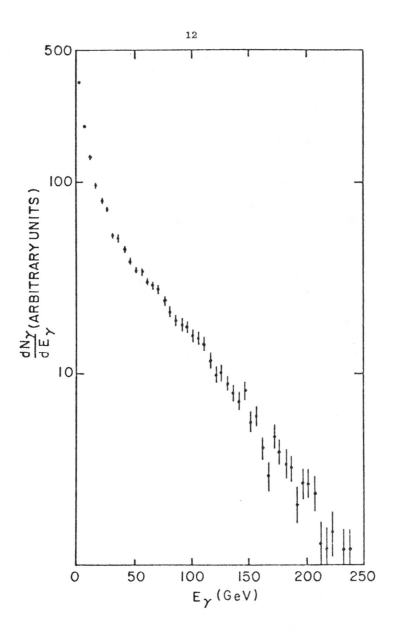

Figure 5

Photon spectrum obtained by the Columbia-Illinois-
Fermilab wide band photoproduction experiment.

266

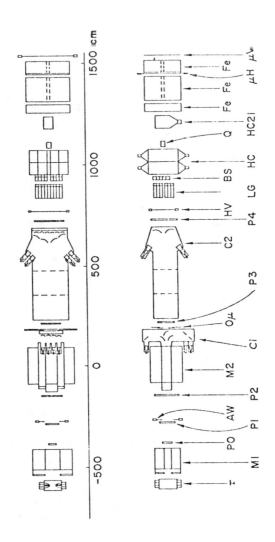

Figure 6

The multiparticle spectrometer employed by the
Columbia-Illinois-Fermilab Collaboration for their study
of photoproduced multihadronic final states. Major
components include two bending magnets (M_1 and M_2),
two Cerenkov counters (C_1 and C_2), and five multiwire
proportional chambers (P_0 through P_4).

267

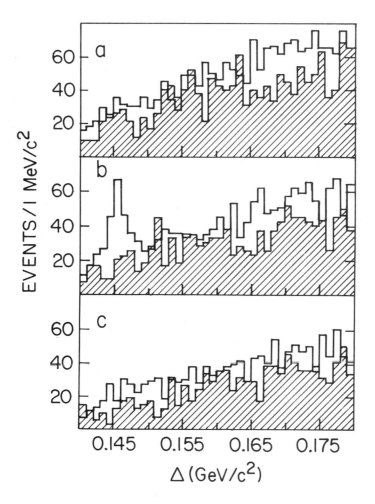

Figure 7

Mass difference distributions ($\Delta \equiv M_{K\pi\pi} - M_{K\pi}$) obtained
by the Columbia-Illinois-Fermilab Collaboration for combinations
with a $K\pi$ mass (a) below, (b) straddling, and (c) above the
known mass of the D^0. Both charm and anticharm states are
included in this plot. The shaded distributions show the
appropriately normalized contributions from hadronic contamination in the photon beam.

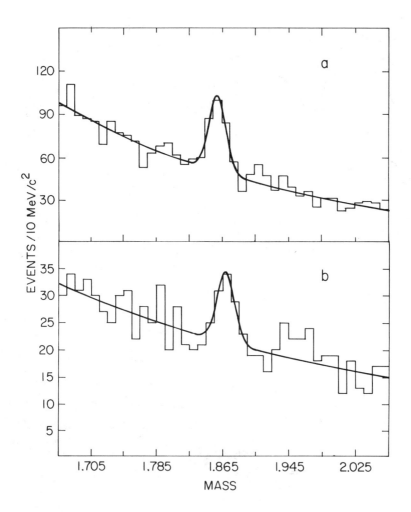

Figure 8

The $K^{\mp}\pi^{\pm}$ (a) and $K^{0}\pi^{+}\pi^{-}$ (b) invariant mass distributions obtained by the Columbia-Illinois-Fermilab Collaboration for combinations within the $D^{*\pm}$ mass difference peak shown in Fig. 7b.

width of the Gaussian peaks are $\sigma = 10$ MeV/c^2 and 12 MeV/c^2 for Fig. 8a and 8b respectively which is consistent in both cases with experimental resolution. After correcting for the relative efficiency of these two final states, the data of Fig. 8 implies the branching ratio:

$$\frac{\Gamma(D^0 \to \bar{K}_0 \pi^+ \pi^-)}{\Gamma(D^0 \to K^- \pi^+)} = 1.7 \pm .8$$

which straddles the Lead Glass Wall Collaboration's value of 1.82 ± .8 and the Mark II Collaboration's value of .96 ± .3.[2]

Figure 9 demonstrates that while there is an enhancement in the $(K^- \pi^+)\pi^+$ channel (9a) corresponding to the D^{*+}, there is no enhancement in the non-exotic $(K^+ \pi^-)\pi^+$ channel (9b). An enhancement in Fig. 9b could arise from either doubly suppressed Cabibbo decay of the D^0, or the conjectured $D^0 - \bar{D}^0$ mixing process. The data of Fig. 9 implies that the fraction of times that a D^0 decays via $K^- \pi^+$ rather than $K^+ \pi^-$ is less than 11% at the 90% confidence level which can be compared to 16% -- the limit obtained by the SPEAR Mark I Collaboration.

In contrast to the WA4 Collaboration, which finds no evidence for the photoproduction of charm = + 1 D mesons, we find that the ratio D^{*+} to D^{*-} in our data is 1.4 ± .4. Our data thus favors a picture where charmed mesons are pair produced by photons rather than produced in association with charmed baryons. Using the Monte Carlo deduced average efficiency of 6% we obtain a spectral averaged inclusive D^{*+} photoproduction cross section of:

$$\sigma_{\gamma P \to D^{*+}_x} \; Br(D^{*+} \to \pi^+ D^0) \; Br(D^0 \to K^- \pi^+) = 1.8 \pm .6 \text{ nb/nucleon}$$

where we have assumed a linear A dependance. Included in the error is a ± 20% uncertainty in detection and trigger efficiency due to model uncertainty. Using the values[2] $Br(D^{*+} \to \pi D^0) = .6 \pm .15$, $Br(D^0 \to \bar{K} \pi^+) = .026 \pm .004$ we obtain the result

$$\sigma_{\gamma P \to D^{*+}_x} = 118 \pm 49 \text{ nb/nucleon}.$$

In order to estimate the ratio of D^{*+}'s to D^0's in our data (and thus estimate the photoproduction cross section for D^0 mesons), we fit the uncut, inclusive $K^- \pi^+$ invariant mass distribution. We see an $\simeq 3\sigma$ excess consisting of 660 ± 230 events over a background of $\simeq 33,000$ events with a mass and width compatible with that of Fig. 8a. Correcting for the relative D^{*+}, D^0 efficiency we find that $.26\vert^{+.13}_{-.06}$ of photoproduced D^0's come from D^{*+} decay in good agreement with .25 ± .09, the fraction obtained in the higher energy $e^+ e^-$ annihilation data reported in Ref. 7. Correcting for the $D^{*+} \to \pi^+ D^0$ branching ratio we find that the D^{*+}/D^0 ratio in our data is $.4\vert^{+.22}_{-.10}$. We thus deduce that the inclusive photoproduced D^0 cross section is 295 ± 130 nb/nucleon. An equal amount of $\bar{D}^0$ cross

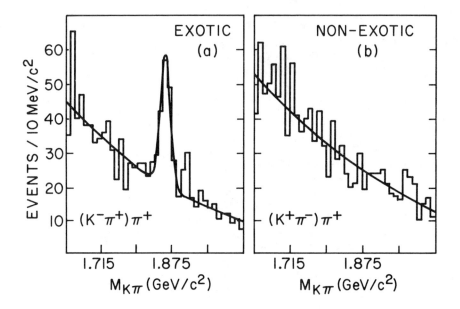

Figure 9

The Kπ invariant mass distribution for exotic (a) and
non-exotic (b) combinations satisfying a tight Δ cut. The
events containing these combinations have seven or less tracks.
This data sets a limit on D^0-$\bar{D}^0$ mixing of less than 11% (90%
CL). Data is from the Columbia-Illinois-Fermilab Collaboration.

section is expected to exist. We thus are seeing D^0 mesons at a level comparable to the WA4 group who quote a $\bar{D}^0$ inclusive cross section of 524 ± 140 nb/nucleon.

RECOIL PARTICLE STUDIES

If D^{*+} mesons are photoproduced in pairs, as our data suggests, one expects to see strange particles recoiling against them. In 143 ± 20 photoproduced D^{*+} events we observe:

Table I

33 ± 9 produced against a			K^+
4 ± 4 "	"	"	K^-
2 ± 2 "	"	"	K_s
3 ± 3 "	"	"	Λ
2 ± 3 "	"	"	$\bar{P}$
4 ± 4 "	"	"	μ^-

We combine statistics for particle and antiparticle states in Table I, but the charge correlations are as implied. The 33 ± 9 D^{*+} events with an additional K^+ candidate include cases where the heavy particle is Cerenkov identified as either a kaon or proton as well as a smaller sample where it is unambiguously identified as a kaon. Table I is not corrected for acceptance or false particle identification probability. Because each entry of Table I is obtained from a separate fit to the $K^-\pi^+$ invariant mass spectrum, Table I is background subtracted. Although we see that the D^{*+} is photoproduced against an assortment of provocative particles, statistical considerations limit us to discussing production against the K^+.

Figure 10 a) shows the $K^-\pi^+$ mass distribution for events with a recoil K^+ or $(K/P)^+$ candidate subject to a cut on the D^*, D mass difference. The fit to this spectrum yields the 33 ± 9 events previously mentioned. Figure 10 b) shows an analogous $K^-\pi^+$ distribution for events with a recoil K^- or $(K/P)^-$ candidate, while Fig. 10 c) shows the $K^-\pi^+$ mass distribution for events with a definite P^- candidate. No discernible D^{*+} or D^{*-} signal observed in our data for events with a recoil K or P with the same charge as the kaon produced from the D^{*+} decay. We thus differ completely with the data of the WA4 Collaboration who find that the requirement of a recoil $(K/P)^+$ enhances their $\bar{D}^0$ signal.

THE PHOTON-GLUON FUSION MODEL

Our data is consistent with models where the D^{*+} recoils against another D or D^* via the decay of a diffractively photoproduced low mass (4 to 6 GeV/c^2) parent. This picture emerges from consideration of the following experimental facts:

1) The D^{*+} photoproduction cross section appears to be independent of photon energy from 50 to 200 GeV.

2) The D^{*+} tends to have $\simeq$ 1/2 of the total visible event energy. The average visible energy fraction carried by the D^{*+} is

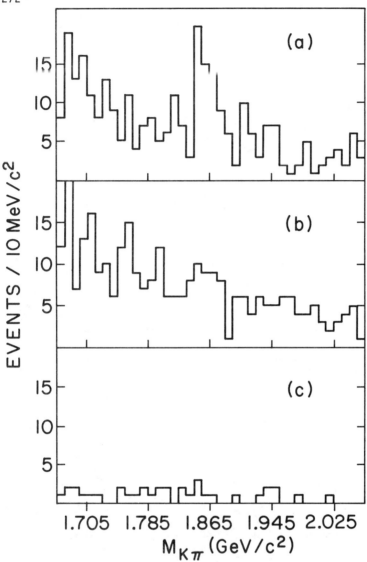

Figure 10

The $K^{\mp}\pi^{\pm}$ invariant mass distribution for combinations subject to a cut on D^{*+} D^0 mass difference for events with:
    a) a recoil $K^{\pm}_{\mp}$ or $(K/P)^{\pm}_{\mp}$
    b) a recoil $K^{\mp}$ or $(K/P)^{\mp}$
    c) a recoil $P^{\mp}$
These results from the Columbia-Illinois-Fermilab Collaboration indicate that photoproduced D^{*+}'s are produced in association with charmed mesons rather than charmed baryons.

.6 ± .1. This number systematically overestimates the true D^{*+} energy fraction because of missing energy.

3) About 3/4 of the D^{*+}'s have a $P_\perp$ (with respect to the incident photon direction) of less than 1 GeV/c. The low average D^{*+} $P_\perp$ serves to limit both the "t" spread of the parent as well as the parent's mass.

4) The multiplicity of events containing a D^{*+} is well matched by a diffractive Monte Carlo where all additional tracks in the event arise from D^{*+} or D^- decays via the Quigg-Rosner[8] statistical model. Hence charmed particles are photoproduced with few extra pions, as in the case in threshold e^+e^- annihilation.[9]

The mass and "t" distribution of the parent can be chosen to match the data. In particular, the use of the photoproduced ψ "t" distribution and a parent mass distribution falling above DD^* threshold like $1/M^N$ with $N > 10$ adequately fits the D^{*+} energy fraction distribution and the D^{*+} $P_\perp$ distribution of the data. Monte Carlo calculations based on this diffraction model have supplied the efficiencies used throughout this discussion.

We have also begun to investigate the applicability of the photon-gluon fusion model for D^{*+} photoproduction. Unfortunately, our limited statistics put most of this discussion in the qualitative sector. In photon-gluon fusion, a $c\bar{c}$ pair is created via diagrams such as that of Fig. 11. The gluon is assumed to lie on mass shell and carry a fraction "x" of the nucleon momentum given by the gluon distribution function $F_{gluon}(x)$. The cross section for the elementary process $\gamma g \rightarrow c\bar{c}$ can be directly taken over from the process $\gamma\gamma \rightarrow e^+e^-$ except for additional QCD vertex factors and the use of the charmed quark mass "M_c" rather than the electron mass. The cross section for the gluon fusion process can then be written as:

$$\frac{d\sigma}{dM^2_{c\bar{c}}} = \frac{\sigma_{\gamma g \rightarrow c\bar{c}}(M^2_{c\bar{c}})}{s} F_{gluon}\left(\frac{M^2_{c\bar{c}}}{s}\right) \qquad \text{(Eqn 1)}$$

Because of the assumed "softness" of the gluon distribution and the peaking of $\sigma_{\gamma g \rightarrow c\bar{c}}$ towards low $M_{c\bar{c}}$, the gluon fusion model tends to produce D^{*+} mesons which look "diffractive" and match the four experimental properties previously listed.

Once created, the $c\bar{c}$ pair can either emerge as a psionic bound state or as a pair of charmed particles depending on whether the $c\bar{c}$ pair mass lies above or below $D\bar{D}$ threshold. In either case we assume the $c\bar{c}$ pair will shed its color through the emission of a soft gluon with unit probability. Charmed mesons resulting from unbound $c\bar{c}$ production are assumed to retain a fraction "z" of the produced quark's energy where "z" is distributed according to the dressing function $D(z)$.

The charmed meson system can presumably acquire $P_\perp$ through the primordial $P_\perp$ of the incident gluon, through higher order QCD corrections, and through the dressing process. Photoproduced D^{*+} mesons acquire $P_\perp$ from both the poorly understood $P_\perp$ of the charmed meson system as well as from the mass of the $c\bar{c}$ system.

274

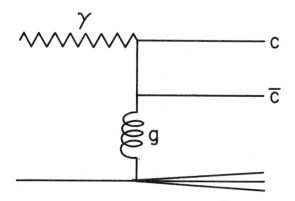

Figure 11

    Charm photoproduction by photon-gluon fusion. The gluon is considered as a free, on mass shell constituent of the nucleon. The $c\bar{c}$ system must subsequently radiate a gluon in order to emerge as a color singlet.

Unfortunately, there is a considerable experimental latitude
in many of the ingredients of the gluon fusion model. The gluon
distribution function is assumed to be of the form:

$$F_{gluon}(x) \propto \frac{(1-x)^N}{x}$$

where the value N = 5 is obtained from constituent counting rules[11]
and the value N = 10 is preferred from fits to ep and μp scattering
data[12] the charmed mass "M_c" is frequently chosen to be 1.65 GeV/c^2--
the value which reproduces the ψ' and ψ masses in a particular
potential theory[13]. There is no compelling experimental evidence to
our knowledge which unambiguously determines this value, however.

In Fig. 12 we compare the prediction of the gluon fusion model
for bound charm photoproduction to a fit[14] of the ψ photoproduction
cross section versus "s" for the value N = 5 and various charmed
quark masses. Values of the charmed quark mass from 1.5 to 1.65
GeV/c^2 appear to roughly reproduce the threshold behavior of the
data with the value M_c = 1.65 giving a good fit to the absolute
cross section as well. One should not disregard values of M_c whose
cross section overshoots the data, however, since we assume in the
gluon fusion calculation that <u>all</u> $c\bar{c}$ bound states emerge as psions.
Changing the power in the gluon x distribution from N = 5 to N = 10
produces a family of cross section curves which approach the
asymptotic cross section slower than the data. A fit by Weiler[15]
to the world's photoproduction data based on a similar analysis also
obtains values of N near 5.

As we have previously discussed, the $c\bar{c}$ pair created by gluon
fusion may acquire a $P_\perp$ through a variety of dimly understood
processes. One can estimate the $c\bar{c}$ pair $P_\perp$ by studying the $P_\perp$ of
high mass hadronically produced dimuon pairs. Experimentally[16]
the dimuon $<P_\perp>$ is found to grow from $\simeq$ .6 GeV/c for dimuon masses
near 2 GeV/c^2 to $\simeq$ 1 GeV/c at dimuon masses near 4 GeV/c^2. For D^{*+}
photoproduction with $c\bar{c}$ masses near threshold, one might expect
$<P_\perp (c\bar{c})>$ of $1/\sqrt{2} \simeq$ .7 GeV/c since in photoproduction we probe one
nucleon constituent rather than two.

A final ingredient in the model is the nature of the charmed
quark dressing function D(z). There are theoretical arguments[17]
that charm quarks dress themselves by giving nearly all their energy
to a single charmed meson [i.e. D(z) $\simeq$ δ(z-.84)]. The D^0 and D^+
inclusive momentum distribution measured in e^+e^- annihilation[18], on
the other hand, favors a much softer dressing function of the form:

$$D(z) \propto e^{-5.5z} \qquad \text{(Eqn 2)}$$

This softer form is also favored by ν induced dilepton data[19].

In Fig. 13 we compare several aspects of the D^{*+} photoproduction
data to Monte Carlo calculations based on the gluon fusion model.
The D^{*+} data, represented by the points with error bars, is obtained
from separate fits to the Kπ invariant mass distribution for every
bin of the indicated distribution. The histograms summarize the
Monte Carlo calculations which incorporates trigger and detection
efficiency and are averaged over the photon energy spectrum. We

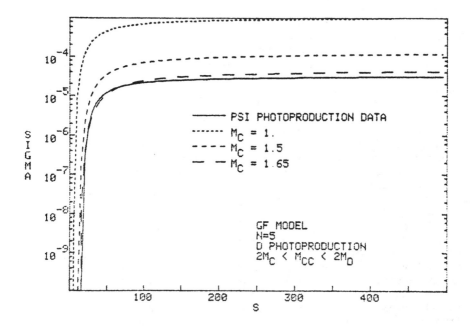

Figure 12

Comparison of the predictions of the photon-gluon fusion model to a fit of the world's ψ photoproduction data for three values of the charmed quark mass. We use a gluon distribution function of the form:

$$F_{gluon}(x) \propto \frac{(1-x)^5}{x}$$

The fit is from Reference 14. The vertical scale is the cross section for $\gamma P \to \psi P$ in millibarns.

note that the D^{*+} energy distribution and the visible event energy distribution are sculpted by D^{*+} acceptance on the low end and the falling photon spectrum on the high end.

The specific gluon fusion calculation shown in Fig. 13 uses the parameters M_c = 1.5 GeV/c^2, N = 10, and the dressing function $D(z) = \delta(z - 1)$. The illustrated distributions, however, are relatively insensitive to this particular choice of parameters. A far more sensitive variable to study is the $P_\perp$ of the D^{*+}. This variable has the attractive property that the D^{*+} acceptance in our detector is nearly independent of $P_\perp$ in the relevant $P_\perp$ range.

In Fig. 14 we compare the data (error bars) to the model (histogram) for three different values for "N" of the gluon distribution function. For the model calculation shown in this figure we assume $D(z) = \delta(z - 1)$ and assume that the $c\bar{c}$ system acquires no $P_\perp$. We note that the data appears more peaked than the Monte Carlo calculations for the value N = 5 which is obtained in fits to the "s" dependance of the ψ photoproduction cross section. Including the additional expected $P_\perp$ of the $c\bar{c}$ system will only make this situation worse. The inclusion of a softer dressing function, on the other hand, makes the situation better.

In Fig. 15 we use N = 5, the dressing function $D(z) = e^{-5.5z}$, and vary the $\langle P_\perp \rangle$ of the $c\bar{c}$ pair. Very good fits to the $P_\perp(D^{*+})$ distribution are obtained for $c\bar{c}$ pair $\langle P_\perp \rangle$'s in the range .7 GeV/c to 1 GeV/c.

Hence we see that the gluon fusion model can be made to fit all aspects of our data with the appropriate choice of gluon structure function, quark mass, primordial $P_\perp$, and dressing function. In particular, our data prefers a relatively soft dressing function for the D^{*+}.

A final check of the model is the predicted level of the total unbound charm cross section. Averaging over our photon spectrum we compute a gluon fusion charm photoproduction cross section of $\simeq$ 230 nb/nucleon, which is comparable with our measured inclusive D^0 cross section but does not leave much room for $D^\pm$ production, charmed baryon production, or $F^\pm$ production. However, higher order corrections to the photon gluon fusion diagram of Fig. 11 can change the predicted cross section by a factor of 2 or so.

THE $D \rightarrow K^-K^+$ DECAY MODE

We have obtained a photoproduced $D \rightarrow K^-K^+$ signal. This Cabibbo suppressed decay mode of the D was first observed by the Mark II Collaboration at SPEAR[20] who obtained the anomolously large ratio:

$$\frac{D \rightarrow K^-K^+}{D \rightarrow K^-\pi^+} = .11 \pm .03$$

where one would naively expect a ratio of $\simeq$ .046.

In order to obtain a relatively large K^-K^+ signal we use neutral two track combinations with at least one track identified by the Cerenkov system as either a kaon or K/P ambiguous candidate. We then combine the two track combination with an additional charged track in order to search for the process:

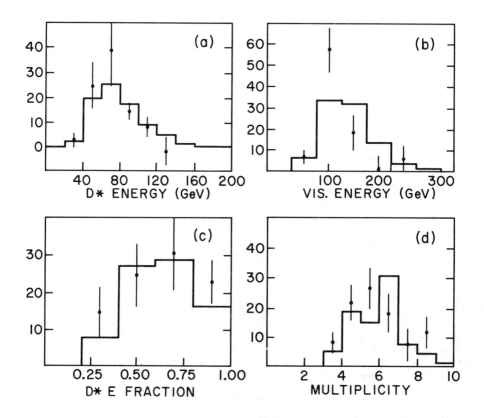

Figure 13

Inclusive properties of the Columbia-Illinois-Fermilab
photoproduced D[*+] signal. Because this experiment uses an
untagged broad band photon beam, we use the total visible energy
of the event (both charged and neutral) as a biased estimator
of the incident photon energy. This quantity is plotted in
(b) for events containing a D[*+]. The points with error bars are
the background subtracted data. The histograms are the results
of a Monte Carlo calculation based on the photon-gluon fusion
model.

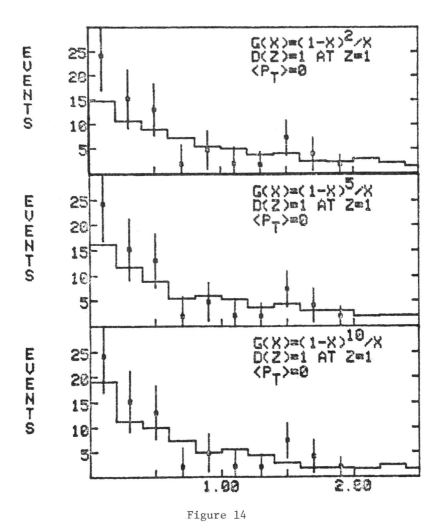

Figure 14

Comparison of the $P_\perp^2$ distribution of the D^{*+} in the data to that predicted by the photon-gluon fusion model for three different gluon distribution functions. The data from the Columbia-Illinois-Fermilab Collaboration is shown by the error bars. The model calculation assumes no $P_\perp$ for the $c\bar{c}$ pair and a dressing function of the form:

$$D(z) = \delta(z - 1)$$

280

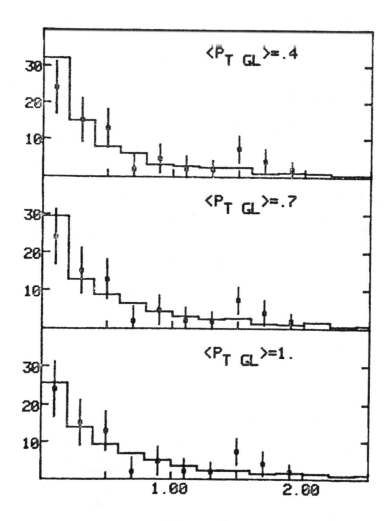

Figure 15

Comparisons of the $P_{\perp}^{2}$ distribution of D^{*+}'s in the
data to those predicted by the photon-gluon fusion model
for three different values of average $c\bar{c}$ system primordial
$P_{\perp}$. All three calculations assume a gluon distribution
function of the form:

$$F_{gluon}(x) \propto \frac{(1-x)^5}{x}$$

and a dressing function of the form:

$$D(z) = e^{-5.5z}$$

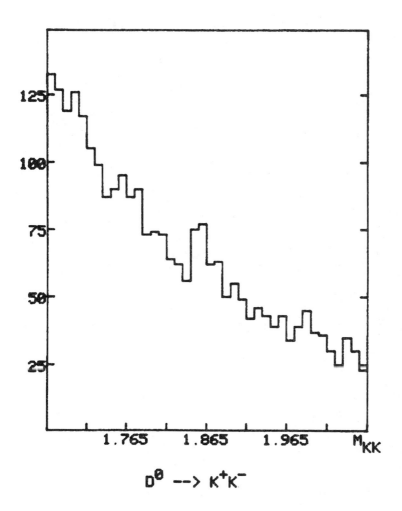

Figure 16

The D → K⁻K⁺ signal obtained by the Columbia-Illinois-
Fermilab Collaboration. One kaon must be identified by the
Cerenkov counter system. There is a tight cut on the
$K^-K^+\pi$, K^-K^+ mass difference around the known D^*, D mass
difference. We plot the number of events per 10 MeV/c²-

$$D^{*+} \rightarrow \pi^+ D^0$$
$$\hookrightarrow K^- K^+$$

Figure 16 shows the $K^- K^+$ invariant mass distribution for combinations satisfying the mass difference cut:

$$.1445 < M_{KK\pi} - M_{KK} < .1465 \text{ GeV/c}^2$$

A fit to the data of Fig. 16 reveals a $\sigma = 7.5$ MeV/c^2 wide signal consisting of 38 ± 15 events at a mass of 1.858 GeV/c^2. Comparing this signal to the $K\pi$ signal of Fig. 8a and correcting for the relative efficiencies, we obtain:

$$\frac{D \rightarrow K^- K^+}{D \rightarrow K\pi} = (20 \pm 9)\%$$

which is higher than but consistent with the SPEAR data. We have not looked for the $D \rightarrow \pi^+\pi^-$ decay mode because of the large backgrounds present in our experiment for this state.

ACKNOWLEDGEMENTS

It is a pleasure to acknowledge the work of my colleagues with the Columbia-Illinois-Fermilab Collaboration which includes: M. S. Atiya, M. Binkley, J. Butler, J. Cumalat, I. Gaines, G. Gladding, M. C. Goodman, M. Gormley, S. D. Holmes, B. C. Knapp, W. Lee, R. L. Loveless, T. O'Halloran, J. Peoples, J. J. Russell, A. Wattenberg, and W. J. Wisniewski. Special thanks go to Paul Avery of this collaboration. I would also like to thank Prof. L. M. Jones of the University of Illinois for useful discussions.

REFERENCES

1. C. Baltay, et al., PRL 41, 73 (1978).
2. J. Kirkby in Proceedings of the 1979 International Symposium on Lepton and Photon Interactions at High Energies, p. 107.
3. P. Schreiner in Proceedings of the 1979 International Symposium on Lepton and Photon Interactions at High Energies, p. 291.
4. J. Blietschau, et al., Phys. Lett. 86B, 108 (1979).
5. D. Aston, et al., submitted to Phys. Lett.
6. P. Avery, et al., PRL 44, 1309 (1980).
7. G. Feldman, et al., PRL 38, 1313 (1977).
8. C. Quigg and J. L. Rosner, Phys. Rev. D17, 239 (1978).
9. G. Goldhaber, et al., Phys. Lett. 69B, 503 (1977).
10. L. M. Jones and H. W. Wyld, Phys. Rev. D17, 759 (1978).
    H. Fritzsch, and K.-H. Streng, Phys. Lett. 72B, 385 (1978).
11. S. J. Brodsky and G. R. Farrar, Phys. Rev. D11, 1309 (1975).
    R. Blankenbecker and S. J. Brodsky, Phys. Rev. D10, 2973 (1974).
12. A. J. Buras and K. J. F. Gaemers, Nucl. Phys. B132, 249 (1978).
13. A. DeRujula and S. L. Glashow, Phys. Rev. D12, 147 (1975).
14. A. R. Clark, et al., PRL 43, 187 (1979).
15. T. Weiler, PRL 44, 304 (1980).
16. J. E. Pilcher in Proceedings of the 1979 International Symposium on Lepton and Photon Interactions at High Energies, p. 185.

17. M. Suzuki, Phys. Lett. $\underline{71B}$, 139 (1977).
18. G. Feldman in Proceedings of the 19th International Conference on High Energy Physics, Tokyo (1978), p. 777.
19. V. Barger, T. Gottschalk, and R. J. W. Phillips, Phys. Lett. $\underline{70B}$, 51 (1977).
20. G. S. Abrams, et al., PRL $\underline{43}$, 481 (1979).

VI International Conference on Experimental Meson Spectroscopy
Brookhaven National Laboratory, Upton, New York
April 25-26, 1980

<u>J. Wiss</u>: Talk Title: "D Meson Production by Photons and Neutrinos"

Answer to question after talk:

Question from L. Littenberg: "Why don't you get a π-K mis-attribution
 peak in your D $\rightarrow$ K$^+$K$^-$ spectrum the way that storage ring experiments
 do?"

Answer: "The main reason is that the Kπ reflection peak is expected
 to be broad (FWHM = 30 MeV) and centered at $\simeq$ 1.94 GeV. It is thus
 lost in the background of Figure 16."

CHARM PRODUCTION IN HADRONIC INTERACTIONS

M. S. Witherell

Princeton University, Princeton N. J. 08544

ABSTRACT

This paper reviews recent results on charm production in hadronic interactions. Charm cross sections inferred from prompt lepton signals are compared with experiments designed to observe specific charm decay modes. All results at FNAL-SPS energy are consistent with a total cross section for D production of 15-35 µb and with a central production mechanism, as seen in ψ production.

INTRODUCTION

From other speakers at this session we have heard about charm production by photons and neutrinos, and in e^+e^- annihilations. Those of us who seek out greater challenges in life search for charm the hard way -- in hadron production. Never has the metaphor of a needle in a haystack been so appropriate.

The first question one must ask about hadronic production of charm is, "What will we learn from these experiments?" It is unlikely that the spectroscopy of charmed particles will be improved by the results. An important point is that the traditional mechanisms, such as one particle exchange, which are responsible for most of ordinary hadron production, are likely to be a small fraction of the charm cross section. Because of the large quark mass and the large momentum transfers required to create the quarks, the production should be dominated by hard collisions -- direct interactions of quarks and gluons. The only such production mechanism which is well-known is the Drell-Yan process, which we know is not important for $c\bar{c}$ production (from the ψ). Models which use ψ-production to try to predict the $D\bar{D}$ production level differ from each other by factors of 10. (At least they did differ by that much before the data began to restrict the possible range.) We need a few precise measurements of charmed particle production to begin testing the models. Of course if we gain some understanding of $D\bar{D}$ production and its relation to ψ production, we can probably apply it to $b\bar{b}$ production using appropriate scaling laws.

To be able to compare various types of experiments, I need to answer two questions. I am making my estimates explicit so that you can see what goes into the cross sections listed later. (1) What is the ratio D^0/D^+ produced in hadronic reactions? It is not unity because the D^* is just at threshold for strong decay. The fraction of D^{*+} that decays to D^0 is 60%, and for D^{*0} the fraction is 100%. Assuming the relative direct production of D^* and D follows 2J+1 weighting, the ratio D^0/D^+ would be 2.5. If equal numbers of D^* and D were directly produced, the ratio would be 2. These numbers set a reasonable range.

(2) What is the semileptonic branching ratio? One has to be careful about exactly what one wants, the branching ratio for D's or the ratio for all charmed particles. Two experiments at SPEAR yield average numbers for the branching ratios BR ($D^O \to X\mu\nu$) = 3% and BR ($D^+ \to X\mu\nu$) = 18%. Using the answer to question (1) above gives BR ($D \to X\mu\nu$) = 7–8%, weighted according to production. If one wants the branching ratio for all charmed particles, one must include Λ_c and F^+, which may have semileptonic branching ratios as low as or lower than the D^O. (Decays in which both quarks participate in the weak vertex are allowed for the D^O, F^+, and Λ_c, but not for the D^+. Preliminary results on the lifetimes support short lifetimes and therefore small semileptonic branching ratios for F^+ and Λ_c.[2]) If Λ_c and F^+ have cross sections comparable to the D^+ and low branching ratios, the branching ratio (BR ($C \to X\mu\nu$)= 5–6%, where C represents all charm weak decays. We don't yet know if this is right, but such a low branching ratio is quite possible.

LEPTON AND LIFETIME EXPERIMENTS

The experiments that look for a charm signal in hadronic interactions are of three types, each type taking an advantage of a different property of the weak charm decay. Experiments look for either prompt leptons, decays with long lifetimes, or a narrow peak in a mass spectrum. An example of the lepton experiments is the Caltech-Stanford experiment which used a target calorimeter followed by an iron toroidal muon spectrometer to look for prompt muons.[3] The experiment required muons of transverse momentum greater than 0.8 GeV/c by requiring the muons to remain in the same quadrant. They separate single muon events from dimuon events, and also observe dimuon events with missing energy, indicating undetected neutrinos. Non-prompt decays are empirically substracted by varying the effective density of the target-calorimeter and extrapolating to infinite density. Figure 1 shows the extrapolation plots for 1μ events and 2μ events with missing energy. A prompt muon signal is seen in each case, although the 2μ signal is less dependent upon the extrapolation procedure. To extract charm cross sections, some dependence of the cross sections on the kinematic variables must be

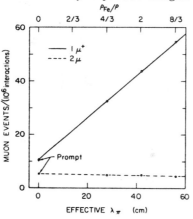

Fig. 1. Measured 1μ and 2μ rates as a function of calorimeter density (from reference 3).

assumed. A good fit to the muon spectra was obtained using the
invariant cross section

$$E \frac{d^3\sigma}{dp^3} = M^{-3} c^{-\alpha p_T} (1 - x)^\beta e^{-\gamma M/\sqrt{s}} \qquad (1)$$

where M is the mass of the composite $D\bar{D}$ system, and with constants
in the following ranges: $\alpha = 1.2-2.7$ GeV^{-1}, $\beta = 3-6$, $\gamma = 15-30$.
The authors quote cross sections assuming an 8% branching ratio,
but get lower cross sections from their 2μ data. What they actu-
ally measure is $\sigma \cdot B_\mu$ in the 1μ data and $\sigma \cdot B_\mu^2$ in the 2μ sample,
where B_μ is the branching ratio for all charm to muons. The
results, then are

$$\sigma \cdot B_\mu = 2.0 \pm 0.4 \ \mu b \qquad (1\mu b) \qquad (2)$$

$$\sigma \cdot B_\mu^2 = .10 \pm .03 \ \mu b \qquad (2\mu b) \qquad (3)$$

where the errors represent systematic uncertainties due to the
range of the constants used. These two measurements determine
σ and B_μ independently, giving $\sigma = 40 \pm 10$ μb and $B_\mu = 4.5 \pm 1\%$.
The cross section here refers to all charm, not just $D\bar{D}$. The low
value of the branching ratio is not too surprising, as was noted
in the last section, if there is substantial contribution from F
or Λ_c. A more recent version of this experiment, by the Caltech-
Fermilab-Rochester-Stanford collaboration looks at a broader kinematic range including small p_T.[4] A similar analysis yields $\sigma B_\mu =$
1.8 ± 0.7 μb, which is consistent with the earlier result, but less
dependent on production model assumed.

Originally the CERN beam dump experiments quoted cross sections
in the range 40-400 μb for charm production by 400 GeV/c protons.[5]
A review by Wachsmuth at the Lepton-Photon conference reported
results of more recent runs.[6] The resulting value of $\sigma \cdot B_\mu$ varies
from 1.4-2.3 μb for these experiments, consistent with the muon
results. The only possible problem is that the BEBC and CHARM
experiments see electron/muon ratios of ½ rather than 1. At the
ISR a CERN-ETH-Saclay group observed electron-muon coincidences in
p-p collisions at $\sqrt{s}$ of 53 and 63 GeV.[7] They measure $\sigma \cdot B_\mu \cdot B_e =$
$.20 \pm .05$ μb, which is somewhat larger than the results at FNAL
energies, as expected.

To compare the results from various experiments and for application to other experiments, I have used two hypotheses for the
source of the charm-produced lepton signal: (a) Only $D\bar{D}$ pairs are
produced, with a branching ratio to muons of 7.5%, and (b) 65% of
charmed particles are D or $\bar{D}$, and the branching ratio for all such
particles is 5%. Table I presents the results in this way. $\sigma_{(a)} =$
24 ± 6 μb and $\sigma_{(b)} = 36 \pm 8$ μb, where the errors reflect differences among experiments. Even under assumption b, however, the cross
section for D's $= .65 \cdot \sigma_{(b)}$ is about 24 μb. This simply reflects

the fact that for the 1μ experiment one is seeing predominantly D's and the results are insensitive to the presence of other charmed species. It is only in the 2μ correlations that it matters whether the D's are being produced in DD pairs or in, say, associated production with a Λ_c of low semileptonic branching ratio.

Table I Cross sections from lepton experiments, in μb.

Group	σB	σB^2	σ$_{(a)}$	σ$_{(b)}$	σ$_D$	(hypothesis b)
C - S	2.0	.10	18-27	40	25	
CFRS	1.8		22	36	22	
BEBC	1.7		21	34	21	
CDHS	1.4		19	28	19	
CHARM	2.3		31	46	28	
C-E-S (ISR)		.20	36	80	50	

Some conclusions can be drawn from the prompt lepton experiments. There is remarkably good agreement between various experiments at the same energy. The cross section for D's is 25 ± 6 μb at $\sqrt{s}$ = 28 GeV, 40 ± 10 μb at $\sqrt{s}$ = 55 GeV, if they are in fact the predominant source of prompt lepton. The total charm cross section is somewhat higher, probably 30-50 μb, but it is not as well measured by the lepton experiments. The results are consistent with a central production mechanism of the type seen in ψ-production.

The second property of charmed particles which can be used to identify them is their lifetime in a range of 10^{-13} - 10^{-12} seconds. Since at x = 0 for 400 GeV/c proton collisions γ = 14, the decay length is about 1 mm. The emulsion technique has been used successfully in neutrino experiments, but it is more difficult in a charged hadron beam. One experiment[8] observed 60000 events, scanning to a distance of 150 μm, and found no event with two decays in that distance. Assuming τ_{Do} = 1.5-10^{-13} sec and τ_{D+} = 10^{-12} sec, consistent with preliminary results[2] of the neutrino emulsion experiment, the limit on $\sigma_{D\bar{D}}$ is about 50 μb for 300 GeV/c pN collisions, which does not contradict other results. The Yale-FNAL streamer chamber experiment also tried to observe the decay of charmed particles. They see an excess of events with decay length of a few mm or more.[9] Making the same assumption on lifetime as above they get $\sigma_{D\bar{D}}$ = 35 ± 15 μb, which is consistent with lepton experiments. The experiment is sensitive to the longer-lived D$^+$; for other lifetimes the observed cross section is proportional to $(\tau_D+)^{-1}$.

SEARCHES FOR MASS PEAKS

The third type of charm experiment looks for mass peaks in specific decay channels. Only in such experiment does one identify the type of charmed particle seen. Experiments done just above charm threshold report upper limits of a few microbarns.[10] Two experiments using the MPS at Brookhaven with restricted charm triggers are now in the analysis stage. At higher energies a number of experiments have been done looking for the D^0 peak in the $K\pi$ mass spectrum. The most sensitive of these is the FNAL - Michigan - Purdue experiment[11] which used a double-arm spectrometer in a 400 GeV/c proton beam. The 4σ upper limit is $d\sigma/dy < 12$ μb, for a total cross section $\sigma < 50$ μb. This experiment collected 10^4 events per 10 MeV and marks the limit of how far one can push the limit without enhancing the signal-to-background ratio. It is difficult to know the background to better than 1% simply by fitting to a polynomial.

A recent experiment by the Princeton-Saclay-Torino-BNL group, of which I am a member, looks at the reaction

$$\pi^- + N \to D^{*+} + X, \quad D^{*+} \to \pi^+ + D^0, \quad D^0 \to K^- + \pi^+ \qquad (4)$$

(and charge conjugate.) Requiring the pion from the D^* decay in coincidence with the D^0 suppresses the non-resonant background more than the signal. The branching ratio for $D^{*+} \to \pi^+ + D^0$ is 60%, so one does not lose much of the signal. The Q-value, defined as $Q = M_{D^*} - M_D - M_\pi$, is measured to be 5.7 ± 0.5 MeV, so the pion is almost at rest in the D center of mass. This means that only pions in a small region of phase space contribute to the background. Figure 2 shows the plan view of the experimental apparatus, most of which is a double-arm spectrometer to measure the K^- and π^+. The low momentum pion from the D^* decay points along the beam with a momentum of 1.0-2.5 GeV/c, and is bent in the vertical plane by the 34 D 48 magnet near the target. Figure 3 shows the elevation view of the arm which detects this pion. For D^0's accepted by the double-arm spectrometer, the associated pions are accepted with 100% efficiency because of the low Q-value.

Figure 4 shows the Q-value spectrum for events with $K\pi$ mass between 1.835 and 1.875 ($K^+ \pi^- \pi^-$ and $K^- \pi^+ \pi^+$ events are combined.) A fit to the data using a Gaussian with $\sigma = 0.6$ MeV and a smooth background gave a peak of 60 ± 25 events with a central value of 5.8 ± 0.3 MeV. Figure 5 shows a plot of $M_{K\pi}$ for events with Q near 5.8 MeV. A fit using a smooth background plus a Gaussian with $\sigma = 14$ MeV yielded a peak of 71 ± 24 events with mass 1851 ± 6 MeV. To use all of the information available we also made a joint fit to the two-dimensional spectrum, Q vs $M_{K\pi}$. These two variables were chosen because their errors are almost uncorrelated. The resulting peak contained 56 ± 21 events above background with $M_{K\pi} = 1851 \pm 9$ MeV and $Q = 5.9 \pm 0.3$ MeV. No other peak of comparable significance exists for any combination of $K\pi$ mass and Q-value. Converting 56 ± 21 events to a cross section gives

290

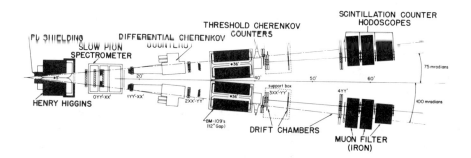

Fig. 2. Plan view of the apparatus

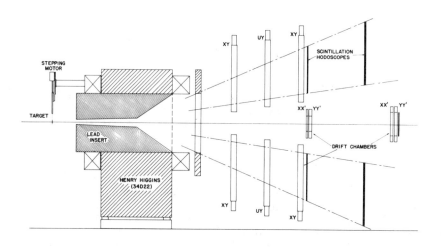

Fig. 3. Elevation view of target region.

$$\sigma\ (D^*) = [\sigma(D^{*+}) + \sigma\ (D^{*-})]/2 = 3.3 \pm 1.2\ \mu b\ \text{or} \qquad (5)$$

$d\sigma/dy = 1.0 \pm 0.4\ \mu b$ at $y = 0$. For comparison, putting in a $(1-x)^3$
dependence and a ratio of D^0/D^{*+} of about 2,[12] the cross section
for D^0 is $6.6 \pm 2.4\ \mu b$, for an upper limit of about 16 µb. (To
compare sensitivity with the earlier experiments, the "upper limit"
calculated as for these experiments would be about 10 µb). The
lepton experiments predict a D^0 cross section of from 6 - 25 µb, and
our result is consistent with that range. It is, however, low
enough that if the D^* cross section is not that large, an inconsis-
tency results.

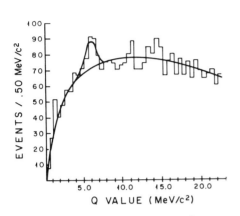 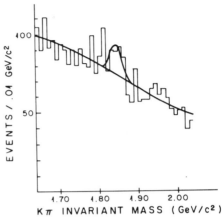

Fig. 4. Q-value spectrum for
events with 1.835<$M_{K\pi}$<1.875.
Fit is described in text.

Fig. 5. Kπ mass spectrum for
events with Q near 5.8 MeV.
Fit is described in text.

The only published result to date showing a D peak in hadronic
production is from the CERN-College de France-Heidelberg-Karlsruhe
group at the ISR.[13] Figure 6 shows the $K^-\pi^+\pi^+$ mass spectrum they
obtain, which shows no evidence of structure. Figure 7 shows a similar plot with the requirement that one K $\pi^!$ combination have a mass
consistent with the K^* (890), along with other cuts on the recoiling
particles detected. There are 92 ± 18 events above the background
shown. This corresponds to B dσ/dy = 2-3 μb at y = 2, where B is
the branching ratio for $D^{*+} \to \bar{K}^{*o} \pi^+$. Assuming that 2/3 of all
$D^+ \to K^- \pi^+\pi^+$ goes through $\bar{K}^*$ they calculate dσ/dy = 50 μb for D^*
production. However, the Mark II group at SPEAR measured the
Dalitz plot for the K$\pi\pi$ decay and it is flat, with an upper limit of
15% for the fraction that is K^*.[14] This upper limit implies there
should be a peak of 10 or more standard deviation in the uncut K$\pi\pi$
plot. The 15% upper limit also changes the cross section to
dσ/dy > 200 μb at y = 0. If such a y-dependence existed at $\sqrt{s}$ = 28
GeV it would be easily seen by the lepton experiments.
Figure 8 shows on one plot the results that have most to say
about the hadronic production of D's. The curves were obtained by
assuming that $\sigma_{D\bar{D}}(\sqrt{s})$ = A $\sigma_\psi(\sqrt{s}$ x $M_J/M_{D\bar{D}})$, where σ_ψ is the cross
section for ψ production. This formula assumes that the dependence
on ($\sqrt{s}$/M) is the same as for ψ production, where a D$\bar{D}$ mass of 2 M_D
is used. The solid line represents proton production with the normalization set to agree with the lepton results; the curved line
represents pion production, using the same normalization. The round
points are the cross sections for D's at FNAL and ISR energies as

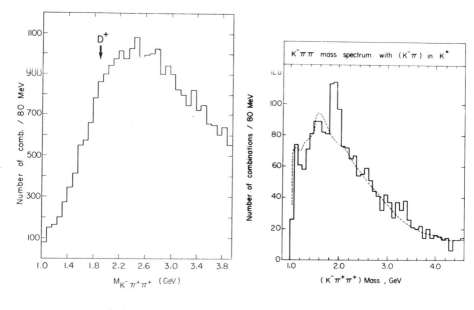

Fig. 6. $K^-\pi^+\pi^+$ mass plot without K^* cut (from ref. 13)

Fig. 7. $K^-\pi^+\pi^+$ mass plot with K^* cut (from ref. 13)

inferred from prompt lepton results, with errors which include the sensitivity of the result to branching ratios and production models. The lines, then, give a rough expectation for dependence on energy and beam particle type. The boxes show upper limits from mass peak searches, open boxes for pion production and closed boxes for proton production. It can be seen that the upper limits of the mass peak experiment are just at the level expected from the lepton experiments, and in fact a cross section of 7 μb for pion production at $\sqrt{s}$ = 20 GeV is at the lower end of the expected range. The triangle point represents the CCHK result <u>if</u> the Mark II branching ratio and the same production model is used.

As has been said before today, "This is a meson conference, so I shouldn't discuss baryons, but..." There have been observations of mass peaks at the ISR which are identified with the Λ_c. Three experiments see peaks in $pK\pi$ or $\Lambda(3\pi)^+$ mass spectra corresponding to $Bd\sigma/dy$ = 2-7 μb at y = 2-3.[15] Mark II measures $BR(\Lambda_c \to p + K^- + \pi^+)$ = 2%[16] which gives $d\sigma/dy$ = 150 μb for Λ_c production at y = 2-3. This is large, but not in contradiction with other results if the semileptonic branching ratio for Λ_c is less than 2% or so, and if they are not produced predominantly with D's.

Many important questions are unanswered, such as dependence of the cross section on s, x, and p_T, and the relative amounts of D, Λ_c and F. Table II lists some of the experiments now in progress to answer these questions, and their present status.

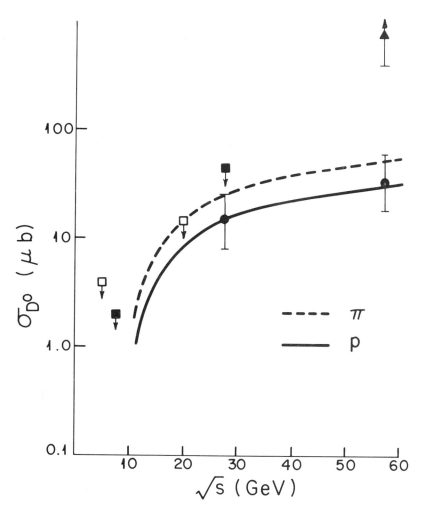

Fig. 8. Results of the experiments which bear on the level of D⁰ production. The round points are cross sections inferred from prompt lepton experiments. The lines give energy dependence for π's and p's expected from ψ production, fixing the normalization with the lepton experiments. The boxes show upper limits, open for pion production and closed for proton production. The triangle represents the D⁺ signal at the ISR if the same, central production mechanism is used.

294

Table II Experiments now in progress

Group	Accelerator	Energy (GeV)	Detection	Status
NYU	AGS	13 $\pi^{\pm}$	D^{o} Σ_c (missing mass)	Prep
CMU–NW–ND	FNAL	200 π^-	μ trigger	Data
A–B–C–C–M–N–R	SPS	200 π^-	e trigger	Data
Yale–FNAL	FNAL	200 π^-	streamer chamber	Prep
P–S–T–B	FNAL	200 π^-	D^*	On hold
C–F–R–S	FNAL	350 p, 200 π^-	leptons	Data

SUMMARY

If we assume that the prompt leptons are from the semileptonic decay of the D, then the level of D production is set to about a factor of 2. The streamer chamber result is consistent with this level. Further work now going on will give a similar result for 200 GeV/c pions. The mass peak experiments must first seek to give conclusive evidence that the prompt lepton level is just from D's. They are just reaching this level of sensitivity. One then needs to study the things one cannot do in the lepton experiments, such as the relative amounts of various charmed particles. More work is needed to fit the large D signal seen by the CCHK group into the picture.

The clearest conclusion is that after five years, this is still a hard business. No easy way of doing the experiments has been found.

REFERENCES

1. V. Lüth, in Proceedings of the International Symposium on Lepton and Photon Interactions at High Energies, Fermilab, 1979, edited by T. B. W. Kirk and H. B. I. Abarbanel, p. 78; J. Kirkby, ibid., p. 107.

2. J. Prentice, this conference.

3. K. W. Brown et al., Phys. Rev. Lett. 43, 410 (1979); and A. Bodek, Rochester report UR-730 (Oct. 1979).

4. J. Ritchie et al., Phys. Rev. Lett. 44, 230 (1980).

5. P. C. Bosetti et al., Phys. Lett. 74B, 143 (1978); P. Alibran et al., ibid., 134; and T. Hansl et al., ibid., 139.

6. H. Wachsmuth, in Proceedings of the International Symposium on Lepton and Photon Interacting at High Energies, Fermilab, 1979, edited by T. B. W. Kirk and H. B. I. Abarbanel, p. 541.

7. A. Chilingarov et al., Phys. Lett. 83B, 136 (1979).

8. G. Coremans-Bertrand et al., Phys. Lett. 65B, 480 (1976).

9. J. Sandweiss et al., Phys. Rev. Lett. 44, 1104 (1980).

10. R. Cester et al., Phys. Rev. Lett. 40, 139 (1978); and J. J. Aubert et al., Phys. Rev. Lett. 35, 416 (1975).

11. W. R. Ditzler et al., Phys. Lett., 71B, 451 (1977).

12. This estimate is made for the assumptions used earlier to estimate the D^0/D^+ ratio. The assumption of 2J+1 weighting for direct production gives a ratio of 2 for D^0/D^{*+}. Assuming equal direct production of D and D^* only changes this ratio to $2\frac{1}{2}$. Recently a photoproduction experiment measured the ratio to be 2.5 ± 1.0. See P. Avery et al., Phys. Rev. Lett. 44, 1309 (1980).

13. D. Drijard et al., Phys. Lett. 81B, 250 (1979); and W. M. Geist, CERN/EP 79-129 (1978).

14. G. Goldhaber, this conference, for example.

15. K. L. Giboni et al., Phys. Lett. 85B, 437 (1979); W. Lockman et al., ibid., 443; and D. Drijard et al., ibid., 452.

16. See paper of V. Lüth, reference 1.

MEASUREMENTS OF CHARMED MESON LIFETIMES

J.D. Prentice
University of Toronto, Toronto, M5S 1A7 Canada

ABSTRACT

Measurements of charmed meson lifetimes are reviewed.
Bare emulsion and track chamber experiments which indicated
that charm lifetimes were of the order of 10^{-13}sec. are briefly
discussed. Fitted decay events from four hybrid emulsion
experiments give lifetime values of $\gamma_{D^0} = .94 \times 10^{-13}$sec
$\gamma_{D^+} = 10.3 \times 10^{-13}$sec and $\tau_{F^+} = 2.0 \times 10^{-13}$sec.

INTRODUCTION

As the lightest charmed mesons must decay through flavour
changing currents the weak interaction is dominant and their
lifetimes are long enough to yield visible decays in sensitive
detectors. That the production and decay of charmed hadrons
could be observed in emulsions was predicted[1] even before the
discovery of the J/ψ. This and other achievements of the
standard model of electroweak interaction might lead us to
expect detailed and accurate predictions for the properties of
these decays. It is clear, however, that the success of the
theory in describing purely leptonic interactions such as muon
decay cannot be extended to the weak decays of hadrons. The
effects of strong interactions in the weak decays of strange
particles are only now, after 2 decades, beginning to be understood.

It is equally clear that the charmed meson decays provide
an interesting new window on the interplay of Q.F.D. and Q.C.D.
weak decays of hadrons composed of u and d quarks involve only
one generation and binding energies are in general large compared to quark mass differences. For the lighter member of each
succeeding weak doublet most decays involve the small components
of the quark mixing matrix and thus almost all the strange
particle decays are Cabibbo forbidden. For charmed particles
however the larger mass of the c quark allows a study of many
more channels both favoured and unfavoured and should also make
the QCD corrections more tractable.

Except for the approximate magnitude of the lifetimes, however, almost all the early theoretical predictions for charm
decays[1]-[5] have been revised in the light of the experimental
data. The decays do not appear to be dominated by the decay of
the charmed quark almost uninfluenced by a light spectator;[6]-[8]
simple colour suppression factors must be modified by gluon
effects; and helicity suppression of two fermion decays of pseudoscalar mesons, which is so clearly observed in pure leptonic

π + K decays, is eliminated by gluon emission in the quark antiquark annihilation and W-exchange diagrams for F and D
decays.[8]-[11] These theoretical developments are well reviewed
in the paper of L.L. Wang in these proceedings.

Since both QED and QCD are in principle fully calculable
theories it should be possible to calculate amplitudes for
individual exclusive decay channels. The present state of non-
perturbative QCD, however, still requires model dependent assumptions. As long as experimental comparisons are limited to the
ratios of partial widths that can be obtained from spectroscopic
measurements of branching ratios many of the assumptions remain
untested. The addition of lifetime measurements transforms
branching ratios to decay rates for exclusive channels and thus
provides far more stringent restraints on the models.

A common feature of the techniques used in the lifetime
measurements surveyed here is the direct observation of the
charmed decay. In many cases this produces a virtually background free sample of charmed particle decays. Other experiments
that depend on effective mass to identify the charmed hadrons
such as those using photoproduction or hadroproduction spectrometers can produce high statistics samples of charmed hadrons but
only in association with large backgrounds. It is clear that
the former extremely low background, albeit low statistics
experiments may be equally useful in exploring the strong decays
of higher mass charmed hadrons and some of the details of the
production mechanisms.

EMULSIONS AND TRACK CHAMBERS WITHOUT FULL DECAY RECONSTRUCTION

Many experiments have recently observed the production and
decay of short lived particles whose decay lengths indicate decay
times $\sim 10^{-13}$ seconds and which appear to have masses too large to
be associated with strange particle decays. To convert an
observed decay length into a decay time requires a measurement of
both the momentum and mass of the decaying particle. To establish
the lifetime of a particular charmed hadron requires a set of
events for which the decaying particles are well identified, their
masses and momenta measured and for which the detection efficiencies as a function of decay length are well understood.

The earliest measurements of decay lengths for charmed
particles were made in emulsions exposed to cosmic rays[12] and
by 1975 ten possible decays had been reported[13]. Only approximate lifetimes can be estimated from these and the other events
reported from emulsions exposed to accelerator hadron beams[14].
Lifetimes of a few times 10^{-13} seconds were indicated by these
results.

In a conventional exposure of the Fermilab 15 foot bubble chamber to the quadrupole triplet neutrino beam (E546) a sample of events with a μ^- and e^+ identified in the final state were carefully examined for visible, short decay tracks. Four events were found consistent with the e^+ originating from a secondary vertex of which one was charged two neutral and one either neutral or charged. The semi leptonic decay implied by the presence of the e^+ requires an accompanying neutrino which prevents the determination of the masses of the decaying particles. The authors however assume the charged decays are D^+ and the neutrals D^0 and quote lifetimes[15].

$$\tau_{D^+} = 2.5 \; {}^{+ \; 3.5}_{- \; 1.5} \; \times \; 10^{-13} \text{sec}$$

$$\tau_{D^0} = 3.5 \; {}^{+ \; 3.5}_{1.7} \; \times \; 10^{-13} \text{sec}$$

Two track chambers constructed specifically to observe short tracks have reported the observation of a few decays from which some information about production cross sections and lifetimes can be deduced.

The Little European Bubble Chamber (LEBC) filled with liquid hydrogen was exposed to a 340 GeV π^- beam. From the 110,000 pictures taken in their first run they obtained 48,000 events which were scanned for examples of 2 short decays indicating associated production of charmed hadrons. The observed sample of 20 events with two visible decays is estimated to include 8 background and 12 associated charm production events. Assuming a lifetime of 10^{-12} seconds and that their charged decays are $D^\pm$ a cross section $\sigma_{D^\pm}$ (inclusive) ~ 35 μb is obtained[16].

The achievement of high bubble densities (70 bubbles/cm) and small bubble size (40-50μ diameter) clearly demonstrated the value of this technique. A recent run ending in June 1980 with a downstream spectrometer and particle identification has produced $> 1.2 \times 10^6$ pictures and it is hoped that ~ 200 well measured and reconstructed charmed decays will result[17].

A high pressure streamer chamber operated in a 350 GeV/c proton beam at Fermilab, with a μ trigger was used by a Yale-Fermilab group to obtain 11 charm decay candidates with an estimated background of 2.6 events. Track widths obtained were from 150 - 200 μ but considerable improvement is expected in the future. The present results permit a lifetime estimate of 10^{-13} sec to 2×10^{-12} for unidentified charmed hadron decays and an inclusive cross section estimate of 20 - 50 μb/nucleon for 350 GeV incident protons[18].

HYBRID EMULSION EXPERIMENTS

The fitted events that can be used for charmed lifetime measurements have been obtained in hybrid emulsion experiments. In this technique the emulsion serves as a high resolution track sensitive target and a downstream detector serves both to locate the events in the emulsion and to provide the momentum measurements and particle identification necessary for full reconstruction of the decay kinematics.

As the emulsion records all charged particle tracks that traverse it a successful experiment requires a high fraction of the events to contain charm decays. In 5 of the 6 experiments described here, this was accomplished by using a neutrino beam and the other employed a tagged photon beam in which each pellicle was exposed for only a few pulses. Thus in all cases a large volume of emulsion, typically 20 - 30 litres, was required. A complete scan of this volume of emulsion would require hundreds to thousands of man years and the downstream detector must therefore localize the events within a small fraction of the total volume.

Once interactions have been found and a search downstream for charged and neutral decays has located charmed candidates it is still necessary to convert decay lengths to proper decay times and sets of decay times to charmed hadron lifetimes. Since the proper decay time

$$\tau = \frac{\ell}{\beta \gamma c}$$

where ℓ is the decay length and $\beta \gamma = \frac{p}{m}$ for the decaying particles, the downstream detector must provide measurements of both p the decaying particles momentum and m, its mass. Identification and momentum measurement of all the charged and neutral decay products provides a complete solution and kinematic constraints can be used in cases where a small number of quantities are unmeasured.

The first hybrid experiment at Fermilab used spark chambers, shower counters and emulsion in the neutrino beam and found one event which could not be fully fitted[19].

Another Fermilab experiment (E553) used a magnetic spark chamber spectrometer and a flash tube calorimeter. A novel event finding technique was based on metal film spark chambers, placed close to the emulsion, that retained a precise permanent record of the spark positions. One neutral and one charged decay candidate are being analyzed.

The other four experiments, two at CERN and two at Fermilab, have each fitted one or more decay candidates that will be used in calculating lifetimes.

Using the Big European Bubble Chamber (BEBC) its associated
external muon identifier and 31.5 litres of emulsion placed just

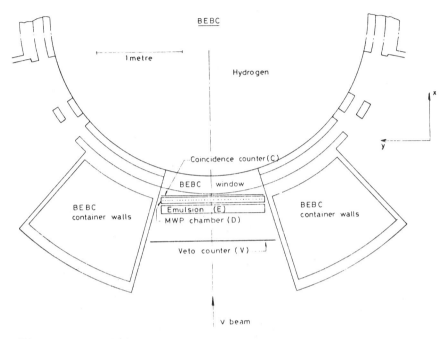

Fig. 1. Experimental apparatus for CERN (WA17) experiment.

upstream of the bubble chamber the CERN (WA17) experimenters
found eight charm decay events induced by neutrinos in the emulsion.

One of the 5 charged decays is uniquely fitted with the
hypothesis $\Lambda_c^+ \to pK^-\pi^+$ with a mass $M(\Lambda_c^+) = 2.26 \pm .02$ GeV/c^2
and a proper decay time of $(7.3 \pm .1) \times 10^{-13}$sec.

As the liquid hydrogen filled bubble chamber had a low
efficiency for neutral particle detection, the other events, which
all had missing neutral decay particles could not be fitted.
Possible hypotheses for these events have been published[20]. A
more recent analysis of this experiment makes use of a relationship between the distance of closest approach of the decay tracks
to the primary vertex and the proper decay time that is independent of momentum of the decaying particle. On the assumption that
four of the charged tracks are D^+ and the three neutrals are D^0
a maximum likelihood fit yields

$$\tau(D^+) = (2.5 \begin{array}{c} + 2.2 \\ - 1.1 \end{array}) \times 10^{-13} \text{sec.}$$

$$\tau(D^0) = (0.53 \begin{array}{c} + 0.57 \\ - 0.25 \end{array}) \times 10^{-13} \text{sec.}$$

There is, however, no way of knowing whether the charged sample contains any F^+ or Λ_c^+ decays and as will be seen below their lifetimes are substantially different from that of the D^+. For the neutral sample there is a danger of contamination from neutral charmed strange baryons (quark content cds and css) that are expected to decay weakly and possible weakly decaying charmed stranged exotic qq $\bar{q}\bar{q}$ states[21] which would invalidate the D^0 lifetime measurement.

Another CERN experiment makes use of the Ω spectrometer Fig. 2. to analyse the downstream tracks from interactions of tagged photons in single emulsion pellicles that are mechanically

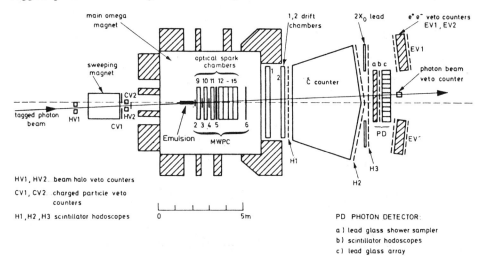

Fig. 2. Omega Spectrometer

placed in the beam for only a few pulses. One well identified $\bar{D}^0 \rightarrow K^+\pi^+\pi^-\pi^-$ with a proper decay time of $\tau = (0.226 \pm .005) \times 10^{-13}$ sec. The invariant mass of this particle is measured to be $M_{\bar{D}^0} = 1866 \pm 8$ MeV/c^2.[22] A second run of this experiment has already produced several examples of decays from associated production events. These will be reported at the xx[th] International High Energy Physics Conference in Madison Wisconsin, July 1980[23].

The Fermilab 15' Bubble chamber, filled with liquid D_2 has been used as a downstream detector for a hybrid experiment (E564). In this case the emulsion, sensitive at liquid hydrogen temperatures was placed inside the bubble chamber, enclosed in a stainless steel box. One decay from the first run has been fully fitted as an

$F^+ \to \pi^+ \pi^+ \pi^- \pi^0$ decay with M_{F^+} = 2017 ± 25 MeV and proper decay time τ = 1.5 x 10^{-13} sec [24].

The other Fermilab experiment (E531) has obtained 10 fitted events, 12 zero constraint but identified fits and 11 other multi-prong charm decay candidates which are too ambiguous to be used in the lifetime fits. This experiment is described in somewhat more detail as an example of the hybrid emulsion technique.

Twenty-three litres of Fuji emulsion placed at the upstream end of the spectrometer shown in Fig. 3. was exposed to the Fermilab horn focussed neutrino beam. An integrated flux of 7 x 10^{18}, 350 GeV/c protons was incident on the neutrino production target during the first run.

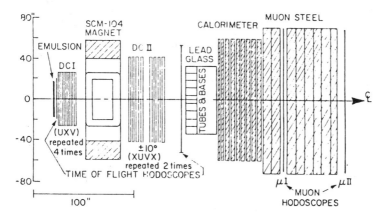

Fig. 3. E531 Target and Spectrometer

The spectrometer was used both to locate the neutrino events in the emulsion and also to measure momenta of the decay products from weak decays and of the other particles produced in the neutrino interactions. The measurements provide the momentum of the decaying particle and its mass thus allowing, in conjunction with the emulsion measurement of its decay length, a calculation of its proper decay time.

In order to obtain an unbiased sample of neutrino inter-actions, a very simple trigger was used which required a neutral incident particle and two or more outgoing charged particles. The veto counter, upstream time of flight (T of F) counter and downstream time of flight hodoscope identified these conditions.

The charged particle momenta were measured by 12 upstream and 8 downstream drift chambers and a large angular acceptance magnet with a central field of 0.6 Tesla.

Neutral pions and η^0 were reconstructed from the momenta of pairs of gamma rays measured in an array of 68 lead-glass blocks, 11 and 13 radiation lengths long and of 19 cm x 19 cm transverse dimensions.

A rudimentary hadron calorimeter with four columns of vertical counters each five layers deep and separated by 10 cm thick iron layers, provided a check on the total hadronic energy in the neutrino events and crude detection of missing neutral hadrons.

Two banks of muon counters, a horizontal hodoscope behind 1.2 m of steel and a vertical hodoscope behind a further 1.2 m of steel gave excellent muon identification above 4 GeV/c and reasonable muon-hadron separation down to 2 GeV/c.

The front Time of Flight counter and the 30 rear T of F counters, with photomultipliers on both ends, gave 130 pico sec resolution and were useful for separating π and K up to about 2.2 GeV/c and identifying protons up to about 4.5 GeV/c.

Information on the resolution of the spectrometer is summarized in Table I.

TABLE I

Quantity Measured	Error ($\pm\sigma$)
charged particle momenta	upstream-downstream tracks $$\frac{\Delta p}{p} = \sqrt{(.013)^2 + (.005p)^2}$$ upstream only (fringe field) $$\frac{\Delta p}{p} > .3p$$
γ momenta	$$\frac{\Delta E}{E} = \frac{0.14}{\sqrt{E(GeV)}}$$
γ position	±5 cm
Time of Flight	±130 pico sec

LOCATION OF NEUTRINO INTERACTIONS IN THE EMULSION

Reconstruction of the drift chamber tracks from our neutrino triggers had yielded a sample of 2200 events with well defined vertices. 1743 of these lie in the region of the emulsion target and the remainder are associated with the support frame.

The results of the counter fits are used to search for the neutrino events in the emulsion by two techniques. In the first the emulsion is scanned in a small volume surrounding the predicted vertex location. The second technique depends on finding an individual track from the event where it leaves the downstream face of the emulsion stack and then following the track back to the vertex. Results of the search for 501 events are summarized in Table II.

TABLE II

Emulsion orientation	"Horizontal"	"Vertical"
Number of events searched for	718	699
Number found	302	551
Efficiency	42%	79%

Two sets of emulsion stacks composed the target. In the first, the emulsion was divided into 27 modules in which emulsion layers of 330µ thickness were deposited on each side of a 70µ polystyrene sheet. 68 such composite sheets in each module gave a thickness of 5 cm along the beam direction. These stacks, in which the neutrinos were incident normal to the plane of the emulsion layers are referred to as "vertical". The other 12 modules containing slightly less than half of the total emulsion volume were composed of 600µ thick, pure emulsion pellicles 5 x 14 cm^2 in area exposed with the beam parallel to the 5 cm dimension. For historical reasons these are referred to as "horizontal" emulsion.

The differences in the event finding efficiencies in the two samples (see Table II) is mainly due to the different search techniques. In the "horizontal" emulsion, almost all the events were found by volume scanning whereas in the "vertical" emulsion almost all the scanning was done by track following.

The background of high momentum tracks with angles within 20^0 of the neutrino beam direction is so great ($\sim$20,000/cm^2) that there are several emulsion tracks which match, within position and angle errors, the prediction for a particular track from the upstream chambers. Our ability to select the correct track to follow and to find the neutrino event depends on changeable sheets of 800µ polystyrene coated front and back with 75µ layers of emulsion which covered the downstream face of all the emulsion stacks in the target. The troublesome background in the emulsion target is due mainly to beam associated muons and partly to nearly horizontal cosmic ray muons. The former background was worst at the beginning of the run and improved substantially later. The target, of course, integrated the background from both sources during the whole period between pouring and development. The changeable sheets however, were poured, mounted for 2 days, and developed immediately and the background was thus reduced to $\sim$ 1000 tracks/cm^2. This allows unique identification of at least one spectrometer track in the changeable sheet for almost every event. The construction of a vertical module and the method of relating the changeable sheet position to that of the main emulsion stack is illustrated in fig. 4.

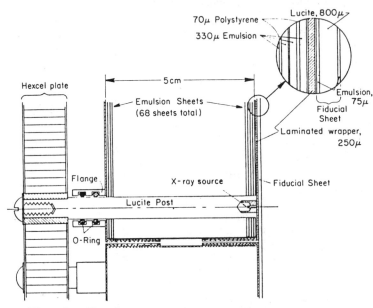

Fig. 4. Vertical emulsion and changeable sheet

Four small, well collimated Fe^{55} sources permanently mounted in each emulsion stack produced small exposed spots on the changeable sheet allowing relative location of the two emulsions to better than 100μ accuracy. The ability to locate and follow individual tracks from the spectrometer to the emulsion target is important, not only for increasing the event finding efficiency by about a factor of two but also for calibrating the efficiency of the search for neutral decays.

The search for charm decays was made by following the charged tracks from the neutrino interaction and by volume scanning, at high magnification, a cylinder downstream of the neutrino interaction for neutral decays. Charged tracks were followed for a length of 6 mm and the neutral search was made in cylinders of 0.6 mm diameter and 1 mm along the beam direction. 12 neutral, 21 3-prong and 27 single-prong kink candidates have been found. Only one of the last group (kink events) has been fitted as a charmed hadron decay.

CHARMED MESON LIFETIMES

Kinematic fits have been attempted for all the charm decay candidates found in this experiment (E531).[25]. When all the decay particles are measured a three constraint fit can be made by assigning a hypothetical mass for the decaying particle or a two constraint fit can be used to determine the decaying mass and a corresponding error. For events with one missing decay track the assumed particle is shown in brackets. By assigning a mass to the decay particle a zero constraint fit can be obtained that normally leads to a quadratic ambiquity in the missing particle momentum. In some cases one of these solutions can be excluded by the absence of a signal in the lead glass or calorimeter. The fits for 10 D^0 events five charged D events and two F decays are shown in Table III. The underlined particles ($\underline{\pi}^+$) in the decay hypotheses

TABLE III

CHARM DECAY EVENTS

Event	Decay Length (μm)	P_μ (GeV/c)	Hypothesis	P (GeV/c)	Mass (MeV)	Decay Time (10^{-13} sec)
1	325	-18.7	$D^0 \to \underline{\pi}^+ \pi^- (k^0)$	19.27		1.05
2	126	- 4.8	$D^0 \to \pi^- \underline{\pi}^- \underline{\pi}^+ \pi^+ \pi^- \pi^+ (\pi^0)$	9.1		0.86
3⁺	256	not seen	$D^0 \to K^- \underline{\pi}^+ \pi^- \pi^+ (\pi^0)$	12.8		1.24
4	27.2	+ 11	$\bar{D}^0 \to K^+ \pi^- \underline{\pi}^+ \pi^- \pi^- \pi^0$	9.2	1766 ±43	0.18
5	116	- 4	$D^0 \to \underline{\pi}^+ K^- \pi^0 \pi^0$	30.1	1935 ±132	0.24
6⁺	41	- 10	$D^0 \to \underline{\pi}^- K^+ \pi^- \pi^+ \pi^0$	15.4	1855 ±43	0.17
7⁺	67	- 30	$D^0 \to \underline{\pi}^+ \pi^- (K^0_{.})$	11.3		0.37
8	6.5	- 4	$D^0 \to \underline{\pi}^+ \pi^+ K^- \underline{\pi}^- \pi^- \pi^+$	19.2	1923 ±46	0.021
9	2647	- 26	$D^0 \to K^- \underline{\mu}^+ (\nu)$	22.8 / 38.7		7.20 / 4.24
10⁺	137	+ 34	$\bar{D}^0 \to K^+ \underline{\pi}^- (\pi^0)$	6.8 / 8.8		1.71 / 1.22
11	457	>150	$D^+ \to K^- \underline{\pi}^+ \pi^+ \pi^0$	10.1	1829 ±35	2.82
12	2145	- 7	$D^+ \to K^- \underline{\pi}^+ \mu^+ (\nu)$	16.1		8.33
13	2307	+ 7	$D^- \to \pi^- \underline{K}^- \pi^+ e^- (\nu)$	9.4		15.3
14	1802	- 11	$D^+ \to K^- K^- \underline{\pi}^+ \pi^+ \pi^0$	17.0	1860 ±25	6.60
15	13000	150	$D^+ \to K^- \underline{\pi}^+ e^+ (\nu)$	118		6.86
16	670	+ 30	$F^- \to \pi^- \pi^- \pi^- \underline{\pi}^+ \pi^0$	12.25	2026 ±56	3.70
17	130	not seen	$F^+ \to \underline{K}^+ \underline{\pi}^+ \underline{\pi}^- \underline{\pi}^+ K^0_L$	9.70	2089 ±121	0.91

⁺These events have a D* with mass ∿2008 ± 3 MeV

have been identified in the spectrometer or by ionization in the emulsion. None of the charged D, zero constraint fits is consistent with a Cabibbo favoured $F^{\pm}$ or Λ^{+}_{c} hypothesis.

For the neutral decays the situation is slightly less clear. Our data contain an event with an identified proton among the decay products. It is a strong candidate for the decay of a neutral charmed baryon and points up the danger of assuming that all short neutral decays are D^{0}. It is true that we cannot exclude baryon hypotheses for events 1, 3, 7 and 9. However they each have an acceptable D^{0} hypothesis and events 3 and 7 contain a π^{+} track from the production vertex which, when combined with the D^{0} hypothesis give an invariant mass within 3 MeV of D* mass.

In order to convert these decay times to lifetimes for the charmed mesons we have made a maximum likelihood fit. This requires a measurement of the scanning efficiency over the whole range of decays. The efficiency drops as one approaches the primary vertex because the primary tracks can obscure the decay vertex. 200 neutrino events were examined to estimate the efficiency for finding secondary vertices as a function of distance from the primary vertex. The efficiency over the whole range was independently checked by following back all spectrometer tracks that were not matched at the primary emulsion vertex for 400 events. More details of the efficiencies and likelihood fits are given elsewhere[25].

The results of the maximum likelihood fits are given in column 3 of Table IV. The probability that each set of decay

TABLE IV

LIFETIME SUMMARY

Particle	E531 Events	Lifetime 10^{-13} sec	Other Events	Combined Lifetime 10^{-13} sec
D^{0}	10	$1.01 \begin{array}{c} + 0.43 \\ - 0.27 \end{array}$	1(WA34) FITTED 3(WA17) UNIDENTIFIED	0.89
$D^{\pm}$	5	$10.3 \begin{array}{c} + 10.5 \\ - 4.1 \end{array}$	4(WA17) UNIDENTIFIED	8.1
$F^{\pm}$	2	$2.2 \begin{array}{c} + 2.8 \\ - 1.0 \end{array}$	1(E564) FITTED	2.0

times are a statistically acceptable sample of a single lifetime
has been calculated. The probability for the D^0 events is $\sim 25\%$
if the high momentum solution for event #9 is selected. The
other choice reduces the probability to less than 5% and has
therefore been rejected.

It is not possible to include events from other experiments
in our likelihood fit since the scanning efficiencies have not
been published. If, however, one takes a weighted average using
a weight of 1 for fitted events and 0.5 for unidentified events
one obtains the values given in the last column of the table.

The results clearly indicate the marked difference between
the $D^\pm$ lifetime and the rather similar short values of the D^0
and F^+ lifetimes. In view of the wide diversity of opinion
concerning the most important contributions to the decay amplitudes[1]-[11] and the substantial errors on the lifetime measurements detailed comparisons with theoretical models seem premature.
It is perhaps interesting to note that an average of the SLAC
Mark II and Delco results from $BR(D^+ \to e^+\nu_e X) = 0.20$ together with
the D^+ lifetime from column 2 of Table IV gives $\Gamma(D^+ \to e^+\nu_e X) =$
$1.9 \begin{array}{c} + 1.4 \\ - 1.0 \end{array} \times 10^{11} sec^{-1}$ in good agreement with several recent
calculations (e.g. ref (3)). If we assume that the D^+ and D^0
have equal semi-electronic partial widths then the branching
ratio $BR(D^0 \to e^+\nu_e X) = 0.019 \begin{array}{c} + .021 \\ - .009 \end{array}$.

A substantial improvement in the accuracy of the lifetime
values can be expected in the next year. As mentioned above the
LEBC has completed a successful data taking run and analysis is in
progress[17]. The CERN photoproduction experiment (WA58) is
expected to obtain 50 - 100 events from their current analysis[23].
The second run of Fermilab (E564), scheduled to start in November
1980 will have greatly improved π^0 detection owing to heavy
Ne - H_2, replacing the D_2 of their first run, in the 15' bubble
chamber. Fermilab E531 is also scheduled for a second run with
a larger emulsion target and improvements to the spectrometer to
provide better charged particle identification and better neutral
particle position resolution.

ACKNOWLEDGEMENTS

I am very grateful to my colleagues for the opportunity to
discuss our data and to all those who provided information and
data from the other experiments discussed herein.

REFERENCES

1. M.K. Gaillard, B.W. Lee, J. Rosner, Rev. Mod. Phys. 47, 277 (1975).
2. J. Ellis, M.K. Gaillard, D.V. Nanopoulos, Nucl Phys B100 313 (1975).
3. N. Cabibbo and L. Maiani, Phys. Lett. 79B, 109 (1978).
4. Altarelli, N. Cabibbo, K. Maiani, Phys. Rev. Lett. 35, 635 (1975).
5. D. Fakirov and B. Stech, Nucl. Phys. B133, 315 (1978).
6. S.P. Rosen, Phys. Rev. Lett. 44, 4 (1980 and Phys. Lett. 89B, 246 (1980).
7. V. Barger, J.P. Leveille, P.M. Stevenson, Phys. Rev. Lett. 44, 226 (1980).
8. M. Bander, D. Silverman and A. Soni, Phys. Rev. Lett. 44, 7 (1979).
9. V. Barger, J.P. Leveille, P.M. Stevenson, Phys. Rev. Lett. 44, 226 (1980) and Phys. Rev. to be published.
10. H. Fritzsch and P. Minkowski, Phys. Lett. P.L. 90B, 455, (1980).
11. B. Guberina, S. Nussinov, R. Peccei, R. Ruckl, Phys. Lett. 89B, 111 (1979).
12. K. Niu, E. Mikumo, Y. Maeda, Prog. Theo. Phys. 46, 1644 (1971).
13. K. Hoshino et al, Proc. 14th International Cosmic Ray Conference 7, 2442 (1975).
14. N. Ushida et al, Lett. Nuovo Cim. 23, 577 (1978).
    H. Fuchi et al, Phys. Lett. 85B, 135 (1979).
    See also review by L. Voyvodic, International Symposium on Lepton and Photon Interactions, Fermilab 1979.
15. H.C. Ballagh et al, Phys. Lett. 89B 423 (1980).
    U.C. Berkeley, Fermilab, Hawaii, Washington Wisconsin Collaboration.
16. W. Allison et al, Phys. Lett. to be published Brussels, CERN, Oxford, Padova, Rome, Rutherford, Trieste collobaration.
17. C. Fisher and J. Mulvey (private communications).
18. J. Sandweiss et al, Phys. Rev. Lett. 44, 1104 (1980.
19. E.H.S. Burhop et al, Phys. Lett. 65B, 299 (1976).
    A.L. Read et al, Phys. Rev. D19, 1287 (1979).
20. C. Angelini et al, Phys. Lett. 80B, 428 (1979) and Phys. Lett. 84B, 150 (1979).
    D. Allasia et al, Phys. Lett. 87B, (1979).
    D. Allasia et al, Nucl. Phys. to be published.
21. H. Lipkin, Phys. Lett. 70B, 113 (1977).
22. M.I. Adamovich et al, Phys. Lett. 89B, 427 (1980).
23. G. Diambrini-Palazzi (Private Communication).
24. R. Ammar et al, Phys. Lett. (to be published).

25.   The list of E531 collaborators is as follows:
N. Ushida, T. Kondo, G. Fujioka, H. Fukushima, Y. Homma,
O. Minakawa, J. Orimoto, Y. Takayama, S. Tatsumi,
Y. Tsuzuki, S.Y. Bahk, T.G. Choi, C.O. Kim, S.N. Kim,
J.N. Park, D.C. Bailey, S. Conetti, J.-R. Fischer,
J.M. Trischuk, H. Fuchi, K. Hoshino, K. Niu, K. Niwa,
H. Shibuya, Y. Yanagisawa, S.M. Errede, M.J. Gutzwiller,
S. Kuramata, N.W. Reay, K. Reibel, T.A. Romanowski,
R.A. Sidwell, N.R. Stanton, K. Moriyama, H. Shibata, T. Hara,
O. Kusumoto, Y. Noguchi, Y. Takahashi, M. Teranaka,
J.-Y. Harnois, C.D.J. Hebert, J. Hebert, B. McLeod, K. Okabe,
J. Yokota, S. Tasaka, P.J. Davis, J.F. Martin, D.B. Pitman,
J.D. Prentice, P. Sinervo, T.-S. Yoon, J. Kimura, Y. Maeda.
More details of this experiment can be found in N. Ushida
et al., Phys. Rev. Lett., to be published and in J. Trischuk
"Measurement of Charmed Particle Lifetimes" Proceedings of
XVth Recontre de Moriord (1980) to be published.

RADIATIVE TRANSITIONS TO AN η_c(2980) CANDIDATE STATE AND
THE OBSERVATION OF HADRONIC DECAYS OF THIS STATE*

Elliott D. Bloom
Stanford Linear Accelerator Center
Stanford University, Stanford, California 94305

ABSTRACT

Preliminary results from the Crystal Ball and Mark II experiments at SPEAR are presented on radiative transitions from ψ'(3684) and J/ψ(3095) to an η_c(2980) candidate state. In addition to the inclusive photon signals reported previously by the Crystal Ball, both detectors now see exclusive hadronic final state signals at masses consistent with the states inclusively determined mass of 2981 ± 16 MeV.

INTRODUCTION

The existence of the $^1S_0$ pseudoscalar partner of the J/ψ, the η_c, and its detailed properties yield important tests of the basic charmonium model.[1] However, the difficult history of the path to the discovery of a likely candidate for the η_c leads one to proceed with some caution before finally pronouncing the task completed. In previously reported preliminary results[2,3] from the Crystal Ball collaboration[4] at SPEAR the existence of a state which I will call "η_c(2980)" was demonstrated by the examination of inclusive photon spectra from ψ'(3684) and J/ψ(3095). Though seen as a signal of more than five σ in the ψ'(3684) inclusive spectrum, no signal was seen initially in exclusive final states by the Crystal Ball (or other detectors). Clearly, confirmation of the states existence through the observation of exclusive hadronic final state decays was needed.

In this review I will report evidence from the Mark II[5] and Crystal Ball detectors for such final state hadronic decays. In addition, I will review the parameters of the state derived from the inclusive photon spectra of the Crystal Ball.

THE INCLUSIVE PHOTON SPECTRA FROM THE CRYSTAL BALL

Figure 1 shows the inclusive photon distribution from the ψ'(3684). The transitions to the well-established χ states are indicated in the figure as are the transitions from χ(3550) and χ(3510)

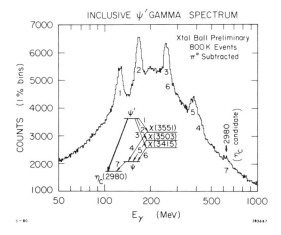

Fig. 1. The inclusive photon spectrum obtained from the decay of 800 K ψ'(3684)'s. The analysis leading to this spectrum and that of Fig. 2 is described in Ref. 2.

to J/ψ(3095). Also clearly seen, but not relatively so large, is a greater than 5 σ signal for a state at $E_\gamma = 634 \pm 13$ MeV. The corresponding mass of the state is 2983 ± 16 MeV as obtained from this spectrum alone.[2]

Figure 2 shows the inclusive photon spectrum from J/ψ(3095). The most obvious structure seen in this spectrum are the radiative transitions to η, η' and some additional structure (labeled "Glue"?), all at the endpoint of the spectrum. These structures are discussed in detail in another report to this conference.[6] In addition an enhancement above background can be seen at a photon energy (after fitting) of about 112 MeV corresponding to a mass of 2981 MeV. This is another indication for the state "η_c(2980)". Also seen in the spectrum is an apparent excess of photons at low energy. This region has been examined, using Monte Carlo simulation and it is found that much of the photon signal below 50 MeV is due to hadronic energy from hadronic interactions in

Fig. 2. The inclusive photon spectrum obtained from the decay of 900 K J/ψ(3095)'s. The data is plotted vs. ℓn E since the resolution, ΔE/E, is slowly varying in E.

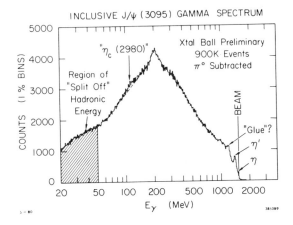

the NaI(Tℓ). This energy has separated from the main hadronic track and is erroneously identified as photons by the Crystal Ball analysis software. It is expected that in the near future, using Monte Carlo codes, one will be able to subtract this background from the photon spectra and so find the true photon yield at low energy. This hadronic "split off" energy also causes some problems in the Crystal Ball when exclusive states are considered, as I will discuss later in this report.

In order to extract the best information possible regarding the mass and Γ, the natural width, of "$\eta_c(2980)$", a 9-parameter fit was simultaneously made to the $\psi'(3684)$ and $J/\psi(3095)$ inclusive spectra in the region of the state. The 9 fit parameters are: 3 for a background quadratic for the $\psi'(3684)$ near 634 MeV, 3 for a background quadratic for $J/\psi(3095)$ near 112 MeV, the amplitude for a Breit-Wigner folded with a Gaussian resolution of 43 MeV (FWHM) at the $\psi'(3684)$, the amplitude for a Breit-Wigner folded with a Gaussian resolution of 11 MeV (FWHM) at the $J/\psi(3095)$, and the mass of the assumed resonance. The natural line width, Γ, of the Breit-Wigner shape was also varied externally to the fit and the dependence of χ^2 on Γ was determined. Figures 3 and 4 show preliminary results of this fitting procedure.[7] In Fig. 3 is shown $\chi^2(\Gamma)$ and $A(\Gamma)$ where $A(\Gamma)$ is the number of counts in the extracted signal at the $J/\psi(3095)$ as a function of Γ. A broad minimum in χ^2 is seen centered at Γ = 20 MeV.

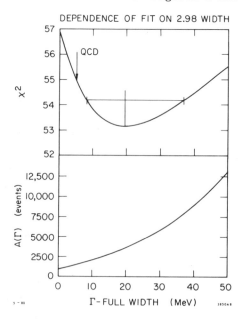

DEPENDENCE OF FIT ON 2.98 WIDTH

Fig. 3. Results of the 9 parameter fit discussed in the text as a function of the separately varied line with Γ. The theoretically expected[1] width is also shown: a) $\chi^2(\Gamma)$; χ^2 for the 9 parameter fit b) $A(\Gamma)$; the number of counts in the extracted "$\eta_c(2980)$" signal from the $J/\psi(3095)$ inclusive spectrum.

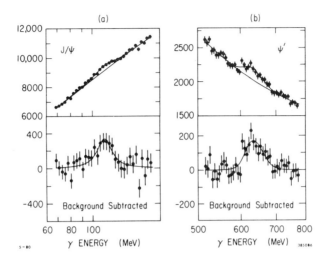

Fig. 4. Blowups of Figs. 1 and 2 in the region of "$\eta_c(2980)$" signal, with minimum χ^2-9 parameter fit results overplotted ($\Gamma = 20$ MeV). In both (a) and (b), unsubtracted, and background (from fit) subtracted spectra are shown.

However, the 1 σ limits $\left(\chi_\pm^2 = \chi_{min}^2 \pm 1\right)$ of Γ are +16 and −11 MeV. Thus, the expected QCD value[1] of ~5 MeV for the width of the standard η_c is only about 1.5 σ away from the experimentally preferred value of 20 MeV. Indeed, $\Gamma = 0$ is less than 2 σ from the experimentally preferred value. Figures 4(a),(b) show the common best fit with M = 2981 (±15) MeV and $\Gamma = 20\left(^{+16}_{-11}\right)$ MeV ($\chi^2 = 53.2$ for 66 degrees of freedom) overplotted on the data for the J/ψ(3095) and ψ'(3684) inclusive photon spectra respectively. The error on the mass of ±15 MeV is predominantly systematic. The error on Γ, however, is purely statistical. No attempt has yet been made to realistically estimate the systematic errors of Γ from the Crystal Ball experiment. Clearly, uncertainty in the form of the background for the J/ψ(3095) inclusive spectrum fit might influence the derived Γ.

Figure 5 shows an angular distribution of the photons in the region of "$\eta_c(2980)$" obtained from the ψ'(3684) inclusive spectrum by dividing the data into bins of $|\cos\theta|$ and fitting the inclusive spectra of each $|\cos\theta|$ bin to a folded Breit-Wigner plus quadratic background.[7] If "$\eta_c(2980)$" is a spin-0 particle one expects the resulting

316

Fig. 5. The angular distribution
of the radiated photon in the
decay $\psi;(3684) \to \gamma"\eta_c(2980)"$. The
angle is the polar angle of the
photon to the e^+ beam direction.
Overplotted is the expected
distribution, Eq. 1 text, for
a spin-0 particle.

angular distribution to be

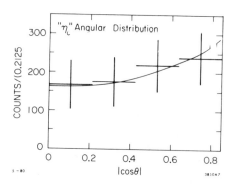

$$\left(\frac{d\sigma}{d\Omega}\right)_{spin-0} \propto \left(1 + \cos^2\theta\right) \qquad (1)$$

The data of Fig. 5 are clearly consistent with (1); however, the data
are also consistent with a flat angular distribution which would be
inconsistent with a spin-0 assignment for the state.

HADRONIC FINAL STATES OF $"\eta_c(2980)"$ FROM THE MARK II

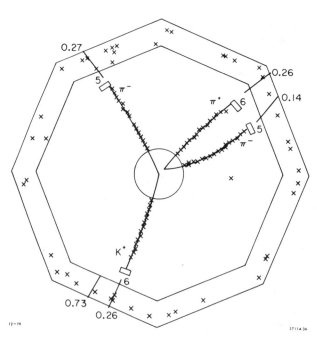

First reports
of the observation of
hadronic decays of
$"\eta_c(2980)"$ came from
the Mark II collabo-
ration at SPEAR.[8]
Figure 6 shows such
a decay in the
Mark II detector

$$\psi'(3684) \to \gamma"\eta_c(2980)"$$
$$\hookrightarrow \pi^\pm K^\mp K_S^0$$

(2)

Fig. 6. The decay $\psi'(3684) \to \gamma"\eta_c(2980)"$,
$"\eta_c(2980)" \to \pi^\pm K^\mp K_S$ in Mark II detector.[8]
The K_S travels several centimeters before
decaying into $\pi^+\pi^-$.

Additional work has
been done since the
publication of
Ref. 8, and I will
report here the pre-
liminary results of
this recent analysis.[9]

Only data from ψ' are presented. The final states studied were,

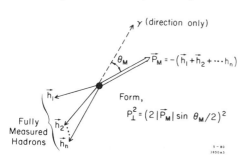

$$\psi' \rightarrow \gamma + \begin{bmatrix} K^{\mp}K_S^{\pm}\pi^{\pm} \\ K^+K^-\pi^+\pi^- \\ \pi^+\pi^-\pi^+\pi^- \\ p\bar{p} \\ p\bar{p}\pi^+\pi^- \end{bmatrix} \qquad (3)$$

Fig. 7. Diagrammatic description of Mark II technique for separating radiative decay signal. For a true radiative decay, with only one photon in the event, $\theta_M \approx 0.0$.

where only the photon's direction was measured and the hadronic decay products angles and momenta were measured. From the fully measured hadrons one constructs a missing momentum, $\vec{P}_M$. As shown in Fig. 7, the direction of $\vec{P}_M$ is obtained relative to the photons direction obtaining θ_M. A quantity,

$$P_\perp^2 = \left(2|\vec{P}_M| \sin\theta_M/2\right)^2 \qquad (4)$$

is then calculated for each event.

Figure 8 shows the resulting $P_\perp^2$ distribution for $M < 3.35$ GeV where M is the O-C mass obtained from the hadrons. The mass cut is made to remove contributions from the χ states.

The major background in this analysis is the process

$$\psi' \rightarrow n\pi^0 + X \quad , \qquad (5)$$

where all but one of the γ's from the π^0's are not seen.

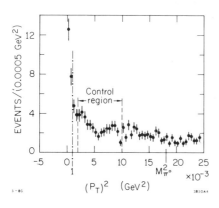

Fig. 8. $P_\perp^2$ vs. events/5×10^{-4} GeV^2, for the final states of Eq. 3 text. The radiative peak region has $P_\perp^2 \leq 1\times10^{-3}$ GeV^2, the control region has $2\times10^{-3}\leq P_\perp^2 \leq 1\times10^{-2}$ GeV^2. A mass cut of $M < 3.35$ GeV on the reconstructed hadron mass is made to remove contamination from the well known χ states.

318

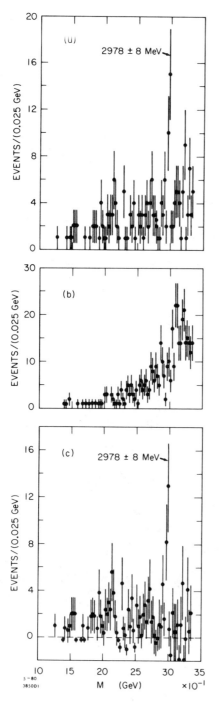

For π^0 decays it can be shown that,

$$dN_\gamma/d\left(P_\perp^l\right) \propto \frac{2\,m_{\pi^0}^2}{\left(m_{\pi^0}^2 + P_\perp^2\right)^2} \approx \text{const.} \ ,$$

$$P_\perp^2 << m_{\pi^0}^2 \quad . \qquad (6)$$

This yields a rather flat smoothly varying $P_\perp^2$ distribution for Fig. 8. However, for a radiative decay of ψ' to a particular mass, one expects a peak at small $P_\perp^2$ as in the case for the data of Fig. 8. This peak results from reaction (3) if only one γ is present in the final state.

Figure 9(a) shows the mass (M) distribution obtained from the data of Fig. 8 cut at $P_\perp^2 < 0.001$ GeV^2. There is a rather clear peak at a mass of 2978 ± 8 MeV in excellent agreement with the results of the Crystal Ball. In Fig. 9(b) is shown the mass distribution obtained from a "control region", 0.002 GeV$^2 \leq P_\perp^2 \leq 0.01$ GeV^2.

Fig. 9. The Mark II signal for exclusive final states of "$\eta_c(2980)$". (a) The reconstructed hadron mass M (GeV) vs. events/2.5× 10^{-2} GeV obtained for $P_\perp^2 \leq 1 \times 10^{-3}$ GeV^2. (b) M (GeV) vs. events/2.5 × 10^{-2} GeV^2 obtained from the control region of Fig. 8. (c) The background subtracted distribution for M (GeV) vs. events/2.5 × 10^{-2} GeV^2. A clear signal is seen at M = 2978 ± 8 MeV.

There is no evidence for an enhancement in this distribution at the "$\eta_c(2980)$" mass. Using the distribution of Fig. 9(b) to give the shape of the π^0 background for 9(a), one can subtract 9(b) from 9(a) after optimally normalizing the distribution; one obtains Fig. 9(c) which shows a somewhat enhanced signal at 2978 ± 8 MeV.

Table I shows a summary of the preliminary results from the Mark II collaboration. As shown in the Table, an upper limit has

TABLE I. MARK II Preliminary Results for $\psi' \rightarrow \gamma"\eta_c(2980)"$
$\rightarrow f$

$M_{"\eta_c"} = 2978 \pm 8$ MeV, $\Gamma_{"\eta_c"} < 30$ MeV (90% C.L.)

Final State - f	$Br\left(\psi' \rightarrow \gamma"\eta_c"\right) * Br\left("\eta_c" \rightarrow f\right)$
$K_s\, K^{\pm}\, \pi$	$(1.5 \pm 0.6) \times 10^{-4}$
$K\, \bar{K}\, \pi$ (from I-spin conservation)	$(4.5 \pm 1.8) \times 10^{-4}$
$p\bar{p}$	$\left(0.8\, {}^{+0.8}_{-0.4}\right) \times 10^{-5}$
$2\,\pi^{+}\, 2\,\pi^{-}$	$\left(4.5\, {}^{+3}_{-2}\right) \times 10^{-5}$

been placed on Γ for the state of $\Gamma < 30$ MeV (90% C.L.). This result is comparable to the mass resolution of the detector in this mass range. It should be noted, however, that the best fit to the peak has $\Gamma = 0$.

HADRONIC FINAL STATES OF "$\eta_c(2980)$" FROM THE CRYSTAL BALL

The Crystal Ball collaboration has recently reported[10] preliminary results on observation of the decay "$\eta_c(2980)$" $\rightarrow \eta\pi^{+}\pi^{-}$. The results were obtained using a 3-constraint fit in the Ball for the process,

$$J/\psi(3095) \rightarrow \gamma + "\eta_c(2980)"$$
$$\rightarrow \text{hadrons} \hspace{3cm} (7)$$

No indication for an exclusive signal has been seen yet at $\psi'(3684)$;

however, preliminary estimates of inclusive branching fractions (see below) yield a larger number of "$\eta_c(2980)$" from the $J/\psi(3095)$ data sample by a factor of about 3 as compared to the $\psi'(3684)$ data sample.

Exclusive hadronic final states are reconstructed in the Crystal Ball by measuring both the energy and angles of the photons, while only measuring the angles of the charged hadrons. Thus for exclusive final states of the type,

$$J/\psi, \psi' \rightarrow \gamma + n\pi^0 + m\eta^0 + c^+ + c^-, \quad m,n=1,2,\ldots \quad (8)$$

the π^0's and η's can be completely reconstructed as is the radiative γ. The four constraints of energy-momentum conservation are reduced to two by the loss of information of $E_{c\pm}$. Various assumptions are made for the masses of $c^\pm$ and 2-C fits are made for each mass assumption. Particle identification for $c^\pm$ is thus made through the fitting process. Fits with C.L. < 0.10 are discarded. Additional constraints are added by assuming π^0 or η^0 mass assignments to the correctly paired photons. These additional constraints improve the mass resolution obtained from the fit.

Presently the Crystal Ball has an anomalous loss of about a factor of 2 in the efficiency for reconstructing exclusive final states like (8). This loss of efficiency is due to the "split off" hadronic energy mentioned previously. The split off energy fakes extra low energy photons in the event, and so confuses the topology routines, i.e., events which should be classified as having $2n+2m+1$ photons, are found with addition photons and thrown out of the correct topology class. Work with the Monte Carlo codes is progressing toward a solution of this problem. It's worth mentioning that the Mark II collaboration had a similar problem with split off energy in the Lq Argon and has solved it quite successfully.

In order to estimate the efficiency for detecting states like (7) quantitatively, a known process was examined in detail. The exclusive state chosen was,

$$J/\psi(3095) \rightarrow \gamma\eta'$$
$$\hookrightarrow \gamma\rho$$
$$\hookrightarrow \pi^+\pi^- \quad (9)$$

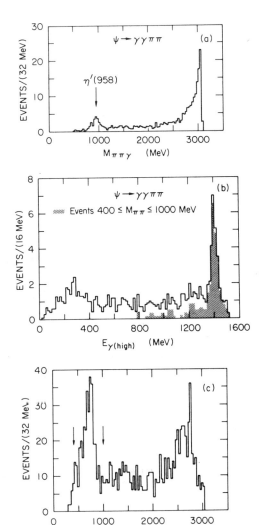

Fig. 10. Examination of the processes of Eq. 9 and 10 text in the Crystal Ball; 2-C fit results are shown.
(a) $M_{\pi\pi\gamma}$ (MeV) vs. events/32 MeV, an
η' signal is seen at $M_{\pi\pi\gamma}=956\pm65$ MeV.
(b) E_γ^{hi} (MeV) vs. events/16 MeV, an η'
signal is seen at 1400±20 MeV. The
shaded histogram shows events with
$400 \le M_{\pi\pi} < 1000$ MeV. (c) $M_{\pi\pi}$ (MeV)
vs. events/32 MeV, a ρ signal is
seen at about 770 MeV.

The general topology of these
events is,

$$J/\psi \to \gamma\gamma C^+C^- . \qquad (10)$$

Figure 10(a) shows the results of 2-C fits to events of
topology (10) when the minimum
χ^2 fit preferred $C^+C^- = \pi^+\pi^-$.
The confidence level for all
events shown is greater than
0.10. A clear indication of
an η' at the mass of 965±65 MeV
is observed. The mass error is
due entirely to the uncertainty in the proton energy measurement. The unshaded histogram of Fig. 10(b) shows the
corresponding distribution in
the high proton energy with
the η' peak at 1400±20 MeV. A
cleaner η' signal is obtained
by cutting on the $\pi\pi$ mass distribution about the ρ mass.
Figure 10(c) shows the $\pi\pi$ mass
distribution obtained from the
events of Fig. 10(a). A clear
indication of a ρ is seen. On
cutting at $400 \le M_{\pi\pi} \le 1000$ the
photon energy distribution
shown as the shaded histogram
of Fig. 10(b) results. Using
the η' signal from the shaded
histogram of Fig. 10(b), we
obtain 365 ± 30 η' events resulting from a sample of 800 K
$J/\psi(3095)$ decays.

322

Using previously measured branching fractions of[11]

$$Br\left(J/\psi(3095) \rightarrow \gamma\eta'\right) \sim 7 \times 10^{-3} \qquad (11)$$

and[12]

$$Br(\eta' \rightarrow \rho\gamma) = 0.298 \pm 0.017 , \qquad (12)$$

plus a Monte Carlo estimate of geometrical efficiency of 0.53 ± 0.1, we expect 905 ± 120 η' events. The efficiency for correctly identifying topology (9) is thus estimated as 0.4 ± 0.1. One should note that (11) was obtained through measurement of the process

$$J/\psi \rightarrow \gamma\eta'$$
$$\qquad\qquad \rightarrow \gamma\gamma , \qquad (13)$$

while (12) was obtained by direct measurement of the $\rho\gamma$ final state. So far the two decays of η' into $\gamma\gamma$ and $\rho\gamma$ have not been measured well in the same detector. The Crystal Ball hopes to accomplish this in the near future and so possibly reduce the systematic errors on the inclusive η' measurements. Of course, in order to obtain superior measurements, resolution of the split off problem is needed.

Figure 11 shows the preliminary $K^+K^-\pi^0$ mass distribution obtained from 3-C fits to the topology

$$J/\psi(3095) \rightarrow \gamma\gamma\gamma K^+K^- \qquad (14)$$

The branching fraction of "$\eta_c(2980)$" is, using I-spin conservation, a factor of two smaller for the $K^+K^-\pi^0$ final state than for the $K_s^0\pi^\pm K^\mp$ final state observed by the Mark II. As is indicated in Fig. 11, no signal is seen yet by the Crystal Ball. Assuming $K^\pm$ split off effects are the same as $\pi^\pm$, an upper limit on the produced branching ratio is obtained.

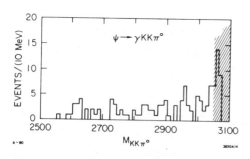

Fig. 11. $M_{KK\pi^0}$ (MeV) vs. events/10 MeV for events in the mass range of "$\eta_c(2980)$". No signal is evident. The region above 3045 MeV is contaminated by split off photons as described in the text; 3-C fit results of the Crystal Ball are shown.

$$Br\left(J/\psi(3095)\rightarrow\gamma\text{"}\eta_c(2980)\text{"}\right)*Br\left(\text{"}\eta_c(2980)\text{"}\rightarrow K^+K^-\pi^0\right)<1.5\times10^{-4}\ (90\%\ C.L.)\quad(15)$$

As shown in the summary section, this upper limit is consistent within error with the Mark II measurement.

Figure 12 shows the preliminary $\eta\pi^+\pi^-$ mass distribution obtained from 3-C fit to the topology

$$J/\psi(3095)\rightarrow\gamma\gamma\gamma\pi^+\pi^-\quad(16)$$

A signal is seen at a mass of $M_{\eta\pi^+\pi^-}=2972\pm15$ MeV in excellent agreement with previously reported masses. The mass error is primarily due to the poor statistics of the measurement; 14 ± 6 events are observed above background. Indicated in both Fig. 11 and 12 is the region

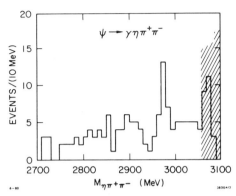

Fig. 12. Evidence for an exclusive final state signal for $J/\psi(3095)$ radiative decay to $\eta\pi\pi$. $M_{\eta\pi\pi}$ (MeV) vs. events/10 MeV is shown for events in the mass range of "$\eta_c(2980)$". A signal is evident at $M_{\eta\pi\pi}=2972\pm15$ MeV. The region above 3045 MeV is contaminated by split off photons as described in the text; 3-C fit results of the Crystal Ball are shown.

where fake split off photons become a serious background. These regions are excluded from consideration.

Using the previously determined estimates of efficiency for topology (15) the Crystal Ball collaboration obtains a preliminary product branching fraction.

$$Br\left(J/\psi(3095)\rightarrow\gamma\text{"}\eta_c(2980)\text{"}\right)*Br\left(\text{"}\eta_c(2980)\text{"}\rightarrow\eta\pi^+\pi^-\right)=(2.7\pm1.5)\times10^{-4}\quad(17)$$

Figure 13 shows the angular distribution of the radiated photon obtained from events in the region of the peak in Fig. 12.

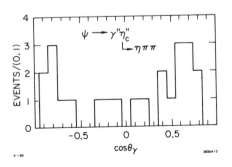

Fig. 13. The angular distribution of the radiated photon obtained from the region of the peak in Fig. 12. θ_γ is the polar angle of the radiated photon relative to the e^+ beam direction.

Though the statistics are poor, the distribution is consistent with the expected distribution (1) for spin-0.

SUMMARY

(a) A candidate η_c state has been observed in radiative transitions from the $\psi'(3684)$ and $J/\psi(3095)$. The mass obtained from the inclusive spectra of the Crystal Ball is,

$$M_{"\eta_c"} = 2981 \pm 15 \text{ MeV} \tag{18}$$

the width obtained is,

$$\Gamma = 20\left(^{+16}_{-11}\right) \text{ MeV (statistical error only).} \tag{19}$$

rough estimates of the branching fractions are[3] (very preliminary),

$$Br\left(\psi'(3684) \rightarrow \gamma"\eta_c(2980)"\right) \sim 0.2 - 0.5\% \tag{20}$$

$$Br\left(J/\psi(3095) \rightarrow \gamma"\eta_c(2980)"\right) \sim 1\% \tag{21}$$

where $J/\psi(3095)$ branching fraction is strongly correlated to Γ.

(b) The "$\eta_c(2980)$" has been observed in exclusive decays from the $\psi'(3684)$. The mass obtained from the exclusive fits of the Mark II is,

$$M_{"\eta_c"} = 2978 \pm 8 \text{ MeV} \tag{22}$$

an upper limit has been obtained for the width,

$$\Gamma < 30 \text{ MeV (90\% C.L.)} \quad . \tag{23}$$

The final states observed and their product branching fractions are given in Table I. Using the Crystal Ball value of,

$$Br\left(\psi'(3684) \rightarrow \gamma"\eta_c(2980)"\right) \sim 0.35\% \tag{24}$$

The "$\eta_c(2980)$" branching fractions of Table II(a) are obtained.

(c) The "$\eta_c(2980)$" has been observed in exclusive decays from the $J/\psi(3095)$. The mass obtained from the exclusive fits of the Crystal Ball is consistent with the inclusively obtained mass. The poor statistics of the present measurement don't allow a significant measurement of the width.

Using the value (21) (1%) for the radiative branching fraction, the "$\eta_c(2980)$" branching fractions of Table II(b) are obtained.

TABLE II(a).   Mark II Branching Fractions Assuming

$$Br\left(\psi'(3684) \to \gamma''\eta_c(2980)''\right) \approx 0.35\%$$

Final State – f	$Br\left(''\eta_c(2980)'' \to f\right)\%$
$K_s K^{\pm} \pi$	4.3 ± 1.7
$K \bar{K} \pi$ (from I-spin conservation)	12.9 ± 5.1
$p\bar{p}$	$0.2\binom{+0.2}{-0.1}$
$2\pi^+ 2\pi^-$	$1.3\binom{+0.9}{-0.6}$

TABLE II(b).   Crystal Ball Branching Fractions Assuming

$$Br\left(J/\psi(3095) \to \gamma''\eta_c(2980)''\right) \approx 1\%$$

Final State – f	$Br\left(''\eta_c(2980)'' \to f\right)\%$
$K^+ K^- \pi^0$	< 1.5 (90% C.L.)
$\eta \pi^+ \pi^-$	3 ± 1.5
$\gamma\gamma$	< 0.5 (90% C.L.)

(d) Determination of the spin-parity of the "$\eta_c(2980)$" is central to the assignment of this candidate as the theoretically desired η_c. So far no decays of the type,

$$''\eta_c(2980)'' \to \pi^+\pi^-, \quad \text{or } K^+K^- \tag{25}$$

have been reported. However, the limits on these decays should be improved before one can state with confidence that they are substantially smaller than the branching fractions into presently observed states. The lack of these decays (25) imply that

$$J^P_{''\eta_c''} = 0^-, 1^+, \dots \tag{26}$$

326

There is evidence (Figs. 5,12) that the radiative photon's angular distribution with respect to the incident e^+ beam is consistent with,

$$\frac{d\sigma}{d\Omega} \propto (1 + \cos^2\theta) \quad . \tag{27}$$

If measurement eventually provides stronger evidence for (27), and if the radiative transition can be shown to be a magnetic dipole transition,[13] then the 0^- assignment will be established.

CONCLUSIONS

A candidate η_c state at M = 2980±10 MeV with radiative transitions from $\psi'(3684)$ and $J/\psi(3095)$ has been firmly established by the results of two experiments and with inclusive and exclusive evidence from both $\psi'(3684)$ and $J/\psi(3095)$. The challenge remains to unambiguously identify this candidate state with the theoretically desired η_c, the $^1S_0$ pseudoscalar partner of the $J/\psi(3095)$. The present sample of data which has led to the establishment of the candidate state is about 10^6 $\psi'(3684)$ decays for both the Crystal Ball and Mark II experiments, and about 10^6 $J/\psi(3095)$ decays for the Crystal Ball experiments. Clearly, in order to make further progress toward uniquely assigning J^P for this state, at least 4×10^6 decays must be gathered by one experiment having at least the capabilities of the Crystal Ball or Mark II detectors.

ACKNOWLEDGMENT

Work supported in part by the Department of Energy under contracts number DE-AC03-76SF00515 (SLAC), DE-AC03-79ER0068 (Caltech), and EY-76-C-02-3064 (Harvard); and by the National Science Foundation contracts number PHY 78-00967 (HEPL) and PHY 78-07343 (Princeton).

REFERENCES

1. Two complete reviews of the charmonium model and its comparisons to the experiment are: T. Appelquist, R. M. Barnett, K. D. Lane, "Charm and Beyond," Ann. Rev. Nucl. Part. Sci. 28 (1978), also SLAC-PUB-2100; and E. Eichten, K. Gottfried, T. Kinoshita, K. D. Lane and T. M. Yan, "Charmonium: Comparison with Experiment," Phys. Rev. D21, 203 (1980), also CLNS-425 (1979).
2. E. D. Bloom, Proc. of the 1979 Int. Symp. on Lepton and Photon Interactions at High Energies, Aug. 23-29, 1979, Fermi-Lab., Batavia, Ill., eds. T. B. W. Kirk and H. D. I. Abarbanel, also SLAC-PUB-2425 (1979). This reference and the following reference contain details of the Crystal Ball apparatus, energy and angular resolutions, and analysis procedures.

3. C. W. Peck, <u>Proc. of the 1979 Annual Meeting of the Div. of Particles and Fields of the APS</u>, Oct. 25-27, 1979, Montreal, Quebec (to be published); also, Caltech report, CALT 68-753.

4. Members of the Crystal Ball collaboration. California Institute of Technology, Physics Department: R. Partridge, C. Peck and F. Porter. Harvard University, Physics Department: A. Antreasyan, W. Kollmann, M. Richardson, K. Strauch and K. Wacker. Princeton University, Physics Department: D. Aschman, T. Burnett, M. Cavalli-Sforza, D. Coyne, M. Joy and H. Sadrozinski. Stanford Linear Accelerator Center: E. D. Bloom, F. Bulos, R. Chestnut, J. Gaiser, G. Godfrey, C. Kiesling, W. Lockman and M. Oreglia. Stanford University, Physics Department and High Energy Physics Laboratory: R. Hofstadter, R. Horisberger, I. Kirkbride, H. Kolanoski, K. Koenigsmann, A. Liberman, J. O'Reilly and J. Tompkins.

5. Members of the SLAC-LBL Mark II collaboration: G. Abrams, M. Alam, C. Blocker, A. Boyarski, M. Breidenbach, D. Burke, W. Carithers, W. Chinowsky, M. Coles, S. Cooper, W. Dieterle, J. Dillon, J. Dorenbosch, J. Dorfan, M. Eaton, G. Feldman, M. Franklin, G. Hollebeek, W. Innes, J. Jaros, P. Jenni, D. Johnson, J. Kadyk, A. Langford, R. Larsen, V. Lüth, R. Millikan, M. Nelson, C. Pang, J. Patrick, M. Perl, B. Richter, A. Roussarie, D. Scharre, R. Schindler, R. Schwitters, J. Siegrist, J. Strait, A. Taureg, M. Tonutti, G. Trilling, E. Vella, R. Vidal, I. Videau, J. Weiss and H. Zaccone.

6. The report of D. Scharre to this conference.

7. T. Burnett, Invited talk, Conference on Color, Flavor and Unification, Irvine, CA, Nov. 30-Dec. 1, 1979. Unpublished.

8. T. M. Himel, SLAC Report No. SLAC-223 (Ph.D. Thesis), (1979).

9. G. Feldman, to be published in the <u>Proc. of the XV Rencontre de Moriond</u>, Les Arcs, France, March 15-21, 1980.

10. D. G. Aschman, to be published in the <u>Proc. of the XV Rencontre de Moriond</u>, Les Arcs, France, March 15-21, 1980.

11. R. Partridge et al., Phys. Rev. Lett. <u>44</u>, 712 (1980).

12. Review of particle properties, Particle Data Group, LBL-100 (1978).

13. A discussion of the measurements needed to determine the multipolarity of this transition is given in G. Karl et al., Phys. Rev. <u>D13</u>, 1203 (1976).

Question: The relation between the mass that you find for your pur-
ported η_c and the rate, does that make sense theoretically
now?

Bloom: We have rough measurements of ~1% for the inclusive branching ratio from the J/ψ and 0.2-0.5% from the ψ. Theoretically, the ψ' transition is a forbidden magnetic transition, and presently the theoretical calculations agree well
with our value. In the case of the J/ψ agreement is not so
good. Theory expects 2%-3% while we presently obtain ~1%.
However, given the present uncertainty of our values, I see
no conflict with theory. A true comparison of experiment
and theory awaits a full understanding of our detection
efficiencies.

Question: Do you think the inclusive branching fraction to J/ψ could
be as large as 2.5%?

Bloom: It might get as big as that.

Question: What would that do to the other branching ratios?

Bloom: Which other branching ratios?

Question: The hadronic branch ratios, or do they all come from the ψ'.

Bloom: The Mark II hadronic branching fractions come from examination of ψ' radiative decays while the Crystal Ball hadronic branching ratios come from examination of J/ψ radiative
decays. The absolute hadronic branching fractions of
"$\eta_c(2980)$" decays depend on Crystal Ball inclusive photon
branching fractions from either J/ψ or ψ'. As is said in
the text, the inclusive branching fractions are very rough
numbers.

Question: So the Crystal Ball hadronic branching fractions would
tend to get smaller?

Bloom: If we finally measure ~2.5% for J/$\psi \to \gamma"\eta_c(2980)"$, it is
likely that the hadronic decays measured by the Crystal
Ball would get smaller. However, what the final result
will be is truly uncertain to about a factor of two or so
at this time.

Question: The Crystal Ball has quoted an upper limit of the branching fraction of the $\chi(3.41)$ (spin 0^+) to J/$\psi + \gamma$. Is this
branching fraction consistent with an E1 transition of the
spin 2^+ χ and the QCD predictions of the hadronic widths
of the two states. [Since $\chi(3.41)$ has spin 0 parity +, its
radiative decay from ψ' and to J/ψ must be E1.]

Bloom: No! The crux of the problem is the apparent large
$\Gamma_{hadronic}$ of $\chi(3.41)$. This width is broader relative to
$\Gamma_{hadronic}$ of $\chi(3.55)$ than lowest order QCD indicates by
about a factor of[2] 3-4. Most theorists I have questioned
about this indicate that higher order QCD calculations
need to be done to check and perhaps augment the lowest
order predictions.

RADIATIVE TRANSITIONS FROM THE ψ(3095) TO ORDINARY HADRONS*

D. L. Scharre
Stanford Linear Accelerator Center,
Stanford University, Stanford, Ca. 94305

ABSTRACT

Preliminary results from the Mark II and Crystal Ball experiments on radiative transitions from the ψ to ordinary hadrons are presented. In addition to the previously observed transitions to the η, η'(958), and f(1270), both groups observe a transition to a state which is tentatively identified as the E(1420).

I. INTRODUCTION

This talk is the second of two reviewing charmonium results from SPEAR. The first talk[1] reviewed the status of the $\eta_c(2980)$ which has now been observed in radiative transitions from both the ψ(3095) and the ψ'(3684). In this talk, I will review the status of radiative transitions from the ψ to ordinary hadrons, where ordinary hadrons are defined to be those which, to first order, do not contain charmed quarks. As in the previous talk, results from both the Mark II[2] and Crystal Ball[3] experiments will be presented.

I will begin with a brief discussion of inclusive photon production at the ψ. This leads naturally into a discussion of the four exclusive radiative transitions which constitute the main part of this talk. Three of these transitions, to the η, η'(958), and f(1270), have been previously observed. The new results are in reasonable, but not perfect, agreement with the previous measurements. The fourth observed transition is to a state which is tentatively associated with the E(1420). This transition has not been previously observed. I will conclude with a review of some recent results on hadronic production of the E, along with an explanation for the relevance of this digression.

---

*Work supported by the Department of Energy, contract DE-AC03-76SF00515.

330

II.   INCLUSIVE PHOTON PRODUCTION

Measurements of inclusive photon production at the $\psi(3095)$ by the
Mark II[4] and Lead-Glass Wall (LGW)[5] collaborations have shown that
there is a sizable direct-photon component in the momentum spectrum.
However, because of the relatively poor photon energy resolutions of
the liquid argon (LA) shower counter system employed by the Mark II
($\delta E/E \approx 12\%/E^{1/2}$, E in GeV) and the lead-glass counters in the LGW
($\delta E/E \approx 9\%/E^{1/2}$), neither experiment was able to observe any narrow
structure in the inclusive photon momentum distribution.

The Crystal Ball detector[6] was designed to provide good energy
resolution for electromagnetic showers. The use of NaI(Tℓ) for shower
detection presently allows a resolution of $\delta E/E \approx 2.8\%/E^{1/4}$ (E in GeV)
to be obtained.

Figure 1 shows a preliminary measurement of the inclusive γ energy
distribution at the ψ from the Crystal Ball.[7] It is plotted as a function
of the logarithm of the γ energy (E_γ in MeV) so that the bin width is roughly
proportional to the energy resolution at all energies. This distribution is based on a sample of
approximately 900,000 events obtained during approximately two weeks of running near the peak of the ψ. Details of the analysis can be found in Ref. 6.

The structure observed in Fig. 1 is evidence for exclusive processes of the type

$$\psi \to \gamma + X .$$

There is clear evidence for the radiative transitions to the η,[8-10] $\eta'(958)$,[8-11] and a new state which I will refer to as the E(1420) which has recently been observed by the Mark II collaboration.[12] [Although I

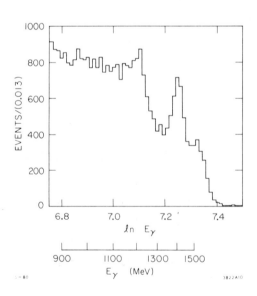

Fig. 1.   Inclusive γ distribution
at the ψ as a function of the
logarithm of the γ energy (in MeV).
(Crystal Ball)

refer to this state as the E(1420), this assignment is still in question.] An additional transition which has been previously observed is
$\psi \to \gamma$ f(1270).[13,14] Because of the relatively small branching fraction
for this transition, it is not observed in this inclusive distribution.
Each of these four transitions will be discussed in turn in the following sections.

III. $\psi \to \gamma\eta$, $\gamma\eta'$

As the η and η' are members of the same SU(3) nonet, it makes
sense to discuss the radiative transitions to these two states at the
same time, along with the transition
to the π°. I will take the extremely
naive approach that it is possible
to understand these processes in
terms of leading-order QCD diagrams.
Thus, one can imagine that the radiated photon is produced either from
the outgoing quark line (assumed to
be u, d, or s) as in Fig. 2(a) or
from the initial charmed quark line
as in Fig. 2(b). In the first case,
the minimal coupling between the
charmed quark line and the ordinary
quark line requires three gluons.
In the second case, two gluons is
sufficient.

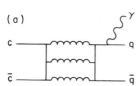
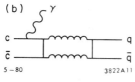

5-80   3822A11

Fig. 2. Leading-order diagrams
for radiative transitions from
the ψ with (a) photon emission
from the final-state quark line
and (b) photon emission from
the initial-state charmed quark
line.

Let me first consider only the process shown in Fig. 2(a) and
assume it is the dominant one. By invoking vector-meson dominance,
I can relate the $\gamma\pi^\circ$ and $\rho^\circ\pi^\circ$ decay widths

$$\Gamma(\psi \to \gamma\pi^\circ) = (\alpha\pi/\gamma_\rho^2) \, \Gamma(\psi \to \rho^\circ\pi^\circ) \; .$$

This leads to a prediction for the $\gamma\pi^\circ$ branching fraction $B(\psi \to \gamma\pi^\circ) \approx$
2×10^{-5} from the measured $\rho^\circ\pi^\circ$ branching fraction.[11,15,16] This is
consistent with the experimental measurement[9] $B(\psi \to \gamma\pi^\circ) = (7 \pm 5) \times 10^{-5}$.

332

The next step is to relate the widths of the $\gamma\eta$ and $\gamma\eta'$ transitions to the width of the $\gamma\pi^\circ$ transition. The η and η' have the following SU(3) singlet and octet components

$$\eta = \eta_8 \cos\theta + \eta_1 \sin\theta$$

$$\eta' = -\eta_8 \sin\theta + \eta_1 \cos\theta \ ,$$

where θ is the standard octet-singlet mixing angle. If one assumes SU(3) invariance, only the octet components contribute to the process shown in Fig. 2(a) and one obtains (up to phase space corrections)

$$\Gamma(\psi \to \gamma\pi^\circ):\Gamma(\psi \to \gamma\eta):\Gamma(\psi \to \gamma\eta') = 3:\cos^2\theta:\sin^2\theta \ .$$

Using the experimentally determined mixing angle $\theta = -11^\circ$, one calculates

$$\Gamma(\psi \to \gamma\pi^\circ):\Gamma(\psi \to \gamma\eta):\Gamma(\psi \to \gamma\eta') = 3:0.96:0.04 \ ,$$

which grossly contradicts the experimental measurements.[8-11] The $\gamma\eta'$ branching fraction has been experimentally determined to be larger than the $\gamma\eta$ branching fraction, and both are at least an order of magnitude larger than the π° transition. The conclusion is that the process in Fig. 2(b) is the dominant one.

One can proceed with similar calculations for the second process [shown in Fig. 2(b)]. Assuming SU(3) invariance (now only the singlet components contribute) and ignoring phase space corrections, one obtains

$$\Gamma(\psi \to \gamma\pi^\circ):\Gamma(\psi \to \gamma\eta):\Gamma(\psi \to \gamma\eta') = 0:\sin^2\theta:\cos^2\theta \ .$$

This is qualitatively in better agreement with the data. However, the ratio $\Gamma(\psi \to \gamma\eta')/\Gamma(\psi \to \gamma\eta)$ is much larger than the experimentally measured ratio.

If one allows for SU(3) symmetry breaking, these results are modified. Fritzsch and Jackson[17] have calculated the relative widths of the $\gamma\eta$ and $\gamma\eta'$ transitions by considering gluon-mediated mixing between the three isoscalar states η, η', and $\eta_c(2980)$. Based on the experimental masses of these states, they find the following admixture of η and η' in the η_c:

$$\eta_c = \bar{c}c + \epsilon \cdot \eta + \epsilon' \cdot \eta' \ ,$$

where $\varepsilon \approx 10^{-2}$ and $\varepsilon' \approx 2.2 \times 10^{-2}$. The decay widths (for M1 transitions) for $\gamma\eta$ and $\gamma\eta'$ are

$$\Gamma(\psi \to \gamma\eta) = \varepsilon^2 \frac{4\alpha}{3m_c^2} \left(\frac{2}{3}\right)^2 k^3 \Omega^2$$

$$\Gamma(\psi \to \gamma\eta') = (\varepsilon')^2 \frac{4\alpha}{3m_c^2} \left(\frac{2}{3}\right)^2 (k')^3 \Omega^2 ,$$

where m_c is the charmed quark mass, $k(k')$ is the momentum of the $\eta(\eta')$, and Ω is an overlap integral. If it is assumed that the overlap integral is the same for both the η and η' transitions, one finds for the ratio of the two partial widths

$$\frac{\Gamma(\psi \to \gamma\eta')}{\Gamma(\psi \to \gamma\eta)} = \left(\frac{k'}{k}\right)^3 \left(\frac{\varepsilon'}{\varepsilon}\right)^2 \approx 3.9 .$$

By estimating the overlap integral $\Omega^2 \approx 0.1$, Fritzsch and Jackson also make predictions for the absolute values of the widths, $\Gamma(\psi \to \gamma\eta) \approx 60$ eV and $\Gamma(\psi \to \gamma\eta') \approx 220$ eV.

Branching fractions for the transitions $\psi \to \gamma\eta$ and $\psi \to \gamma\eta'$ have recently been published by the Crystal Ball collaboration.[10] The measurements were based on a sample of decays $\psi \to 3\gamma$. Figure 3 shows the Dalitz plot for this sample of events. Two distinct bands associated

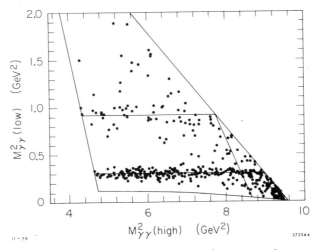

Fig. 3. Dalitz plot for $\psi \to 3\gamma$. Boundary includes effects of both kinematics and $\gamma\gamma$ opening angle cuts. (Crystal Ball)

334

with the $\gamma\eta$ and $\gamma\eta'$ transitions are observed.[18] The projection of the low-mass $\gamma\gamma$ combination, in Fig. 4, clearly shows peaks at the η and η' masses. The branching fractions for these transitions were determined from a fit to the Dalitz plot. They are $B(\psi \to \gamma\eta') = (6.9 \pm 1.7) \times 10^{-3}$ and $B(\psi \to \gamma\eta) = (1.2 \pm 0.2) \times 10^{-3}$.

The Mark II has measured the branching fraction for the process[19]

$$\psi \to \gamma\eta' \; , \; \eta' \to \pi^+\pi^-\gamma \; .$$

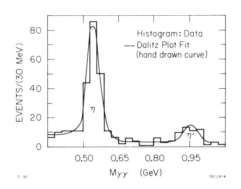

Fig. 4. Low-mass $\gamma\gamma$ invariant mass combinations for $\psi \to 3\gamma$ events. (Crystal Ball)

The data sample used in the Mark II analyses discussed in this talk is basically the same as the Crystal Ball data sample, as both experiments were running at SPEAR simultaneously.[20] Previous publications can be referred to for details on the detector and the analysis.[4,21]

Events with two oppositely charged tracks identified as pions and two or more photons[22] observed in the LA shower counter modules were fit to the hypothesis

$$\psi \to \pi^+\pi^-\gamma\gamma \; . \tag{1}$$

Events in which the fitted $\gamma\gamma$ invariant mass was between 0.12 and 0.15 GeV (i.e., consistent with the π° mass) were eliminated. The $\pi^+\pi^-\gamma$ invariant mass distribution for the events remaining after the χ^2 and π° cuts is shown in Fig. 5. From Monte Carlo calculations of the detection efficiency (which include an assumed $1 + \cos^2\theta$ dependence for the ψ decay, where θ is the angle between the photon and the beam direction), the Mark II measures the branching fraction $B(\psi \to \gamma\eta') = (3.4 \pm 0.7) \times 10^{-3}$.

Due to the bias imposed by the trigger requirement,[23] the Mark II is unable to observe the reaction

$$e^+e^- \to \psi \to 3\gamma \; .$$

Table I. Branching fractions for radiative transitions
from the ψ to the η and η'.

decay	mode	branching fraction	experiment
$\psi \to \gamma\eta'$	$\rho°\gamma$	$(3.4 \pm 0.7) \times 10^{-3}$	Mark II
	$\gamma\gamma$	$(6.9 \pm 1.7) \times 10^{-3}$	Crystal Ball
	$\gamma\gamma$	$(2.2 \pm 1.7) \times 10^{-3}$	DASP[a]
	$\rho°\gamma$	$(2.4 \pm 0.7) \times 10^{-3}$	DESY-Heidelberg[b]
		$\sim3.3 \times 10^{-3}$	theory[c]
$\psi \to \gamma\eta$	$\gamma\gamma$	$(0.9 \pm 0.4) \times 10^{-3}$	Mark II
	$\gamma\gamma$	$(1.2 \pm 0.2) \times 10^{-3}$	Crystal Ball
	$\gamma\gamma$	$(0.8 \pm 0.2) \times 10^{-3}$	DASP[a]
	$\gamma\gamma$	$(1.3 \pm 0.4) \times 10^{-3}$	DESY-Heidelberg[b]
		0.9×10^{-3}	theory[c]

[a] Ref. 9
[b] Refs. 8, 11
[c] Ref. 17

Table II summarizes the measurement of the ratio $B(\psi \to \gamma\eta')/B(\psi \to \gamma\eta)$. The measured values range from approximately 2 to 6, and the theoretical prediction is 3.9. Thus, I think it is fair to say that we have a reasonable understanding of the M1 transitions from the ψ to the ordinary pseudoscalar meson states.

In order to further explore the properties of the charmonium system, the Crystal Ball and Mark II collaborations have begun similar studies of radiative transitions from the ψ'. Naively, one would expect these branching fractions to be approximately an order of magnitude smaller than the corresponding branching fractions at the ψ.[25] Presently, no evidence for $\gamma\eta$ or $\gamma\eta'$ production from the ψ' has been observed, with preliminary 90% confidence level upper limits from the Crystal Ball of $B(\psi' \to \gamma\eta') < 8 \times 10^{-4}$ and $B(\psi' \to \gamma\eta) < 10^{-4}$. As these limits are

Table II. $B(\psi \to \gamma\eta')/B(\psi \to \gamma\eta)$.

ratio	experiment
3.8 ± 1.9	Mark II
5.9 ± 1.5	Crystal Ball
2.8 ± 2.3	DASP[a]
1.8 ± 0.8	DESY-Heidelberg[b]
3.9	theory[c]

[a] Ref. 9
[b] Refs. 8, 11
[c] Ref. 17

only a factor of eight below the measured ψ branching fractions, there is no reason to worry about the absence of these signals at this time.

IV. $\psi \to \gamma f(1270)$

In order to understand the radiative transition to the $f(1270)$, I will once more consider the two processes shown in Fig. 2. The measured branching fraction for the process $\psi \to \omega f$ is approximately 3×10^{-3}.[26] Invoking VMD for the process shown in Fig. 2(a), one is led to expect a rate for the γf transition which is considerably less than the measured value.[13,14] Thus, even in this case, where the final state has $J^P = 2^+$ rather than 0^-, it appears that the process in Fig. 2(b) is dominant.

New measurements of the γf transition have been made by the Mark II.[27] Figure 7 shows the $\pi^+\pi^-$ invariant mass distribution (data points with error bars) for events which satisfy a fit to the hypothesis

$$\psi \to \pi^+\pi^-\gamma \tag{3}$$

with $\chi^2 < 15$. Two structures are evident in the mass distribution, one at the ρ mass and the other at the $f(1270)$ mass. Since the decay $\psi \to \rho^\circ\gamma$ does not conserve charge conjugation parity (C-parity), it is assumed that the events in the ρ° mass region resulted from $\rho^\circ\pi^\circ$ decays

338

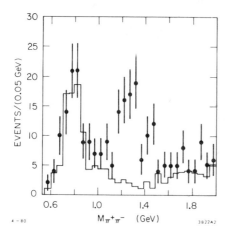

4 – 80   3822A2

Fig. 7. $\pi^+\pi^-$ invariant mass
distribution for events satisfying (3). Histogram shows
the expected feeddown from
the $\pi^+\pi^-\pi^0$ final state as
determined by Monte Carlo.
(Mark II)

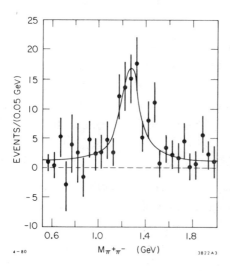

4 – 80   3822A3

Fig. 8. $\pi^+\pi^-$ invariant mass
distribution after subtraction
of $\pi^+\pi^-\pi^0$ feeddown. Curve is
described in text. (Mark II)

in which an asymmetric decay of the
π^0 led to an acceptable fit to (3).
A Monte Carlo was used to determine
the $\pi^+\pi^-\pi^0$ feeddown into the $\pi^+\pi^-\gamma$
channel. The resulting distribu-
tion (including production of both
$\rho^0\pi^0$ and $\rho^\pm\pi^\mp$) is compared with the
data in Fig. 7 and can clearly
account for the observed ρ^0 peak.

Figure 8 shows the $\pi^+\pi^-$ mass
distribution after subtraction of
the $\pi^+\pi^-\pi^0$ background. The distribution is dominated by the f. An
expression consisting of a Breit-
Wigner resonance term plus a flat
background term was fitted to this
distribution. The curve in Fig. 8
shows the best fit which gave
M = 1280 MeV and Γ = 180 MeV for
the resonance parameters. The
branching fraction for (3) was
found to be B($\psi \rightarrow \gamma$f) =
$(1.3 \pm 0.3) \times 10^{-3}$. This branching fraction is consistent with
the previously measured values of
B($\psi \rightarrow \gamma$f) = $(2.0 \pm 0.3) \times 10^{-3}$
from PLUTO[13] and B($\psi \rightarrow \gamma$f) between $(0.9 \pm 0.3) \times 10^{-3}$ and
$(1.5 \pm 0.4) \times 10^{-3}$ (depending on
the helicity of the f in the final
state) from DASP.[14]

As pointed out in the previous
section, we seem to have a fairly
good understanding of the transitions to the $I_z = 0$ members of the

$J^P = 0^-$ nonet. If measurements of the radiative transitions to the f'
and A_2^o could be made, we would have an additional check on the theoretical ideas discussed previously. The Mark II has preliminary results which show no evidence for transitions to either of these two
states. They give 90% confidence level upper limits of $B(\psi \rightarrow \gamma f') \times$
$B(f' \rightarrow K\overline{K}) < 10^{-3}$ and $B(\psi \rightarrow \gamma A_2^o) < 10^{-3}$. Unfortunately, these limits
are not yet small enough to provide meaningful constraints on models.
As in the case of the $\gamma\pi^\circ$ transition, one expects to see a very small
branching fraction for γA_2^o because of isospin conservation. However,
the $\gamma f'$ transition should be observable. Based on a naive calculation
assuming SU(3) invariance (similar to the $\eta-\eta'$ calculation described
earlier),[28] one expects

$$\frac{B(\psi \rightarrow \gamma f')}{B(\psi \rightarrow \gamma f)} = \frac{1}{2} .$$

The Mark II limit is not yet inconsistent with this prediction.

V. $\psi \rightarrow \gamma E(1420)$

As the E(1420) is a fairly obscure resonance, I will briefly
review what was known about the E as of the last (1978) Particle Data
Group tables[29] before discussing the results on the γE radiative transition. The E is a fairly narrow resonance with width estimates
ranging from 40 to 80 MeV. Measurements of the mass lie between 1400
and 1440 MeV. None of the quantum numbers of the E have been firmly
established. The isospin is believed to be zero as no charged E has
ever been observed; the C-parity is believed to be even; and analyses
of the decay Dalitz plot favor an abnormal spin-parity assignment.
$J^P = 0^-$ and 1^+ are the preferred values. The principally observed
decay mode is $K\overline{K}\pi$, but there is some evidence for an $\eta\pi\pi$ decay mode.
Finally, up until 1978, the best signals for the E were observed in
$\overline{p}p$ annihilations at rest. I will mention only one of these experiments here. Baillon et al.[30] studied a sample of $\overline{p}p$ annihilations in
the CERN 81-cm hydrogen bubble chamber. They did a spin-parity analy-
sis of the E observed in the reaction $\overline{p}p \rightarrow E\pi\pi$ and determined $J^P = 0^-$.
The Mark II sees evidence for the process[12]

$$\psi \rightarrow \gamma E , \quad E \rightarrow K_S K^\pm \pi^\mp . \tag{4}$$

Observation of this transition establishes C = + for the E. Figure 9(a) shows the $K_S K^\pm \pi^\mp$ invariant mass for events satisfying the 5-constraint (5C) fit to (4) with $\chi^2 < 15$.[31] The constraints are the normal ones of energy-momentum conservation with an additional constraint for the K_S mass. A peak is seen near the mass of the E(1420). One is not compelled to interpret this structure as the E(1420), but due to the similar characteristics of this structure and the previously observed E, I will make this tentative assignment.

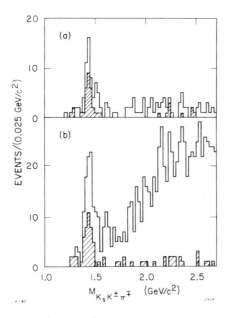

Fig. 9. $K_S K^\pm \pi^\mp$ invariant mass distributions for events satisfying (a) 5C fits and (b) 2C fits (i.e., observation of the photon is not required) to (4). Shaded regions have the additional requirement $M_{K\overline{K}} < 1.05$ GeV. (Mark II)

The parameters of the resonance were obtained by fitting the invariant mass distribution to a Breit-Wigner[32] plus a smooth background. The Mark II finds $M = 1.44^{+0.01}_{-0.015}$ GeV and $\Gamma = 0.05^{+0.03}_{-0.02}$ GeV. These errors include systematic uncertainties due to the functional form used in the fit. The branching fraction product, based on 47 ± 12 observed events, is $B(\psi \to \gamma E) \times B(E \to K_S K^\pm \pi^\mp) = (1.2 \pm 0.5) \times 10^{-3}$.[33] With the assumptions that the E is an isoscalar and that K_S and K_L production are equal in the decay of the E, one can relate the $K^+ K^- \pi^0$, $K^\circ \overline{K}^\circ \pi^0$, and $K^\circ K^\pm \pi^\mp$ branching fractions and determine the branching fraction product $B(\psi \to \gamma E) \times B(E \to K\overline{K}\pi) = (3.6 \pm 1.4) \times 10^{-3}$.

Previous experiments[29] have found the decay of the E to be associated with a low mass $K\overline{K}$ enhancement which is also observed by the Mark II. If a cut requiring $M_{K\overline{K}} < 1.05$ GeV is imposed on the data, the shaded region in Fig. 9(a) is obtained.

Since the signal is quite clean, it is possible to relax the requirement that the photon be observed. The resulting 2C fit to (4) is shown in Fig. 9(b). Although there is an improvement in statistics,[34] there is also an increase in the background level. However, as shown by the shaded region, the $\overline{KK}$ mass cut again substantially reduces the background.

The Dalitz plot for the sample of events shown in Fig. 9(b) with masses between 1.375 and 1.500 GeV (the signal region) is shown in Fig. 10. The curves show the low-mass and high-mass kinematic boundaries and the dashed lines show the nominal K*(890) mass values. The points are plotted as functions of the $(K\pi)^{\circ}$ invariant mass squared vs. the $(K\pi)^{\pm}$ invariant mass squared. The $\overline{KK}$ axis, if it were shown, would be at an angle approximately bisecting the two $K\pi$ axes. One sees an excess of events in the upper right-hand corner of the Dalitz plot.[35] It is not clear whether these events correspond to a low-mass $\overline{KK}$ enhancement (spread out by the movement of the kinematic boundary as the $\overline{KK}\pi$ mass changes), or to constructive interference where the K* bands overlap.

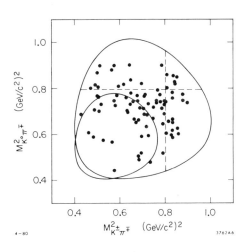

Fig. 10. Dalitz plot for events with $1.375 \leq M_{\overline{KK}\pi} < 1.500$ GeV. Curves show low-mass and high-mass kinematic boundaries. Dashed lines show nominal K* mass values. (Mark II)

Figure 11(a) shows the $K_S K^{\pm}$ invariant mass distribution for events in the signal region and Fig. 11(b) shows the corresponding distribution for events outside the signal region. There is evidence for a low-mass $\overline{KK}$ enhancement for events in the signal region which is absent for events outside the signal region. One possible interpretation of this enhancement is the $\delta(980)$.

In an attempt to understand the decay mechanism of the E, fits were made to the Dalitz plot which included $K^*\overline{K}$ (the inclusion of both this state and the charge conjugate state are implied by this notation), $\delta\pi$, and phase space contributions. These three contributions

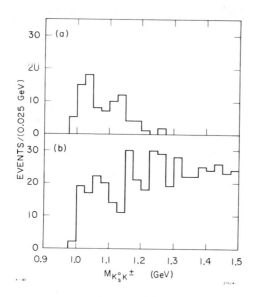

Fig. 11. $K_S K^{\pm}$ invariant mass distributions for events (a) in the signal region and (b) outside of the signal region. (Mark II)

were added incoherently, but the $K^* \bar{K}$ contribution included components from both the charged and neutral K^* states, which were assumed to interfere constructively where they cross on the Dalitz plot (as demanded by the even C-parity of the E). The best fit favors $\delta \pi$ as the primary component of the decay with

$$\frac{B(E \to \delta \pi) \times B(\delta \to K \bar{K})}{B(E \to K \bar{K} \pi)} = 0.8 \pm 0.2$$

The quoted error does not include possible systematic errors. One has to be careful in interpreting this result, as the best fit to the Dalitz plot does not completely simulate the $K \bar{K}$ invariant mass distribution. This indicates that the decay mechanism is not completely understood.

An attempt has been made to determine the spin of the E by analysis of the double decay angular distribution for events consistent with

$$\psi \to \gamma E \ , \ E \to \delta \pi \ .$$

However, the limited statistics do not allow a statistically significant determination of the spin.

Preliminary results from the Crystal Ball also show evidence for the transition $\psi \to \gamma E$.[7] Figure 12 shows the $K^+ K^- \pi^\circ$ invariant mass distribution[36] for events which satisfy the 2C fit to

$$\psi \to \gamma K^+ K^- \pi^\circ \ , \tag{5}$$

with $M_{K \bar{K}} < 1.1$ GeV. Although the Crystal Ball detector has excellent energy resolution for photons, the absence of a magnetic field does not allow a momentum measurement for charged particles. This reduces the constraint class for (5) from 4 to 2. Evidence for an E signal is seen in this distribution.

As the Crystal Ball efficiency calculations are still in a very preliminary state, estimates of the branching fraction are only good to a factor of two at best. When corrections are made for the K^+K^- mass cut[37] and the unobserved decay modes of the E, they find $B(\psi \to \gamma E) \times B(E \to K\bar{K}\pi) \approx 2 \times 10^{-3}$.

As was mentioned earlier, there is some evidence for the decay of the E into $\eta\pi^+\pi^-$. Fig. 13 shows the $\eta\pi^+\pi^-$ invariant mass distribution (from the Crystal Ball) for events satisfying fits to

$$\psi \to \gamma\eta\pi^+\pi^- \tag{6}$$

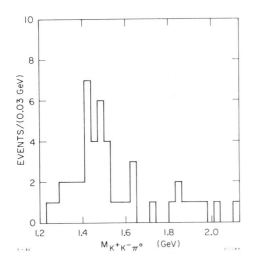

Fig. 12. $K^+K^-\pi^\circ$ invariant mass distribution for events satisfying (5) with $M_{K\bar{K}} < 1.1$ GeV. (Crystal Ball)

In addition to the η' signal, there is evidence for a peak in the E mass region. A preliminary estimate of the branching fraction product $B(\psi \to \gamma E) \times B(E \to \eta\pi\pi)$ finds it to be smaller than the corresponding number for $K\bar{K}\pi$, but a firm number will have to wait until calculations of the efficiencies are made.

In summary, the E is observed very strongly in radiative transitions from the ψ. The only other transition that has been observed with a comparable branching fraction is the $\gamma\eta'$ transition. The possible significance of this will be discussed in the next section. Observation of this transition

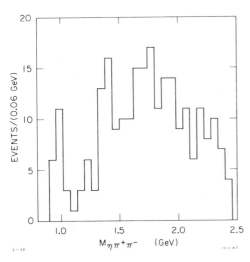

Fig. 13. $\eta\pi^+\pi^-$ invariant mass distribution for events satisfying (6). (Crystal Ball)

has established the C-parity of the E as even. Unfortunately, a determination of the spin is impossible with the present statistics. Finally, the Mark II finds the $\overline{K}K\pi$ decay mode of the E to be predominantly δπ. The consequences of this will also be discussed in the next section.

VI. REVIEW OF THE STATUS OF THE E(1420)

I was asked by the organizers of this conference to include a review of the status of the E(1420) in my talk. Although this is somewhat outside the original scope of the talk, namely charmonium studies, I agreed as I think an understanding of the E could have important consequences in regard to understanding the charmonium system. As I have already given a brief introduction to the status of the E as of 1978, I will confine my discussion to two recent hadronic experiments which observe the E, and a comparison of their results with those of the Mark II.

The first results are from a high statistics (90 events/μb) bubble chamber experiment in which the reaction

$$\pi^- p \rightarrow K_S K^{\pm} \pi^{\mp} n \qquad (7)$$

was studied at 3.95 GeV/c.[38] Figure 14 shows the $K_S K^{\pm} \pi^{\mp}$ invariant mass for events which satisfy the 1C fit (the neutron was not observed) to (7). Evidence is seen for both D(1285) and E(1420) production. A fit to the $K_S K^{\pm} \pi^{\mp}$ invariant mass distribution yields values for the E mass and width of M = 1426 ± 6 MeV and Γ = 40 ± 15 MeV.[39] These errors are statistical only.

The Dalitz plot for events in the region $1.39 \leq M_{\overline{K}K\pi} \leq 1.47$ GeV is shown in Fig. 15. As was observed in the Mark II data, there is evidence for an enhancement in the upper right-hand corner of the Dalitz plot. However, in this case, there is also clear evidence for K*(890) production. A partial-wave analysis of the data determined the spin-parity of the E to be $J^P = 1^+$, and also determined the branching fraction ratio

$$\frac{B(E \rightarrow K^* \overline{K})}{B(E \rightarrow K^* \overline{K} + \delta\pi)} = 0.86 \pm 0.12 \ .$$

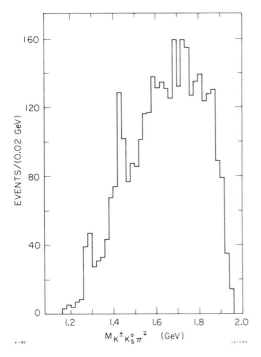

Fig. 14. $K_S K^{\pm} \pi^{\mp}$ invariant mass distribution for events satisfying (7). Data is from Ref. 38.

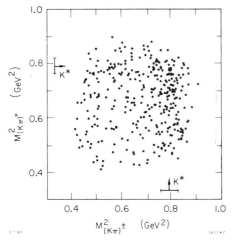

Fig. 15. Dalitz plot for events with $1.39 \leq M_{K\bar{K}\pi} \leq 1.47$ GeV. Data is from Ref. 38.

However, it should be pointed out that the E signal is over a relatively large background which has a significant $K^{*}\bar{K}$ component, so that one should regard this result with caution.

In another experiment, the reaction

$$\pi^- p \to K_S K^{\pm} \pi^{\mp} + X \qquad (8)$$

was studied at 50 and 100 GeV/c.[40] The $K_S K^{\pm} \pi^{\mp}$ invariant mass distribution for this sample of events, in Fig. 16, shows no evidence for an E signal. However, if a δ cut is applied, $M_{K\bar{K}} < 1.04$ GeV, both the D and the E become quite prominent, as shown in Fig. 17. If instead of a δ cut, a cut is applied requiring one of the $K\pi$ invariant mass combinations to be in the K^* mass region ($0.84 < M_{K\pi} < 0.94$ GeV), one still sees an E signal, but with considerably worse background. A fit to the $K_S K^{\pm} \pi^{\mp}$ mass distribution in Fig. 17 yielded values of the resonance parameters of $M = 1440 \pm 6$ MeV and $\Gamma = 110 \pm 27$ MeV. (The curve in Fig. 17 represents the best fit to the data.) The errors are statistical only. The systematic errors, especially for the

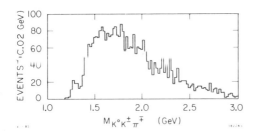

Fig. 16. $K_S K^{\pm} \pi^{\mp}$ invariant mass distribution for events satisfying (8). Data is from Ref. 40.

width, are probably large. Another fit made to a similar spectrum (after subtraction of the estimated background due to $K-\pi$ misidentification) yielded $M = 1440 \pm 5$ MeV and $\Gamma = 62 \pm 14$ MeV. On the surface, this data seems to indicate a preference for the $\delta\pi$ decay mode of the E over the $K^*\overline{K}$ decay mode. However, questions of kinematic overlap in the Dalitz plot and phase space boundaries have not been considered in detail. Thus, this preference should be considered only as an indication until a more sophisticated analysis is done.

Despite all the new information on the E from recent experiments, the situation is not much clearer than it was in 1978. One point of controversy is whether the E decays predominantly into $\delta\pi$ or $K^*\overline{K}$. The Mark II (and possibly also the Fermilab experiment of Bromberg et al.[40]) seem to favor the decay $E \to \delta\pi$. On the other hand, Dionisi et al.[38] see little evidence for $\delta\pi$ and find the predominant decay of the E is into $K^*\overline{K}$. As for the spin, Dionisi et al. find $J^P = 1^+$ which agrees with some earlier results, but disagrees with others. However, their determination of the spin goes hand-in-hand with the determination of the predominance of the $K^*\overline{K}$ decay mode. Since this predominance is not firmly established,

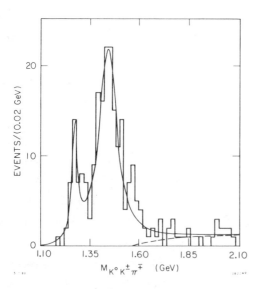

Fig. 17. $K_S K^{\pm} \pi^{\mp}$ invariant mass distribution with $M_{\overline{KK}} < 1.04$ GeV. Solid curve shows fit to mass spectrum. Dashed curve shows background distribution determined from fit. Data is from Ref. 40.

I think that one should still consider the spin of the E to be an open question until the decay mechanism is understood better.[41]

To understand my reasons for this excessive interest in the quantum numbers of the E, let me refer for the last time to Fig. 2(b). As discussed yesterday by Donoghue,[42] if gluonium states[43] exist, the process shown in Fig. 2(b), after elimination of the outgoing quark lines, would be an ideal process for production of such states. I would like to suggest the possibility that the E might be such a gluonium state, rather than an ordinary $q\bar{q}$ resonance. Although there is certainly no real evidence for this hypothesis, there are a few peculiarities associated with the γE radiative transition from the ψ which I would like to point out.

First, the branching fraction for $\psi \to \gamma$E is larger than the corresponding branching fractions for transitions to other ordinary hadrons, with the possible exception of the η'. This is in contrast to hadronic experiments where E production is in general small compared to the production of other resonances. This would lead one to infer a connection between the E and the 2-gluon intermediate state in Fig. 2(b). Whereas the production of gluonium states is expected to be significant in ψ radiative transitions, there is no reason to expect significant production of such states in hadronic reactions.

Second, whereas in most hadronic experiments in which an E is observed to decay into $K\bar{K}\pi$, one observes roughly comparable D(1285) production, neither the Mark II nor the Crystal Ball see much evidence for D production. The Mark II gives an upper limit for D production of $B(\psi \to \gamma D) \times B(D \to K\bar{K}\pi) < 0.7 \times 10^{-3}$ at the 90% confidence level. This might be taken as strong evidence for a difference in the production mechanisms involved in the two different processes, and hence an indication of a large gluonium component in the E. However, if one assumes that the D and E are both members of the standard $J^{PC} = 1^{++}$ nonet, and the E is the primarily singlet state and the D is the primarily octet state,[44] one would expect D production to be suppressed relative to E production because of SU(3) symmetry arguments. Thus, this suppression may not be relevant to the gluonium question at all.

In my opinion, the most important question which should be resolved regarding the E is its spin. If the E can be firmly established as an axial vector state, there is no reason not to make the standard $q\bar{q}$ meson interpretation and put it in the same nonet as the D(1205), A_1, and Q_A. If, on the other hand, the E is finally established as a pseudoscalar, it is difficult to interpret it within the standard quark model. The $J^P = 0^-$ nonet is complete, and one would have to consider the existence of another 0^- nonet, possibly a radial excitation of the ground state, in order to accommodate the E. However, I think it is equally plausible to interpret the E as a gluonium state.

VII. CONCLUSIONS

The first part of this talk dealt with radiative transitions from the ψ to the η, η', and f. The new results from the Mark II and Crystal Ball collaborations are basically compatible with previous results (ignoring minor factor-of-two problems with the η'). I tried to emphasize that these transitions can be understood in terms of minimal gluon-coupling ideas, with mixing between the different isoscalar states. Further work is being done to extend our understanding of these processes. The Mark II is in the process of studying the radiative transitions to the other tensor states, the f' and A_2°. Another direction which is being pursued by both the Mark II and the Crystal Ball collaborations is an analysis of similar radiative transitions from the ψ'. As mentioned previously, these transitions are expected to have branching fractions approximately an order of magnitude smaller than the corresponding ψ transitions. This should be verified, and may lead to surprises.

The rest of the talk dealt with the E(1420). As I discussed in detail in the previous section, it is interesting to entertain the possibility that the E is a gluonium state. If this were true, it would open up a whole new field of spectroscopy. However, let me emphasize that even if the spin-parity of the E were determined to be 0^-, there would be no compelling reason to believe that it is a gluonium state.

Although it was not emphasized during the talk, there has been some effort by the Mark II collaboration to look for other radiative transitions from the ψ. All states with reasonable acceptance in the Mark II detector (i.e., states decaying into combinations of $\pi^{\pm}$, $K^{\pm}$, K_S, p, and $\bar{p}$), and even some with poor acceptance (e.g., states with π°'s or η's in the final state), have been considered. No statistically significant signals aside from those shown today have been observed. Thus, if the E is not a gluonium state, neither the Mark II nor the Crystal Ball has any evidence for such a state.[45]

Let me conclude by remarking that in addition to the understanding of the charmonium system that can be gained by studying ψ decays (in particular, radiative transitions), there is also the possibility of being able to study ordinary (i.e., non-charmed) hadrons in a cleaner environment than can be obtained in typical hadronic interactions.

REFERENCES

1. E. D. Bloom, invited talk this conference.
2. Members of the SLAC-LBL Mark II collaboration: G. Abrams, M. Alam, C. Blocker, A. Boyarski, M. Breidenbach, D. Burke, W. Carithers, W. Chinowsky, M. Coles, S. Cooper, W. Dieterle, J. Dillon, J. Dorenbosch, J. Dorfan, M. Eaton, G. Feldman, M. Franklin, G. Gidal, G. Goldhaber, G. Hanson, K. Hayes, T. Himel, D. Hitlin, R. Hollebeek, W. Innes, J. Jaros, P. Jenni, D. Johnson, J. Kadyk, A. Lankford, R. Larsen, V. Lüth, R. Millikan, M. Nelson, C. Pang, J. Patrick, M. Perl, B. Richter, A. Roussarie, D. Scharre, R. Schindler, R. Schwitters, J. Siegrist, J. Strait, H. Taureg, M. Tonutti, G. Trilling, E. Vella, R. Vidal, I. Videau, J. Weiss, and H. Zaccone.
3. Members of the Crystal Ball collaboration. California Institute of Technology, Physics Department: R. Partridge, C. Peck and F. Porter. Harvard University, Physics Department: D. Andreasyan, W. Kollmann, M. Richardson, K. Strauch and K. Wacker. Princeton University, Physics Department: D. Aschman, T. Burnett, M. Cavalli-Sforza, D. Coyne, M. Joy and H. Sadrozinski. Stanford Linear Accelerator Center: E. D. Bloom, F. Bulos, R. Chestnut, J. Gaiser, G. Godfrey, C. Kiesling, W. Lockman and M. Oreglia.

Stanford University, Physics Department and High Energy Physics
Laboratory: R. Hofstadter, R. Horisberger, I. Kirkbride,
H. Kolanoski, K. Koenigsmann, A. Liberman, J. O'Reilly and
J. Tompkins.

4. G. S. Abrams et al., Phys. Rev. Lett. 44, 114 (1980); D. L. Scharre
    et al., Stanford Linear Accelerator Center Report No. SLAC-PUB-2513
    (1980), to be submitted for publication.

5. M. T. Ronan et al., Phys. Rev. Lett. 44, 367 (1980).

6. Details of the experimental apparatus can be found in E. D. Bloom,
    in Proceedings of the Fourteenth Rencontre de Moriond, Vol. II,
    edited by Trân Thanh Vân (R.M.I.E.M. Orsay, 1979), p. 175; and
    C. W. Peck et al., California Institute of Technology Report No.
    CALT-68-753 to be published in the Proceedings of the Annual
    Meeting of the American Physical Society, Division of Particles
    and Fields, McGill University, Montreal, Canada, October 25-27,
    1979.

7. D. G. Aschman, to be published in the Proceedings of the Fifteenth
    Rencontre de Moriond, Les Arcs, France, March 15-21, 1980.

8. W. Bartel et al., Phys. Lett. 66B, 489 (1977).

9. W. Braunschweig et al., Phys. Lett. 67B, 243 (1977).

10. R. Partridge et al., Phys. Rev. Lett. 44, 712 (1980).

11. W. Bartel et al., Phys. Lett. 64B, 483 (1976).

12. D. L. Scharre et al., Stanford Linear Accelerator Center Report
    No. SLAC-PUB-2514 (1980), to be submitted for publication.

13. G. Alexander et al., Phys. Lett. 72B, 493 (1978).

14. R. Brandelik et al., Phys. Lett. 74B, 292 (1978).

15. W. Braunschweig et al., Phys. Lett. 63B, 487 (1976).

16. B. Jean-Marie et al., Phys. Rev. Lett. 36, 291 (1976).

17. H. Fritzsch and J. D. Jackson, Phys. Lett. 66B, 365 (1977).

18. Because of the six-fold symmetry of the final state, the Dalitz
    plot has been folded. This results in the observed folding of
    the η and η' bands at the boundary.

19. D. L. Scharre, in Proceedings of the Fourteenth Rencontre de
    Moriond, Vol. II, edited by Trân Thanh Vân (R.M.I.E.M. Orsay,
    1979), p. 219.

20. However, due to problems with the Mark II liquid argon shower counter system, there was no photon detection for approximately half of the running time. Thus, any analyses which required photon detection were based on half of the total data sample.

21. G. S. Abrams et al., Phys. Rev. Lett. 43, 477 (1979); G. S. Abrams et al., Phys. Rev. Lett. 43, 481 (1979); G. S. Abrams et al., Phys. Rev. Lett. 43, 1555 (1979).

22. Due to noise in the LA electronics, spurious photons were occasionally reconstructed by the tracking program. In order not to lose good events, extra photons were allowed in candidate events. When analyzing these events, separate fits were attempted for each two-photon combination.

23. In general, two or more charged tracks were required to trigger the detector.

24. Approximately one million ψ' events were taken. From this sample 92,000 ψ decays were identified by missing mass from the $\pi^+\pi^-$ system.

25. This ratio is expected to be roughly the same as the ratio of the leptonic branching fractions of the ψ' and ψ.

26. J. Burmester et al., Phys. Lett. 72B, 135 (1977); F. Vanucci et al., Phys. Rev. D 15, 1814 (1977).

27. C. Zaiser et al., in preparation.

28. It is expected that the mixing due to gluon exchange will not affect this ratio as severely as in the case of the pseudoscalar nonet.

29. Particle Data Group, Phys. Lett. 75B, 1 (1978) and references therein.

30. P. Baillon et al., Nuovo Cimento 50A, 393 (1967).

31. This distribution includes a few additional events corresponding to the process $\psi' \to \pi^+\pi^-\psi$, $\psi \to \gamma E$. Details can be found in Ref. 12.

32. The mass resolution of these constrained events is considerably smaller than the natural line width of the resonance and is ignored in the fit.

33. The efficiency used in the determination of this branching fraction was based on a Monte Carlo analysis which assumed all decay distributions were isotropic. If the spin-parity of the E were 0^-, which results in a $1 + \cos^2\theta$ distribution for the angle of the photon with respect to the beam axis, the branching ratio product should be increased by 19%.

34. The increase in sample size arises principally from the fact that the sample of data in which the LA system was not operational could be used.

35. Monte Carlo analysis shows the acceptance to be roughly flat over the entire Dalitz plot. Hence, the observed structure is not the result of variations in the acceptance.

36. This decay mode is expected to be half the $K_S K^{\pm} \pi^{\mp}$ mode by isospin conservation.

37. The correction was based on the $K_S K^{\pm}$ mass distribution for the sample of $E \to K_S K^{\pm} \pi^{\mp}$ events from the Mark II.

38. C. Dionisi et al., CERN Report No. CERN/EP 80-1 (1980), submitted to Nucl. Phys. B.

39. The actual fit was made to a distribution which required at least one $K\pi$ combination within the K^* mass region, with some additional kinematic cuts to reduce the background.

40. The data are from experiment E110 using the Fermilab multiparticle spectrometer [C. Bromberg et al., California Institute of Technology Report No. CALT-68-747 (1980)].

41. It has been suggested that the resonance seen in hadronic experiments and known as the E(1420) is not the same state that has been observed by the Mark II in radiative transitions from the ψ. However, because of the consistency of the parameters of the states seen in these two processes and the outward similarity of the Dalitz plots, I think it is logical to consider them to be the same state until some evidence to the contrary is produced.

42. J. Donoghue, invited talk this conference.

43. A gluonium state is a bound state of two or more gluons.

44. Unfortunately, the masses of the other members of the nonet (the A_1 and the Q_A) are not well enough known to provide a reliable estimate of the mixing angle.

45. This does not mean that none exist. Neither the masses nor the widths are well defined theoretically, and the decay modes expected for such states are often such that their detection by the Mark II would be difficult. For a brief but excellent review of the possibilities, see J. D. Bjorken, Stanford Linear Accelerator Center Report No. SLAC-PUB-2366, to be published in the Proceedings of the 1979 EPS High Energy Physics Conference, Geneva, Switzerland, June 27-July 4, 1979.

Prentice: There is some recent ηππ phase-shift data which shows a 0^- state sitting under the D. We interpret that as a radial excitation of either the η or the η'. There is probably another radial excitation a little higher up and there is some evidence in the 0^- wave in that phase-shift analysis for another slightly higher state. However, it seems that there is a difference in the ratio of ηππ and $K\bar{K}\pi$ between what you are seeing and what π^-p experiments see.

Scharre: I think that one could interpret the E as a radial excitation if it were 0^-, but I think that the state at 1260 is somewhat wider than the E. It's hard for me to understand the narrow width of the E if it were a radial excitation, although I do not know of any calculations of the expected width.

Question: On the subject of the branching ratios of the E, it's fairly important to decide whether it decays mainly into strange particles or not. If it does, you would expect it to be a quark state, for instance a hidden strangeness state, which is what you are looking for in the 1^+ sector. There is a way to get a handle on this. If you look at ηππ in π^-p, you see mostly D and very little E. If you look at $K\bar{K}\pi$, you see mostly E and very little D. Knowing the branching ratios of the $\delta(980)$, which is roughly equal for $K\bar{K}$ and ηπ, you can conclude from those two that the E is decaying more often into strange particles than it would if it were really decaying through the $\delta\pi$ intermediate state. This suggests that the E decays mainly into K*'s.

Scharre: I think that in the Dionisi experiment, the D and E production cross sections are equal within about a factor of two. In regard to the $\eta\pi\pi$ data, I have not seen an E signal that I would regard as conclusive. Finally, in regard to the $K\bar{K}$ vs. $\eta\pi$ decay modes of the δ, I think that the relative branching fractions are essentially undetermined. Therefore, in light of these uncertainties, I am not sure that we can gain any fundamental understanding from this type of comparison.

NEW RESULTS ON THE T-SYSTEM FROM DORIS

H. Schröder
Deutsches Elektronen-Synchrotron DESY, Hamburg,
Germany

presented at the VI International Conference on Experimental Meson
Spectroscopy, Brookhaven, April 25,26,1980

ABSTRACT

Further studies of e^+e^- annihilations in the T region at the
DESY storage ring DORIS have yielded improved results on the properties of the T mesons. For the T-meson the leptonic width Γ_{ee}
and branching ratios $B_{\mu\mu}$ are found to be Γ_{ee} = 1.29 ± 0.07 keV
and $B_{\mu\mu}$ = 3.2 ± 0.8 %. This gives a total width of the T-meson
of Γ_{tot} = 40 $^{+13}_{-8}$ keV. The leptonic width of the T'meson was determined to $\Gamma_{ee}(T')$ = 0.57 ± 0.06 keV.

The T mesons are by now well established[1-5]. They represent
quarkonium states of the heavy quark b and its antiquark. From
the leptonic width of the T mesons the charge of the b quark was
found to be e_b = -1/3. The mass spectrum is well known up to
the T''' mass just above the flavor threshold[6]. It has been shown
that the T decay pattern can be understood by assuming that it
consists mainly of decays into 3 gluons[7-9]. This implies that the
width of the T meson should be rather small, comparable to that of the
J/ψ particle which is 69 keV. Since up to now the upper limit for
$\Gamma_{tot}(T)$ amounted to some few MeV it was highly desirable to improve
this situation. The main point of my talk is therefore to present
an accurate value of this basic quantity.

The total width Γ_{tot} is given by

$$\Gamma_{tot} = \Gamma_{ee} / B_{\mu\mu} \qquad (1)$$

Is is mainly due to the fact that $B_{\mu\mu}$ has large errors[10,11] that
Γ_{tot} is poorly known. Thus the experiments carried out at the DESY
storage ring DORIS aimed for a better determination of $B_{\mu\mu}$ (T). Besides this result refined values for $\Gamma_{ee}(T)$ and $\Gamma_{ee}(T')$ will be
given.

The DORIS machine was operating in a single-ring-single-bunch
mode. It delivered a peak luminosity of 10^{30} cm^{-2} s^{-1} at beam currents of 18 mA. With a beam lifetime of about 4 hours and a refill
time of few minutes the average luminosity amounted to 30 nb^{-1} day^{-1}.
During the last months of 1979 and in the beginning of 1980 an integrated luminosity of about 1000 nb^{-1} was taken both in the T and
T' region.

The experiments were carried out with the DASP detector (fig. 1) and the NaI-LG detector (fig. 2), which are well known for many years. The DASP-detector was operated by the DASP2-group and the NaI-LG-detector by the LENA-group

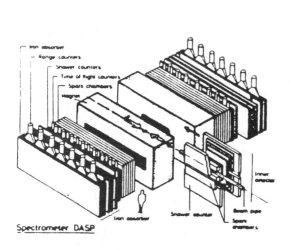

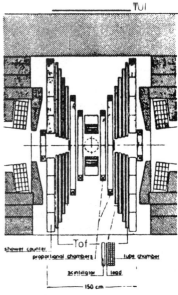

Fig. 1
DASP - detector

(a) spectrometer   (b) inner detector

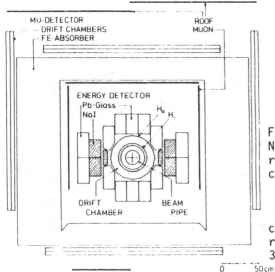

Fig. 2
NaI-Lead Glass detector with roof and bottom-time-of-flight counters

The measured hadronic cross section for each experiment is displayed in figs. 3 and 4.

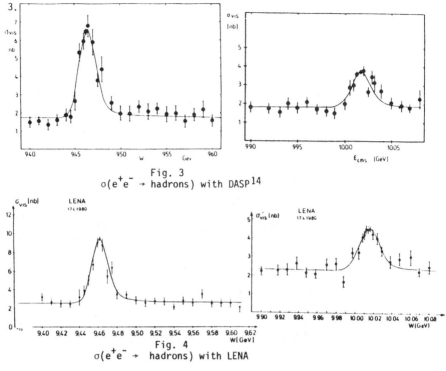

Fig. 3
$\sigma(e^+e^- \to$ hadrons) with DASP[14]

Fig. 4
$\sigma(e^+e^- \to$ hadrons) with LENA

It was expressed in terms of a continuum contribution and a Breit-Wigner function

$$\sigma_{had} = \frac{3\pi}{s} \left(\frac{4\alpha^2}{9} \cdot R_{had} \cdot \eta_1 + \frac{\Gamma_{ee} \, \Gamma_{had}}{(M-\sqrt{s})^2 + \Gamma^2/2} \cdot \eta_2 \right) \quad (2)$$

This expression was folded with the gaussian energy resolution of the machine and radiative corrections were applied. Knowing the efficiencies η_1 for two-jet events and η_2 for mainly 3-jet-events the following quantities could be extracted (table I):

TABLE I: T and T' resonance parameters*

	DASP2	LENA	AVERAGE	
M (T)	9463.1 ± 0.7±10	9461.6 ± 0.6±10	9462.4 ± 0.5±10	MeV
M (T')	10016.8 ± 1.5±10	10014.2 ± 0.9±10	10014.9 ± 0.8±10	MeV
M (T')-M(T)	553.7 ± 1.7±10	552.6 ± 1.0±10	553.1 ±0.06±10	MeV
$\Gamma_{ee}(1-3B_{\mu\mu})$(T)	1.23± 0.09±0.2	1.10 ± 0.07±0.11	1.15±0.06	keV
$\Gamma_{ee}(1-3B_{\mu\mu})$(T')	0.61± 0.11±0.11	0.55 ± 0.07±0.06	0.57±0.06	keV

* The first error is the statistical error whereas the second one is the systematic error.

The muonic branching ratio is given by

$$B = \frac{\sigma(e^+e^- \to \Upsilon \to \mu^+\mu^-)}{\sigma(e^+e^- \to \Upsilon \to had) + 3 \cdot \sigma(e^+e^- \to \Upsilon \to \mu^+\mu^-)} \tag{3}$$

The muonic resonance cross section is determined by measuring the muon cross section in the resonance and subtracting the QED-contribution from the continuum. Due to a lack of statistics only the $B_{\mu\mu}$ for the Υ meson could be determined.

Muon pairs are required to be collinear. They are recognized by their ability to penetrate large amounts of material without interacting (see figs. 1 and 2) and by time-of-flight (TOF) measurements in order to separate them from cosmic rays. This discrimination is illustrated fro the DASP inner detector in fig. 5 where the bunch crossing time is plotted versus the TOF difference between two opposite TOF counters. Muon pairs from the e^+e^- annihilation should be correlated with the bunch crossing at 16 ns and should have a zero TOF difference (see arrows in fig. 5). Cosmic rays are not correlated with the bunch signal and have a TOF difference of -6 ns. In fig. 5 one clearly sees an enhancement for muon pairs in

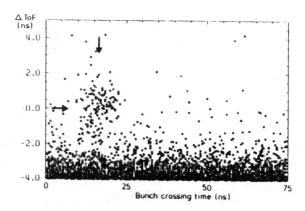

Fig. 5
Bunch crossing time versus TOF-difference
between two opposite counters

the right region. Fig. 6 shows a projection onto the TOF axis for muon pairs correlated (a) and uncorrelated with bunch crossing (b). The subtraction of both spectra yields the spectrum in fig. 6 c which exhibits a clear μ-pair signal. The angular distribution of these μ pairs shows the expected $(1 + \cos^2\theta)$ distribution (fig. 7).

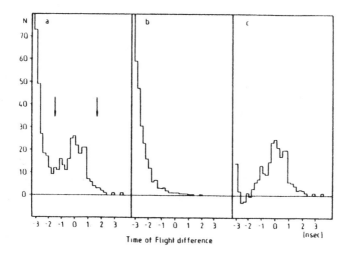

Fig. 6
TOF-difference for collinear, non showering particles

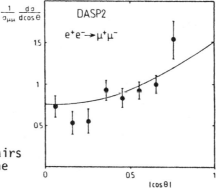

Fig. 7
Angular distribution of μ-pairs
from e⁺e⁻ annihilation at the
Υ-resonance energy

The resonance cross section for μ-pair production is then
given by:

$$\sigma(e^+e^- \to \Upsilon \to \mu^+\mu^-) = \sigma_{\mu\mu}^{QED} \left(\frac{N_{\mu\mu}^{on}}{N_{\mu\mu}^{cont}} \frac{L_{cont}}{L_{on}} - 1 \right) \qquad (4)$$

where $N_{\mu\mu}$ are the number of μ - pairs on the peak of the resonance
and in the continuum and the L's the corresponding luminosities.
$\sigma_{\mu\mu}^{QED}$ is the QED cross section $\sigma(e^+e^- \to \mu^+\mu^-)$ at the resonance energy.

From the measured numbers one then gets for the DASP inner
detector:

$$B_{\mu\mu} (\Upsilon) = 2.5 \pm 1.8 \%$$

and for the LENA detector

$$B_{\mu\mu}^{LENA}(T) = 3.5 \pm 1.4 \pm 0.5 \%$$

In the analysis of the DASP outer detector, which has an acceptance of only 5 % of 4π but an excellent particle discrimination, the number of μ-pairs coming from the decay of the T-meson is determined in the following way: There are 24 μ-pairs at the resonance energy whereas from 135 e^+e^- pairs in the outer detector one would expect only 14.3 ± 2 from the QED process. This excess of about 10 μ-pairs leads to a branching ratio of

$$B_{\mu\mu} = 5.6 \pm 3.3 \%.$$

Averaging the DASP2 values including the result from 1978[10] of $B_{\mu\mu}$ = 2.5 ± 2.1 % yields

$$B_{\mu\mu}^{DASP2}(T) = 2.9 \pm 1.3 \pm 0.5 \%$$

where the first error is statistical and the second one is systematic.

The PLUTO group has evaluated the Bhabha cross section from the first T-measurement at DORIS in 1978[15]. From the excess of e^+e^- pairs compared to the Bhabha cross section at backward angles (fig. 8) a branching ratio

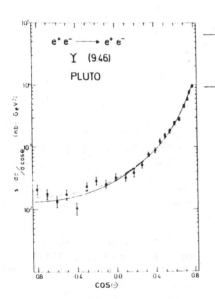

Fig. 8

Angular distribution of Bhabha pairs at the T - resonance energy

$$B_{ee} = (5.1 \pm 3.0) \%$$

is deduced. Averaging this number with the already published value of $B_{\mu\mu}$[11] yields

$$B_{ee}^{PLUTO}(T) = 3.1 \pm 1.7.$$

In Table II the final results are summarized.

TABLE II: DORIS results on $B_{\mu\mu}$ and Γ_{ee} of the T-mesons

	DASP2	LENA	PLUTO 78	AVERAGE ***
$B_{\mu\mu}(T)(\%)$	2.9 ±1.3 ± 0.5	3.5 ± 1.4±0.4	3.1 ± 1.7	3.2 ± 0.8
$\Gamma_{ee}(T)$ (keV)	1.35 ±0.11 ± 0.22	1.23±0.13±0.14 (1.04±0.28)*	1.33 ± 0.14	1.29 ± .0.07
$\Gamma_{ee}(T')$(keV) **	0.61 $^{+0.12}_{-0.11}$ ± 0.11	0.55±0.07±0.06		0.57 ± 0.06
$\Gamma_{ee}(T')/\Gamma_{ee}(T)$	0.45 ±0.09 ± 0.05	0.45±0.06±0.02		0.45 ± 0.05

*   Result from the DESY-Hamburg-Heidelberg-München-group[5]
** $B_{\mu\mu}(T') \lesssim 2\%$ assumed

*** Only statistical errors quoted

By using relation (1) one can now deduce the total width of the meson:

$$\Gamma_{tot}(T) = 40 ^{+13}_{-8} \text{ keV.}$$

This value is very close to that of the J/ψ particle suggesting that their nature is very similar. In both cases one expects that the direct hadronic decay into 3 gluons is the dominant decay mode. This decay width can be calculated using the following relation:

$$\Gamma_{dir} = \Gamma_{tot} - (R + N) \Gamma_{ee} \qquad (5)$$

where N is the number of leptonic decay modes and R takes into account the decay into a q$\bar{q}$ pair also through a virtual photon. In lowest order QCD the direct hadronic decay width is given by[16]:

$$\Gamma_{dir} = \Gamma_{3g} = \frac{10(\pi^2 - 9)}{81\pi} \frac{\alpha_s^3}{e_Q^3 \alpha^2} \cdot \Gamma_{ee} \qquad (6)$$

and thus is a measure of the strong coupling constant α_s.

Neglecting for the moment the possible corrections to (6) one gets the following results (see Table III):

TABLE III:

	T(9.46)	J/ψ
Γ_{tot} (keV)	40 $^{+13}_{-8}$	67 ± 12
Γ_{3g} (keV)	31 $^{+13}_{-8}$	48 ± 12
α_s	0.16 ± 0.02	0.19 ± 0.02

This result shows that as expected α_s is indeed small. Furthermore, both values show that α_s is only a very weak function of energy. A comparison with the values of α_s deduced from gluon bremsstrahlung events shows also a remarkable agreement[17]. If this is not just an accident then it strongly suggests that the corrections to (6) are not substantial.

ACKNOWLEDGEMENTS

I wish to thank my colleagues from the DASP2-group who helped in preparing this talk. I thank the LENA- and PLUTO-group for allowing me to present their results.

REFERENCES

1. S. Herb et al., Phys. Rev. Lett 39,252 (1977)
2. Ch. Berger et al., Phys. Lett. 76B, 243 (1978)
3. C.W. Darden et al., Phys. Lett. 76B,246 (1978) and 78B,364 (1978)
4. J.K. Bienlein et al., Phys. Lett. 78B,360 (1978)
5. D. Andrews et al., Phys. Rev. Lett. 44,1108 (1980)
    T. Böhringer et al., Phys. Rev. Lett. 44,1111(1980)
6. C. Bebek, 'Results on the τ Region from Cleo', at this conference
    J. Lee-Franzini, 'Results on the τ Region from the Columbia-Stony Brook Experiment at CESR', at this conference
7. W.Schmidt-Parzefall in Proc. of the XIX International Conference on High Energy Physics, Tokyo, 1978
8. PLUTO-Collaboration, Phys. Lett. 82B,449 (1979)
9. F.H. Heimlich et al., Phys. Lett. 86B,399 (1979)
10. C.W. Darden et al., Phys. Lett. 80B, 419 (1979)
11. C. Berger et al., Z. Physic C1,343 (1979)
12. R. Brandelik et al., Phys. Lett. 56B,483 (1975), 67B,243 (1977)
13. W. Bartel et al., Phys. Lett 66B,483(1977), 77B,331 (1978)
14. H. Albrecht et al., A determination of the total width of the $\tau(9.46)$ meson, DESY 80/30 and submitted to Phys. Lett.
15. PLUTO-collaboration, DESY 80/15, March 1980
16. T. Appelquist and D.H.Politzer, Phys. Rev. Lett. 34,43 (1975)
17. M. Chen, G. Knies: 'Recent results from PETRA' at this conference
    TASSO-collaboration, DESY 80/40, May 1980

RESULTS ON THE ϒ REGION FROM CLEO AT CESR

C. J. Bebek
Harvard University, Cambridge, MA 02138

ABSTRACT

The total hadronic cross section in the mass region 9.41 GeV to 10.65 GeV has been measured by the CLEO collaboration at CESR. Three narrow and one broad resonance have been observed. The latter resonance is used to place a limit on the mass of the B meson.

INTRODUCTION

This talk presents the results of the first data taking with the CLEO detector at the Cornell Electron Storage Ring, CESR. The data were taken during two running periods, November-December and February-March. CLEO is the collaborative effort of the following institutions: Cornell, Harvard, Ithaca College, LeMoyne College, Rochester, Rutgers, Syracuse and Vanderbilt. A description of the detector will be given first and then the preliminary results will be presented for four resonances in the ϒ region, 9.41 GeV to 10.65 GeV. Data will be presented that indicate the fourth resonance to the above the B-$\bar{\text{B}}$ threshold.

DETECTOR

Figure 1 is a beam view and Figure 2 is a side view of the CLEO detector. Surrounding the interaction region is a PWC used for localizing the event vertex along the beam axes. It is an important element in the elimination of stray beam particles that interact in the beam pipe. Surrounding this is a 17 layer cylindrical drift chamber for charged particle tracking; the present achieved resolution is 250 microns. Next comes a magnetic coil capable of generating a 0.5T solenoidal field. Mounted on each magnet pole tip is a lead-proportional-tube shower counter that can also track charged particles. Just outside each pole tip is a luminosity monitor. The detector outside the coil is divided into octants that contain devices for particle identification. Immediately outside the coil is a drift chamber that is used to enhance the inner drift chamber longitudinal information. At the rear and two sides of each octant are lead-proportional-tube shower counters. In front of the rear shower counter is a layer of TOF scintillators. The remaining space in the octants is occupied by Cerenkov counters in 6 octants and dE/dx counters in 2 octants. Surrounding the whole detector is a hadron filter for identifying muons. For the data taken during December only 6 of the octants were installed; the February running was done with all 8 octants. Figure 3 is a display of a typical hadronic event in CLEO. The results presented here come from an analysis of the inner PWC, the cylindrical drift chamber, the time-of-flight system, and the octant

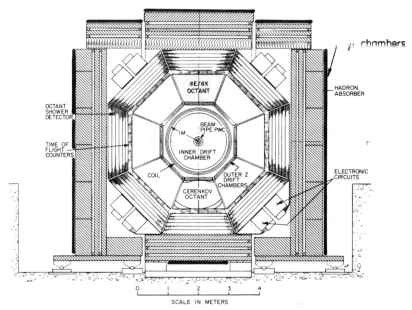

Figure 1. A beam view of the CLEO detector.

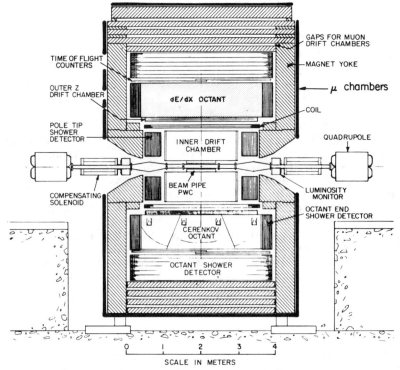

Figure 2. A side view of the CLEO detector.

shower counters. Different triggers were used during the two
running periods. The December trigger was:

(≥2.5 charged tracks .OR. 2 octant energies >2 GeV)

.AND.

(≥2 TOF counters)

The February trigger which implemented a neutral trigger was:

(≥2 charged tracks .OR. 2 octant energies >2 GeV)

.AND.

(≥2 TOF counters .OR. total octant energy >3 GeV)

EVENT SELECTION

Hadronic event selection proceeded as follows:
1. Events were checked to see if they met the hardware trigger requirements.
2. The inner-PWC was analized to reject beam-pipe wall inter-actions.
3. Octant shower counter energy ≥0.25 GeV
4. ≥3 charged tracks with a common vertex.
5. Visible energy (charged+neutral) ≥0.15E (c.m.)
6. Vertex z position ≤8 cm (alonq the beam axis)
/. Vertex x-ray position <2.5 cm (transverse to the beam axis)

Figure 4 shows a z-vertex distribution with all the data cuts
except the vertex cut. The above cuts leave a 6% background of beam-
pipe interactions and reject 3% of the hadronic events. A statistical subtraction using data from the wings of the z-vertex
distribution is performed to eliminate the remaining background from
beam pipe interactions, beam gas interactions, and cosmic rays. The
subtraction is <9%. A Monte Carlo calculation for the detection
efficiency has been made incorporating the detector geometries and
efficiencies. It is found to be 75% for jet-like events and 85% for
resonance-like events (this assumes 8 octants).

RESULTS

The existence of the b quark brings with it a rich spectrum of
excited states of the b-$\bar{b}$ system. The first two S-wave states have
been observed already at Fermilab[1] and Doris.[2] Two additional
resonances which we believe to be the 3S and 4S states of the T
family have been observed in CLEO.

Figure 5 shows the cross section for the 1S and 2S from 66
nb^{-1} and 55 nb^{-1}, respectively. The third peak which we associate
with the 3S state comes from 320 nb^{-1}. The first two resonances

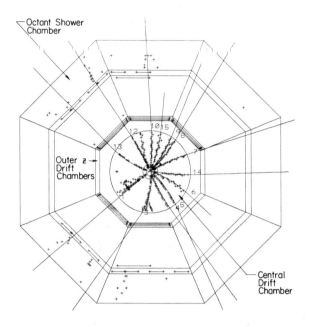

Figure 3. An on-line display of a hadronic event in CLEO.

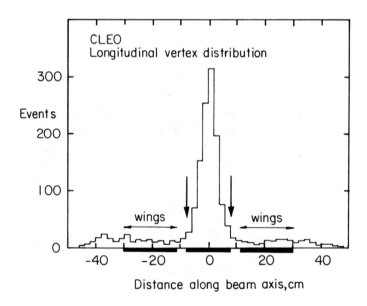

Figure 4. Reconstructed vertex distribution along the beam axis.

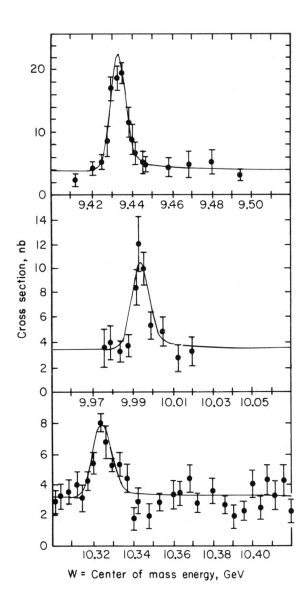

Figure 5. Corrected cross-sections for the $\Upsilon(1S)$, $\Upsilon(2S)$, and $\Upsilon(3S)$. The curves are the result of a simultaneous fit to all three resonances.

agree with the Fermilab and Doris resonances. The 3S state may
have been seen in the Fermilab experiment and is resolved for the
first time at CESR. All three peaks are narrow and consistent with
the beam energy spread of CESR.

The curves in Figure 5 represent a simultaneous fit to all
three peaks assuming an S^{-1} background behavior, an S dependence of
the beam energy spread, and three radiated gaussian (Jackson-
Scharre)[3]. The χ^2/DOF that results is 0.94 and the implied beam
energy spread agrees with the 4.1 MeV expected in CESR at 10 GeV.
Table I summarizes the fitted mass differences between the
resonances. The absolute mass scale is uncertain by 30 MeV and
the mass differences are uncertain by 3-4 MeV due to a 0.3%
uncertainty in the ∫B.dl of CESR. Table II summarizes the leptonic
widths extracted from the fit according to

$$\int \sigma . dW = 6\pi^2 \Gamma_{ee}^2/M^2$$

Theoretical calculations[4-6] that take the 1S-2S mass splitting as
input predict a 1S-3S splitting between 889 and 898 MeV in good
agreement with the experimental number of 891 MeV. The ratios of
leptonic widths are also in good agreement.[4]

The largest amount of CLEO running has been done in the energy
range 10.45 GeV to 10.65 GeV. Figure 6a shows the measured cross
section from 1060 nb^{-1} of integrated luminosity. An enhancement
near 10.55 GeV can be seen. To demonstrate that this bump behaves
like a resonance, data on the bump will be compared with the data
on either side of the bump using the Fox-Wolfram[7] shape variable
R_2 defined by:

$$H_n = \Sigma[p_i \ p_j \ P_n(\theta_{ij})/W^2]$$
$$R_2 = H_2/H_0$$

P_n are the Legendre polynomials, $p_{i,j}$ are the momenta magnitudes of
the particles i and j, and θ_{ij} is the angle between them. R_2 is
used instead of H_2 directly to take into account the use of only
charged tracks in the calculation; H_0 is essentially a measure of
the charged energy fraction detected in the event. An R_2 distribution will peak at small values of R_2 for spherical events. Figure
7 shows an R_2 distribution for the 1S peak. The resonance data is
skewed to small values of R_2 while the continuum is shifted to
larger values.

Figure 8a shows the differential cross section in R_2 for
data on and off the peak of Figure 6a. The on-peak data favors
smaller values of R_2 as would be expected for the decay of a
resonance. Figure 8b shows the subtraction of the off-peak data
from the on-peak data. An excess at low values of R_2 is seen.
Returning to Figure 6b, the cross section is replotted for $R_2 < 0.3$.
The signal to noise is seen to be improved. Thrust and sphericity
distributions also show a resonance-like contribution in this region.

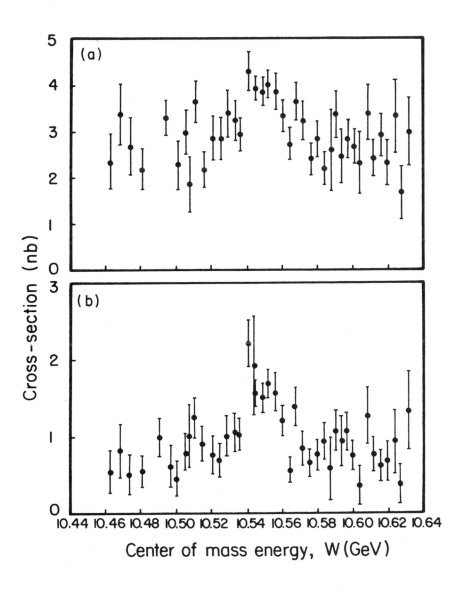

Figure 6. a) Corrected cross-section in the $\Upsilon(4S)$ region.
b) Corrected cross-section for $R_2 < 0.3$.

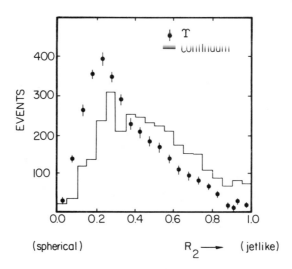

Figure 7. R_2 distributions for the $\Upsilon(1S)$ resonance and continuum region. The two distributions are normalized to an equal number of events.

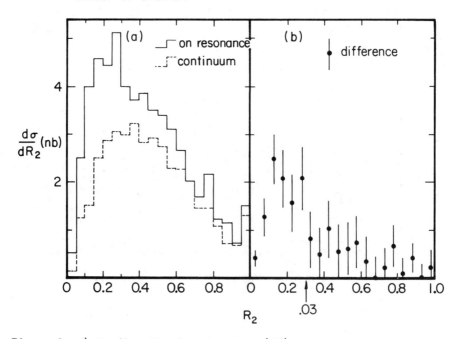

Figure 8. a) R_2 distribution for the $\Upsilon(4S)$ resonance and continuum region. b) R_2 distribution for the difference of the two distributions in a).

A fit to the data in Figure 6a using an S^{-1} background and a radiated gaussian gives a mass difference relative to the 1S state of 1112 $\pm$ 2 $\pm$ 4 MeV and a leptonic width relative to the 1S state of 0.21 $\pm$ 0.06. After unfolding the beam energy spread, we find Γ_{tot}=21.5$\pm$5.7 MeV. The fit has a χ^2/DOF of 42/39. The width is much broader than the beam energy spread; if a width corresponding to the known energy spread of 4.6 MeV at this energy is forced, the χ^2/DOF becomes 70/40.

It should be mentioned that there is no good reason for fitting the 4S peak to a gaussian or a Breit-Widner. The shape of the resonance is a critical function of the mass difference between the resonance and the threshold for B meson production.

CONCLUSION

We have shown the existence of four resonances in the region 9.41 GeV to 10.65 GeV. The widths of the three lower mass resonances are consistent with the energy spread of the beams and are in good agreement with the members of the Υ family observed elsewhere. The fourth resonance is seen for the first time at CESR and has a width inconsistent with the energy spread of the beams. If this is interpreted as the 4S state being above the B-$\overline{B}$ threshold, the B meson mass is limited to the range 5.16 GeV to 5.26 GeV.

REFERENCES

1. S. W. Herb et al., Phys. Rev. Lett. 39, 252 (1977)
2. C. Berger et al., Phys. Lett. 76B, 243 (1978)
3. J. D. Jackson and D. L. Scharre, Nucl. Instr. 128, 12 (1975)
4. G. Bhanot and S. Rudaz, Phys. Lett 78B, 119 (1978)
5. J. L. Richardson, Phys. Lett. 82B, 272 (1979)
6. E. Eichten, K. Gottfried, T. Kinoshita, K. Lane and T. M. Yan, Cornell report CLNS-425
7. G. C. Fox and S. Wolfram, Phys. Rev. Lett. 41, 1581 (1978) and Nucl. Phys. B149, 413 (1979)

QUESTIONS

D. Hitlin (Caltech): For a given luminosity what is your trigger rate?

Answer: For the data presented here, the trigger was 2 Hz at $10^{30} cm^{-2} sec^{-1}$. For our most recent running we have made progress in controlling the stray beam particles and the corresponding trigger rate is 0.7 Hz.

D. Hitlin (Caltech): Is there any evidence for an enhancement of direct electrons at the 4S state?

Answer: I have no results to present at this time. As you might imagine, an enhancement of electrons and kaons on the 4S is the focus of a major analysis effort at this moment.

TABLE I: Upsilon Masses
Mass differences are relative to the T(1S).
The first error is statistical, the second is systematic.
The DORIS results are from the 1980 Moriand Conference.

Resonance	CESR		DORIS	
	CLEO	CUSB	LENA	DASPII
T(1S)	9.4331	9.4345	9.4616	9.4630
(mass in GeV)	±.0003	±.0004	±.0005	±.0005
	±.0300	±.0300	±.0100	±.0100
ΔMT (2S)	0.5607	0.5590	0.5527	0.5537
(mass relative	±.0008	±.0010	±.0012	±.0017
to T(1S)	±.0030	±.0030	±.0010	±.0010
ΔMT (3S)	0.8911	0.8890		
(mass relative	±.0007	±.0010		
to T(1S)	±.0050	±.0050		
ΔMT (4S)	1.112	1.114		
(mass relative	±.0020	±.0020		
to T(1S)	±.0050	±.0050		

TABLE II. Upsilon Electronic Widths
T(1S) width in KeV, others relative to T(1S)

Resonance	CESR		DORIS	
	CLEO	CUSB	LENA	DASPII
T(1S)	1.15		1.23	1.35
(Γ in KeV)	±.08		±.09	±.11
	±.17			±.22
T(2S)	0.44	0.39	0.46	0.48
Γ/Γ(1S)	±.06	±.06	±.07	±.10
	±.04			
T(3S)	0.35	0.32		
Γ/Γ(1S)	±.04	±.04		
	±.03			
T(4S)	0.21	0.25		
Γ/Γ(1S)	±.06	±.07		
	±.03			

RESULTS ON THE ϒ REGION FROM THE
COLUMBIA-STONY BROOK EXPERIMENT AT CESR

J. Lee-Franzini
SUNY at Stony Brook, Stony Brook, N.Y. 11794

ABSTRACT

Using the CUSB layered NaI detector at CESR, we have observed three narrow $(ϒ, ϒ', ϒ'')$ and one broad enhancements $(ϒ''')$ in the $\sigma(e^+e^- \to \text{hadrons})$ over the c. of m. energy interval of 9.4 to 10.6 GeV. Their mass spacings are $M(ϒ')-M(ϒ)=559\pm1(\pm3)$ MeV, $M(ϒ'')-M(ϒ)=889\pm1(\pm5)$ MeV, $M(ϒ''')-M(ϒ)=1114\pm2(\pm5)$ MeV and their relative leptonic widths are $\Gamma_{ee}(ϒ')/\Gamma_{ee}(ϒ)=0.39\pm.06$, $\Gamma_{ee}(ϒ'')/\Gamma_{ee}(ϒ)=0.32\pm.04$ and $\Gamma_{ee}(ϒ''')/\Gamma_{ee}(ϒ)=0.25\pm.07$, where errors in parenthesis represent systematic uncertainties. These values allow us to identify these enhancements as the four lowest triplet S levels of the bound $(\bar{b}b)$ quark-anti quark system. The broad natural width (~ 13 MeV) of the 4^3S_1 state indicates that it lies above the threshold for $B\bar{B}$ production.

INTRODUCTION

Towards the end of November of 1979, CESR initiated its first physics running period which lasted about five weeks, yielding about 1000nb^{-1} of integrated luminosity. For two-thirds of that period (corresponding to $\sim 600\text{nb}^{-1}$) data were collected around the c. of m. energies of 9.45 GeV, 10 GeV and 10.3 GeV, i.e., around the mass positions of the first three upsilons. The remainder of that period was devoted to an energy scan around the c. of m. energy of 10.6 GeV, where the fourth ϒ was expected to exist. In fact, the existence of this last resonance was not proved till the next running period, which occured during February-March of 1980, where data were collected in the energy region from 10.46 GeV to 10.6 GeV for ~ 1100 nb^{-1} of integrated luminosity. In the following, I will report on the level spacing and leptonic width determinations of these first four upsilon resonances with data from the CUSB detector, as well as on their interpretations as the four lowest triplet S states in the bound $(b\bar{b})$ quark-anti-quark system. [1,2]

APPARATUS AND DATA COLLECTION

The CUSB layered NaI-Pb glass detector is located in the 18'x21' "North Area" of CESR. It was designed primarily to study the upsilon family spectroscopy and inclusive photon processes at CESR, i.e., able to measure with accuracy the soft photons (of 50-200 MeV) arising from the radiative transitions and the hard photons (of order GeV) when one of the upsilons possibly decay into a gamma and one of the new particles such as Higgs Bosons. However, as the present report will show, this detector also functions extremely well as a hadron detector in total σ(e⁺e⁻→hadrons) measurements. In its final design, the CUSB detector consists of a system of drift chambers for tracking, which is followed by layered NaI strip chamber arrays which localize photon conversion vertices and measure over 70% of its energy, and is finally backed by Pb glass $\hat{c}$ counters to contain the remainder of the electromagnetic showers. The NaI is divided into 32 azimuthal sectors and 2 polar sectors, thus providing complete azimuthal coverage in the region $45^{\circ} < \theta < 135^{\circ}$. Radially, the inner four NaI layers are 1 r. l. thick, interspersed with low mass cathode readout proportional chambers, followed by one four r. l. NaI layer, and surrounded by 260 8 r. l. thick Pb glass blocks. This system will have good spatial resolution (1 mrad in θ, 5 mrad in ϕ for track position, 0.5 cm for shower conversion point position) and good energy resolution (Δe/e∼7% at 0.1 GeV and ∼2.5% at 10 GeV). Figure 1 illustrates the end and cut away side view of the NaI components of the CUSB detector.

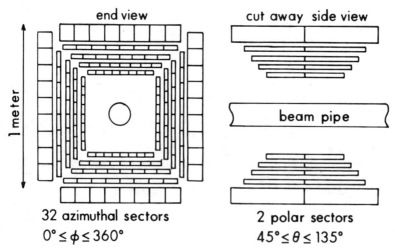

end view   cut away side view

1 meter

beam pipe

32 azimuthal sectors   2 polar sectors
0° ≤ φ ≤ 360°   45° ≤ θ ≤ 135°

Coverage: ∼67% of 4π
Crystals : 324

Fig. 1. CUSB detector (NaI).

For the Nov-Dec run only one of the two polar sections was

used and it was centered optimally over the interaction point. In the Feb-March run, both polar sections were used, and the detector was centered along the interaction point.

For the Nov-Dec run, four small strip chambers placed around the beam-pipe, localized the beam-beam interaction point. In the Feb-March run, drift chambers were used instead. Charged particles originating from the beam interaction point deposit energy in each of the NaI layers, thus yielding five independent dE/dx measurements. Also, charged particles whose path is confined to within one of the 64 NaI sectors subtending a solid angle of 1% of 4π, point to the beam-beam interaction point. Therefore, even though drift and strip chambers were in place during the present runs, the track information obtained from them were used only to confirm the cleanliness of the hadron selection based solely on NaI information.

The absolute energy scale for each NaI is set using radioactive sources: Cs^{137} sources, which yield 0.66 MeV γ's, are interspersed thru the thin NaI planes, and Co^{60} sources whose sum peak gives 2.5 MeV gammas, are placed outside the 4" thick NaI crystals. Photomultiplier tube (PM) stability was monitored with light from a spark in argon. The PM gains were stable to within 5% throughout the running periods. All our analog signals are differentially transmitted and received. All signals from the PM's were integrated every beam crossing (every 2.56μs) while a trigger decision was made. If no trigger was present, all integrators were reset to be ready for the next crossing. Only a total energy trigger was used for the present runs: requiring $\geq$ 420 MeV to be deposited in the outer three layers of the one polar sector of the NaI array for the Nov-Dec run and $\geq$ 700MeV to be deposited in both sectors for the Feb-March run.

Once a trigger was produced, all signals were digitized and recorded on tape. This trigger gave an event rate of 0.1 Hz for a luminosity of 1 $\mu b^{-1}s^{-1}$. Approximately 15% of the triggers are hadronic events at the Υ and about 10% of the triggers are large angle Bhabha scattering events in our detector. The integrated luminosity for each run was measured by detecting and counting small angle (40 to 80 mrad) collinear Bhabha scatters with lead-scintillator sandwich shower detectors. Hadronic and large angle Bhabha scatter yields in the detector are also monitored on-line by counting on scalers events satisfying appropriate criteria. The Bhabha scattering criterion requires that the total energy deposited in NaI (E_{NaI}) be greater than 5.4 GeV and that two collinear octants each have $\geq$ 100 MeV in the outer four layers. These online "Bhabhas" comprise approximately 80% of those found later in off line analysis and serve as an online consistency check of the small angle Bhabha lumi-

nosity monitor. The long term stability of the luminosity
monitor is confirmed by the yield of the large angle Bhabha events recorded on tape which are also used for absolute normalization. A typical run of CESR lasts 3 to 5
hours, yielding an integrated luminosity of up to 15nb^{-1}.
During such a run, the CUSB R-Meter, a scaler counting
events which satisfy the hadron criteria (1.4 GeV $\leq E_{NaI}$
$\leq$ 4.2 GeV in the outer three layers, that two or more
octants in each half have $>$ 100 MeV deposited in the
outer four layers, and that two such octants be collinear
see fig. 2a) shows clear Υ , Υ " signals. The visible

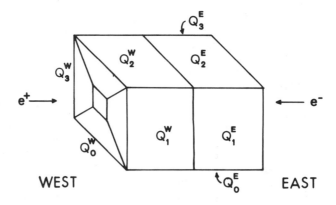

WEST   EAST

R Meter requires :

1) Discriminate : $E_a > 100$ MeV

2) Logic : (2 out of 4 in Q^E)•(2 out of 4 in Q^W)•

(at least 1 collinear E-W pair)•

(1.4 GeV $\leq E_{total} \leq 4.2$ GeV)

Fig. 2a CUSB On-Line R-Meter Logic
cross section, multiplied by the factor $(M/M_\Upsilon)^2 \approx (1/s)$,
as a function of mass (i. e. $2E_{beam}$) as obtained from the
on-line R-meter scaler for the Feb-March run is shown in
figure 2b. As will be seen later, the accuracy in mass
position and relative leptonic width determination was
good and in agreement with the off line analyses. The
live time of our detection system is 99%.

DATA ANALYSIS AND RESULTS

Electromagnetic showers, with their characteristic
energy deposition pattern are easily identifiable in our
layered and segmented active converter array. Charged
particles from beam-beam interactions also give a unique
signature in our detector. For example, at normal inci-

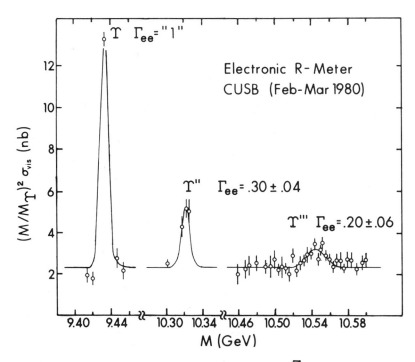

Fig. 2b CUSB On-Line R-Meter Results. Γ_{ee} are given relative to Γ_{ee} (Υ).

dence, minimum ionizing particles deposit 15 MeV in the first four NaI layers and approximately 68 MeV in the last layer of a single sector,(defined as "track"). A hadronic event usually includes at least one track, some showers, and a balanced deposition of energy in the two polar sectors. In order to develop hadronic event selection computer algorithms, one or more physicists have examined over 95% of the hadronic event candidates. The various algorithms developed for the Feb-March data have efficiencies for continuum events of from 65% to 75% and with contamination from non beam-beam background ranging from 1 to 5% respectively. These efficiencies include losses due to detector solid angle, hence are considerably smaller for the Nov-Dec data. The background estimates are obtained from single beam runs, and from reconstructed vertex positions for those events having drift chamber information. The efficiency for detecting upsilon events is higher than that for continuum events because the former has higher multiplicity and sphericity than the latter. For the Nov.-Dec. data the net efficiency of our detector was 28% for continuum events and 37% at the Υ peak; for the Feb.-March data the corresponding numbers are 73% and 82%.

The main aim of CESR's Nov-Dec run was to locate the position of the upsilon for calibrating the machine energy scale, to quickly scan the Υ' already established at DORIS and to search for higher states in the Upsilon family. Our half detector performed efficiently and well in this task, collecting 214Υ, 53Υ', and 133Υ'' events above the continuum and 272 events from the continuum around the three Υ's. The hadronic yield is presented in figure 3, plotted in arbitrary units proportional to the ratio of detected events to small angle Bhabha yield. In this way, the energy dependence ($\propto 1/s$) of the single photon process is removed. The horizontal scale is $M(e^+e^-)$, twice the nominal machine energy. Mass values of the

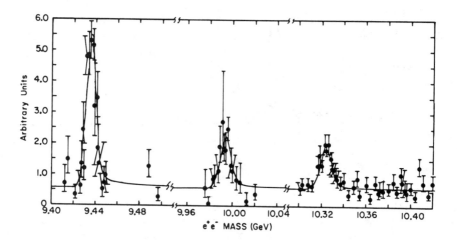

Fig. 3 Hadron yield/Bhabha event vs e^+e^- mass

three resonances are determined by fitting the data with a constant continuum plus three radiatively corrected Gaussians[3] whose width represent the machine energy spread with however only a single beam width parameter used for all three resonances, scaling with energy as s, according to machine theory. The machine total energy spread at 9.5 GeV is determined to be 4 $\pm$ 0.3 MeV rms from the Nov-Dec data, and 3.7 $\pm$ 0.3 MeV from the Feb.-March data. These values are consistent with each other and with the value expected from CESR design parameters.

From the fit to the Nov-Dec data we found m(Υ)= 9.4345 $\pm$ 0.004 GeV, m(Υ')= 9.993 $\pm$ 0.0010 GeV and m(Υ'')= 10.3232 $\pm$ 0.0007 GeV. The mass scale at CESR appears at present to be displaced by 2.7 parts in a thousand with respect to Doris[4]. The determination of which scale is correct will have to await future CESR energy calibrations. The quantities which are relevant to the phenomenology of the bound states (which had been discovered at Fermilab[5] as a narrow enhancement in the dimuon spectrum near 10 GeV) of the b quark are the level splittings and ratios of leptonic widths, for which we obtain $M(\Upsilon')-M(\Upsilon)=559\pm1(\pm3)MeV$ and $M(\Upsilon'')-M(\Upsilon)=889\pm1(\pm5)MeV$, where errors in parenthesis represent the aforementioned systematic uncertainties in CESR energy calibration, $\Gamma_{ee}(\Upsilon')/\Gamma_{ee}(\Upsilon)=.39\pm.06$ and $\Gamma_{ee}(\Upsilon'')/\Gamma_{ee}(\Upsilon)=0.32\pm.04$. The $\Upsilon'-\Upsilon$ mass level spacing had been measured at Doris and agrees well with our results. Our ratio for the Υ' to Υ leptonic width is different from and is more accurate than Doris's published results. New results from Doris reported at this conference are now in agreement with ours. The Υ'' has been isolated and seen at e^+e^- machine for the first time at CESR, by CUSB and CLEO simultaneously. The errors on the ratios are statistical only; however, they are larger than our estimates of the systematic errors in these ratios due to possible scanning inefficiencies for the three resonances. Our results are in good agreement with many quarkonium model predictions,[6] reinforcing the interpretation that these first three Υ's, $(\Upsilon, \Upsilon', \Upsilon'')$ are the three lowest triplet S states of the $b\bar{b}$ bound system.

In such quarkonium models, the number of quasistable radial excited states increase with the mass of the quark. In particular, for the b quark (m~5GeV), the 4^3S_1 state should exist with an excitation energy of~ 1.15 GeV. The 4^3S_1 state is expected to be very close to the threshold for $B\bar{B}$ production, where B is a pseudo scalar bound system of b and u quarks. If the 4^3S_1 state lies below the $B\bar{B}$ threshold, its natural width would be well below 1 MeV, giving observed width dominated by machine width, as for $\Upsilon, \Upsilon', \Upsilon''$, whereas if it lies above the $B\bar{B}$ threshold, the opening up of decay channels would result in a natural width which increases rapidly with $M(4\,^3S_1)-2M(B)$.[7] As has been stated in the introduction, the search for the 4^3S_1 state began in the Nov-Dec run;indeed, a hint of the structure was seen by our group by Christmas, but the definitive results did not come until the Feb-March run. Five thousand $1^3S_1(\Upsilon)$ events and 450 $3^3S_1(\Upsilon'')$ events were also collected during the Feb-March running period. Three thousand events were collected in the region 10.46 to 10.6 GeV. The presence of an enhancement around 10.54 is quite evident, even on the on-line R meter results as was shown in figure 2. In figure 4, the visible cross section in our detector for e^+e^- hadrons, multiplied by the scaling variable $k=(M/M_\Upsilon)^2$, in the region of the Υ'' and in the energy range between 10.46 and 10.6 GeV. Both the presence of a sizeable enhancement over the continuum and its broadness relative to its neighbor (Υ'') are apparent. To enhance discrimination between resonance and continuum events, we make use of a property first discovered at Do-

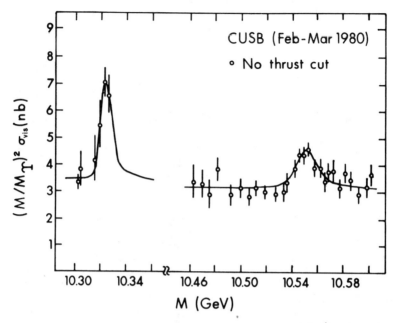

Fig. 4. The observed cross-section for (e^+e^-) hadrons multiplied by $k=(m/M)^2$. M is the e^+e^- invariant mass.

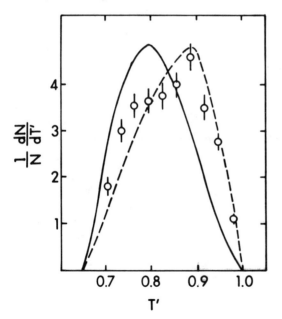

ris[8] about the Υ, that the spatial distributions of the continuum and decay events are different. We chose to use a simplified thrust variable, T', defined as the maximum of $\Sigma|\vec{E}_{NaI}\cdot\hat{n}|/\Sigma E_{NaI}$ over all possible $\hat{n}$ perpendicular to the beam axis. The distributions of T' for Υ events (solid line), continuum events (dashed line), and events in the enhancement peak (data points) are shown in Figure 5.

Fig. 5. Pseudo thrust (see text) distribution for Υ events (solid line), continuum events (dashed line), events in the enhancement peak (data points).

The difference between the solid and dashed line figures is quite obvious, confirming the conjecture that continuum events have a two jetlike structure, while resonance decays have a more spherical distribution. It is also clear that events in the enhancement region have contributions from both types of events. A cut at T'<0.85 has been made for all the data, and the result is shown in Fig. 6.

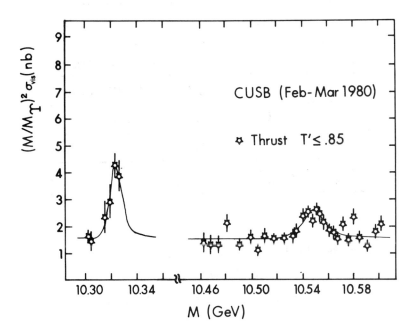

Fig. 6. Observed cross-section for $e^+e^- \rightarrow$ hadrons multiplied by $k=(M/M)^2$ after removing events with T'>.85.

We determine the parameters of the Υ''' by fitting the data sample with T'<.85 to a constant continuum plus a Gaussian with radiative corrections, finding a central position of $M(\Upsilon''')=10.547\pm.002$MeV and an apparent width of 19 ± 4 MeV. The non-thrust cut data yielded similar results. Scaling our fitted stored beam energy spread at the Υ of 3.7 ± 0.3 MeV by the expected $(E_{beam})^2$ dependence yields an anticipated beam spread FWHM at 10.54 GeV of $10.8\pm.9$ MeV, about half of the observed value. We therefore assume a Breit-Wigner resonance shape for the enhancement, fold in the machine energy spread and radiative corrections, and fit the Υ''' data with the resulting curve. This results in a natural width of 12.6 ± 6.0 MeV. If we constrain the natural width to be much smaller than the machine energy spread, χ^2 increases by 8.3, from 40.3 for 30 degrees of freedom to 48.6 for 29 degrees of freedom. Thus, our value for the natural width is inconsistent with the expected width of less than 1 MeV for a resonance below $B\bar{B}$ threshold. In table I we list the properties of the Υ'''.

Table I. Summary of Υ''' properties

σ_{beam} (at Υ mass)	3.7	± 0.3	MeV
$M(\Upsilon)$		9.433 ± 0.001	GeV
$M(\Upsilon''')$		10.547 ± 0.002	GeV
$M(\Upsilon''') - M(\Upsilon)$		1.114 ± 0.002	GeV
$\Gamma(\Upsilon''')$	12.6	± 6.0	MeV
$\Gamma_{ee}(\Upsilon''')/\Gamma_{ee}(\Upsilon)$	0.25	± 0.07	

Note: Masses shown use local CESR

energy scale which has a

systematic uncertainty of 0.3%

The mass values are again given in the nominal CESR energy scale. The mass difference is $M(\Upsilon''')-M(\Upsilon)=1114\pm2$ MeV, with again a systematic uncertainty of 5 MeV. The ratio of leptonic widths calculated from the fitted areas is $\Gamma_{ee}(\Upsilon''')/\Gamma_{ee}(\Upsilon)=.25\pm.07$. Both the mass difference and the ratio of leptonic width are in excellent agreement with many phenomenological calculations of the 4^3S_1 state of $b\bar{b}$[6]. Therefore we conclude that the enhancement we see at M=10.547 is most likely that state. The natural width of ~13 MeV implies that the 4S is above $B\bar{B}$ threshold and that the mass of the B is less than 5.274 GeV. CLEO at the South Area has very similar results and conclusions as we do with regards to all four members of the Upsilon family seen so far at CESR.[9] With the anticipated enhanced production rate for B mesons around the Υ''' region, both groups plan an intense program of B meson study in the next running periods.

385

ACKNOWLEDGEMENT

The author (J. L-F) and all the CUSB collaborators
(T. Böhringer, F. Costantini, J. Dobbins, P. Franzini,
K. Han, S. W. Herb, D. M. Kaplan, L. M. Lederman, G.
Mageras, D. Peterson, E. Rice and J. K. Yöh of Columbia
University; G. Finocchiaro, G. Giannini, R. D. Schamber-
ger Jr., M. Sivertz, L. J. Spencer and P. M. Tuts of
SUNY at Stony Brook) are supported in part by the National
Science Foundation. I also thank Paula Franzini for ty-
ping this manuscript.

REFERENCES

1. T. Bohringer et al., Phys. Rev. Lett. 44, 1111(1980).
2. G. Finocchiaro et al., Submitted to Phys. Rev. Lett.
3. J. D. Jackson and D. L. Scharre, Nucl. Instrum.
   Methods 128, 13 (1975).
4. Ch. Berger et al., Phys. Lett. 76B, 243 (1978); C. W.
   Darden et al., Phys. Lett. 76B, 246 (1978); Phys.
   Lett. 78B, 364(1978); J. K. Bienlein et al, Phys.
   Lett. 78B, 360(1978).
5. S. W. Herb et al., Phys. Rev. Lett. 39, 252(1977);
   W. R.Innes et al., Phys. Rev. Lett. 39, 1240, 1640(E)
   (1977); K. Ueno et al., Phys. Rev. Lett. 42, 486(1979).
6. C. Quigg and J. L. Rosner, Phys. Lett. 71B, 153(1977);
   E. Eichten et al., Phys Rev. D17, 3090(1978), D21,
   203 (1980); G. Bhanot and S. Rudaz, Phys. Lett. 78B,
   119 (1978).
7. C. Quigg and J. L. Rosner, Phys. Rpts. 56, #4 (1979).
8. Ch. Berger et al, Phys. Lett. 78B, 176(1978); 82B,
   449(1979); F. H. Heimlith et al., Phys. Lett. 86B,
   399(1979).
9. D. Andrews et al., Phys. Rev. Lett. 44, 1108(1980),
   and submitted to Phys. Rev. Lett.

DISCUSSION

Q. (Roy Weinstein, Northeastern Univ). On your recent
(post Christmas) data, just below the Υ''' there seemed
to be a hint of structure. Could this be the $B\bar{B}^*$
threshold?
A. That hint of structure, while tantalizing, is only one
standard deviation above the continuum at the moment,
and appears to be only as wide as the machine energy
spread. Therefore, while it could be a D-state or
vibrational state candidate, it is too narrow to be a
threshold onset.
Q. (M. Chen, DESY/MIT). Can you show me the measured energy spectra at the Υ''' and off the resonance to see
whether you see any diffence in the spectra?

A. In our detector we see only a part of the total energy from the emitted particles. For that fraction we see no difference between the energy spectrum on and off resonance.

RECENT THEORETICAL DEVELOPMENTS FOR HEAVY
QUARK-ANTIQUARK SYSTEMS[*]

E. Eichten[†]
Lyman Laboratory of Physics
Harvard University
Cambridge, MA 02138

ABSTRACT

The status of potential models of the heavy quark-antiquark ($Q\bar{Q}$) interaction is summarized and some recent developments are discussed.

1. INTRODUCTION

There is now a wealth of data on the $c\bar{c}$ system of resonances and new data[1] on the even heavier $b\bar{b}$ system. With so much interesting data I cannot possibly give a detailed comparison with theory here; rather I will concentrate on the present status (in general terms) of the phenomenological potential models of the heavy quark-antiquark ($Q\bar{Q}$) systems and point out some aspects of the theory of $Q\bar{Q}$ systems which may be tested or clarified within the next few years. In particular we should learn about: (1) the coupling of $Q\bar{Q}$ systems to decay channels, (2) relativistic corrections to the $Q\bar{Q}$ potential, (3) hadronic transitions between $Q\bar{Q}$ narrow resonances, and (4) applications of Q.C.D. to $Q\bar{Q}$ hadronic decay processes.

2. POTENTIAL MODELS

Let me begin with a summary of the present status of potential models. There are two aspects of Q.C.D. which are essential for building potential models of a heavy-quark-antiquark system. First, the short distance behaviour of the potential is calculable.[2] The small R behaviour is basically Coulombic but modified by calculable logarithms; i.e. $R V(R) \rightarrow -4\alpha_s(R)/3$ as $R\rightarrow 0$ where $\alpha_s(R)$ is the strong coupling constant, a weak function of R for small R. Explicitly, as $R\rightarrow 0$,

$$\alpha_s(R) \rightarrow \frac{2\pi}{9 \ln\left(\frac{1}{\Lambda R}\right)}, \Lambda \sim 400 \text{ MeV}$$

where the contribution of the three light quarks have been included. The second property of Q.C.D. which is not yet proven but generally

[*]Talk presented at the VI International Conference on Experimental Meson Spectroscopy, Brookhaven National Laboratory, Upton, N.Y., April 25,26 (1980).
[†]Research supported in part by the National Science Foundation under Grant No. PHY77-22864, and the Alfred P. Sloan Foundation.

accepted is confinement[3]; in particular, linear confinement at
large distance. Thus $V(R) \to a^{-2}R$ as $R \to \infty$ where a^{-2} is a string
tension related phenomenologically to the slope of Regge trajectories for the mesons. A phenomenological potential must contain
an interpolation between the behaviour expected in the two extremes
of long and short distance.

There is one further property of Q.C.D. that should be
incorporated into a potential model: the static energy between
quarks is independent of the quark mass. Of course there are
relativistic corrections to the potential which depend on $1/m_Q^2$. I
will discuss these corrections in section 4. In addition to
relativistic effects associated with $1/m_Q^2$, contributions to the
interaction due to light quarks (u,d,s) exist even when $m_Q \to \infty$.
These light quarks slightly modify the potential. For example,
the string tension, a^{-2}, will differ from the value in a theory
without light quarks where the string tension arises purely from
pure Yang-Mills interactions. It is likely that this is not a
large effect. At least this is true in one model--the Cornell
Model.[4] Here light quarks are introduced by adding to the phenomenological Hamiltonian which gives the potential between Q and $\bar{Q}$
pair creation and annihilation terms for light quarks. The effect
on the value of a^2 in the effective potential for $Q\bar{Q}$ resonances
below threshold for Zweig allowed decays is to reduce a^2 by ~15%.[5]

However, as well as modifying the effective potential the
introduction of light quarks (ℓ) allows the $Q\bar{Q}$ states to decay
strongly into two meson states ($Q\bar{\ell} + \ell\bar{Q}$) carrying the new flavor.
The potential picture must break down at large distance as the
physical states of lowest energy contain the two mesons rather than
simply a $Q\bar{Q}$ pair. Physical states which have a large radius in the
potential models will have large decay widths and decay rapidly, so
are difficult to study experimentally; thus the large distance
behaviour of the phenomenological potential is difficult to study
experimentally. The structure of the potential that is probed
experimentally[6] is therefore limited to distances $R \lesssim 1$ fermi and
is also limited at short distances by the Compton wavelength of the
quark, $R \gtrsim 1/m_Q$.

Now consider four particular phenomenological potentials all
seemingly very different. The potentials are as follows:

(1) The Cornell Model[4] A three-parameter model which has the
     right large R behaviour and only approximately the correct
     short distance behaviour.

$$V_1(R) = -\frac{K}{R} + \frac{R}{a^2} + V_0. \qquad (2.2a)$$

(2) The Logarithmic Potential.[7] This two-parameter model does not
     have the appropriate behaviour at large or small R but fits
     the spectrum of $c\bar{c}$ excitations and the $T' - T$ mass splitting
     reasonably well.

$$V_2(R) = A \ln (R/R_0). \qquad (2.2b)$$

(3) The MIT Bag Model. This potential arises from the MIT Bag for

heavy quarks under certain reasonable assumptions.[8] Treated as a phenomenological potential it has three parameters and the correct large and small R behaviour.

$$V_3(R) = -\frac{4}{3}\frac{\alpha(R)}{R} + \frac{32\pi\ B\ \alpha(R)^{1/2}}{3} R + C[B\ \alpha^3(R)]^{1/4} \quad (2.2c)$$

and finally

(4) Richardson's Potential.[9] A two-parameter model. The correct short distance behaviour is built in exactly and by simply replacing the usual $\ell n(Q^2/\Lambda^2)$ in the definition of the effective charge by $\ell n[(Q^2+\Lambda^2)/\Lambda^2]$ the correct large R behaviour is also insured.

$$V_4(R) = -\frac{4}{3}\int\frac{d^3Q}{(2\pi)^3}\ e^{i\vec{Q}\cdot\vec{R}}\ \frac{1}{\vec{Q}^2}\ \frac{16\pi^2}{9\ \ell n[(Q^2+\Lambda^2)/\Lambda^2]} \ . \quad (2.2d)$$

The parameters of these potentials and the charm quark mass, m_c, are determined by using the following experimental inputs:[5,10]

$$m(\Psi') - m(\Psi) = 590 \text{ MeV} \quad (2.3a)$$

$$r \equiv \frac{m(^3P_J \text{ center of gravity}) - m(\Psi)}{m(\Psi') - m(\Psi)} = .72 \quad (2.3b)$$

If additional constraints are needed, the condition $m_c = 1.85$ GeV $\simeq m_B$ is used. Finally zero of each of the potentials was arbitrarily chosen so that $V_i(R_0) = .1$ GeV (2-i) i=1,...,4 at $R_0 = .4$ fermi (for ease of simultaneous display). The resulting potentials are shown in Figure 1. The agreement between the potentials is impressive

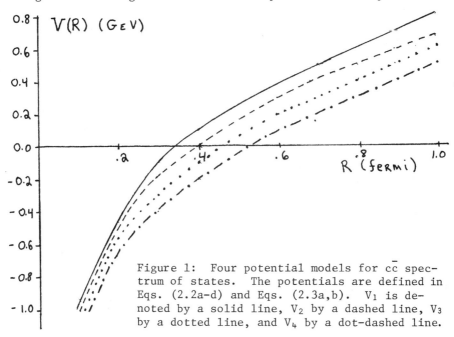

Figure 1: Four potential models for $c\bar{c}$ spectrum of states. The potentials are defined in Eqs. (2.2a-d) and Eqs. (2.3a,b). V_1 is denoted by a solid line, V_2 by a dashed line, V_3 by a dotted line, and V_4 by a dot-dashed line.

390

in the region $.2 \lesssim R \lesssim 1$. The only appreciable deviation occurs for the logarithmic potential for $R \gtrsim .8$ fermi. Thus for V_1, V_3, and V_4 (and somewhat less completely for V_2) in spite of the very different looking analytic forms presented in Eqs. (2.2a-d) they are numerically very similar in the range of R probed by experiments within $c\bar{c}$ system. For example, the spectrum of resonances E1 and M1 transition rates, and wavefunctions at the origin will be roughly similar. I conclude simply that we know the phenomenological potential to reasonable accuracy although it may parameterize in many different ways. The constraints of proper small and large distance behaviour are quite powerful for a smoothly interpolating potential.

Since $1/m_b \ll 1/m_c$ we can probe smaller distances in the (b$\bar{b}$) system than the (c$\bar{c}$) system. Furthermore, we expect that the same potential function should work for this even heavier Q$\bar{Q}$ system. The small R behaviour of the four potentials $V_1,\ldots,V_4$ is shown in Figure 2. The Richardson potential, V_4, and the MIT Bag Potential, V_3, have exactly the short distance behaviour of Q.C.D. (see Eq. (2.1); while the Cornell Model, V_1, has a Coulombic behaviour with constant coupling K. This variation in the small R behaviour of the potentials can actually be differentiated through the spectrum of (b$\bar{b}$) resonances. In the case of the Cornell Model the splitting $T' - T$ is 580 MeV for $K = .52$ (the value determined for the c$\bar{c}$ system). The value of m_b is determined by independent considerations to be 5.17 GeV.[5] The observed value of the mass splitting $T' - T \simeq 560$ MeV requires modifying K to .48 for the b$\bar{b}$ system.

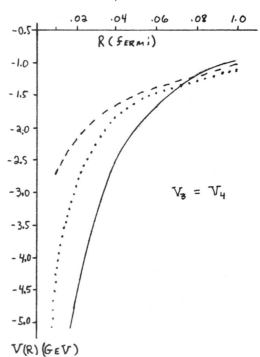

Figure 2: Small R behaviour of V_1, V_2, V_3, V_4. Notation is given in Figure 1. Zero of potential readjusted so that $V_i(R_0) = 0$ for $R_0 = .4$ fermi, V_i.

Evidently the effective coupling of the Coulomb interaction decreases with increasing mass m_Q as predicted in Q.C.D. In fact, potentials V_3 and V_4 which have the short distance behaviour of Q.C.D. (Eq. (2.1) fit the T' - T mass splitting without any adjustment of parameters. Table 1 gives a comparison of the modified Cornell Model and Richardson's Model with the available data for the $b\bar{b}$ resonances.

TABLE 1: Comparison of potential models with recent data on the excitation energy and the ratio of electronic width $\Gamma_n(e^+e^-)$ to that of the $T(\Gamma_1(e^+e^-)$ for some of the $n\,{}^3S_1(b\bar{b})$ resonances.

TABLE 1

	Cornell Model[(5)]	Richardson Potential	Experiment
$M(T') - m(T)$	560 MeV (input)	555 MeV	560 MeV
$\Gamma_2/\Gamma_1(e^+e^-)$	.38	.42	0.44 ± .06
$m(T'') - m(T)$	898 MeV	886 MeV	891 MeV
$\Gamma_3/\Gamma_1(e^+e^-)$	.28	.34	0.35 ±0.04
$m(T''') - m(T)$	1,172 MeV	1,160 MeV	1,113 MeV

These potential models work extremely well at determining the gross structure of heavy $Q\bar{Q}$ systems. Furthermore, there is some evidence for the expected short distance behaviour of the potential. In the remaining sections I want to consider some aspects of heavy quark systems which can provide important new theoretical insights in the next few years.

3. COUPLING OF $Q\bar{Q}$ SYSTEMS TO DECAY CHANNELS

The Cornell Model provides a detailed model[4,5] of the decay amplitudes $Q\bar{Q} \to (Q\bar{\ell}) + (\bar{\ell}Q)$ with $\ell = u,d,s$ and $Q = c$. I won't go into the details here but simply point out that the resulting decay amplitudes have a strong energy dependence; for example, the decay amplitudes of radially excited $Q\bar{Q}$ systems have nodes. This model was applied with some success to the $(c\bar{c})$ resonances above threshold and in particular ΔR_{charm}. One can analyze the $(b\bar{b})$ system using the same model. In fact, there are no new parameters except the masses of the ground state mesons carrying the bottom flavor: the pseudoscalars $B^-(b\bar{u})$ and $B^0(b\bar{d})$, and the vector states $B^{*-}(b\bar{u})$ and $B^{*0}(b\bar{d})$. It is expected[11] than $m(B^0) - m(B^-) = 4$ MeV and $m(B^*) - m(B) = 50$ MeV. The only remaining unknown is the mass of the $B^-(m_B)$ and hence the threshold $(2m_B)$ for Zweig allow decays of $b\bar{b}$ resonances. The details of this analysis are contained in a separate publication.[11] The complicated nature of the resulting decay amplitudes is illustrated in Figure 3. The evidence

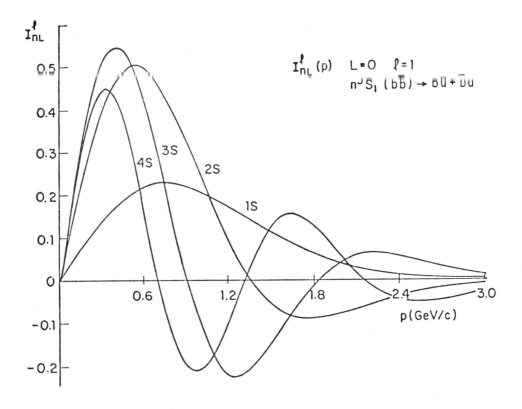

Figure 3: Invariant decay amplitudes for $n^3S_1(b\bar{b})$ resonances to decay into B^+B^- pair of relative three momentum p. More details may be found in Reference 11.

from CESR[1] indicates that the T''' resonance is considerably broader than experimental resolution with a FWHM ~15 MeV and a peak ΔR ~ 1-2. This strongly suggests that the threshold for Zweig allowed decays lies between the 3S and 4S $(b\bar{b})$ states. The behaviour of ΔR_b in the region of the 4S resonance can also be calculated in the Cornell Model. The shape is sensitive to the exact position of B^-B^+ threshold. Figures (4a,b) show ΔR_b for two reasonable choices of $m(T''') - 2m_B$. The values of FWHM and peak ΔR obtained in this model agrees well with the premininary evidence on the T'''. I emphasize there are no new adjustable parameters (except $M(4S) - 2m_B$) in this calculation.

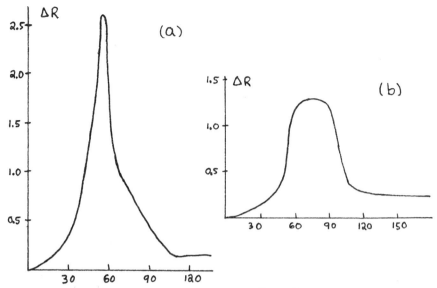

Figure 4: $\Delta R(E_{cm})$ in the region of the $4^3S_1(b\bar{b})$ resonance for two values of $M(4S) - 2M_B$: (a) 50 MeV and (b) 70 MeV. The horizontal axis is $E_{cm} - 2M_B$ (in MeV).

There is further information on energy dependence of the contribution from each exclusive channel I have not displayed[11] which can be checked experimentally in the region of the 4 resonance. In the more immediate future the detailed behaviour of the exclusive charmed meson channels in the 4 GeV region will be measured and comparison made with the predictions of the Cornell Model. The results of these experiments should confirm the complicated energy behaviour of the decay amplitudes for radially excited meson states; and hopefully lead to the construction of a more realistic model of the interplay between quark confinement and hadronic strong decays.

4. RELATIVISTIC CORRECTIONS IN $Q\bar{Q}$ SYSTEMS

As discussed in Section 2, the form of the non-relativistic phenomenological potential is reasonably well established. It is therefore natural to turn to the relativistic corrections. The spin independent relativistic corrections, e.g. higher order radial kinetic energy and orbital angular momentum ($\vec{L}^2$) corrections are difficult to determine without a better theory of the non-relativistic potential. The spin-dependent corrections are more easily measured experimentally since a multiplet of states of given L and radial quantum number (n) are only split in mass by the spin dependent forces. Furthermore, the $b\bar{b}$ system contains many multiplets of narrow resonances for which detailed measurements of the spin splittings are possible.

The short distance behaviour of the spin dependent forces can be calculated in perturbation theory[12] but little is known about their long distance behaviour. The most general form for the lowest order spin dependent corrections in a meson system consisting of a quark with mass m_1 and spin $\vec{s}_1$ and an antiquark with mass m_2 and spin $\vec{s}_2$ is given by:

$$V_{spin}(R) = \left[\frac{\vec{S}_1 \cdot \vec{L}}{2m_1^2} + \frac{\vec{S}_2 \cdot \vec{L}}{2m_2^2} \right] \frac{1}{R} \frac{dV_1(R)}{dR}$$

$$+ \frac{1}{m_1 m_2} \vec{L} \cdot (\vec{S}_1 + \vec{S}_2) \frac{1}{R} \frac{dV_2(R)}{dR} + \frac{2}{3m_1 m_2} \vec{S}_1 \cdot \vec{S}_2 \nabla^2 V_4(R)$$

$$+ \frac{1}{3m_1 m_2} [3(\vec{S}_1 \cdot \hat{R})(\vec{S}_2 \cdot \hat{R}) - (\vec{S}_1 \cdot \vec{S}_2)] \left[\frac{1}{R} \frac{dV_3(R)}{dR} - \frac{d^2 V_3(R)}{dR^2} \right]$$

$$(4.1)$$

In Q.E.D. there is a connection between the spin independent potential (the Coulomb potential $V_c(R)$) and the spin dependent interaction summarized in the form of the Breit-Fermi equation.[13] This relation yields $V_1(R) = V_2(R) = V_3(R) = V_4(R) = V_c(R)$ in Eq. (4.1). What is the analogy of the Breit equation in Q.C.D.? This question has been analyzed by F. Feinberg and myself[14]; and we conclude that without any additional assumptions only a part of the spin-dependent potential can be determined from the static(non-relativistic)potential $\varepsilon(R)$. The relations obtained are

$$V_4(R) = V_2(R) \tag{4.2a}$$

and

$$V_1(R) = \varepsilon(R) + \tilde{V}_1(R) \tag{4.2b}$$

The remaining potentials $\tilde{V}_1(R)$, $V_2(R)$ and $V_3(R)$ are independent. The relation (4.2b) is actually non-trivial only in the sense that $V_1(R)$ is determined by a gauge magnetic field correlation while $V_1(R)$ has both gauge magnetic and gauge electric field correlations terms. In fact, all the uncalculable potentials $\tilde{V}_1(R)$, $V_2(R)$ and $V_3(R)$ involve only the gauge magnetic field correlations. This . suggested a simple additional assumption that allows these potentials to be calculated. Since confinement is likely to be associated with long range gauge electric field correlations[14] it is tempting to assume that the long range components of the spin dependent interactions are also only electric. Then the gauge magnetic correlations might sensibly be calculated in perturbation theory. In lowest order they would be determined by single gluon exchange.[15] A parameter free form for the spin dependent potential in Q.C.D. results:

$$V_{spin}^{Q.C.D.} = \left[\frac{\vec{S}_1 \cdot \vec{L}}{2m_1^2} + \frac{\vec{S}_2 \cdot \vec{L}}{2m_2^2}\right] \frac{1}{R} \frac{d\,\varepsilon(R)}{dR} + \left(\frac{4}{3}\alpha_s\right)\frac{1}{m_1 m_2} \frac{\vec{L} \cdot (\vec{S}_1 + \vec{S}_2)}{R^3}$$

$$+ \left(\frac{4}{3}\alpha_s\right)\frac{1}{m_1 m_2}(3\vec{S}_1 \cdot \hat{R}\,\vec{S}_2 \cdot \hat{R} - \vec{S}_1 \cdot \vec{S}_2)\frac{1}{R^3}$$

$$+ \left(\frac{4}{3}\alpha_s\right)\frac{2}{3m_1 m_2}\,\vec{S}_1 \cdot \vec{S}_2\, 4\pi\,\delta(\vec{R}) \qquad (4.3)$$

where α_s is the running coupling constant and $\varepsilon(R)$ is the non-relativistic potential itself. The results of using Eq. (4.3) for the $c\bar{c}$ system for $\alpha_s = .38$, $m_c = 1.84$ GeV, and $\varepsilon(R)$ given by the Cornell Model is given in Table 2:

TABLE 2: Comparison of Spin-Splittings With Data For Charmonium Resonances

Splittings	Theory[12]	Experiment
$1^3S_1 - 1^1S_0$	129.8 MeV	116 ± 9 MeV[22]
$^3P_2 - ^3P_J\,(COG)$	36.5 MeV	30 ± 2 MeV[10]
$^3P_1 - ^3P_J\,(COG)$	-29.4 MeV	-17 ± 2 MeV[10]
$^3P_0 - ^3P_J\,(COG)$	-94.1 MeV	-90 ± 3 MeV[10]

The agreement is quite good considering the corrections expected from higher order perturbation theory (~30%) and higher order relativistic corrections (~20%). These corrections are systematically smaller for the $(b\bar{b})$ system; and therefore, if the short range magnetic correlation assumption is valid, the spin splitting in the $b\bar{b}$ system should be in good agreement with the predictions derived from Eq. (4.3) $\alpha_s = .26$ and $m_b = 5.17$ GeV.

Some of the results[12] are presented in Table 3:

TABLE 3:   Predictions of Spin Splittings for the $b\bar{b}$ System

	S States $^3S_1 - {}^1S_0$	P State Splittings (MeV) $^3P_2$	$^3P_1$	$^3P_0$	D State Splittings (MeV) $^3D_3$	$^3D_2$	$^3D_1$
n=1	95 MeV	14	-11	-36	3.5	-3.5	-12.5
n=2	41 MeV	10	- 8	-26	7	-3	-11

Measurement of the spin splittings in the $(b\bar{b})$ system will provide a critical test of this approach.

5.   HADRONIC TRANSITIONS

Only the $J^{PC} = 1^{--}$ states of a $(Q\bar{Q})$ system can be directly produced in e^+e^- annihilation through a single virtual photon; however the excited 1^{--} states which are below threshold for Zweig allowed decay have large branching rate to make photonic and hadronic transitions to lower mass $(Q\bar{Q})$ states. For the $(b\bar{b})$ system there is a rich variety of possible transitions of the form $(b\bar{b}) \rightarrow (b\bar{b})' +$ light hadrons[16] as shown in Figure 5. There has been considerable theoretical work in the last few years on the structure of these hadronic transitions in Q.C.D. In terms of the underlying field theory a hadronic transition results from the emission of gluons by the initial quark (Q) and antiquark ($\bar{Q}$) which gives rise to a final quark and antiquark in a different final state and the gluons produce the light hadrons. This underlying picture suggests an analog with photonic transitions in atomic (or nuclear) physics where a multipole expansion is physically justified. K. Gottfried proposed such a multipole expansion for gluon emission in Q.C.D.[17] There are two scales in the problem: the separation of Q and $\bar{Q}$ in the initial state (a), and the momentum transfer to the gluons in the particular process $k \equiv 1/r$ (equivalently r is the size of light hadrons). For (ka) << 1 we have a multipole expansion. The actual details of the expansion is complicated because of the non-Abelian nature of the emission.[18] It has been demonstrated recently by T. M. Yan that a formal expansion can be generated in a way similar to Q.E.D. by making full use of the non-Abelian gauge invariance.[19] Yan then applies the results of multipole expansion to the hadronic transitions in the Υ system

(Figure 5). These transitions involve strong dynamics so the

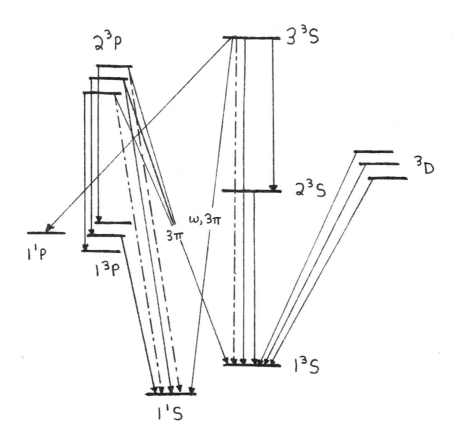

Figure 5: Some allowed hadronic transitions in the $b\bar{b}$ system. Solid lines denote the emission of two pions. The dot-dashed line emission of an η. Other transitions indicated explicitly. See References (16,19) for details.

actual rates cannot yet be calculated (although interesting m_Q scaling law can be deduced[17,19]). However, the Wigner-Eckart Theorem can be used (as in atomic transitions) to obtain relations between rates. For example, Yan derives a set of relations among the transition $m \ ^3P_J \rightarrow n \ ^3P_{J'} + \pi\pi$: (m>n)

$$\Gamma(1\rightarrow1) = \Gamma(0\rightarrow0) + \frac{1}{4} \Gamma(0\rightarrow1) + \frac{1}{4} \Gamma(0\rightarrow2) \qquad (5.1a)$$

$$\Gamma(1\rightarrow2) = \frac{5}{12} \Gamma(0\rightarrow1) + \frac{3}{4} \Gamma(0\rightarrow2). \qquad (5.1b)$$

$$\Gamma(2\to2) = \Gamma(0\to0) + \frac{3}{4}\,\Gamma(0\to1) + \frac{7}{20}\,\Gamma(0\to2) \qquad (5.1c)$$

where $\Gamma(J\to J')$ is the partial rate for $m^3P_J \to n^3P_{J'} + \pi\pi$. These and many more relations are discussed in detail by Yan. Clearly the measurement of the hadronic transition rates in the $(b\bar{b})$ system will provide important tests of this approach to hadronic transitions as well as give information on the strong dynamics which determines the actual rates.

6. APPLICATIONS OF PERTURBATIVE Q.C.D.

Finally, let me turn briefly to the applications of perturbative Q.C.D. and renormalization group methods to sufficiently heavy $(Q\bar{Q})$ systems. In order to eliminate the wavefunction at the origin which appears in decay rates I look at ratios of decays:

$$\frac{\Gamma(^3S_1\to 3\text{ gluons})}{\Gamma(^3S_1\to e^+e^-)} = \frac{10(\pi^2-9)\,\alpha_s^3}{81\pi\,e_Q^2\alpha_{EM}^2} + 0(\alpha_s^4) \qquad (6.1a)$$

$$\frac{\Gamma(^3S_1\to 3\text{ gluons})}{\Gamma(^3S_1\to\text{photon} + 2\text{ gluons})} = \frac{5}{36}\,\frac{\alpha_s}{e_Q^2\alpha_{EM}} + 0(\alpha_s^2) \qquad (6.1b)$$

$$\frac{\Gamma(^1S_0\to 2\text{ gluons})}{\Gamma(^1S_0\to 2\text{ photons})} = \frac{2}{9}\,\frac{\alpha_s^2}{e_Q^4\alpha_{EM}^2} + 0(\alpha_s^3) \qquad (6.1c)$$

and for the spin dependent interaction in Eq. (4.3)

$$\frac{m(^3S_1)-m(^1S_0)}{\Gamma(^3S_1\to e^+e^-)} = \frac{8}{9}\,\frac{\alpha_s}{e_Q^2\alpha_{EM}^2} + 0(\alpha_s^2). \qquad (6.1d)$$

The gluons produce light hadrons with assumedly unit probability and hence the rates appearing in Eqs. (6.1a-d) are experimentally measurable. However, the value of α_s determined by each of these four processes need not agree. This is because in Q.C.D. the coupling constant α_s is conventionally defined by the value of some off-shell Green's function at momenta of scale μ. Of course physical quantities such as the rates in the above equations do not depend on the value of μ (the normalization point). This is true in Eqs. (6.1 a-d) since dependence on μ is formally higher order in α_s; however, it also means it is important to compute the next order corrections in α_s for this process to insure that, for the particular choice of α_s used, the corrections are small. It may also be that some corrections are physically large, i.e. for no choice of the definition of α_s can corrections to all the above processes be made simultaneously small. In particular, it is likely that with any reasonable definition of α_s the corrections to $\Gamma(^3S_1 \to e^+e^-)/|\Psi(0)|^2$ are large.[20] Comparison of the ratio in Eqs. (6.1 a-d) for the $(c\bar{c})$ and $(b\bar{b})$ systems will test the quark mass dependent part of the higher order perturbative corrections.

There is also an exciting new development in the application of perturbative methods to heavy quark systems. It has been shown recently[21] that the methods of the renormalization group can be extended to the calculation of any heavy $Q\bar{Q}$ annihilation process in which the momentum of every gluon appearing in the lowest order connected perturbative diagram is forced by kinematics to be sufficiently large. These techniques can therefore be applied to total hadronic decays, π or K inclusive decays, or even exclusive hadronic final states. For example, the branching ratio

$$\frac{\Gamma(^3P_J \to \pi\pi)}{\Gamma(^3P_J \to \text{light hadrons})} \qquad \text{is calculable.}$$

$$(6.2)$$

In fact, the leading behaviour (as $m_Q \to \infty$) has no unknown parameters (the only strong interaction effects are expressible in terms of f_π). These methods should lead to many predictions which can be tested in the $(b\bar{b})$ system.

7. SUMMARY

The basic assumption underlying the phenomenology of heavy $(Q\bar{Q})$ systems (the existence of an non-relativistic effective potential) is strongly supported by the $(c\bar{c})$ and $(b\bar{b})$ data and furthermore the form of the potential is reasonably well determined. There is still much to learn from these systems: (1) the structure of the Zweig allowed decay amplitudes; (2) the detailed structure of the spin dependent interactions; (3) relations among the hadronic transition rates: $(Q\bar{Q}) \to (Q\bar{Q})' + \text{light hadrons}$; and (4) comparison of the observed rates of hadronic decays to light meson (and baryon) final states with calculations in perturbative Q.C.D. All these aspects are presently under both experimental and theoretical investigation. Of course, there may also be surprises waiting to be discovered through the study of the decays of the $(b\bar{b})$ ground state mesons (glueballs, the Higg meson, Pseudo Goldstone Bosons, etc.).

REFERENCES

1. D. Andrews , et al., Phys. Rev. Lett. 44, 1108 (1980) and T. Bohringer, et al, Phys. Rev. Lett. 44, 1111 (1980); and the references contained therein.
2. A. Duncan, Phys. Rev. D 13, 2866 (1976); F. L. Feinberg, Phys. Rev. Lett. 39, 316 (1977); T. Appelquist, M. Dine, and I. Muzinich, Phys. Lett. 69B, 231 (1977; and W. Fischler, Nucl.

Phys. B129, 157 (1977).

3. For a recent review of confinement mechanisms see: M. Bander, "Theories of Quark Confinement", to appear in Physics Reports.

4. E. Eichten, et al., Phys. Rev. D 17, 3090 (1978); and Errata Phys. Rev. D 21, 313 (1980).

5. The phenomenological applications of the model discussed in Reference 4 are presented in: E. Eichten, et al., Phys. Rev. D 21, 203 (1980).

6. The potential may be constructed by inverse scattering methods from the spectrum and leptonic widths of the $J^{PC} = 1^{--}$ ($c\bar{c}$) and ($b\bar{b}$) states. See, for example, C. Quigg and J. Rosner, "Inverse Scattering and the T Family", University of Minnesota Preprint (1980) and references contained therein.

7. C. Quigg and J. Rosner, Phys. Lett. 71B, 153 (1977); M. Machacek and Y. Tomozawa, Ann. Phys. (N.Y.) 110, 407 (1978).

8. W. C. Haxton and L. Heller, Los Alamos Preprint LA-UR 80-855 (1980).

9. J. L. Richardson, Phys. Lett. 82B, 272 (1979).

10. The most recent data on the ^3P_J ($c\bar{c}$) masses from the Mark II Detector Group at SPEAR is reported in T. M. Himel, et al., Phys. Rev. Lett. 44, 920 (1980) while the results of the Crystal Ball Group are presented by E. Bloom in Proceedings of the International Symposium on Lepton and Photon Interactions at High Energies, Batavia, Illinois, 1979 (to be published).

11. E. Eichten, Harvard Preprint Hutp-80/A027 (1980).

12. M. Dine, Yale Preprint COO-3075-204 (1978) and Yale Thesis (unpublished). Also, see Reference 2.

13. A. Akhiezer and V. Berestetskii, "Quantum Electrodynamics", translated by G. Vokkoft, Wiley, New York (1965), c.p. 528.

14. E. Eichten and F. Feinberg, Phys. Rev. Lett. 43, 1205 (1979); and Harvard Preprint HUTP-80/A053 (1980).

15. The single gluon exchange assumption was first proposed by A. De Rújula, H. Georgi, and S. Glashow, Phys. Rev. D 12, 47 (1975).

16. E. Eichten and K. Gottfried, Phys. Lett. 66B, 286 (1977).

17. K. Gottfried, Phys. Rev. Lett. 40, 598 (1978).

18. G. Bhanot, W. Fischler, and S. Rudaz, Nucl. Phys. B155, 208 (1979); M. E. Peskin, Nucl. Phys. B156, 391 (1979).

19. T. M. Yan, Cornell Preprint, CLNS 80/451 (1980).

20. R. Barbieri, et at., Nucl. Phys. B105, 125 (1976); W. Celmaster, Phys. Rev. D 19, 1517 (1979).

21. A. Duncan and A. Mueller, Columbia Preprint, CU-TP-175 (1980).

22. T. M. Himel, et al., Phys. Rev. Lett. 45, 1146 (1980); R. Patridge, et al., Phys. Rev. Lett. 45, 1150 (1980).

Comment made by Leon Heller, Los Alamos, after the talk by Eichten
in the second morning session on April 26.

W. Haxton and I have derived the central part of the heavy quark-
antiquark potential [1] by treating the MIT bag model in Born-Oppenheimer
approximation. For fixed q-$\bar{q}$ separations the color electrostatic fields
and bag shapes are determined by solving numerically the Yang-Mills
equations with bag boundary conditions to lowest order in the quark-gluon
coupling constant. As mentioned by Dr. Eichten, this calculation gives
the absolute scale of the energy, and not just energy differences.

The only two adjustable parameters in our work (besides quark masses)
refer to the variation of the running coupling constant with q-$\bar{q}$ separation, for which we assume the following simple form

$$\alpha(r) = \frac{4\pi}{11 - \frac{2}{3}n_f} \frac{1}{\ln\left[\frac{1}{\Lambda^2 r^2} + \gamma\right]} \quad .$$

Λ is the scale parameter of asymptotic freedom, and tells how the
coupling constant goes to zero as r goes to zero. In addition it
is necessary to specify the value of α at some finite separation,
and this is achieved with the parameter γ. The figure shows the
curve of α vs. r for the values of the parameters which were found
to give the best fit to the $c\bar{c}$ and $b\bar{b}$ spectra and leptonic decay
widths. (Rather than quote the value of γ, we give the value of α
at 1 Fermi.) The dashed curve is the pure asymptotic freedom formula,
i.e., that obtained by putting $\gamma=0$ for the same value of Λ.

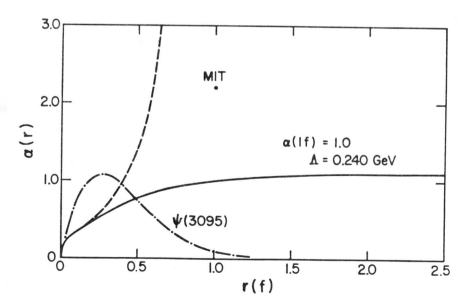

The dash-dot curve is the wave function of the ground state of charmonium. It is clear from the figure that there is a very large
variation in the coupling constant over the range of distances for
which this wave function has significant values. It is also clear
that the departure of the coupling constant used in our work from
the pure asymptotic freedom formula is very important.

The point labelled 'MIT' shows the value of the coupling constant
which was used by the MIT group [2] in their fit to the masses of
hadrons composed of light quarks, using the 'fixed bag' approximation.
These bags had radii and interquark separations which were typically
1f. If we had insisted that the curve of α vs. r pass through
that point, we could successfully fit the low lying states of $c\bar{c}$
and $b\bar{b}$ (by altering the value of Λ and changing the quark masses),
but the higher $c\bar{c}$ states would appear much too high in energy. So
there appears to be a discrepancy in this regard between the present
work on heavy quark systems and the earlier work on light quark
systems.

We have also applied our $q\text{-}\bar{q}$ potential to the strange quark, and
obtain consistency between the masses of the ϕ and F^* mesons
with a quark mass $m_s = 0.64$ GeV. These systems are becoming
relativistic, however, and this suggests that the strange quark may
require a description which is intermediate between the 'fixed bag'
approximation and the Born-Oppenheimer approximation.

1. W. C. Haxton and L. Heller, "The Heavy Quark-Antiquark Potential
in the MIT Bag Model," Los Alamos preprint LA-UR-80-855, submitted to
Physical Review D.

2. T. DeGrand et al., Phys. Rev. D12 (1975) 2060.

FLAVOR MIXING AND QUARK DECAY

Ling-Lie Chau Wang

Brookhaven National Laboratory, Upton, New York   11973

Since this is an experimental conference I shall begin my talk with that spirit. We can view that the subject of my talk as a result of "the ORY Collaboration" with more than fifty theorists involved. The topics covered are the results of four task forces: I. the Mixing Matrix Task Force, II. the D-decay Task Force, III. the Boredom-Escaping Group and IV. the Far-and-Beyond Group.

I. THE MIXING MATRIX

Here I shall give a brief historical review of the development of the mixing among quarks, i.e, the quark states contributing to the weak currents are mixtures of the quark mass eigenstates of strong interactions. In the weak theory of Cabibbo[1] of 1963, the weak interacting quark states are a left-handed doublet and a left-handed singlet:

$$\begin{pmatrix} u \\ d' \end{pmatrix}_L, \quad s_L, \tag{I.1a}$$

where

$$d' = \cos\theta_c d + \sin\theta_c s. \tag{I.1b}$$

In this theory there is the strangeness-changing neutral current $\bar{d}s$ term from $\bar{d}'d'$. However this is not consistent with the very small branching ratio of $B(K_L \to \mu\bar{\mu}) = (9.1\pm1.8)\times10^{-9}$. This led to the introduction, after the renormalizability of the weak interaction theory had been realized, of the charm quark to form another left-handed doublet by Glashow-Iliopolous-Miani[2] in 1970,

$$\begin{pmatrix} u \\ d' \end{pmatrix}_L, \quad \begin{pmatrix} c \\ s' \end{pmatrix}_L, \tag{I.2a}$$

where

$$(d', s') = (d, s) \begin{pmatrix} \cos\theta_c & -\sin\theta_c \\ \sin\theta_c & \cos\theta_c \end{pmatrix}. \tag{I.2b}$$

The orthogonality of the mixing matrix guarantees the absence of strangeness flavor changing neutral current $\bar{d}s$ term from $(\bar{d}'d' + \bar{s}'s')$. Then it was noticed in 1973 by Kobayashi and Maskawa[3] that the reality of the matrix in Eq. (I.2b) would not allow CP violation in the standard $SU(2)_L \times U(1)$ theory with a single Higgs doublet. Purely based upon this theoretical observation they introduced the

404

third pair of left-hand quark doublets,i.e.

$$\begin{pmatrix} u \\ d' \end{pmatrix}_L, \quad \begin{pmatrix} c \\ s' \end{pmatrix}_L, \quad \begin{pmatrix} t \\ b' \end{pmatrix}_L,$$ (I.3a)

where

$$(d', s', b') = (d,s,b) \begin{pmatrix} V_{ud} & V_{cd} & V_{td} \\ V_{us} & V_{cs} & V_{ts} \\ V_{ub} & V_{cb} & V_{tb} \end{pmatrix}$$ (I.3b)

The V_{ij}'s characterize the coupling of the quarks q_i, q_j to the weak intermediate boson $W^\pm$. Again the unitarity of the V matrix, $VV^+ = I$ insures the absence of flavor changing neutral current. Now the V_{ij}'s are complex and characterized by three angles and one phase. This complexity via the diagram in Fig. 1 gives imaginary part in the

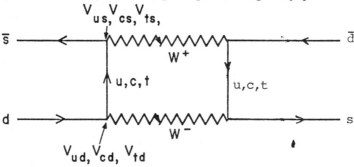

Fig. 1: The box graph for calculating
the $K^0 - \bar{K}^0$ transition matrix.

$K^0 \leftrightarrow \bar{K}^0$ transition matrix, thus the CP violations. The important point is that in this model CP violation is tied with the finiteness of some of the matrix elements in the third column or the third row, i.e. the b and t flavored particles must have pure hadronic decays. Models with CP violation coming from the Higgs couplings[4], by having more Higgs doublets than the standard $SU(2)_L \times U(1)$ model, have no such correlation. Actually in many of these models, the b-flavored particles have only semileptonic decays though this is not imposed on by any first principles.

The matrix element $|V_{ud}|$ can be determined from the $0^+ \to 0^+$ nuclear β-decay of ^{14}O, $^{26m}A_\ell$ in comparison with the μ decay rate (here the effect of lepton mixing is assumed to be absent).The matrix element $|V_{us}|$ can be determined from semileptonic decays of the hyperons and Ke_3. The results of Shrock and Wang[5] were

$$|V_{ud}| = .9739\pm.0025, \qquad (I.4)$$

$$|V_{us}| = .219\pm.003, \qquad (I.5)$$

$$|V_{ud}|^2 + |V_{us}|^2 = .996\pm.004. \qquad (I.6)$$

The reader is referred to the original paper for the discussions of errors and uncertainties involved. The significance of Eq. (I.6) is that the central value of $|V_{ud}|^2 + |V_{us}|^2$ is less than one, indicating that the old Cabibbo theory was not exactly true and there is "leakage" from the first two doublets. It allows the third doublet to decay, i.e. the b can decay into u.

The other two parameters are determined from the CP violation property of the K^0's, i.e. the small parameter ε and the K_L, K_S mass difference. However the determination of these two parameters relies upon a theoretical estimate of the matrix element

$$\mathcal{M} \equiv \langle \bar{K}^0 | \bar{s}\Gamma_\mu d \bar{s}\Gamma_\mu d | K^0 \rangle, \qquad (I.7)$$

which is still rather uncertain. The vacuum insertion and the MIT-bag calculations[6] can differ by a factor of two. Allowing such a factor of two uncertainty in comparison with data Shrock, Treiman, Wang obtained the central values of the V matrix:

$$V = \begin{array}{ccc} u & c & t \\ \left(\begin{array}{ccc} .97 & -.20 & -.11 \\ .22 & .95-.75\times10^{-3}i & .20+1.3\times10^{-3}i \\ .068 & -.22+2.4\times10^{-3}i & .97-4.1\times10^{-3}i \end{array} \right) & \begin{array}{c} d \\ s \\ b \end{array} \end{array} \qquad (I.8)$$

Using different estimates of Eq. (I.7), or different tolerance in comparison with experiments, results of different ranges certainly can be reached. However the result is rather insensitive to the t quark mass as long as it is not comparable to the intermediate boson mass.

It is interesting to note that the matrix elements in the V of Eq. (I.8) get smaller as they move away from the diagonal, i.e. though there are flavor mixing, the flavors like to keep their original identity. Also it implies that b and t prefer to decay in a cascade fashion, $t \to b \to c \to s$.

Another way to bound the angle, independent of Eq. (I.7) is to consider $K_L \to 2\mu$. Shrock and Voloshin[7] obtained $|V_{ts}V_{td}^*| \lesssim .06$, which is satisfied by the V matrix of Eq. (I.8).

Here I want to emphasize that the V matrix analysis is purely based upon the six-quark model. So far the sixth member, the t

quark is still eluded from observations. With the observation of
Υ''' and its width as reported in this morning's talk of results from
CESR we may be on the threshold of observing the b-flavor particles.
We must keep an open mind whether b is from a doublet, whether the
flavor changing neutral current exits in B^0 decays, i.e. whether
$B^0 \to \tau^-\tau^+$, $B^0 \to$ hadrons $+ e^+e^-$ exist. It is also preferable if we
can determine the matrix element V_{cd}, V_{cs} more directly from experi-
ment like $|V_{ud}|$ and $|V_{us}|$. This I shall mention in the later sect-
ions.

II. THE D DECAYS (NON-LEPTONIC)

Before I launch into the discussions on the non-leptonic decays
of the D meson, I shall first briefly review the history of our
learning processes of the K non-leptonic decays. It was in 1956,
Birge et al[8] observed that the $K^+ \to \pi^+\pi^0$ decay rate was greatly
suppressed compared to that of $K^0 \to \pi^+\pi^-$, i.e. $\Gamma(K^+ \to \pi^+\pi^0)/$
$\Gamma(K^0 \to \pi^+\pi^-) \approx 1/670$. Since only the isospin I=2 component contri-
butes in $K^+ \to \pi^+\pi^0$, i.e. $\Delta I = 3/2$, it was proposed[9] that the weak
Hamiltonian H_w was dominated by I = 1/2; or in SU(3) classification,
H_w transforms like an 8. However no good dynamical reason was found
for this $\Delta I = 1/2$ rule. Soon after the realization of the renormalizability of the weak interaction[10], some quark diagrams with the
lowest order gluon corrections were pointed out to give $\Delta I = 1/2$
rule[11] (see Fig. 2.) But it gave only a factor of 4 enhancement in

Fig.2: Quark diagram
with one loop of
the W and gluon
exchange

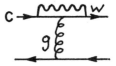

Fig. 3: The
"Penguin" diagram

rate comparing to the experimental number 670. Then in 1977, the
so-called "Penguin" diagram of Fig. 3 was claimed[12] to give enough
enhancement. Afterwards, the bag model was also used to explain
part of the effect.

If it has taken more than 20 years to search for the explanation
of the mechanism of K decay, I am sure that it will take some time
before we really understand the charm non-leptonic decays. Therefore I personally prefer first to take a phenomenological approach.[13]
We classify the charm decays in quark diagrams (Fig. 4, next page).
There are six distinct diagrams. These diagrams are meant to
include all strong interaction effects (the gluon lines), which are
in general not yet calculable. Thus we do not know the magnitude of
each diagram. However we can classify experimental results using
the diagrams, hoping to obtain the sizes and phases of these

diagrams. This attitude is justified by the fact that Nature has surprised, or defied , those who attempted to make some simple dynamical guesses , as I am now going to report.

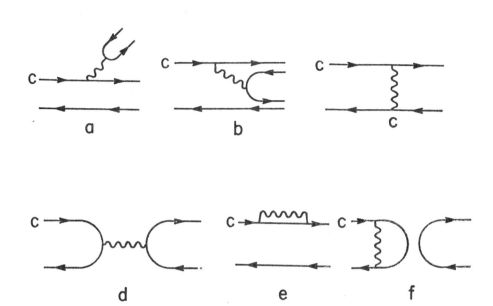

Fig. 4: Diagrams for charm decay
(noting that diagrams e and f
are Cabibbo suppressed)

Since this a very fast developing field, I shall describe events following the sequence in time. April 1979 came the first surprise,[14]

$$R_1 \equiv \frac{\Gamma(D^0 \to K^- K^+)}{\Gamma(D^0 \to K^- \pi^+)} = .113 \pm .030, \qquad (II.1a)$$

$$R_2 \equiv \frac{\Gamma(D^0 \to \pi^- \pi^+)}{\Gamma(D^0 \to K^- \pi^+)} = .033 \pm .015 \qquad (II.1b)$$

(Though the errors are still large and the possibility of $R_1 = R_2$ may still turn out to be true, let's discuss in the context that R_1, R_2 are unequal.)   This was a surprise because calculations were made in 1975 in the 4-quark scheme assuming SU(3) symmetry[15] and the result was $R_1 = R_2 = \tan^2\theta_c \approx .05$.   The historically interesting thing is that though the Kobayashi-Maskawa six-quark model was proposed in 1973, no such calculation was done before the data of Eqs.(II.1a) and (II.1b).

Several explanations have so far been offered:

(1) Rather than the 4-quark model, one ought to generalize the SU(3) symmetric calculations to 6-quark model. Then if $V_{us}/V_{ud} \neq -V_{cd}/V_{cs}$, R_1 and R_2 do not have to be the same and the data shown in Eqs. (II.1a) and (II.1b) can be easily accommodated.[16] If $V_{us}/V_{ud} = V_{cd}/V_{cs}$, one must resort to other reasons.

(2) The inequality of R_1 and R_2 may be from SU(3) breaking,[17] but it is hard to make quantitative statements.

(3) It was suggested that maybe a charged Higgs[18] is involved since Higgs prefers to couple to heavier quarks than the lighter ones, so as to make $R_1 > R_2$. But it was pointed out by Abbot et al[18] such a Higgs boson will give too much CP violation effect in K^0 system via the double Higgs-exchange box diagram.

(4) The "Penguin" diagram[19] may be important for the charm decay too. But it is generally agreed that unlike for the K decay, the "Penguin" diagram cannot give the whole effect.

(5) It has been pointed out that the final state interaction effects[20] are important. But the difficulty is how to incorporate such effects.

The second surprise came during the summer of 1979. The lifetimes[21] of D^+ and D^0 were different:

$$\tau(D^+)/\tau(D^0) > 5.8 \pm 1.5; \quad 3.1^{+4.1}_{-1.3}. \qquad (II.2)$$

This came as a surprise because it was argued[22], considering only the one gluon loop diagram of Fig.2, that, among the Cabibbo favored, only graphs a and b of Fig.4 are important. All other graphs are small because of helicity conservation, just like $\pi^+ \to e^+ \nu_e$ being suppressed. The total decay rate of D^0 decay is given by the diagrams a, b, and c of Fig. 4, and that of D^+ is given by diagrams a and b only. Therefore according to this "Simple QCD" argument diagram c is very small so $\tau(D^+) = \tau(D^0)$. But if we take a more phenomenological approach[13,23] the data of Eq. (II.2) simply implies that the W-exchange diagram c of Fig. 4 is important. Some authors then modified the Simple-QCD[24] and argued that W-exchange diagram with one gluon emission (Fig. 5) though being a tree graph, is important. Then there

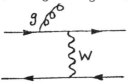

is no helicity conservation arguments. The fact is that two more parameters are introduced. Educated guesses on these parameters do do give reasonable numbers comparing to the data. The interesting question is what is the lifetime of F^+, whose total decay rate is given by

Fig.5: The W-exchange diagram with one gluon emission

the graphs a, b, and d. But the diagram d cannot have one gluon emission tree graph, so diagram d is still small. Thus in this "Modified-Simple-QCD" scheme, the lifetime of F^+ will be more like the D^+.

The third surprise to Simple-QCD argument is that[14]

$$\frac{\Gamma(D^0 \to K^-\pi^+)}{\Gamma(D^0 \to \bar{K}^0\pi^0)} = 1.6\pm0.9. \qquad (II.3)$$

The reason is the following,

$$\Gamma(D^0 \to K^-\pi^+) = |V_{ud}V_{cs}(a+c)|^2 , \qquad (II.4)$$

$$\Gamma(D^0 \to \bar{K}^0\pi^0) = \frac{1}{2}|V_{ud}V_{cs}(b-c)|^2 . \qquad (II.5)$$

As mentioned before, Simple-QCD implies that diagram c is small. By color counting a = 3b, so $\Gamma(D^0 \to K^-\pi^+) = 18\Gamma(D^0 \to \bar{K}^0\pi^0)$, which is certainly violently contradicted by the data. Again, here we can take the naive phenomenological view again that the exchange graph c is important. But it has been also pointed out final state inter-action may be very important in the $\bar{K}^0\pi^0$ channel.[25] Similarly the experimental result

$$B(D^+ \to \bar{K}^0\pi^+) \approx B(D^0 \to K^-\pi^+) \qquad (II.6)$$

is consistent with the conclusion that the W-exchange diagram is important.[23] One interesting conclusion of the W-exchange diagram dominance is that

$$\Gamma(D^0 \to \bar{K}^0\eta^0)/\Gamma(D^0 \to K^-\pi^+) \approx \frac{1}{6} , \qquad (II.7)$$

which is still to be checked by future experimental data.

Status of 6 dominance[15]: the $\Delta c \neq 0$ H_w transform like 20 or 84 in SU(4). However only the 20 (=6+6*+8) contains the $\Delta c=0$ H_w transforming like an 8 in SU(3), which is implied by the $\Delta I = 1/2$ rule in K decay. Thus it was conjectured that the H_w transforms like 20 in SU(4), which contains the 6 and 6* with $\Delta c \neq 0$. So far, the predictions of 6 dominance have been consistent with the experimental results of Eqs. (II.3), (II.6), but not so good with data of Eq. (II.1a) and (II.1b). Other interesting predictions are
$\Gamma(F^+ \to \bar{K}^0 K^+)/\Gamma(D^0 \to K^-\pi^+) \approx 1$, $\Gamma(F^+ \to \eta\pi^+)/\Gamma(D^0 \to K^-\pi^+) \approx 2/3$,
and $\Gamma(D^0 \to \bar{K}^0\eta)/\Gamma(D^0 \to K^-\pi^+) = 3\Gamma(D^0 \to \bar{K}^0\pi^0)/\Gamma(D^0 \to K^-\pi^+)$.

Four years ago L. N. Chang and I looked into the implication of 6 dominance and PCAC for charm decays. One of the results was that one constant characterizes all three body decays, and non-uniformity in the Dalitz plot of $D^\pm \to K^\pm\pi^\pm\pi^\pm$ was predicted, though the data then showed uniform distribution. It is nice to hear from Goldhaber yesterday that the Dalitz plot indeed is not uniform. We shall compare the results. I am sure that such detailed analysis of Dalitz plot will be very fruitful in learning the weak decay mechanism.

Finally let me mention a few "neat" cases of charm decays:
(1) From isospin of the $C \neq 0$ H_w being one, the amplitudes

$$A(D^+ \to \bar{K}^0\pi^+) - A(D^0 \to K^-\pi^+) + \sqrt{2}\, A(D^0 \to \bar{K}^0\pi^0) = 0,$$

$$A(F^+ \to \pi^+\pi^0) = 0.$$

(2) From SU(3) symmetry[16]

$$\Gamma(D^0 \to \bar{K}^0\eta)/\Gamma(D^0 \to \bar{K}^0\pi^0) = \frac{1}{3}, \text{ and}$$

$$\Gamma(D^+ \to \pi^0\pi^+)/\Gamma(D^+ \to \bar{K}\pi^+) = \frac{1}{2}|V_{cd}/V_{cs}|^2.$$

The last one is especially interesting. Both final states are exotic, so no contamination of final state interactions. It provides a good means to measure $|V_{cd}/V_{cs}|$.

Summing up the results of the D-decay analysis,
(1) Phenomenological quark-diagram classification can accommodate all known decays. Indications are that graphs c, and possibly e and f, are important.
(2) "Simple-QCD" ideas need to be modified. Graphs like Fig. 5 is important. One interesting prediction is $\tau(F^+) \sim \tau(D^+)$.
(3) $\underline{6}$ dominance is quite reasonable in the Cabibbo un-suppressed decays. The ration of $\Gamma(D^0 \to K^+K^-)/\Gamma(D^0 \to \pi^+\pi^-) \neq 1$ poses a problem , but may be explained away by final state interaction.
(4) Final state interactions should be properly put in. So far , a systematic method of calculating final state interaction is still lacking.

III. THE BOREDOM-ESCAPING ATTEMPTS

So far I have talked about quark mixing and charm decays in the six quark scheme. Yet the sixth quark, the t quark, has escaped observation. So what are the alternatives?
(1) The Georgi-Glashow model:[26] They proposed that b remains a singlet. Yet they require that there is still no flavor-changing neutral current. Not only the $d\bar{s}$ neutral current is missing which is required by $K_L \to \mu\bar{\mu}$ but also the $d\bar{b}$ and the sb neutral currents are also missing, which is yet to be established by experiment. To achieve this they had to assign the b quark a new quantum number, which then can only decay via a new gauge boson which carries this same new quantum number, into a lepton pair one of which carrying the new quantum number. The quark multiplets are

$$\begin{pmatrix} u \\ d' \end{pmatrix}_{L,} \begin{pmatrix} c \\ s' \end{pmatrix}_{L}, \underset{\sim}{b}, \qquad \text{(III.1)}$$

where $d' = \cos\theta_c d + \sin\theta_c s$, and $s' = -\sin\theta_c d + \cos\theta_c s$, and the wiggle underneath the $\underset{\sim}{b}$ noting this new quantum number, Now the $\underset{\sim}{b}$ can only decay semileptonically via

$$\underset{\sim}{b} \to q\underset{\sim}{W}$$
$$\qquad \hookrightarrow \ell\bar{\underset{\sim}{\ell}} \qquad \qquad . \qquad \text{(III.3)}$$

Thus the definite predictions of this model are that the $\underset{\sim}{b}$ only decays semileptonically and there exists the new lepton $\underset{\sim}{\ell}$.

(2) The George-Machacek[27] model: here the b-flavored meson B can even decay into $\bar{p}\tau$.

(3) The George-Pais[28] model: here they considered a different group $SU(3)_L \times U(1)$. The quarks are in the triplet representation

$$\begin{pmatrix} u \\ d'' \\ b'' \end{pmatrix}_{L,} \begin{pmatrix} c \\ s'' \\ \ell'' \end{pmatrix}_{L.} \qquad \text{(III.4)}$$

Here the d'', s'', b'' ℓ'' are mixtures of d, s, b and ℓ, and ℓ, using the author's notation, is a new charge $-1/3$ quark. Since the flavor b is mainly in the same triplet as u,

$$b \to u \gg b \to c. \qquad \text{(III.5)}$$

In this scheme, the horizontal mixing is constrained not to give flavor changing neutral current, i.e. no $\bar{ds}$, $\bar{b\ell}$; but the vertial mixing allows flavor changing neutral current, i.e. there exist first order weak $b\bar{d}$, $s\bar{\ell}$, type of neutral currents.

(4) A phenomenological model: we may ask if we can have a "minimum bias" model, i.e. a phenomenological model accommodating all existing experimental informations. There has been no sign of the t quark. Maybe nature is just teasing us a bit. The quark hierarchy may well be just the Cabibbo world extended, i.e.

$$\begin{pmatrix} u \\ d' \end{pmatrix}_{L,} \begin{pmatrix} c \\ s' \end{pmatrix}_{L}, b_L \qquad \text{(III.6)}$$

where

$$d' = V_{ud}d + V_{us}s + V_{ub}b,$$

$$\text{(III.7)}$$

$$s' = V_{cd}d + V_{cs}s + V_{cb}b,$$

and, in addition to the orthogonality and unitarity, the constraint of no strangeness changing neutral current is imposed,

$$V_{ud}V_{us} + V_{cd}V_{cs} = 0. \qquad \text{(III.8)}$$

However the b-flavor changing neutral currents $d\bar{b}$, $s\bar{b}$ are allowed. Such a model was discussed by Barger and Pakvasa[29]. Taking the known value of V_{ud} = .97, V_{us} = .23, one can actually solve the mixing elements. One of the results is

$$V_{ub} = .1, \ V_{cd} = .89, \ V_{cs} = -.25, \ V_{cb} = .39. \qquad \text{(III.9)}$$

They calculated the $s\bar{b}$ flavored neutral current decay of the b having a branching ratio $B(b \rightarrow se^+e^-) \approx 2\%$.

IV. THE FAR AND BEYOND

It utilizes exotic things[30] like compactified dimensions higher than four, graded Lie algebra, SU(n/1),... . The claim is that everything can be calculated, $\sin^2\theta_W = 1/4$, Higgs mass $m_H \approx 245$ GeV, number of flavors = 2^{n-5}. But only a special group of physicists understand the theory and lots of work needed to be done.

CONCLUSION

To end the lecture, I would put these challenges to the experimentalists:

(1) direct measurement of V_{cs}, V_{cd} from inclusive semileptonic decays of charm, and $\Gamma(D^+ \rightarrow \pi^+\pi^0)/\Gamma(D^+ \rightarrow \bar{K}^0\pi^+)$.

(2) study the B decay properties: does the B decay only semileptonically? $b \rightarrow c \rightarrow s$ or $b \rightarrow u$? Is there b flavor-changing neutral current, like $B \rightarrow X$ $\ell\bar{\ell}$, $B \rightarrow \tau^+\tau^-$?

(3) CP properties[31] of D^0, $\bar{D}^0$; B^0, $\bar{B}^0$ system. The size of $(N^{++} - N^{--})/(N^{++} + N^{--})$ in e^+e^- reaction, where $N^{++}(N^{--})$ is the number of events with two positively (negatively) charged leptons.

If you think that is tough for the experimental group, let's see the challenges to the ORY group (by now you should have guessed what I meant by the ORY group):

(1) solve the Family problem.

(2) explain the generation mixing.

(3) source of CP violation?

(4) dynamics of hadronic weak decay.

Acknowledgement: I would like to thank Drs. T. Rizzo and T. L. Trueman for many helpful and enlightening discussions during the preparation of this talk.

REFERENCES

1. N. Cabibbo, Phys. Rev. Lett. 10, 531 (1963).
2. S. Glashow, J. Iliopoulos and L. Maiani, Phys. Rev. D2, 1285 (1970).
3. M. Kobayashi and T. Maskawa, Prog. Theor. Phys. 49, 652 (1973).
4. T. D. Lee, Phys. Reports, 9C, 143 (1974); S. Weinberg, Phys. Rev. Lett. 37, 657 (1976).
5. R. Shrock and L.-L. Wang, Phys. Rev. Lett. 41, 1692 (1978).
6. R. Shrock and S.B. Treiman, Phys. Rev. D19, 2148 (1979) and R. E. Shrock, S.B. Treiman and L.-L. Chau Wang, Phys. Rev. Lett. 42, 1589 (1979). See also V. Barger, W.F. Long, and S. Pakvasa, ibid, 42, 1585 (1979).
7. R. E. Shrock, M. B. Voloshin, Phys. Lett. 87B, 375 (1979).
8. B. W. Birge et al. N.C. 4, 834 (1956).
    G. Alexander et al. N.C. 6, 478 (1957).
9. M. Gell-Mann and A. Pais, Proc., Int'l. Conf. on High Energy' Physics, Pergamon Press, (1955).
10. G. 't Hooft and M. Veltman, Nucl. Phys. B50, 318 (1972); B. W. Lee and J. Zinn-Justin, Phys. Rev. D7, 1049 (1973); for a review, see E. Abers and B. W. Lee, Phys. Reports, 9C, 1 (1973).
11. M. K. Gaillard, B. W. Lee, Phys. Rev. Letters 33, 108 (1974); G. Altarelli and L. Maiani, Phys. Letters, 32B, 351 (1974).
12. M. A. Shifman, et.al., JETP Lett. 22, 55 (1975); Nucl. Phys. B120, 316 (1977); Sov. Phys. JETP 45, 670 (1977); J. F. Donoghue and B. R. Holstein, "Dynamical Effects in Two Body Decay", MIT and NSF Preprint (1979).
13. L.-L. Chau Wang, "Flavor Mixing and Charm Decay", Proc. of 1980 Guangzhou (Canton) Conference on Theoretical Particle Physics, Jan. 5 - 14, 1980.
14. See review talk by V. Luth, Proceedings of the International Symposium on Lepton and Photon Interactions at High Energies, Fermilab, August 23 - 29, 1979; see also review talk by G. Goldhaber, at this conference.
15. R. Kingsley, S. Treiman, F. Wilczek, and A. Zee, Phys. Rev. D11, 1919 (1975); G. Altarelli, N. Cabibbo, and L. Maiani, Nucl. Phys. B88, 285 (1975); M. B. Einhorn and C. Quigg, Phys. Rev. D12, 2015 (1975); M. B. Voloshin, V. I. Zakharov, and L. B. Okun, Zh. Eksp. Teor. Fiz. 21, 183 (1975) [Sov. Phys. JETP 21, 403 (1975)].
16. L.-L. Chau Wang and F. Wilczek, Phys. Rev. Lett. 43, 816 (1979); for other related discussions, see M. Suzuki, Phys. Rev. Lett. 43, 818 (1980); C. Quigg, Zeit Phys. C4, 55 (1980).
17. V. Barger and S. Pakvasa, Phys. Rev. Lett. 43, 813 (1979); M. Fukugita, T. Hagiwara and A. I. Sanda, "The Color-Radius Interactions, $\Gamma(D^0 \to K^+K^-)/\Gamma(D^0 \to \pi^+\pi^-)$ and Violations of the $\Delta I = \frac{1}{2}$ Rule", Rutherford Preprint RL-79-052, T. 248.

414

18. G. Kane, SLAC Report No. SLAC-PUB-2326, 1979 (unpublished); L. F. Abbott, P. Sikivie, M. B. Wise, Phys. Rev. D21, 1393 (1980).

19. A. I. Sanda, "Non-Leptonic Decays of Charm Mesons", Rockefeller Univ. Preprint, COO-22323-191; B. Guberina, R. D. Peccei, R. Ruckl, Phys. Lett. 89B, 111 (1980); K. Ishikawa, "Some Dynamical Contributions to Cabibbo Suppressed D^0 Decay", Univ. of Cal. at Los Angeles Preprint, L. F. Abbott, P. Sikivie and M. B. Wise, Phys. Rev. D21, 768 (1980); M. Gluck, Phys. Lett. 88B, 145 (1979).

20. D. G. Sutherland, Phys. Letters, 90B, 173 (1980); I. I. Y. Bigi, Phys. Letters, 90B, 177 (1980).

21. See review talk by J. Kirkby, Proceedings of the International Symposium on Lepton and Photon Interactions at High Energies, Fermilab, August 23 - 29, 1979; see review talks by J. Prentice and G. Goldhaber in this conference.

22. N. Cabibbo and L. Maiani, Phys. Lett. 73B, 418 (1978); B. Stech, Nucl. Phys. B133, 315 (1978).

23. E. Ma, S. Pakvasa, and W. A. Simmons, "Quark-Number Selection Rule for Non-leptonic Weak Decays", U. of Hawaii preprint, UH-511-369-79; H. Fritzsch and P. Minkowski, "The Puzzle of Non-leptonic Decays and Its Resolution", U. of Bern preprint (1979); H. C. Lee, Chalk River preprint, CRNL-TP-80-Jan-04; S. P. Rosen, Phys. Lett. 89B, 246 (1980), Phys. Rev. Lett. 44 41 (1980); T. G. Rizzo and L.-L. Chau Wang, "The Quark-diagram Classification of Charm Decays", Brookhaven National Laboratory preprint, BNL-27950 (1980), and Ref. (16).

24. M. Bander, et al. Phys. Rev. Letters, 44, 7 (1980); Guberina, et al., Phys. Letters 89B, 111 (1980); W. Bernreuther, O. Nachtmann and B. Stech, Z. Physik C, Particle & Fields 4, 257-267 (1980); V. Barger, J. P. Leveille and P. M. Stevensen, "Gluon Enhancement in Charmed Meson Decays", Wisconsin preprint 00-881-124 (1979).

25. H. J. Lipkin, Phys. Rev. Lett. 44, 710 (1980).

26. H. Georgi and S. Glashow, Nucl. Phys. B167, 173 (1980).

27. H. Georgi and M. Machacek, Phys. Rev. Lett. 43, 1639 (1979).

28. H. Georgi and A. Pais, Phys. Rev. D19, 2746 (1979).

29. V. Barger and S. Pakvasa, Phys. Lett. 81B, 195 (1979).

30. D. B. Bairlie, Phys. Lett. 82B, 97 (1979); E. J. Squires, Phys. Lett. 82B, 395 (1979); J. G. Taylor, Phys. Lett. 84B, 79 (1979); Y. Ne'eman, "Unification Through a Supergroup", Tel Aviv Univ. preprint, Taup 132-80 and references therein.

31. Due to lack of time I cannot go into the details of discussing the phenomenological implications of CP violation; the reader is referred to the original publications: A. Pais and S. B. Treiman, Phys. Rev. 12, 2744 (1975); L. B. Okun, V. I. Zakharov, and

32. B. M. Pontecorvo, Lettere Al Nuovo Cimento 13, 218 (1975); A. Ali, Z. Z. Aydin, Nucl. Phys. B148, 165 (1979).

A SEARCH FOR NARROW STATES PRODUCED IN THE
REACTION $\pi^-p \to n + \gamma$'s AT 13 GeV/c*

I-H. Chiang, R.A. Johnson, B. Kwan, T.F. Kycia, K.K. Li,
L.S. Littenberg and A. Wijangco
Brookhaven National Laboratory, Upton, New York 11973

L.A. Garren and J.J. Thaler
University of Illinois, Urbana, Illinois 61801

G.E. Hogan, K.T. McDonald and A.J.S. Smith
Princeton University, Princeton, New Jersey 08544

ABSTRACT

Using a double arm lead-glass lead-scintillator calorimeter
system we have searched for narrow states, such as the η_c, produced
in the exclusive reactions $\pi^-p \to \gamma\gamma n$, $\pi^-p \to \pi^0\gamma n$, and $\pi^-p \to \pi^0\pi^0 n$
at 13 GeV/c. We find a 90% c.l. upper limit $\sigma \cdot BR < 260$ pb for $\gamma\gamma$
states with masses from 2.6 to 3.1 GeV/c^2. Corresponding limits on
narrow $\pi^0\gamma$ and $\pi^0\pi^0$ states are also given.

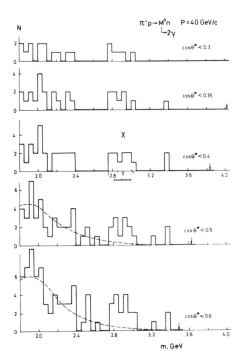

Fig. 1. $\gamma\gamma$ mass spectra from Ref. 2.

This is a brief status report on AGS Experiment 732 which is a search for narrow states produced in the exclusive reaction $\pi^-p \to n + \gamma$'s at 13 GeV. Our primary motivation was the unsatisfactory state of the $^1S_0$ charmonia, at the time of our proposal. One had hyperfine splittings apparently larger than fine splittings, M1 radiative rates too low by orders of magnitude, etc.[1] We hoped to find the real η_c and η_c' in hadronic production. There was also a tantalizing Russian result in 40 GeV $\pi^-p \to \gamma\gamma n$. Apel et al.,[2] claimed to see a $\gamma\gamma$ peak at 2.88 GeV/c^2 (consistent in mass with the DESY η_c candidate). The cross section × branching ratios was ~ 200 picobarns. Fig. 1 shows their data. After a number of cuts were imposed on the data, a signal was

*Presented by L.S. Littenberg

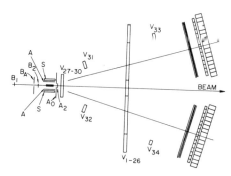

Fig. 2(a). Experimental layout. S and V counters are lead-scintillator shower counters. A are charged particle veto counters which surround the scintillator target.

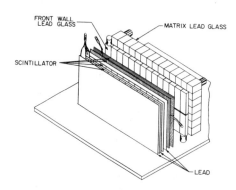

Fig. 2(b). Details of one photon calorimeter. Components of the converter section are shown in an exploded view for clarity.

apparent but not really compelling.

Our apparatus is shown schematically in Fig. 2(a). Briefly, there is a 13 GeV/c beam incident on a live scintillator target, there are two calorimeters at ± 16° and the rest of 4π sr is filled with lead-scintillator shower counters set in veto. The calorimeter shown in Fig. 2(b) is 16 radiation lengths in all, 4 r.l. of lead scintillator followed by 12 r.l. of lead glass in 2 layers. At the end of the lead scintillator section are two crossed fine grain hodoscopes. These serve to identify π^0's which appear as two clusters in each projection. π^0 rejection is really the key to this experiment, and that worked quite well.

Photon energy resolution was $\sigma_E/E \sim 15\%/\sqrt{E}$ which would imply a $\gamma\gamma$ mass resolution of about 120 MeV/c at 2.8 GeV/c^2. However, the fact that we have an exclusive process works in our favor here because we can make use of constrained fitting. This reduces $\sigma_{M_{\gamma\gamma}}$ by more than a factor of 2 to around 45 MeV/c^2.

We took a number of test runs at 6 GeV/c in order to see some common garden variety resonances. In Fig. 3(a) we see a large f^0 in the $\pi^0\pi^0$ spectrum. In the $\gamma\gamma$ spectrum of Fig. 3(b) we see both an η and an η' in the right places. We also see two other peaks which are due to $\pi^0 \rightarrow \gamma$ feed-down, in one case from $\omega \rightarrow \pi^0\gamma$ and in the other case from the f^0, which show how π^0 misidentification can be a problem. It turns out that at the sensitivity we have reached, this is not yet the limiting factor.

We triggered on neutral energy of more than 9 GeV, at least one GeV in each arm, and no vetos. After 10 days running at about $10^7 \pi^-$/burst, not much more than a test run, we collected approximately 3 million triggers, mainly $\pi^0\pi^0$. With the trigger in force

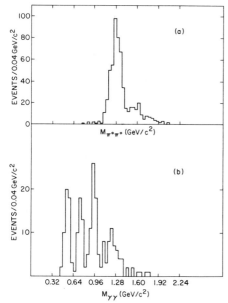

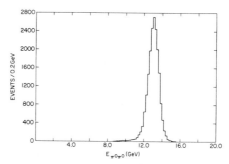

Fig. 4. $E_{\pi^0\pi^0}$ spectrum from 13 GeV/c running.

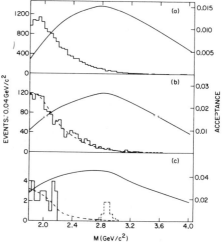

Fig. 3.
a) $M_{\pi^0\pi^0}$ from $\pi^-p \to n + 4\gamma$ at 6 GeV/c.
b) $M_{\gamma\gamma}$ from the same runs.

one can look at the energy distribution of say the $\pi^0\pi^0$ sample (shown in Fig. 4) and see that the events are not all piled up against the cut at 9 GeV but fall into a clean, full energy peak. This indicates that the trigger was successful. The 10 day run allowed us to reach about 10 events per nanobarn for $\gamma\gamma$.

Fig. 5 shows the results in the three modes that we have analyzed thus far: $\pi\pi$, $\pi\gamma$, and $\gamma\gamma$. One doesn't really expect narrow resonances in the $\pi\pi$ spectrum and none are seen. The solid curve is the shape of the acceptance. In the $\pi\gamma$ spectrum there are fewer events and again there is no apparent structure. The dashed curve

Fig. 5. Histogram of effective mass from 13 GeV/c runs. Solid lines indicate the acceptance. Dashed lines in (b) and (c) indicate the calculated $\pi^0\pi^0$ feeddown.
a) $\pi^0\pi^0$
b) $\pi^0\gamma$
c) $\gamma\gamma$. The dotted peak shows the predicted contribution of the X^0.[2]

418

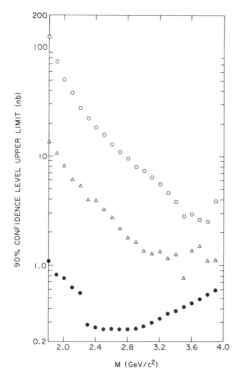

Fig. 6. 90% confidence level upper
limits for narrow resonance produc-
tion in $\pi^0\pi^0$(∘), $\pi^0\gamma$(△), and $\gamma\gamma$(•).

here is the expectation of the Monte Carlo calculation of the $\pi\pi$ feed-down and it accounts for the $\pi\gamma$ that we see quite nicely. In the most interesting process, $\gamma\gamma$, there are very few events, nothing above 2.2 GeV, and again the feed-down from $\pi\pi$ explains what we do see. This translates into a limit on $\sigma\cdot BR$ of ∿ 260 picobarns at 2.88 GeV/c². This is higher than the $\sigma\cdot BR$ claimed by Apel et al.,[2] but in an exclusive reaction one would expect the cross section at 13 GeV/c to be much larger than the cross section at 40 GeV/c. In fact one would expect the η_c to scale like η's which go something like $p^{-1.5}$ or at least like p^{-1}. If the cross-section scales like p^{-1}, this would imply 600 picobarns at 13 GeV/c, which would give 5½ events for our run. The dotted curve in Fig. 5(c) is such a peak, smeared with our resolution, super- imposed upon our spectrum.

I don't think it could have been missed. After this experiment
was proposed the theorists decided that the X(2880) makes a poor
η_c candidate because the $\sigma\cdot BR$ claimed by Apel et al.,[2] is too
high; it implies a very large $\Gamma(\eta_c \to \pi A_2)$.[3,4] There was specula-
tion that the X might be a 4-quark object à la Jaffe in which
case one expects a momentum scaling even faster than for an
η-like object, i.e., $\sigma \propto p^{-2.5}$. This would predict 33 events
which certainly could not have been missed.

Fig. 6 shows the 90% confidence level upper limits for all
three modes as a function of mass from 1.8 to 3.8 GeV/c².

Finally, we are currently proceeding with our run with an improved apparatus. Our aim is to improve the sensitivity by a factor of 20-down to about the 10 picobarn level where the η_c is expected to show up.

REFERENCES

1. K. Gottfried, Proceedings of the International Symposium on
    Lepton and Photon Interactions at high Energies, ed. F.

Gutbrod (DESY, Hamburg, 1977) p. 667.
2. Apel et al., Phys. Lett. 72B, 500 (1978).
3. A. Yu Khodjaminian EΦ – 281 (6)-78 (Yerevan Physics Institute preprint).
4. G. Eilam, B. Margolis and S. Rudaz, Phys. Lett. 80B, 306 (1979).
5. H.J. Lipkin, H.R. Rubinstein and N. Isgur, Phys. Lett. 78B, 295 (1978).

RECENT RESULTS OF MARK J:
PHYSICS WITH HIGH ENERGY ELECTRON-POSITRON COLLIDING BEAMS

(the AACHEN, DESY, MIT, NIKHEF, PEKING Collaboration)

presented by M. Chen

TABLE OF CONTENTS

1.   EXPERIENCE AT PETRA

2.   THE MARK J EXPERIMENT

    2.1.   Physics Objectives

    2.2.   The Detector

3.   PHYSICS RESULTS

    3.1.   Tests of QED and of Universality of Charged Leptons

        a.   Bhabha Scattering

        b.   Muon and Tau Pair Production

    3.2.   Hadronic Final States

        a.   Hadron Identification

        b.   Total Hadronic Cross Section

    3.3.   Jet Analysis

        a.   Thrust Distributions

        b.   Jet Analysis Using Fox-Wolfram Moments

        c.   A Study of Inclusive Muons in Hadronic Events

        d.   Discovery of 3-Jet Events

        e.   Determination of the Strong Coupling Constant α_s

    3.4.   Comparison with other Experiments at PETRA

4.   CONCLUSION

0094-243X/81/670421-91$1.50 Copyright 1981 American Institute of Physics

1. EXPERIENCE AT PETRA

PETRA[1] (Positron Elektron Tandem Ringbeschleuniger Anlage) began operation in the fall of 1978 as the world's highest energy e^+e^- colliding beam machine. Since its commissioning, PETRA beams have been available for physics runs 60% of the time, with the remaining time being devoted to machine development and maintenance periods[2].

The ring, with a circumference of 2.3 kilometers, has eight long straight sections of which two are reserved for the RF accelerating cavities. At present only four of the experimental areas are occupied. The remaining two experimental areas are reserved for second generation experiments.

The original injection scheme utilized both of the existing DESY facilities, DESY and DORIS. Electrons, initially accelerated in LINAC I (see Figure 1) are injected into DESY (Deutsches Elektronen Synchrotron) where they are further accelerated to 6 GeV and injected into PETRA. Positrons follow a somewhat more complicated path: after initial accelertion in LINAC II, positrons are injected via DESY into DORIS (Doppel-Ring-Speicher), where they are accumulated at an energy of 2.2 GeV. Stored positron bunches in DORIS are then transferred back to DESY for further acceleration to 6 GeV, the minimum PETRA injection energy.

With the discovery of the upsilon (T) resonance in 1977 at FNAL[3] and the confirmation in e^+e^- interactions[4], the need to operate DORIS as a storage ring independent of PETRA was realized. Consequently, in the fall of 1977 the decision was made to construct an Intermediate Positron Accumulator (PIA)[5] to free DORIS for physics runs. In this new injection scheme, positrons are accumulated in PIA after acceleration in LINAC II. Twenty successive LINAC bunches are injected into PIA, compressed in phase space, and transferred to DESY for acceleration and injection into PETRA. PIA was assembled in record time and since the summer of 1979 has served as the injector for both DORIS and PETRA.

The average luminosity is $2 \times 10^{30} \text{cm}^{-2} \text{sec}^{-1}$ up to beam energies of 18.3 GeV. It is expected that the luminosity will increase in the near future with more operational experience.

Recently, PETRA has run up to an energy of 36.6 GeV. It has run reliably with very little failure. The stability of the machine was the major reason why all groups at PETRA have been able to perform their experiments satisfactorily.

2. THE MARK J EXPERIMENT

2.1. Physics Objectives

The MARK J detector[6], which identifies and measures the energy and direction of muons, electrons, charged and neutral hadrons with close to uniform efficiency and with ~ 4π acceptance, is capable of fulfilling a broad range of physics objectives. Some of the prime physics goals of the experiment are:

1) To study the various QED processes shown in Figure 2 and to study the universality of the known charged leptons in their electromagnetic interactions. At PETRA the available c.m. energy is $\sqrt{s}$ = 37 GeV (q^2 up to 1400 GeV^2). Since first order QED processes exhibit a 1/s cross section dependence the MARK J can probe the validity of QED with an order of magnitude greater sensitivity than that previously available in earlier colliding beam experiments performed at storage rings at SLAC, DESY, and the CEA in the range of $q^2 \lesssim 50$ GeV^2.

2) To search for new quark flavors by studying the energy and angular distributions of inclusive muon production in hadronic events (Figure 3a).

3) Using the distributions of μe and μh final states shown in Figure 3b to search for the existence of new charged leptons heavier than the tau.

4) To measure the total hadronic cross section (Figure 4) and thereby the structure and energy dependence of the total cross section, in order to search for new thresholds in the hadronic final state continuum, and to search directly for more J-like particles which appear as sharp resonances.

5) To study the topology of hadronic events by measuring the direction and energy of charged and neutral particles. In particular, at PETRA energies, the fragmentation of hard gluons emitted in association with quark-antiquark pairs leads to the creation of additional gluons and quarks, resulting in the production of multi-jet events. Study of the properties of these jets enables us to make a direct comparison with the predictions of QCD[7]. The rate of 3-jet events relative to 2-jet events enables

us to measure directly the strong interaction coupling constant α_s.

6) To measure the charge asymmetry expected from the interference of weak and electromagnetic interactions in the production of $\mu^+\mu^-$ pairs. As shown in Figure 5, diagrams in which a virtual photon is exchanged or in which a Z^0 vector boson is exchanged both contribute to $\mu^+\mu^-$ production. The interference can be understood in terms of a variety of models based on the weak interaction Lagrangian

$$\mathcal{L}_{int} = i\,\bar\mu\gamma^\tau\,(g_V - g_A\,\gamma^5)\,\mu Z_\tau.$$

In the simple V-A model for example, one assumes $g_V = g_A = g$, where $g^2/M_Z^2 = G/\sqrt{2}$, and where G is the Fermi coupling constant. In the now standard Glashow-Weinberg-Salam (GWS) model the couplings are expressed in terms of the single parameter Θ_W, the Weinberg angle:

$$g_V = 1/4\,g\cos\Theta_W\,(3\tan^2\Theta_W - 1),\text{ and } g_A = 1/4\,g\sec\Theta_W.$$

In order to distinguish between theoretical hypotheses, we can use the forward-backward charge asymmetry $A \equiv \dfrac{\sigma_- - \sigma_+}{\sigma_- + \sigma_+}$, where σ_- (σ_+) corresponds to the μ^- (μ^+) appearing in the forward hemisphere. At $\sqrt{s} = 30$ GeV, with a total time-integrated luminosity of 10^{38} cm^{-2}, one obtains $\sim 10^4$ events in a 4π detector, leading to a 10 standard deviation asymmetry effect in the V-A model and a 5 standard deviation effect in the GWS model[6].

Because the expected asymmetry is small and because higher order QED processes also produce sizable charge asymmetry at small angles, the measurement of asymmetry requires attention in reducing and understanding systematic errors in the detector design.

One notes that before the direct observation of the Z^0, the precise determination of the charge asymmetry arising from weak-electromagnetic interference is the most important verification of the idea of the unified electromagnetic and weak theory.

2.2 The Detector

The MARK J detector is shown in Figures 6-10. It is designed to distinguish charged hadrons, electrons, muons, neutral hadrons and photons and to measure their directions and energies. It covers a solid angle of $\phi = 2\pi$ and $\theta = 12°$ to $168°$ (θ is the polar and ϕ is the azimuthal angle). The detector, which consists of five magnetized iron toroids built around a non-magnetic inner detector complemented by end caps, was designed to be insensitive to the effects of synchrotron radiation. Particles leaving the interaction region first pass through a five millimeter thick aluminium beam pipe, with an outer diameter of 190 mm. The aperture of the beam pipe is large enough so that the synchrotron radiation produced in the final PETRA bending magnets and quadrupoles will pass unobstructed through the entire detector. Two thick copper absorbers are located symmetrically around the interaction region, at a distance of 1 meter, to trap synchrotron radiation reflected back towards the interaction region by collimaters just in front of the last PETRA quadrupoles. The detector layer structure is best understood by referring to Figure 11.

During the first nine months of operation, a ring of 2 x 16 lucite Cerenkov counters each covering an azimuthal sector of $22.5°$ and a polar-angle region from $9° < \theta < 171°$ surrounded the beam pipe. These counters are divided at $\theta = 90°$ to permit a crude determination of the momentum balance between the forward and backward hemispheres. The counters are insensitive to the effects of synchrotron radiation and can be used to separate charged from neutral particles.

In the latter part of 1979 the lucite counters were replaced by a four-layer inner track detector of 992 drift tubes. Each tube is 300 mm long and 10 mm wide and has a spatial resolution of 300 microns. The tubes, which are arranged perpendicular to the beam line, distinguish charged from neutral particles in the angular range $30° < \theta < 150°$ and reconstruct the position of the event vertex along the beam line to an accuracy of two millimeters. The distribution of event vertices obtained using the drift tubes is shown in Figure 12. The observed r.m.s. width of 1.27 cm is compatible with that expected from the known bunch length of the machine.

Particles then pass through 18 radiation lengths of shower counters used to identify and measure the energy of electrons, photons, charged and neutral hadrons. This inner calorimeter is divided into three layers of shower counters (labelled A, B and C in Figure 6). Each counter is constructed of 5.0 mm thick pieces of scintillator alternated with lead plates of equal thickness.

The A and B counters are each 3 radiation lengths thick, while the C shower counter is a total of 12 radiation lengths thick (measured normal to the surface of the counter).

The twenty A shower counters are each 2 m long and cover the angular region of $\theta = 12^{\circ}$ to 168°. The 24 B counters are constructed identically to the A counters and cover an angular region from $\theta = 16^{\circ}$ to 164°.

Since every shower counter is viewed by one phototube at each end, the longitudinal (z) position of particle trajectories can be determined by comparing the relative pulse heights from each end of the counter. Timing information provides another measure of the longitudinal position. The trajectory location determined by this method was found to be in excellent agreement with the data from the drift tubes (see Section 2.2a).

Twelve planes of drift chambers (labelled S and T) measure the angles of particles penetrating the inner electromagnetic calorimeter. Each of the sense wires is connected to its own amplifier and time digitizer. Both end cap regions are covered by an additional ten planes of drift chambers (labelled U and V) of similar construction. These chambers are protected from beam backgrounds from the interaction region by the shower counters A, B and C. The energy sampling elements of the calorimeter K, shown in Figure 6, are 192 scintillation counters arranged in four layers. The main body of the calorimeter is composed of the magnetized iron plates which are also used to momentum-analyze muons. These plates range in thickness from 2.5 to 15 cm. Hadrons penetrating the inner shower counter layers, and secondary particles produced by hadronic showers initiated in the inner layers, deposit most of their remaining energy in the calorimeter K. The energy sampled by the K counters is thus used to help distinguish hadrons from electrons, and to help identify minimum-ionizing particles.

Muons are identified by their ability to penetrate the iron of the hadron calorimeter. The low-momentum cut-off is about 1.3 GeV/c at normal incidence. The initial muon trajectory is measured in the S and T (U and V) chambers and in the drift tubes.

The bend angle and position of muons exiting from the calorimeter are measured in 10 planes of drift chambers, labelled R and P in Figure 6. The total thickness of the iron is 87 cm and it has a bending power of approximately 17 kG-meters. The typical bend angle for a 15 GeV muon is 30 mrad.

An additional 2 layers of drift chambers (Q chambers) are situated amidst the iron layers to measure the muon tracks in the bending plane. Adjacent to these chambers are the 32 muon trigger counters marked (D) used to trigger on single and multiple muon events and to reject cosmic rays. Each of these counters is 30 cm wide and 450 cm long and has a phototube at each end. The timing difference between real dimuon events and cosmic rays is about 10 ns. These counters have a timing resolution of about 400 ps.

Covering each of the end cap regions are the E counter hodoscopes. Each of these counters has dimensions 80 cm x 450 cm x 1 cm and they are used to trigger on muons produced in the forward and backward directions as well as to reject cosmic rays and beam-gas background.

One of the prime goals of the MARK J experimental program (see Section 2.1) is to measure the charge asymmetry in the angular distribution of muon pairs produced in e^+e^- annihilation to an accuracy of ~ 1%. This goal can only be achieved if small systematic effects due to variations in chamber efficiency and counter gains, and slight asymmetries in the construction of the magnet and the positions of particle detectors in space, do not influence the overall charge asymmetry measurement. In order to isolate and subsequently eliminate the effects of these systematic errors in the measurement, the supporting structure is designed so that the entire detector can be rotated azimuthally about the beam line by ± 90° and 180° about a vertical axis. The rotation about the vertical axis maps θ into 180° - θ, and is therefore most useful in checking the measurement of the front-back charge asymmetry. The azimuthal rotation, which is used to check for beam polarization, can also be used to aid in the charge asymmetry measurement in the presence of polarized beams.

For the data in this report detectors E and R were not used.

The luminosity monitor consists of two arrays of twenty-eight lead glass blocks[8] (labelled G in Figure 6), each with dimensions of 8 cm x 8 cm x 70 cm located 5.8 m from the interaction point. They are designed to measure Bhabha events at small scattering angles (~ 30 mrad). Scintillators (F) in front of the lead glass define the acceptance and the lead glass counters measure the energy of the electron pairs.

The trigger is arranged in two stages. The first stage is a fast loose trigger generated from the counter hit information with the following requirements:

    i)   For electron pairs we require at least 0.5 GeV total energy deposited in opposite quadrants of the A and D counters.

    ii)   For muon pairs we require at least two A and two B counters in coincidence with a pair of D counters which are coplanar within 50°.

    iii)   For single muon events we require at least two A, two B, two C and one D counters to be triggered.

    iv)   For hadrons we require at least four A and three B counters; each triggered quadrant must be in coincidence with the opposite quadrant.

    All triggers are required to be in coincidence with the beam crossing signal.

    After the fast trigger, a second stage imposes two more selections depending on event type. For electron pairs and hadron events the total energy deposited in the inner calorimeters A, B and C is determined by measuring the pulse area of linearly added signals. We require at least 13% of the total C.M.S. energy for hadrons and at least 10% for electron pair events. For muon pairs, single muon and hadron events, a microprocessor applies a loose track requirement demanding at least three pairs of wires to be hit in the S or T chambers.

3. PHYSICS RESULTS

3.1   Test of Quantum Electrodynamics and of Universality for Charged Leptons

    There have been many experiments testing quantum-electrodynamics (QED) with electrons, muons and photons at electron-positron storage rings. Notable experiments[9] were done by Alles-Borelli et al., Newman et al., Augustin et al., O'Neill et al., and by our group at PETRA[10] up to a center of mass energy of 17 GeV. For a comprehensive review of QED work, see Brodsky and Drell[11]. Much has been learned about the properties of the leavy lepton tau since the original search began at ADONE on $e^+ + e^- \rightarrow \mu e + \ldots$[12]. The discovery of the τ lepton at SLAC[13] and its subsequent confirmation at DESY[14] has inspired further studies. We know it is a spin 1/2 particle which decays weakly[15] and whose properties are very similar to the muon.

    In this experiment we study the reactions $e^+ + e^- \rightarrow \ell^+ + \ell^-$ for all the known charged leptons (ℓ = e, μ, τ) by measuring the dependence of the cross section on center of mass energy or scattering angle over a wide range of PETRA energies. These measurements enable us to compare the data with predictions of quantum electrodynamics, to test the universality of these leptons at very small distances, and to set a limit on the charge radius of these particles. Up to the present time the reactions:

$$e^+ + e^- \rightarrow e^+ + e^- \quad \text{(Bhabha scattering)} \qquad (2)$$

$$e^+ + e^+ \rightarrow \mu^+ + \mu^- \qquad (3)$$

$$e^+ + e^- \rightarrow \tau + \tau \qquad (4)$$
$$\quad\quad \longrightarrow \text{(h's or e)} + \nu\text{'s}$$
$$\quad\quad \longrightarrow \mu + \nu\text{'s}$$

have been measured at the center of mass energies $\sqrt{s}$ = 12, 13, 17, 22, 27.4, 30, 31.6, 35, 35.8, and 36.6 GeV.

a)   Bhabha Scattering

The Bhabha events are identified by requiring two back-to-back showers in the A, B and C counters which are collinear to within 20° in ϕ and θ and with a measured total shower energy greater than 1/3 of the incident beam energy. Photons emitted close to either electron are included in the electron momentum. From the measurement of the acollinearity angle $\Delta\theta$, and the acoplanarity angle $\Delta\phi$, we observe that most of the events are in the region $\Delta\theta < 4^\circ$, $\Delta\phi < 4^\circ$. Because there are few events near the 20° cuts in $\Delta\theta$ we conclude that the background to Bhabha scattering is negligible.

To eliminate most background from hadron jets, the energy in the K counters was required to be less than 7% of the total energy. Because the QED test is most sensitive to background in the large angle region, all events having θ larger than 60° were scanned on graphic displays which showed the distribution of counter hits. On the basis of a Monte Carlo study of hadron events, we conclude that the background from this source is less than 1% of the events. As mentioned above, the acceptance for $e^+e^- \rightarrow e^+e^-$ was computed using Monte Carlo techniques and is defined by the geometry of the first shower counter array A. Both energy and acceptance losses in the corners were found to be small.

The first order QED photon propagator produces an s^{-1} dependence in the $e^+e^- \to e^+e^-$ cross section. Thus when radiative corrections have been taken into account in the data, the quantity $s\frac{d\sigma}{d\cos\theta}$ vs $\cos\theta$ should be independent of s. This distribution is plotted for the data at $\sqrt{s} = 13$, 17 and 27.4 GeV in Figure 13. Excellent agreement with QED predictions is seen. To express this agreement analytically, we compare our data with the QED cross section in the following form (since charge is not distinguished here)[16]:

$$\frac{d\sigma}{d\Omega} = \frac{\alpha^2}{2s} \left\{ \frac{q'^4+s^2}{q^4} F_s^2 + \frac{2q'^4}{q^2 s} \mathrm{Re}(F_s F_T^*) + \frac{q'^4+q^4}{s^2} F_T^2 \right.$$

$$\left. + \frac{q^4+s^2}{q'^4} F_s'^2 + \frac{2q^4}{q'^2 s} \mathrm{Re}(F_s' F_T^*) + \frac{q'^4+q^4}{s^2} F_T^2 \right\} \{1 + C(\theta)\},$$

where

$$F_s = 1 \mp q^2/(q^2 - \Lambda_{s\pm}^2)$$

is the form factor of the spacelike photon, $F_s' = 1 \mp q^2/(q^2 - \Lambda_{s\pm}^2)$,

$$F_T = 1 \mp s/(s - \Lambda_{T\pm}^2)$$

is the form factor of the timelike photon, $q^2 = -s\cos^2(\theta/2)$, $q'^2 = -s\sin^2(\theta/2)$, Λ is the cutoff parameter in the modified photon-propagator model[17] and $C(\theta)$ is the radiative correction term as a function of θ.

The radiative correction to the e^+e^- elastic scattering process was calculated using the program of Berends for these particular event selection criteria[20].

In order to establish lower limits on the cut-off parameters a Monte Carlo program was used to generate e^+e^- pairs which were then traced through the detector with the inclusion of measured θ, ϕ resolutions. A χ^2 fit to all of the 13, 17 and 27.4 GeV data is then made using the Monte Carlo generated angular distribution. The normalization is treated in two ways: (1) the total number of Monte Carlo events in the region $0.90 < \cos\theta < 0.98$ was set equal to the total number of measured events in the same region, (2) the minimum-χ^2 for the entire data sample determined the normalization.

The two methods agree with each other to within 3% and give essentially the same result in the cut-off parameter Λ. The lower limits of Λ at 95% confidence level under various assumptions are shown in Table I. The JADE and PLUTO groups have also analyzed their QED reactions and have obtained similar conclusions with regard to the validity of QED at small distances (see Section 3.4).

b) <u>Muon and Tau Pair Production</u>[18]

The MARK J detector is designed to distinguish muons from electrons and hadrons and to distinguish back-to-back muon pairs from cosmic ray muons. Muon identification is also aided by the short decay path allowed to hadrons before reaching the shower counters. In addition to the cuts described in Section 2.3b, single muons are identified as particles which:

i) are reconstructed in the inner drift chambers to come from the interaction region;

ii) leave minimum ionizing pulse heights in the A, B, C, K, and D counters, a total of seven layers;

iii) leave a track in the outer drift chambers (P) and thus fall into an angular range $45^{o} < \theta_{\mu} < 135^{o}$.

In addition back-to-back muon pairs from reaction (3) are distinguished from cosmic rays by the requirement that:

i) the D counter timing signals are coincident with one another (and not relatively off time as in the case for cosmic rays traversing the detector);

ii) the muons should be collinear and coplanar, and they should pass through the intersection region.

A Monte Carlo study shows that the $\mu^{+}\mu^{-}$ acceptance, which is dominated by the geometrical acceptance of the P drift chambers, is 41% ± 3% independent of beam energy.

Tau leptons from reaction (4) are identified by detecting μ-hadron and μ-electron final states. The cross section is determined using the known branching ratio of $\tau \rightarrow \mu \, \nu\bar{\nu}$ (16%) and $\tau \rightarrow$ (e, lepton + hadron, or multi-hadrons) $+ \nu$ (84%)[15]. The muons, hadrons and electrons are identified as described previously. The total deposited energy of hadrons (or electrons) is required to be greater than 2 GeV.

The major background to reaction (4) is the two-photon process:

$$e^+ + e^- \to e^+ + e^- + \mu^+ + \mu^- \qquad (7)$$

This process becomes important at high energies since the total cross section grows as $\ln(\sqrt{s}/m_e)^2 \ln(\sqrt{s}/m_\mu)$ with increasing beam energy in contrast to reaction (3), the rate for which falls as $1/s$. The observed cross section for the two-photon process is suppressed by a factor $\sim 10^3$ relative to the total cross section by momentum cuts on the muons and by energy and angle cuts on the electrons, but rates of the accepted events remain significant.

In Figure 14 we show the calculated cross section in our detector when the observed particles are

    (a)   two μ's only,

    (b)   only one μ and one e and,

    (c)   two μ's and one e.

The cross section for each of these configurations were computed using the Monte Carlo integration program of Vermaseren[17]. The computations were compared to the cross sections measured for reaction (7) using the following cuts:

$$168^\circ \geq \theta_e \geq 12^\circ \qquad\qquad E_e \geq 2 \text{ GeV}$$
$$135^\circ \geq \theta_{\mu_1} \geq 45^\circ \qquad\qquad P_{t_{\mu_1}} \geq 1.5 \text{ GeV (first muon)}$$
$$147^\circ \geq \theta_{\mu_2} \geq 33^\circ \qquad\qquad P_{t_{\mu_2}} \geq 0.8 \text{ GeV (second muon)}$$

The measured cross sections, also shown in Figure 14, agree well with the calculations in all cases.

The cross section for case (a) above is much larger than that for the process

$$e^+ + e^- \to \tau^+ + \tau^- \to \mu^+ + \mu^- + 4\nu,$$

and the $\mu^+\mu^-$ events from the two processes are hard to distinguish. We therefore exclude case (a) from the sample of $\tau\bar{\tau}$ candidates. In case (b) the electron from the two-photon process is strongly peaked in the small angle region. Typically, $\dfrac{d\sigma_{\mu e}}{d(\cos\theta_e)}$ decreases by two orders of magnitude from $\cos\theta_e = 0.98$ to 0.80. Furthermore, the observed muons and electrons tend to be coplanar because of conservation of transverse momentum. By requiring $30^\circ < \theta_e < 150^\circ$ for the $\bar{\tau}\tau$ sample, and by requiring that the final state μe be collinear within 30°, we were able to reduce the two-photon contribution to a negligible level ($<10^{-3}$ picobarns). The rate for case (c) is small and can readily be separated from the τ events.

The fact that the μ-hadron events produced by reaction (4) in our energy region are almost collinear can also be used to distinguish reaction (4) from μ-hadron events produced by the semi-leptonic decay of particles with c (charm) or b (bottom) quantum number. In the latter cases the muon is accompanied by hadrons emitted close to the muon direction.

The measured muon momentum and hadron energy for the $\bar{\tau}\tau$ candidates is in agreement with calculations based on the known decay properties of the τ lepton[15].

The acceptance is calculated using a Monte Carlo method to generate $\bar{\tau}\tau$ production from reaction (4) including radiative corrections. We obtain a detection efficiency of ~ 10% for τ pairs at various energies, when requiring one decay muon to be detected in association with a single electron, or one or more hadrons.

The resultant $e^+e^- \rightarrow \mu^+\mu^-$ and $\bar{\tau}\tau$ cross sections as a function of s are plotted in Figures 15 and 16 together with the QED prediction. We see that from $q^2 = s = 169$ to $q^2 = 1300$ GeV^2 the data agree well with the predictions of QED for the production of a pair of point-like particles. In particular, Figure 16 represents the first evidence that the τ lepton is a point-like particle over a large range of q^2, and demonstrates that it belongs in the same family as the electron and muon. To parametrize the maximum permissible size (radius) of the particles, we use a form factor:

$$F_\ell = 1 \mp \frac{q^2}{q^2 - \Lambda^2_{\ell\pm}} \qquad (\ell = e, \mu, \tau).$$

By comparing our data with the cross sections including this form factor we find lower limits on the cut off parameters, at the 95% confidence level, summarized in Table II.

Thus, from Heisenberg's Uncertainty Principle, all the known charged leptons are point-like particles in their electromagnetic interactions, with characteristic radii $< 2 \times 10^{-16}$ cm.

3.2 Hadronic Final States

a) Hadron Identification

The final selection for hadrons is made by scanning the events on an interactive graphics system after a preselection which includes energy and momentum balance cuts. In this scanning, hits observed in the drift tubes and in the drift chambers allow an easy determination of the event vertex. Thus beam gas events which do not come from the interaction region are readily recognized. The charged multiplicity observed in the drift tubes and the shower envelope in the counters is used to reject events of electromagnetic origin. For events which have low multiplicity both in the tubes and the counters, we require that a minimum fraction of the total energy is deposited in the hadron calorimeter (K in Figure 6) to further discriminate against purely electromagnetic final states.

The energy spectrum of the events passing the cuts is shown in Figure 17. For the analysis of hadronic final states we employed an additional energy cut of $E_{vis} > 0.5 \sqrt{s}$ for measurement of total cross section (Section 3.2b), for most of the thrust analysis (Section 3.3a) and the study of inclusive muons in hadron events (Section 3.3c). This additional cut also reduces the contamination from beam-gas events, from two-photon processes[19] and $e^+e^- \rightarrow \tau^+\tau^-$ events which yield hadrons in the final state[20].

For reasons discussed in Section 3.3d, we employed an even more restrictive cut of $E_{vis} > 0.7 \sqrt{s}$ for study of gluon effects in thrust distributions (Section 3.3a), for jet analysis using Fox-Wolfram moments (Section 3.3b), for the discovery of 3-jet events (Section 3.3d) and determination of the strong coupling constant α_s (Section 3.3e). Both of these cuts are indicated by arrows in Figure 17.

b) Total Hadronic Cross Section[21, 22, 23]

The total cross section for $e^+e^- \rightarrow$ hadrons was measured over a wide range of center of mass energies from 12 to 35.8 GeV, including results obtained by extended periods of running at a fixed beam energy and by two fine energy scans covering the range of 29.92 to 31.46 GeV and of 35.0 to 35.26 GeV respectively. The results are expressed in terms of R:

$$R = \sigma\,(\,e^+e^- \rightarrow \text{hadrons}) \,/\, \sigma\,(\,e^+e^- \rightarrow \mu^+\mu^-).$$

In the naive quark-parton model, the cross section for the hadron production process is simply given by the sum over flavors of the point-like $q\bar{q}$ pair cross sections. Using this picture with spin 1/2 massless quarks, and with three colors gives

$$R = 3 \sum_q e_q^2 \,,$$

where e_q is the charge of the quark with flavor q. Considering the five known quark flavors (u, d, s, c and b), and correcting the naive model for gluon emission as predicted by QCD, we expect $R \simeq 4$, over the entire PETRA energy range, with only a slight decrease in R with increasing beam energy.

The MARK J results for R are shown in Figure 18 along with the QCD predictions. The data consist of a total of 2300 hadron events. The results are summarized in Table III. Figure 18 shows that the data agree with the QCD predictions for five quark flavors (represented by the lower dashed line), and that there is no sign up to 35.8 GeV of a threshold in the hadron continuum corresponding to a new heavy quark of charge 2/3 such as the top quark (the upper line).

The experimental R-values in Figure 18 have been corrected for initial-state radiative corrections ($\Delta R \simeq -0.3$) for contamination of the sample by hadronic events produced by the two photon process $e^+e^- \rightarrow e^+e^- + \text{hadrons}$ ($\Delta R \lesssim -0.1$), and for the contribution of $e^+e^- \rightarrow \tau^+\tau^- \rightarrow \text{hadrons} + \text{leptons}$ (ΔR typically $= -0.3$). In addition to the statistical errors shown, there is an additional systematic uncertainty due to model dependence of the acceptance of 10%. The accuracy in evaluating the contributions of the two photon and tau-pair contributions to the hadronic event sample is limited by the lack of detailed experimental data on the high energy, high multiplicity states arising from these sources.

In addition to the $t\bar{t}$ contribution to the hadronic continuum, the toponium system should form one or more bound states, with the number of such states depending on the shape of the binding potential[21]. Interpretation of the vector mesons ρ, ω, ϕ, J, ψ', T, T', and T'', as nonrelativistic $q\bar{q}$ bound states, or "quarkonia" leads to the prediction that the gap between the lowest bound state and the continuum is probably ~ 1 GeV, and very likely < 2 GeV.

In order to check for the existence of $t\bar{t}$ bound states lying below 35.8 GeV, the two energy scans mentioned earlier were performed in 20 MeV center of mass energy steps (matching the r.m.s. energy spread of PETRA). The results of the scan are shown in Figures 19a and 19b, along with the results obtained earlier at 30.0 and 31.6 GeV and at 35 and 35.8 GeV respectively. The figures show that the data are entirely consistent with the predictions of QCD for u, d, s, c and b quarks, that is, with a constant value of R over the whole range. No single point lies $\geq 3\sigma$ above the QCD line, and there is no indication of an upward slope with increasing energy signaling the onset of a new contribution to the continuum. The values of R averaged over the individual energy range of the scan are 4.33 ± 0.17 for $30 < E_{cm} < 31.6$ GeV and 4.2 ± 0.3 for $35 < E_{cm} < 35.8$ GeV.

In order to set a quantitative upper bound on the possible production of a narrow resonance, the data in Figure 19 were fitted by a constant plus a gaussian:

$$R = R_o + R_v e^{[-(\sqrt{s} - M)^2/2 \Delta_w^2]}$$

where R_o represents the non-resonant continuum, M is the mass of the resonance, Δ_w is the r.m.s. machine energy width, and R_v is the peak value of the resonant contribution. The largest value of R_v consistent with the data was determined by trying fits with M, the center of the gaussian, fixed at all the center of mass energies at which data were taken. The largest value of R_v was obtained at 31.32 GeV corresponding to an upper limit on the resonance strength

$$\Sigma_v \equiv \int q_v(w)dw, \text{ where } w = \sqrt{s},$$

of 33 MeV-nb at the 90% confidence level. Using the relation between the resonance strength, the width into e^+e^- (Γ_{ee}), the hadronic width (Γ_h), the total width (Γ), and the hadronic branching ratio ($B_h \equiv \Gamma_h/\Gamma$):

$$\Sigma_v \equiv \int \frac{3\pi}{M^2} \frac{\Gamma_{ee}\Gamma_h}{(w-M)^2 + \Gamma^2/4} \, dw = \frac{6\pi^2}{M^2} B_h \Gamma_{ee}$$

we obtained

$$B_h\Gamma_{ee} < 1.3 \text{ KeV (90\% C. L.) for } 30 < E_{cm} < 31.6 \text{ GeV}$$

Similarly, $\quad B_h\Gamma_{ee} < 1.2$ KeV (90% C. L.) for $35 < E_{cm} < 35.28$ GeV.

This upper limit excludes the production of a vector particle consisting of a $q\bar{q}$ bound state where the quark has charge 2/3. On the basis of the experimental fact that Γ_{ee}/e_q^2 is approximately constant for the vector meson ground states ρ, ω, ϕ, J and Γ, as is predicted by duality arguments[21], one expects $\Gamma_{ee} \sim 5$ KeV for the lowest mass meson in the toponium family. A rigorous lower bound of the production cross section of the ground state toponium in e^+e^- annihilation has been calculated by Quigg[21] as a function of the mass of the toponium. The theoretical predictions are shown in Figure 19c. We see the experimental upper limits are below the lower bound both at $E_{cm} = 31$ and $E_{cm} = 35.1$ GeV.

3.3 Jet Analysis

a) Thrust Distributions

Data at lower energies from SPEAR[24] have shown that the final state hadrons from the process $e^+e^- \to$ hadrons are predominantly collimated into two back-to-back "jets" in agreement with the expectations of simple models in which the time-like photon materializes initially into a quark antiquark pair. It is thus necessary to develop kinematic quantities which describe the jet-like nature of the MARK J hadronic events.

A jet analysis of the hadronic events has been devised using the spatial distribution of the energy deposited in the detector. For each counter hit, a vector $\vec{E}^i$ (the energy flow) is constructed, whose direction is given by the position of the signal in the counter, and whose magnitude is given by the corresponding deposited energy. A parameter thrust (T) is defined as[23]

$$T = \max \Sigma \; E_{/\!/}^{i} \; / E_{vis}$$

where $E_{/\!/}^{i}$ is the parallel component of E^{l} along a given axis, and the maximum is found by varying the direction of this axis, and where the sums are taken over all counter hits. The resultant direction is thus the direction along which the projected energy flow is maximized.

For events in which the spatial distribution of energy is isotropic, T is expected to approach 0.5. This would be the situation if, for example, the virtual photon materializes into two very heavy quarks, each with a mass close to the beam energy. Such quarks would be produced almost at rest. On the other hand, pairs of light quarks would move at high speed and the Lorentz boost of their hadronic fragmentation products would result in the hadrons being produced in narrow jets collimated around the initial quark directions. Higher beam energies would result in narrower jets, so that T should approach the value 1. Thus, thrust measurements can be used as a sensitive method to detect the presence of a new threshold due to new heavy quarks.

These expectations are illustrated qualitatively in Figure 20 in which the thrust distribution expected just below and just above the threshold for production of a new heavy quark is shown. Production of a new heavy quark would also result in lowering the average T as the energy is raised and the threshold is passed.

The normalized thrust distributions $\frac{1}{N} \frac{dN}{dT}$ for 13, 17, 22, and the combination of 27.4 and 27.7 GeV data (labelled 27 GeV combined) are shown in Figures 21 a–d along with the Monte Carlo predictions of a quark–parton model with u, d, s, c, and b quarks and no gluon emission. (The individual distributions at 27.4 and 27.7 GeV are in agreement with each other). As expected for production of final states with two jets of particles, the distributions become narrower and shift towards high thrust with increasing energy.

Figure 22 shows the normalized thrust distributions for combined data between 29.9 and 31.6 GeV and for data between 35 and 35.8 GeV where the measured energy of the selected events is at least 70% of $\sqrt{s}$. The curves show the Monte Carlo predictions with and without inclusion of gluons. As can be seen, these data lend support to the necessity of including gluons in the Monte Carlo program. The data is inconsistent with a Monte Carlo calculation which includes a charge 2/3 t quark produced as described previously. We thus conclude that there is no evidence for production of a new heavy quark with charge $q = \frac{2}{3}$ e.

Figure 23 shows the average thrust <T> plotted at the six energies. The solid curves are from Monte Carlo calculations which include u, d, s, c and b quarks with gluon emission. The energy dependences of the data are smooth and show none of the steps which would have appeared at new quark thresholds.

b)   Jet Analysis Using Fox-Wolfram Moments

Another method of jet characterization has been proposed by Fox and Wolfram. The simplest observables characterizing the three-dimensional shape of the energy distribution in a hadronic event are the Fox-Wolfram moments[25].

$$H_\ell = \sum_{i,j} \frac{|\vec{E}_i| \cdot |\vec{E}_j|}{E_{vis}^2} P_\ell(\cos\theta_{ij}), \text{ where } P_\ell \text{ is } \ell\text{th Legendre}$$

polynomial, and

$$\pi_1 = \sum_{i,j,k} \frac{|\vec{E}_i| \cdot |\vec{E}_j| \cdot |\vec{E}_k|}{(E_{vis})^3} (\hat{E}_i \cdot \hat{E}_j \times \hat{E}_k)^2$$

The H's parametrize the shape of the energy distribution by measuring the correlation between pairs of energy flow elements in terms of spherical harmonics. For cigar-shaped events expected from the process $e^+ + e^- \rightarrow q + \bar{q}$ the even moments H_2 and H_4 tend to be large, while they take relatively low values for events broadened on one side by the radiation of a hard non-collinear gluon ($e^+ + e^- \rightarrow q + \bar{q} + g$). Less significant differences are exhibited by the odd moments. The moment π_1, involving a three-particle angular correlation, is sensitive to the planarity of events expected from gluon bremsstrahlung.

Since all these quantities are invariant against a split-up of the particle energy vector E into separately measured components $\vec{E}_n'$ ($\sum_n \vec{E}_n' = \vec{E}_i$) having a small angular spread, they are well suited for detectors not identifying individual particle tracks. Since they are also rotationally invariant, they are easily calculable and do not involve the definition of an event axis by a maximization process.

Figures 24 and 25 show the distributions $\frac{1}{N}\frac{dN}{dH_2}$ and $\frac{1}{N}\frac{dN}{dH_4}$ of the combined data taken at high energies ($\sqrt{s} > 27$ GeV). They are compared to predictions of QCD and of the quark-parton model with $<P_t> = 225$ MeV/c, where P_t refers to the quark transverse momentum in the fragmentation process. The experimental distributions are in excellent agreement with QCD and clearly rule out the simple quark model at this particular $<P_t>$. The same is true for the differential distribution $\frac{1}{N}dN/d\pi_1$, shown in Figure 26. The predictions of the quark model for these shape parameters are, however, sensitive to the $<P_t>$ chosen.

c)   A Study of Inclusive Muons in Hadronic Events

In the framework of the six quark model for the weak decays of heavy quarks, (c, b and t) copious muon production is expected from the cascade decays $t \rightarrow b \rightarrow c$[14, 15]. The onset of production of a new heavy lepton would also lead to an increase in muon production. Thus, in addition to indications based on thrust and R measurements, a measurement of inclusive muon production in hadronic final states should provide a clear indication of the formation of top quarks or new leptons. All the hadron data for $\sqrt{s}$ from 12 to 35.8 GeV has therefore been analyzed and scanned in a search for muons. The sample of events used in the inclusive muon survey is a subsample of that used to measure R.

The main sources of muons in the hadron sample are decay products of bottom and charm quarks. Background contributions to the muon signal, arising from hadron punch through and decays in flight of pions and kaons have been calculated using the Monte Carlo simulation to be ~ 2% at these energies. The contribution of $\tau^+\tau^-$ events to the μ + hadron sample becomes negligible when the total energy cut and the energy balance cut are applied.

. Table IV summarizes the results for the relative production rate of hadronic events containing muons. The table demonstrates once again the absence of toponium up to 35.8 GeV. For $\sqrt{s} \geq 30$ GeV the observed rate of $3.75 \pm 0.57\%$ agrees with the Monte Carlo predictions for five quark flavors, but is approximately 5 standard deviations away from the prediction which includes the top quark.

The relative production rate of hadronic events containing muons as function of energy is shown in Figure 27a, together with the predictions based on the production of u, d, s, c and b quarks (slashed curve) and the production of top quarks (solid curve). We see the data agree with the five quark model and disagree with the six quark hypothesis. The predicted rate based on a theoretical model which assumes there are only five quarks in nature and the b quark decays only semi-leptonically is shown as -x- in Figure 27a. We see the data are lower than the prediction.

Figure 27b shows the thrust distribution of the hadronic events containing muons compared to a QCD calculation containing five quark flavors. There is very good agreement between the data and the Monte Carlo prediction. The scarcity of events at low thrust in the figure also rules out the existence of the top quark.

d)   Discovery of Three Jet Events

In this section we review the detailed topological analysis which was used by the MARK J to unambiguously isolate the 3-jet events arising from the emission of hard non-collinear gluons[26]. Examination of the azimuthal distribution of energy around the thrust axis was used to obtain a sample of planar events. An analysis of the spatial distribution of the energy flow for the planar event sample established the underlying 3-jet structure in agreement with the QCD predictions for $e^+e^- \rightarrow q\bar{q}g$.

In the reaction $e^+ + e^- \rightarrow$ hadrons, the final states have many appearances: spherical, 2-jet like, 3-jet like, 4-jet like, etc. Events which fall into each of these visual categories can be produced by a variety of underlying processes:

$$e^+ + e^- \rightarrow \text{ a phase space like distribution of hadrons}$$
$$e^+ + e^- \rightarrow q + \bar{q} \text{ (quark } <P_t> = 200 - 500 \text{ MeV)}$$
$$e^+ + e^- \rightarrow q + \bar{q} + \text{ gluon}$$
and
$$e^+ + e^- \rightarrow q + \bar{q} + 2 \text{ gluons}$$

These alternatives make any conclusions which may be drawn from the jet-like appearance of individual events of little value in distinguishing between the models, nor can such an appearance provide information about the nature of the basic final-state constituents. Neutral particles carry a large fraction of the total energy. When the statistics are limited it is important to measure both charged and neutral particles. For a consistent analysis, one must collect a statistically significant number of events in a given kinematic region and compare the number of events in the region with specific model predictions on a statistical basis. A meaningful comparison with models must take into account the uncertainties in the models such as the quark P_t distributions, fragmentation functions, etc. Before conclusions can be drawn, background contributions from other processes must be understood and kept small[27].

In order to exclude events where leading particles have escaped down the beam pipe, or where part of a broad jet is missed, we select only those events for which $E_{vis} > 0.7 \sqrt{s}$. This cut also eliminates two-photon events and events where a hard photon is emitted in the initial state. The drift tubes enable us to separate more distinctly the distribution of charged particles from neutrals. Since neutral particles carry away a large portion of the total energy, they will not only affect the axes of the jets, but will also affect the identification of individual jets.

The jet analysis of hadronic events and the search for 3-jet events is based on a determination of the three dimensional spatial distribution of energy deposited in the detector. This method is quite different from the pioneering method used by the PLUTO and TASSO groups[28]. The characteristic features of hard non-collinear gluon emission in $e^+e^- \rightarrow q\bar{q}g$ are illustrated in Figure 28. Because of momentum conservation the momenta of the three particles have to be coplanar. For events where the gluon is sufficiently energetic, and at large angles with respect to both the quark and antiquark, the observed hadron jets also tend to be in a recognizable plane. This is shown in the upper part of the figure where a view down onto the event plane shows three distinct jets; distinct because the fragmentation products of the quarks and gluons have limited P_t with respect to the original directions of the partons. The lower part of the figure shows a view looking towards the edge of the event plane, which results in an apparent 2-jet structure. Figure 28 thus demonstrates that hard non-collinear gluon emission is characterized by planar events which may be used to reveal a

3-jet structure once the event plane is determined.

The spatial energy distribution is described in terms of three orthogonal axes called the thrust, major and minor axes. The axes and the projected energy flow along each axis T_{thrust}, F_{major} and F_{minor} are determined as follows:

(1) The thrust axis, $\vec{e}_1$, is defined as the direction along which the projected energy flow is maximized. The thrust, T_{thrust} and $\vec{e}_1$ are given by

$$T_{thrust} = \max \ \Sigma_i \ \frac{\vec{E}^i \cdot \vec{e}_1}{\Sigma_i |\vec{E}^i|}$$

where $\vec{E}^i$ is the energy flow detected by a counter as described above and $\Sigma_i \vec{E}^i$ is the total visible energy of the event (E_{vis}).

(2) To investigate the energy distribution in the plane perpendicular to the thrust axis, a second direction, $\vec{e}_2$, is defined perpendicular to $\vec{e}_1$. It is the direction along which the projected energy flow in that plane is maximized. The quantity F_{major} and $\vec{e}_2$ are given by

$$F_{major} = \max \ \frac{\Sigma_i \ \vec{E}^i \cdot \vec{e}_2}{E_{vis}} \ ; \ \vec{e}_2 \perp \vec{e}_1$$

(3) The third axis, $\vec{e}_3$ is orthogonal to both the thrust and the major axes. It is found that the absolute sum of the projected energy flow along this direction, called F_{minor}, is very close to the minimum of the projected energy flow along any axis, i.e.,

$$F_{minor} = \frac{\Sigma_i \ \vec{E}^i \cdot \vec{e}_3}{E_{vis}} \sim \min \frac{\Sigma_i \ \vec{E}^i \cdot \vec{e}}{E_{vis}}$$

If hadrons were produced according to phase-space or a $q\bar{q}$ two-jet distribution, then the energy distribution in the plane as defined by the major and minor axes would be isotropic, and the difference between F_{major} and F_{minor} would be small. Alternatively, if hadrons were produced via three-body intermediate states such as $q\bar{q}g$, and if each of the three bodies fragments into a jet of particles with $<P_t^h> \sim 325$ MeV, the energy distribution of these events would be oblate (P_t^h refers to the final state hadrons).

Following the suggestion of H. Georgi, the quantity oblateness, O, is defined as:

$$O = F_{major} - F_{minor}$$

The oblateness is $\sim 2\, P_t^{gluon}\, \sqrt{s}$ for three-jet final states and is approximately zero for final states coming from a two-jet distribution.

Figure 29a shows the event distribution as a function of oblateness for the data at $\sqrt{s}$ = 17 GeV where the gluon emission effect is expected to be small. The data indeed agree with both models, although the prediction with gluons is still preferred.

Figure 29b shows the event distribution as a function of oblateness for part of the data at $27.4 \leq \sqrt{s} \leq 31.6$ GeV as compared with the predictions of $q\bar{q}g$ and $q\bar{q}$ models. Again, in the $q\bar{q}$ model we use both $<P_t^h>$ = 325 MeV and $<P_t^h>$ =425 MeV. The data have more oblate events than the $q\bar{q}$ model predicts, but they agree with the $q\bar{q}g$ model very well. Figure 29b also illustrates a useful feature of the oblateness: it is quite insensitive to the details of the fragmentation process.

To study the detailed structure of the events we also divided each event into two hemispheres using the plane defined by the major and minor axes, and separately analyzed the energy distribution in each hemisphere as if it were a single jet. The jet having the smaller P_t with respect to the thrust axis is defined as the "narrow" jet (n) and the other as a "broad" jet (b). In each hemisphere we calculate the oblateness, $O_n = 2\,(F_{major}^n - F_{minor}^n)$, and $O_b = 2\,(F_{major}^b - F_{minor}^b)$, and thrusts T_n and T_b.

One approach to analyzing the flat events for a possible 3-jet structure is illustrated by Figure 30. The figure shows the energy flow diagram for each of two high energy hadronic events, viewed in the event plane determined by the major and thrust axes. The energy flow diagram is a polar coordinate plot in which we summed the energy vectors E^i in 10^0 intervals. Each point in the plot represents the summed energy in an angular interval, with the radius given by the magnitude, and the azimuth given by the center of the angular interval in the event plane. The two events in the figure both show an apparent 3-jet structure. The second event also shows that one of the jets is completely neutral, emphasizing the importance of measuring charged as well as neutral energy in performing a consistent topological analysis.

As mentioned earlier, however, the examination of individual event appearances cannot be used to establish the underlying 3-jet structure characteristic of $q\bar{q}g$ final states. This is demonstrated by Figure 31, which shows two low thrust, planar events at 12 GeV center of mass energy. The events also show a distinct multi-jet structure. It should be noted that all the measured distributions at low energies (thrust, oblateness, etc.) are well described by a simple $q\bar{q}$ model, so that the suggestive event appearances are unrelated to gluon emission, but are dominated by fluctuations in the quark fragmentation process. The views of the events in the minor-thrust plane (looking at the edge of the event plane) also show that the events are planar.

At the 1979 International Symposium on Lepton and Photon Interactions at Fermilab[29], all DESY groups (JADE, MARK J, PLUTO and TASSO) reported evidence for hard gluon emission. The first statistically relevant results, establishing the 3-jet pattern from $q\bar{q}g$ of a sample of hadronic events were presented by the MARK J collaboration. These are shown in Figure 32 in which a sample of the events with low thrust and high oblateness, where the gluon-emission effect is expected to be relatively large, is selected for detailed examination. The key feature of this figure is that it consists of the superposition of an entire event sample, and thus displays the average behavior of the energy flow for planar events at high energy. The event sample is composed of 40 events with $T < 0.8$ and $O_b > 0.1$ out of 446 hadronic events obtained up to the time of the Fermilab Conference in the energy range of 27-31.6 GeV. Each of the 40 events has been examined by physicists and found to be 3-jet in appearance like those shown in Figure 30. In superposing events the narrow jet of each event points in the direction of the thrust axis, and the other two jets are oriented in the other hemisphere according to their sizes.

The calculated Monte Carlo model predictions in the figure are compatible with the data with $\chi^2 = 67$ for 70 degrees of freedom. The accumulated energy distribution in the left part of the figure, showing the view in the plane defined by the thrust and minor axes, exhibits a flat distribution consistent with the model predictions.

These results can be contrasted to those obtained with a simple phase space model. When viewed in the major-thrust plane, phase-space shows three nearly identical lobes due to the method of selection used. However, at $\sqrt{s} \sim 30$ GeV there lobes are different in appearance from the jets shown in Figure 32. In general, one expects the three jets from $q\bar{q}g$ to become slimmer and easier to distinguish from the phase-space distribution as the center-of-mass energy increases. Using a χ^2 fit of the phase-space energy distribution to the data we found that $\chi^2 = 222$ for 70 degrees of freedom. Therefore, phase-space is inconsistent with the data. Furthermore, large contributions of phase-space distributions are ruled out by our thrust distributions as shown in Figures 21-23.

For the analyses performed in the earlier part of 1979, as shown in Figures 29 and 32, we used a Monte Carlo model implemented by P. Hoyer et al.[30]. In the analysis discussed below, the Monte Carlo of Ali et al.[7] was adopted. This model incorporates higher order QCD effects, the q^2 evolution of the quark and gluon fragmentation functions and the weak decays of heavy quarks. The QCD predictions presented in Figures 29 and 32 are not noticeably different for the two models.

Figures 33 and 34 show the event distribution as a function of O_n and O_b, compared to the predictions of the QCD model[32] and of two quark-parton models with quark $<P_t> = 300$ MeV and 500 MeV respectively[33]. Figure 33 shows that the narrow jet distribution agrees with the various models indicating that it comes from a single quark jet. Figure 34, however, shows that the quark-parton models severely underestimate the number of events with $O_b > 0.3$ while the QCD model correctly predicts the observed distribution.

The T_n and T_b thrust distributions of the flat events in the region $O_b > 0.3$ are shown in Figures 35 and 36 along with the QCD and high P_t $q\bar{q}$ models. The observed distribution is also compared to a "flattened" $q\bar{q}$ model in which the quarks have $<P_t> = 500$ MeV in the thrust-major plane and $<P_t> = 300$ MeV in the thrust-minor plane. The T_n distribution in Figure 35 is in good agreement with the QCD model predictions and has the same general shape as the thrust distribution for high energy $e^+e^- \to q\bar{q}$ reactions shown in Figures 21 and 22. As expected the T_b distribution, however, is much broader than that of T_n and agrees only with the QCD predictions.

The distributions in Figures 35 and 36 demonstrate that the relative yield of flat events, and the shape of these events as measured by T_n and T_b, can only be explained by the QCD model. The distributions T_n and T_b further exclude phase space (which peaks at lower thrust values) as well as $q\bar{q}$ models as possible explanations of the energy flow plots.

The development of the 3-jet structure with decreasing thrust and increasing oblateness, as predicted by QCD[34], is shown in a series of energy flow diagrams in Figures 37-39.

As seen in Figure 37 the events at high thrust values are dominated by a 2-jet structure characteristic of $e^+ + e^- \rightarrow q + \bar{q}$. In Figure 38, where the thrust is lower, we begin to see the appearance of the gluon jet; and in Figure 39 the 3-jet events are predominant. It is important to note that in all three cases the data agree with the QCD model prediction, showing the increased incidence of hard non-collinear gluon emission with decreasing thrust and increasing oblateness. The energy flow in the minor-thrust plane contains only two nearly identical lobes similar to the narrow jet in Figure 32b, in good agreement with QCD predictions. In Figure 40 we unfold the energy flow diagram of Figure 39 to see more clearly the comparison of the data with the predictions of QCD, $q\bar{q}$ ($<P_t> = 500$ MeV), and a "mixed model" consisting of a combination of $q\bar{q}$ and phase-space contributions. All models in Figure 39 are normalized to have the same areas (as the data) before the individual cuts are imposed. The normalization for the mixed model was determined by adjusting the $q\bar{q}$ ($<P_t> = 300$ MeV) and phase-space contributions to agree with the measured thrust distribution as illustrated in Figure 41. As seen in Figure 40, only QCD can describe the observed 3-jet structure. The conclusions obtained by the JADE, PLUTO and TASSO collaborations on their tests of QCD and studies of multi-jet events are in good agreement with us (see Section 3.4).

A very powerful method to eliminate phase-space as a source of the 3-jet event and in fact eventually to study the nature of the individual jets is shown in Figure 42. Here all the data with $27 < E < 35.8$ GeV are included. The thrust distributions of each of the jets, separated according to their position in the energy flow figure, are compared to the expectations of QCD and phase-space. The thrust distributions agree well with the QCD prediction but are in strong disagreement with phase space predictions.

e)   Determination of the Strong Coupling Constant α_s.

Recent experiments on scaling violations in lepton inelastic scatter-ing[35], on high Γ_t events in dilepton production by hadrons[36], and multi-jet events in e^+e^- annihilations[23, 26, 37] all indicate that the results are explained naturally in the quantum chromodynamics (QCD) theory of the strong interactions of quarks and gluons[7]. The strong coupling constant α_s (q^2) between quarks and gluons has been measured indirectly in quarkonium bound states[38], and in deep inelastic experiments[35]. At PETRA, where the q^2 is much larger, computations are expected to be more reliable. In addition, high energy e^+e^- annihilations offer a more direct way of measuring α_s and testing perturbative QCD because it is expected to give rise to multi-jets which can be systematically identified.

The 3-jet events discussed in the previous section, which consist of $q\bar{q}g$ fragmentation products with relatively small backgrounds from fluctuations of phase-space-like processes or quark-antiquark intermediate states, allow us to make further comparisons of the event properties with the predictions of QCD. In particular the relative yield of 3-jet events and the shape distribution gives a way to measure directly α_s, the strong coupling constant.

We used several methods in determining the strong coupling constant α_s, including:

1)   the average oblateness $<O_b>$,

2)   the fraction of events with $O_b > 0.3$,

3)   the relative yield of events with $O_b - O_n > 0.3$ where O_n is constrained to be greater than zero.

For each quantity we allowed α_s to vary in the QCD model, and we then determined the range of α_s values for which the QCD model predictions agree with the data within errors. In particular, the samples obtained using criteria 2) and 3) consist predominantly of 3-jet events from $e^+e^- \to q\bar{q}g$, in which the gluon emitted is both very energetic and at a large angle with respect to both the quark and antiquark. This leads to an event sample where the number of events in the sample is a quasi-linear function of α_s, and in which the influence of non-perturbative effects which are not calculable in QCD is minimal. For criterion 2), for example, we observed 161 events, which matches the QCD model with $\alpha_s = 0.23$. The $e^+e^- \to q\bar{q}$ contribution is calculated to be 21 events. The predominance of $q\bar{q}g$ in a sample with

$O_b > 0.3$ is maintained even if $<P_t>$ is allowed to vary from 225 MeV to 500 MeV in the model. With $<P_t>$ 500 MeV the $e^+e^- \rightarrow q\bar{q}$ contribution is calculated to be 58 events.

The methods described above yield a self-consistent set of α_s values, as illustrated in Figure 43. On the basis of the results of the three methods we obtain

$$\alpha_s = 0.23 \pm 0.02 \text{ (statistical error)}$$
$$\pm 0.04 \text{ (systematic error)}$$

The large systematic error was mostly due to uncertainties in QCD calculations[7]. For method 2) the range of α_s due to variation in $< P_t >$ from 225 to 400 MeV is ± 0.01 and the change in α_s due to different cuts in O_b from $O_b > 0.3$ to $O_b > 0.15$ or cuts in O_n from no cuts to $O_n < 0.1$, is $- 0.01$. For method 2), changing the fragmentation function $zD(z)$ to $1-z$ for u, d and s quarks and $zD(z)$ to z for c and b quarks does not change the α_s value noticeably. Tables V and VI show in detail the change of α_s with respect to the O_b cuts and O_n cuts.

Our value of α_s is consistent with the values of the JADE group obtained with a different Monte Carlo program. (See Section 3.4). The value of α_s[31, 39] is in qualitative agreement with the values obtained in deep inelastic lepton nucleon scattering experiments[35], and in the analysis of the quarkonium states[38]. However, detailed comparison among these results cannot yet be made without accurate higher order QCD calculations.

3.4   Comparison With Other Experiments at PETRA

Our detector uses calorimetric techniques which are very different from other PETRA groups (JADE, PLUTO and TASSO) which use solenoidal magnetic field followed by particle identification devices. These differences in technique imply that the event selection criteria, Monte Carlo analysis programs, and the assignment of systematic errors are quite different. However, despite the different techniques used the results of all the PETRA groups are complementary and supportive of each other in their physics conclusions on the test of QED[40], on the measurement of R, the search for new flavors[37, 41] and of the effects of hard gluon jets[37, 42].

4. CONCLUSIONS

In the first one and one-half years of experimentation with a simple detector, we have obtained the following results:

1)   We have established the validity of quantum electrodynamics to a distance $< 2 \times 10^{-16}$ cm. Quarks, electrons, muons and tau leptons are point-like with sizes smaller than 2×10^{-16} cm.

2)   The relative cross sections and event distributions show that there is no new charge 2/3 quark pair production up to $\sqrt{s} = 35.8$ GeV.

3)   The energy flow of events at high energies is in good agreement with quantum chromodynamics. The quantity of flat events and their distributions disagree with the simple quark antiquark model prediction.

4)   We have discovered 3-jet events; the rate of their production and their distribution agree with the prediction of QCD.

5)   We have measured the strong interaction coupling constant α_s.

There are two reasons which made it possible to obtain these results:

1)   In e^+e^- collisions, the signal is clear and unique. Every event has a definite physical interpretation and can be analyzed in terms of QED or QCD. This is quite different from our previous experience with proton proton collisions where the signal of the virtual photon events is less than one part in 10^6 of the background.

2)   PETRA was reliably constructed and available for use by experimentalists from the beginning.

ACKNOWLEDGEMENTS

We wish to thank Professors H. Schopper, G. Voss, A. N. Diddens, H. Faissner, E. Lohrmann, F. Low, Drs. F. J. Eppling and G. Soehngen for their valuable support and A. Ali, A. DeRujula, H. Georgi, S. Glashow, J. Kouptsidis, T. D. Lee, E. Pietarinin, T. Walsh, L. L. Wang for helpful discussions.

REFERENCES

1.   PETRA Proposal (updated version), DESY, Hamburg, (February, 1976).

2.   G. Voss, The 19 GeV e^+e^- Storage Ring, Internal Report, DESY M/79/16.

3.   S. W. Herb, et al., Phys. Rev. Lett. 39, 252 (1977).

4.   C. W. Darden et al., Phys. Lett. 76B, 246 (1978).

Ch. Berger et al., Phys. Lett. 76B, 243 (1978).

J. K. Bienlein et al., Phys. Lett. 78B, 360 (1978).

C. W. Darden et. al., Phys. Lett. 78B, 364 (1978).

5.   A. Febel and G. Hemmie, "PIA, the Positron Intensity Accumulator for the PETRA Injection," Internal Report, DESY M/79/13.

6.   U. Becker et al., "A Simple Detector to Measure e^+e^- Reactions at High Energy," Proposal to PETRA Research Committee, (March, 1976).

7.   D. J. Gross and F. A. Wilczek, Phys. Rev. Lett. 30, 1343 (1973).

H. D. Politzer, Phys. Rev. Lett. 30, 1346 (1973).

J. Ellis et al., Nucl. Phys. B111, 253 (1976).

T. deGrand et al., Phys. Lett. D16, 3251 (1977).

G. Kramer et al., Phys. Lett. 79B, 249 (1978).

A. DeRujula et al., Nucl. Phys. B138, 387 (1978).

P. Hoyer et al., DESY Preprint 79/21 (unpublished).

A. Ali et al., Phys. Lett. 82B, 285 (1979); also DESY Report 79/54 submitted to Nucl. Phys. B.

8.   P. D. Luckey et al., Proceedings of the International Symposium on Electron and Photon Interactions at High Energies, Hamburg, 1965 (Springer, Berlin 1965) Vol. II, p. 397.

9.   V. Alles-Borelli et al., Nuovo Cimento 7A, 345 (1972).

H. Newman et al., Phys. Rev. Lett. 32, 483 (1974).

J-E. Augustin et al., Phys. Rev. Lett. 34, 233 (1975).

L. H. O'Neill et al., Phys. Rev. Lett. 37, 395 (1976).

10.   D. P. Barber et al., Phys. Rev. Lett. 42, 1110 (1979).

11.   S. J. Brodsky and S. D. Drell, Annu. Rev. Nucl. Sci. 20, 147 (1970).

12.   ADONE Proposal INFN/AE-67/3, (March 1967), ADONE-Frascati (unpublished) and M. Bernardini et al., (Zichichi Group), Nuovo Cimento 17A, 383 (1973).

        S. Orito et al., Phys. Lett. 48B, 165 (1974)

13.   M. Perl et al., Phys. Rev. Lett. 35, 1489 (1975).

        G. Feldman et al., Phys. Rev. Lett. 38, 117 (1977).

14.   J. Burmester et al., Phys. Lett. 68B, 297 (1977).

        J. Burmester et al., Phys. Lett. 68B, 301 (1977).

15.   For a review of our present knowledge of the τ lepton, see Guenter Fluegge, Zeitschr. f. Physik C1, Particles and Fields, 121 (1979), and the references therein.

16.   R. Hofstadter, Proceedings of the 1975 International Symposium on Lepton and Photon Interactions at High Energies (Stanford Linear Accelerator Center, Stanford, California, 1975) 869.

17.   S. D. Drell, Ann. Phys. (N. Y.) 4, 75 (1958).

        T. D. Lee and G. C. Wick, Phys. Rev. D2, 1033 (1970).

18.   D. P. Barber et al., Phys. Rev. Lett. 43, 1915 (1979).

19.   H. Terazawa, Rev. Mod. phys. 45, 615 (1973). We used σ ($\gamma\gamma \rightarrow$ multipion) = 240-270/W^2 nb. W is the energy of the two-photon system.

20.   Y. S. Tsai, Phys. Rev. D4, 2821 (1971).

        H. B. Thacker and J. J. Sakurai, Phys. Lett. 36B, 103 (1971).

        K. Fukijawa and N. Kawamoto, Phys. Rev. D14, 59 (1976).

        See also Reference 28.

21.   C. Quigg, Fermilab Conference 79/74-THY, September 1979.

        J. L. Rosner, C. Quigg, H. B. Thacker, Phys. Lett. 74B, 350 (1978).

        C. Quigg, contribution to the 1979 International Symposium on Lepton and Photon Interactions, Fermilab.

        M. Greco, Phys. Lett. 77B, 84 (1978).

22.   T. Applequist and H. Georgi, Phys. Rev. D8, 4000 (1973).

        A. Zee, Phys. Rev. D8, 4038 (1973).

23.   D. P. Barber et al., Phys. Rev. Lett. 42, 1113 (1979).

        D. P. Barber et al., Phys. Rev. Lett. 43, 901 (1979).

        D. P. Barber et al., Phys. Lett. 85B, 463 (1979).

        For a theoretical discussion on the use of thrust variables see:

454

E. Farhi, Phys. Rev. Lett. 39, 1587 (1977).

S. Brandt et al., Phys. Lett. 12, 57 (1964).

S. Brandt and H. Dahmen, Zeitschr. f. Phys. C1, 61 (1979).

24.   R. Schwitters, Proceedings of the 1975 International Symposium of Lepton and Photon Interactions at High Energies (Stanford Linear Accelerator Center, Stanford, California, 1975) 5.

G. Hansen et al., Phys. Rev. Lett. 35, 1609 (1975).

G. Hansen, Talk at the 13th Recontre de Moriond, Les Arcs, France, (March 12-26, 1978), SLAC-PUB 2118 (1978).

25.   G. C. Fox and S. Wolfram, Phys. Rev. Lett. 41, 1581 (1978).

G. C. Fox and S. Wolfram, Nucl. Phys. B413, (1979).

G. C. Fox and F. Wolfram, Phys. Lett. 82B, 134 (1979).

26.   D. P. Barber et al., Phys. Rev. Lett. 43, 830 (1979).

27.   Ch. Berger et al., Phys. Lett. 86B, 418 (1979), have presented a systematic study in the selection of 3-jet events from $q\bar{q}$ and 3-jet events from $q\bar{q}g$ (see Tables 1 and 2 of their paper).

28.   Ch. Berger et al., Phys. Lett. 76B, 176 (1978).

S. Brandt and H. Dahmen, Zeitschr. f. Phys. C1, 61 (1979).

B. H. Wiik, Proc. Intern. Neutrino Conf. (Bergen, Norway, June 1979).

P. Soeding, European Physical Society Conference Report, July (1979). This report includes five events of 3-jets of charged particles. Each event has a measured energy ~1/2 of the total energy. Their prediction for $q\bar{q}g$ was 9 events with an unstated 3-jet background from $q\bar{q}$. Statistically significant results on planar events have been published (see Table 4) in R. Brandelik et al., DESY-Preprint 79/61 (1979).

29.   H. Newman, paper presented at the 1979 Fermilab Conference (to be published in Proceedings). See Reference 26.

See also S. Orito (the JADE Collaboration) ibid.

Ch. Berger (the PLUTO Collaboration) ibid.

G. Wolf (the TASSO Collaboration) ibid.

For a summary of DESY work up to that time, see H. Schopper, DESY Report 79/79 (1979).

30.   P. Hoyer et al., DESY Report 79/21 (1979).

31.   D. P. Barber et al., Phys. Lett. <u>89B</u>, 139 (1979).

32.   The data in Figures 33 and 34 at $O < O_b$ or $O < O_n$ is due to statistical fluctuations, and the energy resolution $\frac{\Delta E}{E} \sim 20\%$ and possible higher order QCD effects.

33.   In Figures 34–40 if we choose quark $<P_t> \; < 300$ MeV we observe larger deviations between $q\bar{q}$ model and the data.

34.   A. DeRujula et al., Nucl. Phys. B <u>138</u>, 387 (1978).

35.   H. L. Anderson et al., Phys. Rev. Lett. <u>40</u>, 1061 (1978).

         P. C. Bosetti et al., Nucl. Phys. B <u>142</u>, 1 (1978).

         J. G. H. de Groot et al., Zeitschr f. Phys <u>C1</u>, 143 (1979).

         J. G. H. de Groot et al., Phys. Lett. <u>82B</u>, 292 (1979).

         J. G. H. de Groot et al., Phys. Lett. 82B, 456 (1979).

36.   D. Antreasyan et al., (to be published).

37.   W. Bartel et al., (the JADE Collaboration), Phys. Lett. <u>88B</u>, 171 (1979).

         W. Bartel et al., (the JADE Collaboration), Phys. Lett. <u>89B</u>, 136 (1979).

         Ch. Berger et al., (the PLUTO Collaboration), Phys. Lett. <u>81B</u>, 410 (1979).

         Ch. Berger et al., (the PLUTO Collaboration), Phys. Lett. <u>86B</u>, 413 (1979).

         Ch. Berger et al., (the PLUTO Collaboration), Phys. Lett. <u>86B</u>, 418 (1979).

         Ch. Berger et al., (the PLUTO Collaboration), Phys. Lett. <u>89B</u>, 120 (1979).

         R. Brandelik et al., (the TASSO Collaboration), Phys. Lett. <u>83B</u>, 261 (1979).

         R. Brandelik et al., (the TASSO Collaboration), Phys. Lett. <u>86B</u>, 243 (1979).

         R. Brandelik et al., (the TASSO Collaboration), Phys. Lett. <u>88B</u>, 199. (1979).

         R. Brandelik et al., (the TASSO Collaboration), Phys. Lett. <u>89B</u>, 418 (1980).

38.   M. Krammer and H. Krasemann, DESY Report 78/66 (1978).

456

39.   It is not yet clear which q^2 should be used to extract Λ to compare quantitatively with other data.

40.   S. Orito (the JADE Collaboration), invited talk 1979 Symposium at FNAL and DESY Report 79/64, 1979.
        Ch. Berger et al., (the PLUTO Collaboration), DESY Report 80/01.

41.   Ch. Berger et al., (the PLUTO Collaboration), DESY Report 80/02.
        W. Bartel et al., (the JADE Collaboration), DESY Report 80/04.

42.   W. Bartel et al., (the JADE Collaboration), DESY Report 79/80.
        Ch. Berger et al., (the PLUTO Collaboration), DESY Report 79/83.

TABLES

I.   Cut-off Parameters in GeV for photon form factors from Bhabha scattering.

II.   Cut-off parameters in GeV for e^+e^-, $\mu^+\mu^-$ and $\tau^+\tau^-$ production using leptonic form factors.

III.   Results of R measurement. The errors are statistical only.

IV.   Monte Carlo predictions and data for hadronic events which include muons.

V.   Value of α_s with different O_b cuts, without cuts in O_n and $<P_t> = 247$ MeV.

VI.   Value of α_s with different O_n cuts, with $O_b > 0.3$ and $<P_t> = 247$ MeV.

458

TABLE I

Λ	$1 - \cdot\dfrac{q^2}{q^2 - \Lambda_+^2}$	$1 + \cdot\dfrac{q^2}{q^2 - \Lambda_-^2}$
Λ_S	91	152
Λ_T	58	64
$\Lambda_S = \Lambda_T$	97	157

TABLE II

ℓ	electron	muon	tau
Λ_-	142	150	157
Λ_+	94	126	76

TABLE III

$\sqrt{s}$ [GeV]	$R \pm \Delta$ (statistical)
12.0	4.03 ± 0.28
13.0	4.1 ± 0.5
17.0	4.4 ± 0.6
22.0	4.7 ± 0.7
27.57 *)	3.8 ± 0.3
30.0	4.2 ± 0.3
29.90–31.46	4.33 ± 0.17
31.6	4.0 ± 0.5
33.0	2.9 ± 0.6
35.0	3.8 ± 0.4
35.0–35.28	4.2 ± 0.3
35.8	4.4 ± 0.7

*) = combination of 27.4 and 27.7 GeV data

TABLE IV

E_{cm} GeV	Luminosity nb^{-1}	Number of Hadron Events	Number of Muon Events	% of Muon Events	Monte Carlo (no top) ±	Monte Carlo (with top) ±
12	97.7	239	2	0.8 ± 0.6	1.1 ± 0.3	
13	53	95	1	1.05 ± 1.0	1.25 ± 0.3	
17	60	67	2	3.) ± 1.7	2.0 ± 0.3	
27.4	414	188	11	5.35 ± 1.8	3.3 ± 0.4	
30 to 31.6	2804	1147	43	3.75 ± 0.57	4.5 ± 0.5	7.8 ± 0.5
35 to 35.8	1885	640	22	3.5 ± 0.7	5.1 ± 0.4	7.8 ± 0.5

TABLE V

$O_b>$	α_s
0.15	0.22 ± 0.02
0.20	0.22 ± 0.02
0.25	0.22 ± 0.02
0.30	0.23 ± 0.02

TABLE VI

$O_n <$	α_s
No cuts	0.23 ± 0.02
0.24	0.23 ± 0.02
0.20	0.23 ± 0.02
0.16	0.23 ± 0.02
0.12	0.22 ± 0.03
0.08	0.22 ± 0.03

FIGURE CAPTIONS

1. The layout of PETRA e^+e^- Storage Ring at DESY.

2. Electron, muon and tau pair production in lowest order.

3. a. Diagram for production and decay of heavy quarks in e^+e^- annihilation.

    b. Diagram for production and decay of heavy leptons in e^+e^- annihilation.

4. The reaction $e^+e^- \to$ hadrons in lowest order.

5. First order electromagnetic and weak processes contributing to the reaction $e^+e^- \to \mu^+\mu^-$.

6. The MARK J detector in a side view.

7. The MARK J detector in end view. Beam pipe (1), drift tubes (DT), shower counters (A, B, C), inner drift chambers (S, T), calorimeter counters (K), outer drift chambers (Q, P, R), and magnetized iron (2).

8. The MARK J detector showing the outer drift chambers.

9. Aerial view of the MARK J showing movable cable supports.

10. End view of the MARK J showing the inner chamber and calorimeter.

11. The layer structure of the MARK J detector as seen by a particle emerging from the interaction point at a right angle to the beam axis.

12. Distribution of event vertices along the beam direction reconstructed using drift tube tracks.

13. The differential cross section s $\frac{d\sigma}{d\cos\theta}$ for $e^+e^- \to e^+e^-$ at $\sqrt{s}$ of 13, 17, and 27.4 GeV.

14. The observed cross sections of $e^+ + e^- \to e^+ + e^- + \mu^+ + \mu^-$ in our detector as functions of $\sqrt{s}$ when the observed particles are:

    a. two μ's

    b. one μ and one e

    c. two μ's and one e

    The solid lines are Monte Carlo calculations of the yield from two photon diagrams and the points are the measurements.

15. Observed cross section for the reaction $e^+e^- \to \mu^+\mu^-$ compared to predictions of QED.

16. Observed cross section for the reaction $e^+e^- \to \tau^+\tau^-$ compared to predictions of QED.

17.   Energy distribution of hadron events satisfying the cuts described in Section 3.2a.

18.   The total relative hadronic cross section $R = (e^+e^- \to$ hadrons)/ $\sigma(e^+e^- \to \mu^+\mu^-)$ at all energies covered by this experiment. Note that the point at 27.5 GeV corresponds to data taken over the range of 27.4 to 27.7 GeV, and the point at 30.7 GeV represents the energy scan from 29.9 to 31.5 GeV.

19.   a.   R values measured during the energy scan between 29.9 and 31.5 GeV. The line represents the predictions of QCD.

       b.   R values measured during the energy scan between 35.0 and 35.26 GeV. The line represents the predictions of QCD.

       c.   Theoretical predictions of the production cross section of the ground state toponium in e^+e^- annihilation; slashed curve is calculated assuming Γ_e/e_q^2 is constant; solid line is the rigorous lower bound as calculated by Quigg in reference 21; $\downarrow$ indicates the experimental upper limit.

20.   A sketch showing the qualitative change in the thrust distribution expected when the energy is increased and the threshold for production of a new quark flavor is crossed.

21.   Thrust distribution observed at $\sqrt{s} =$ a) 13, b) 17, c) 22, d) 27 GeV (see text). The solid line is the quark model prediction for u, d, s, c and b quarks with no gluon emission.

22.   The thrust distribution with a 70% energy cut for $\sqrt{s} = \sim 30$ GeV. The curves are predictions for various models as described in the text.

23.   Average value of thrust as a function of $\sqrt{s}$ together with the QCD prediction (solid line). The values expected from a phase space distribution and from a QCD model with a top quark are also shown.

24.   The distribution of the combined high energy data ($\sqrt{s} > 27$ GeV) in the Fox-Wolfram moment, H_2 compared to the prediction of QCD at $\alpha_s = 0.23$ and the quark model with $<P_t> = 225$ MeV.

25.   The Fox-Wolfram moment H_4 distribution.

26.   The Fox-Wolfram moment π_1 distribution.

27.   a.   The relative production rate of hadronic events containing muons as function of energy. The slashed curve is the prediction based on the production of five types of quarks. The solid curve

includes the top quark as well. The -x- is calculated assuming that the b quark decays only semi-leptonically.

b. The thrust distribution of high energy hadron events ($\sqrt{s} > 27$ GeV) containing at least one muon. The data (solid points) are compared to the prediction of QCD with five quark flavors.

28. A schematic view of the process $e^+e^- \to q\bar{q}g$, and the three resulting hadron jets showing the axes used to describe the event.

29. Differential oblateness distribution at a) $\sqrt{s} = 17$ GeV and b) at high energies (combined data) compared to the predictions of QCD (solid line) and quark model (dashed lines).

30. Energy flow diagrams for two high energy hadronic events viewed in the major-thrust plane.

31. Two events measured at $\sqrt{s} = 12$ GeV with

    a.   T = 0.71

    b.   T = 0.67

The lines show direction and magnitude of energy deposited in the calorimeter displayed in three projections. The events appear to have a multi-jet structure in the thrust-major plane. The view in the thrust-minor plane shows the events are flat. The top event is a 3-jet event. The bottom event is a planar multi-jet event. The numbers associated with the tracks in the Major-Thrust plane are the numbers of charged tracks in the jets.

32. a. Energy distribution in the plane defined by the thrust and major axes for all events with $T < 0.8$ and $O_b > 0.1$ at $\sqrt{s} = 27.4$, 30 and 31.6 GeV obtained up to the time of the 1979 Fermilab Conference. The radial distance of the data points is proportional to the energy deposited in a 10° bin. The superimposed dashed line represents the distribution predicted by QCD.

    b. Measured and predicted energy distribution in the plane defined by the thrust and minor axes, which shows only 2-jets.

33. The narrow jet oblateness distribution $\frac{1}{N}\frac{dN}{dO_N}$ for all hadron events with measured energy $E_{vis} \geq 0.7\sqrt{s}$. The data are compared to the predictions of the QCD model and to two quark-antiquark models with $\langle P_t \rangle = 300$ MeV and $\langle P_t \rangle = 500$ MeV respectively.

34. The broad jet oblateness distribution $\frac{1}{N}\frac{dN}{dO_b}$ under the same condition as Figure 33.

35. The thrust distribution of the narrow jet events with $O_b \geq 0.3$. Also shown are the various model predictions including a flattened $q\bar{q}$ (500, 300) discussed in the text. Note the narrow jet thrust distribution is consistent with thrust distribution of all the hadron events labelled "ALL DATA." This curve is normalized to the data of Figure 22.

36. The thrust distribution of the broad jets for events with $O_b > 0.3$. The curves are discussed in the text.

37. Energy flow diagram in the thrust-major plane for high energy data (27-31.6 GeV) with $T > 0.9$. The solid line is the prediction of QCD.

38. Same as Figure 37 for events with $0.8 \leq T \leq 0.9$ and broad jet oblateness $O_b > 0.1$.

39. Same as Figure 37 for events with $T < 0.8$ and $O_b > 0.1$.

40. The unfolded energy flow diagram of Figure 39 as compared to QCD, the quark model ($<P_t> = 500$ MeV) and a mixed $q\bar{q}$ and phase space model (see text).

41. Illustration of the method used in determining the maximum permissible admixture of phase-space to $q\bar{q}$ in the mixed model shown in Figure 40.

42. Thrust distribution $\frac{1}{N}\frac{dN}{dT}$ for each individual jet in the 3-jet sample of Figure 39 which were selected by using $O_b > 0.1$ and thrust > 0.8. The corresponding distribution normalized to the same total number of 3-jet events are also shown for QCD (solid curve) and phase-space model (dashed curve).

43. The left graph: The average value of oblateness $<O_b>$ for all events with $E_{vis} \geq 0.7\sqrt{s}$ as a function of α_s, computed by varying α_s in the QCD model.

The right graph: The fraction of hadronic events with $O_b > 0.3$ (σ_{3j}) as a function of α_s computed by varying α_s in the QCD model.

468

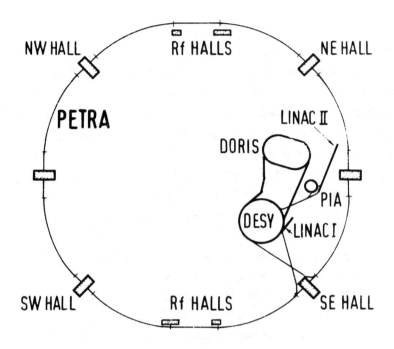

Fig. 1

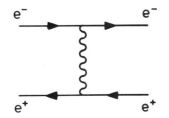

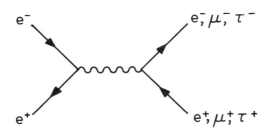

Fig. 2

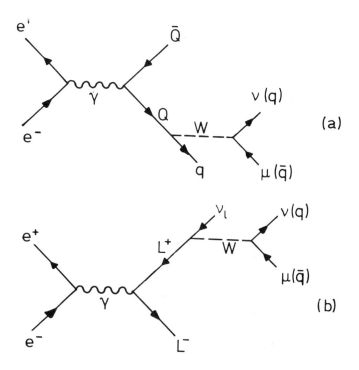

Fig. 3

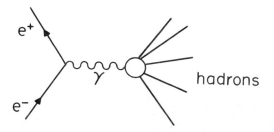

Fig. 4

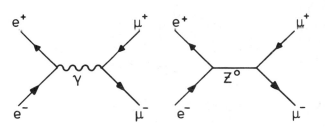

Fig. 5

472

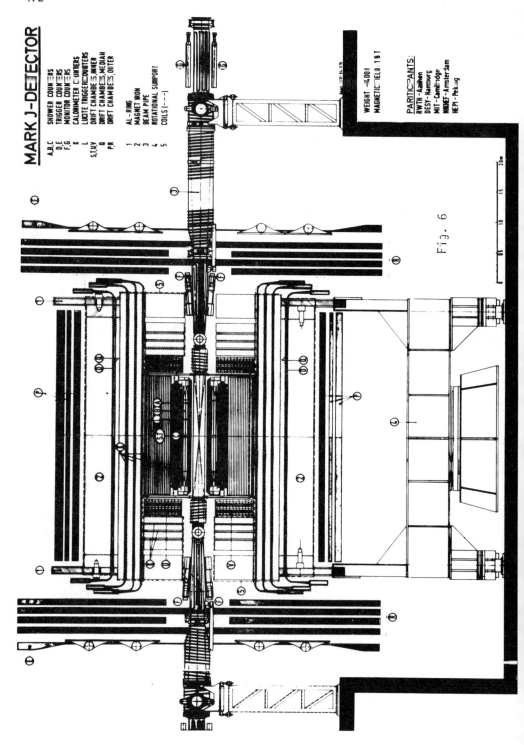

MARK J-DETECTOR

A,B,C	SHOWER COUNTERS
D,E	TRIGGER COUNTERS
F,G	MONITOR COUNTERS
K	CALORIMETER COUNTERS
L	LUCITE TRIGGER COUNTERS
S,T,U,V	DRIFT CHAMBERS, INNER
Q	DRIFT CHAMBERS, MEDIAM
P,R	DRIFT CHAMBERS, OUTER
1	AL-RING
2	MAGNET IRON
3	BEAM PIPE
4	ROTATIONAL SUPPORT
5	COILS (----)

WEIGHT = 4001
MAGNETIC FIELD : 1.9 T

PARTICIPANTS:
RWTH-Aachen
DESY-Hamburg
MIT-Cambridge
NIKHEF-Amsterdam
IHEP-Peking

Fig. 6

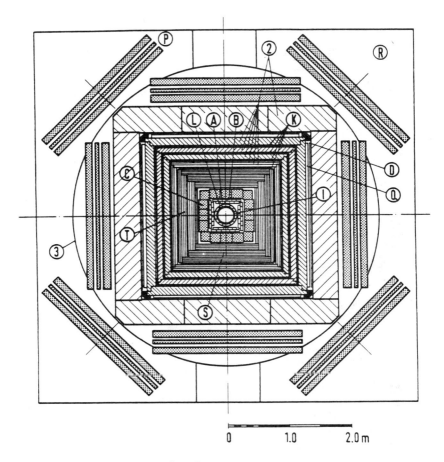

Fig. 7

474

Fig. 8

Fig. 9

476

Fig. 10

477

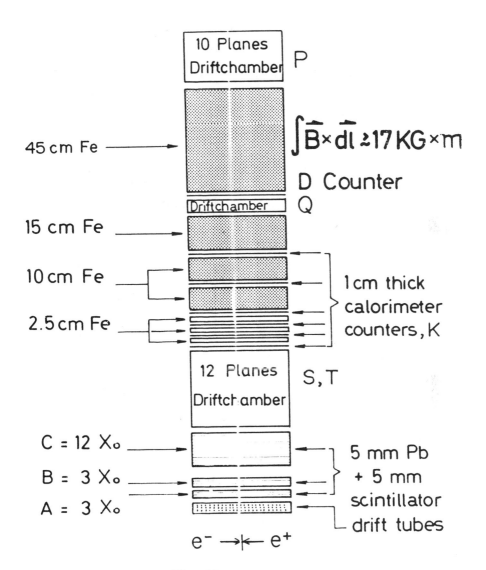

Fig. 11

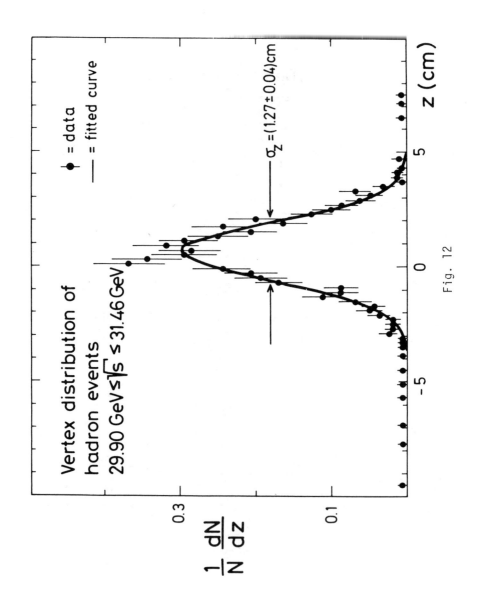

Fig. 12

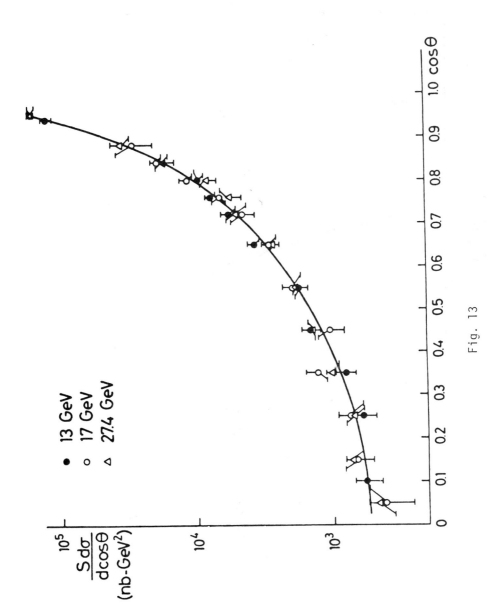

Fig. 13

480

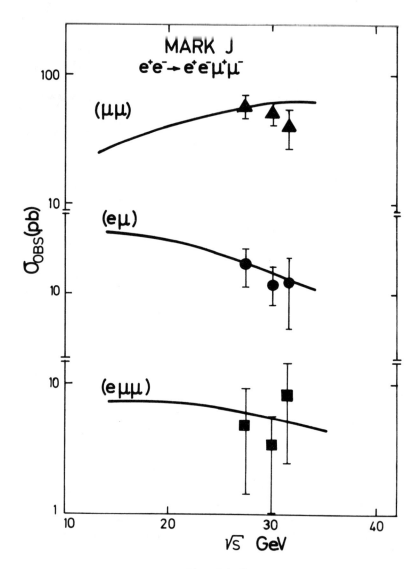

Fig. 14

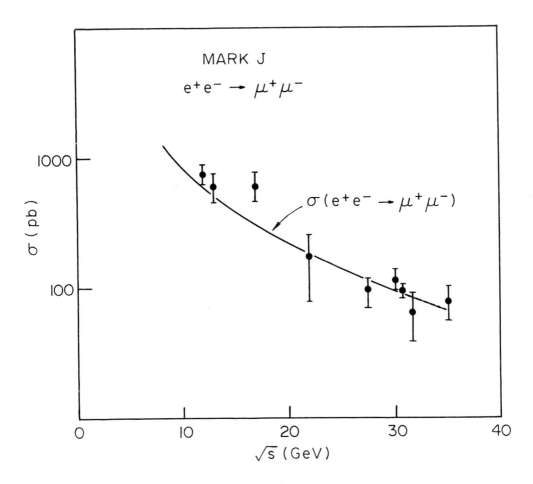

Fig. 15

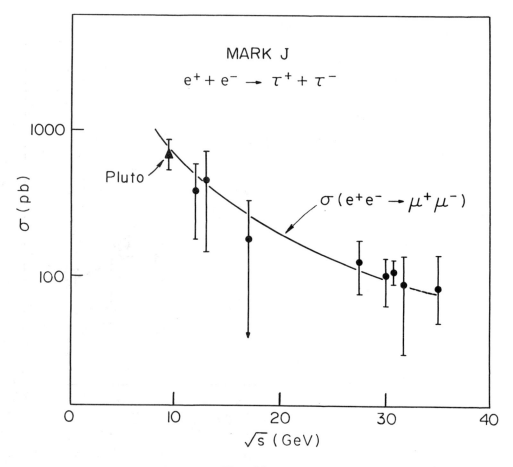

Fig. 16

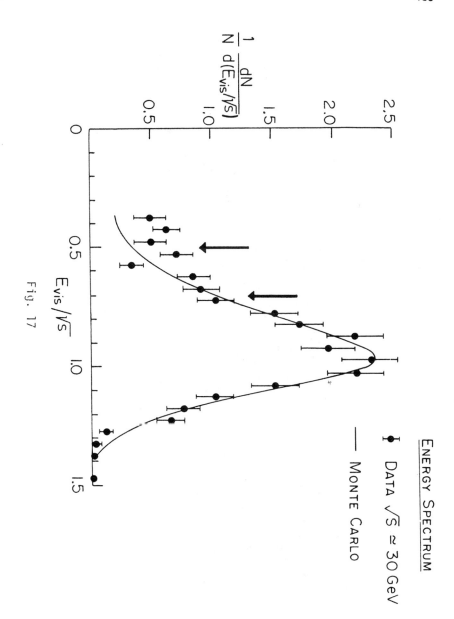

Fig. 17

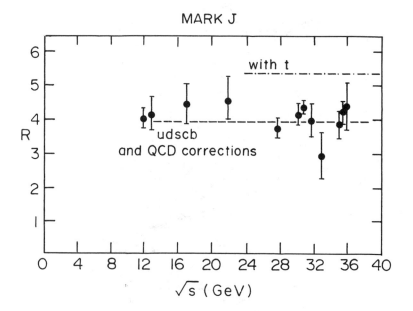

Fig. 18

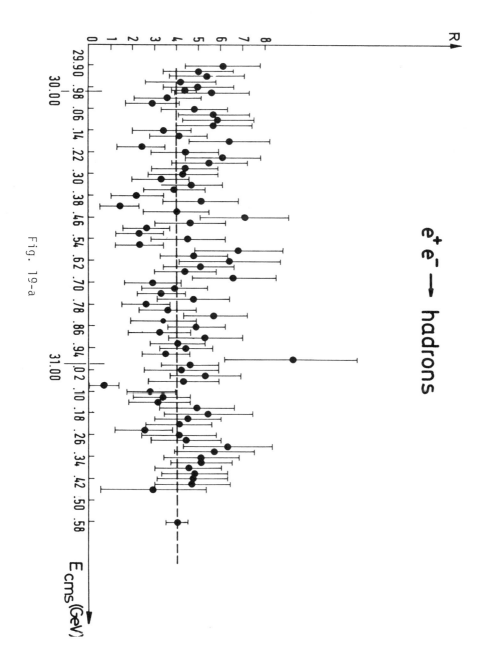

Fig. 19-a

486

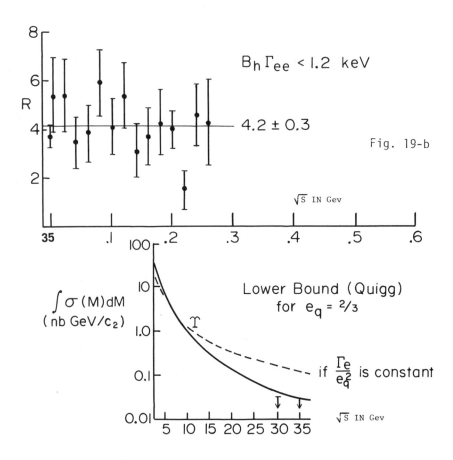

$B_h \Gamma_{ee} < 1.2$ keV

4.2 ± 0.3

Fig. 19-b

$\sqrt{S}$ IN Gev

Lower Bound (Quigg)
for $e_q = 2/3$

$\int \sigma (M) dM$
(nb GeV/c_2)

--- if $\dfrac{\Gamma_e}{e_q^2}$ is constant

$\sqrt{S}$ IN Gev

Fig. 19-c

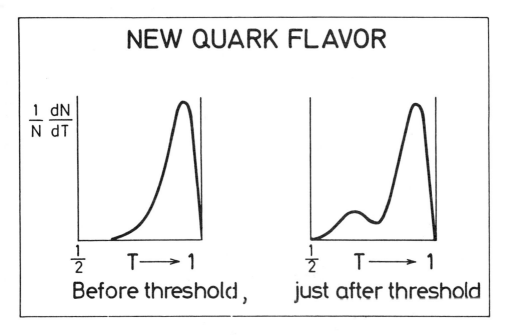

Fig. 20

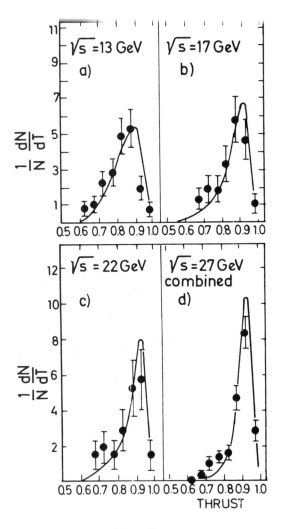

Fig. 21

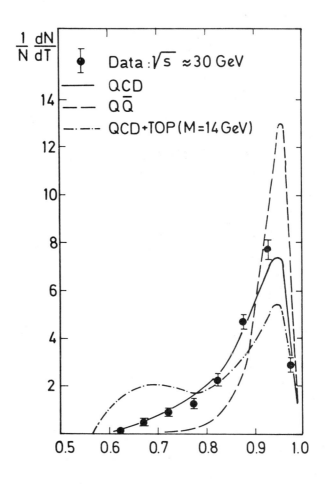

Fig. 22

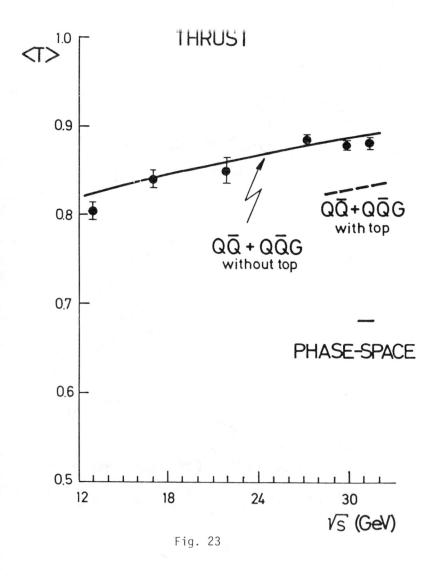

Fig. 23

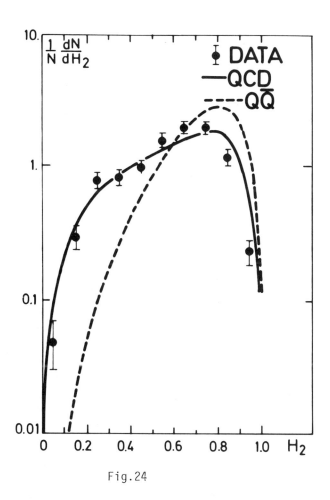

Fig.24

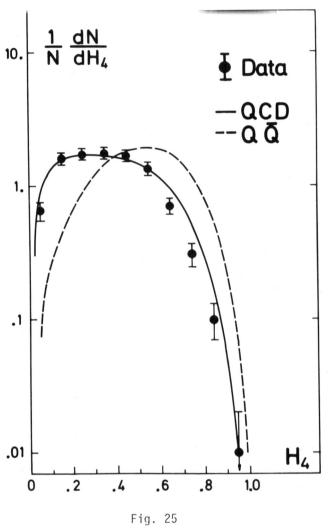

Fig. 25

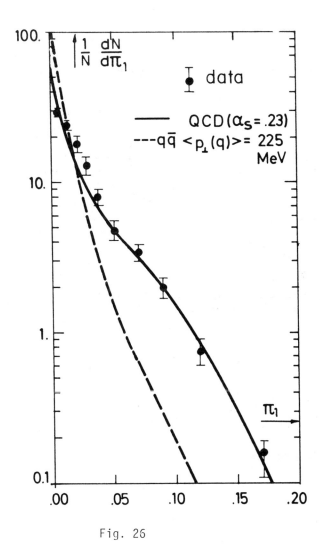

Fig. 26

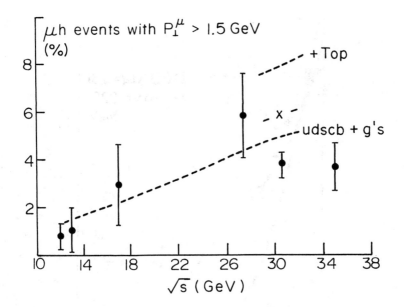

Fig. 27-a

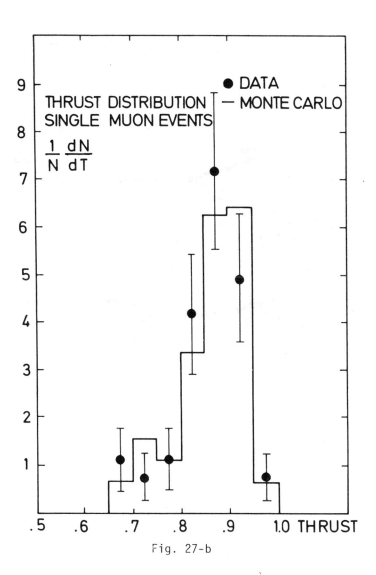

Fig. 27-b

496

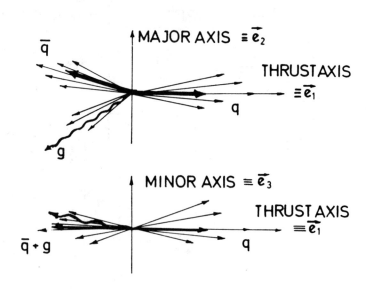

$$e^+ + e^- \longrightarrow \underbrace{q + \bar{q} + g}_{\text{PLANAR}} \longrightarrow \text{HADRONS}$$

Fig. 28

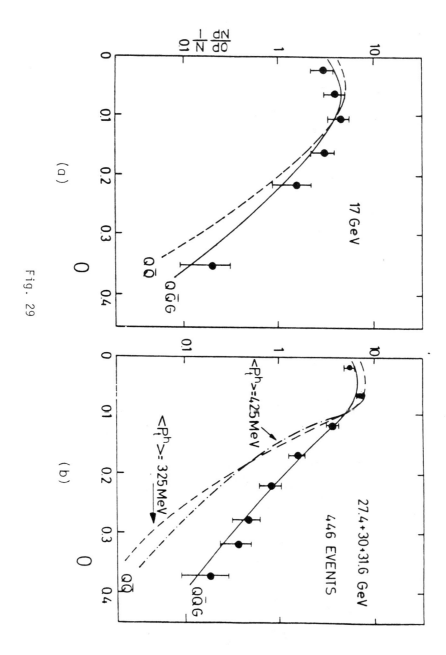

Fig. 29

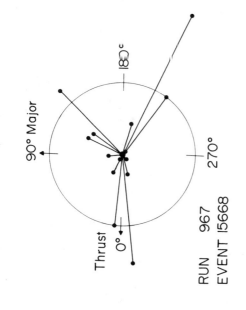

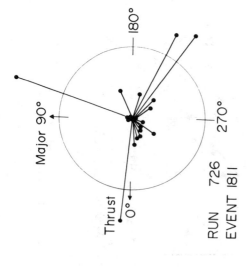

Fig. 30

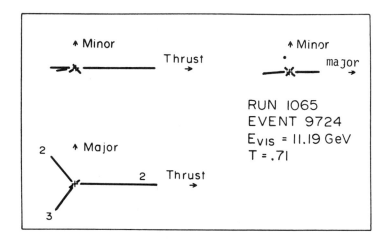

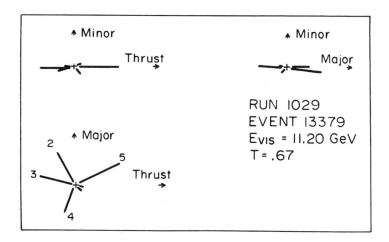

Fig. 31

500

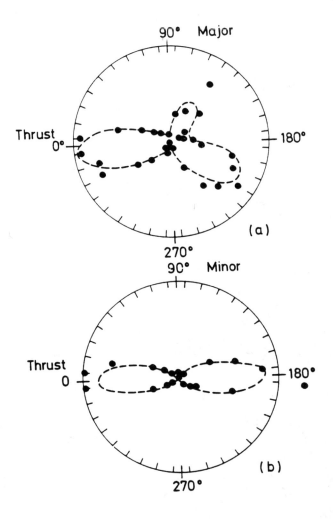

Fig. 32

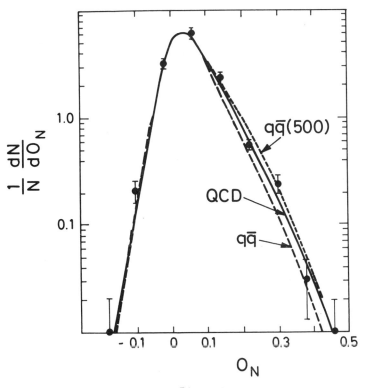

Fig. 33

502

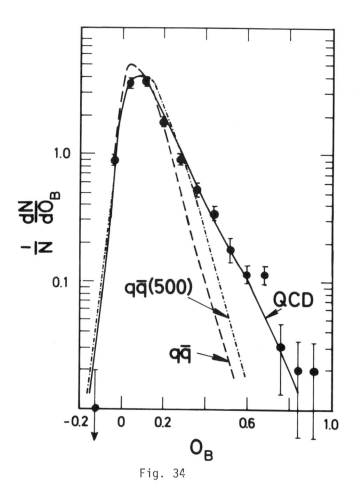

Fig. 34

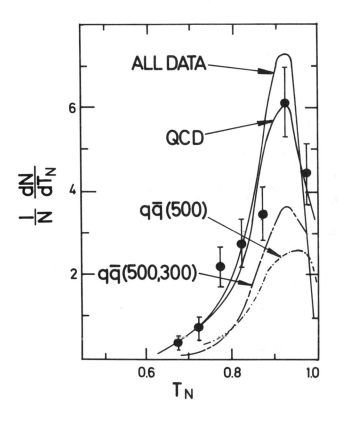

Fig. 35

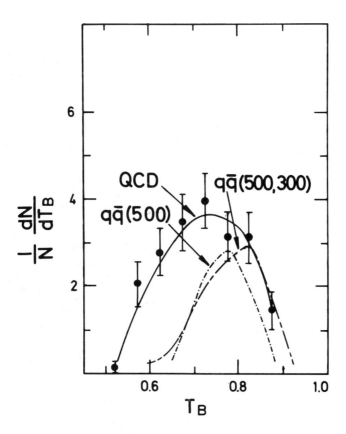

Fig. 36

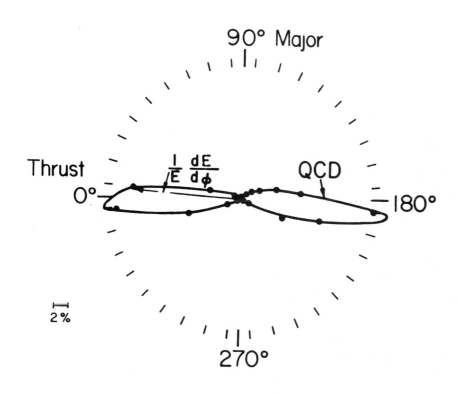

Fig. 37

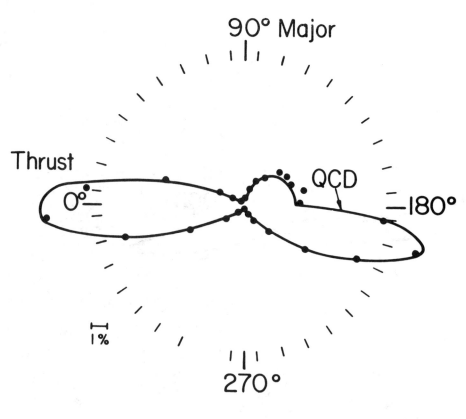

Fig. 38

507

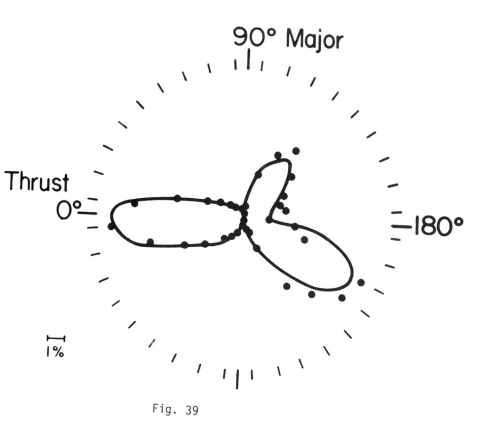

Fig. 39

508

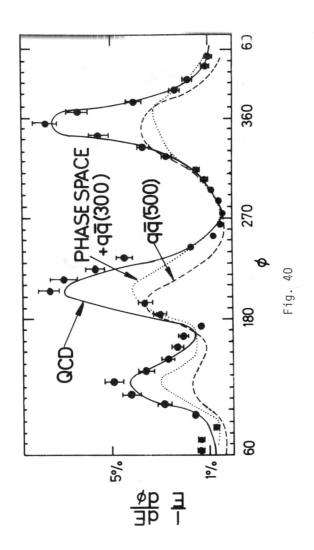

Fig. 40

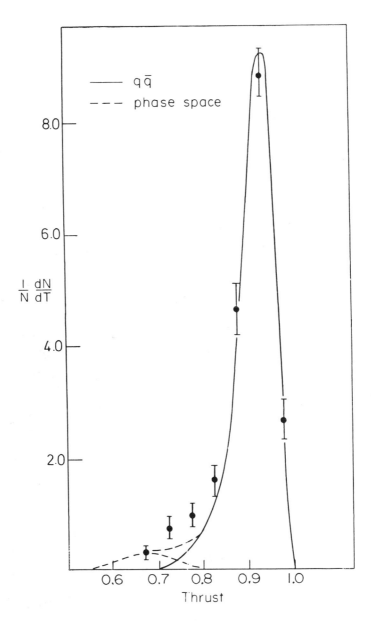

Fig. 41

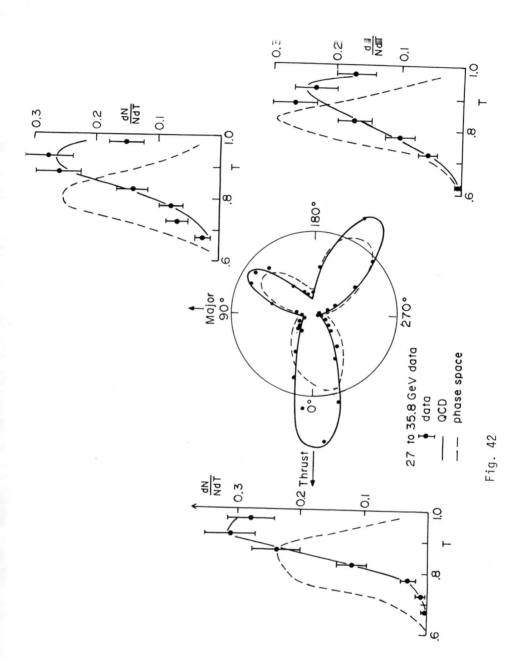

Fig. 42

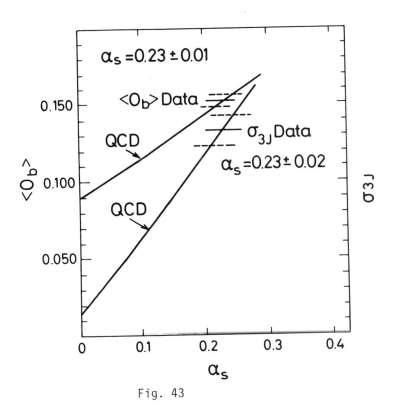

Fig. 43

RECENT RESULTS ON γγ AND e^+e^- ANNIHILATION FROM

THE PLUTO EXPERIMENT

Gerhard Knies

Deutsches Elektronen-Synchrotron, DESY, Hamburg

ABSTRACT

Results on $\gamma\gamma \to f^o$ (1270), σ^{tot} ($\gamma\gamma \to$ hadrons) and $\gamma\gamma \to$ large $p_\perp$ events are presented. The experimental evidence for color and spin of gluons, for first order gluon bremsstrahlung processes and jet evolution as in LLA summations is described.

514

1. INTRODUCTION

In this talk I want to report on results from e^+e^- experiments at high energies, done with the detector PLUTO at PETRA (mainly) and at DORIS. A few results from the TASSO experiment are also included.

The topics I want to address are:

$\gamma\gamma$ interactions,
$e^+e^- \rightarrow$ hadrons and tests of QCD.

The detector PLUTO is shown in Fig. 1.

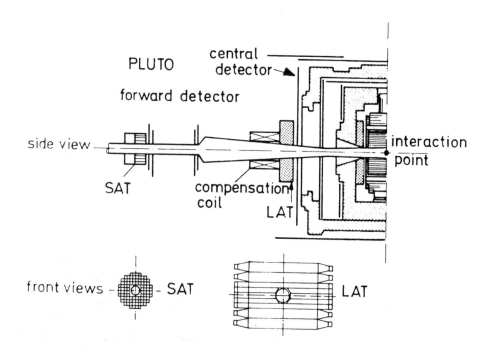

Fig. 1: Cross section of the detector PLUTO, along the beam line (upper part), and front views of the small (SAT) and large (LAT) angle taggers.

More details are given in previous publications of the PLUTO
collaboration, e.g. in Phys. Lett. 81B (1979) 410, Phys. Lett. 86B
(1979) 413 and Phys. Lett. 89B (1980) 120. The members of the PLUTO
Collaboration are:

Ch. Berger, H. Genzel, R. Grigull, W. Lackas, F. Raupach
I. Physikalisches Institut der RWTH Aachen, Germany.

A. Klovning, E. Lillestöl, E. Lillethun, J.A. Skard
University of Bergen, Norway

H. Ackermann, G. Alexander, F. Barreiro, J. Bürger, L. Criegee,
H.C. Dehne, R. Devenish, A. Eskreys, G. Flügge, G. Franke,
W. Gabriel, Ch. Gerke, G. Knies, E. Lehmann, H.D. Mertiens,
U. Michelsen, K.H. Pape, H.D. Reich, M. Scarr, B. Stella,
T.N. Ranga Swamy, U. Timm, W. Wagner, P. Waloschek, G.G. Winter,
W. Zimmermann
Deutsches Elektronen-Synchrotron DESY, Hamburg, Germany

O. Achterberg, V. Blobel, L. Boesten, V. Hepp, H. Kapitza,
B. Koppitz, B. Lewendel, W. Lührsen, R. van Staa, H. Spitzer
II. Institut für Experimentalphysik der Universität Hamburg,
Germany

C.Y. Chang, R.G. Glasser, R.G. Kellogg, K.H. Lau, R.O. Polvado,
B. Sechi-Zorn, A. Skuja, G. Welch, G.T. Zorn
University of Maryland, College Park, USA

A. Bäcker, S. Brandt, K. Derikum, A. Diekmann, C. Grupen,
H.J. Meyer, B. Neumann, M. Rost, G. Zech
Gesamthochschule Siegen, Germany

T.Azemoon, H.J. Daum, H. Meyer, O. Meyer, M. Rössler, D. Schmidt,
K. Wacker
Gesamthochschule Wuppertal, Germany

2. $\gamma\gamma$ INTERACTIONS

2.1 INTRODUCTION

In the framework of classical linear Maxwell equations two
light quanta do not interact. In QED, however, the photon has a
well defined transition probability into a virtual pair of -
pointlike - charged particles like electrons etc. A second photon
can then interact with this virtual pair, thus establishing
$\gamma\gamma$ interactions. Historically this was the first way $\gamma\gamma$ interactions were predicted [1]:

516

A second approach to γγ interactions, in particular for the process γγ → hadrons came from the discovery of a hadronlike behaviour of the photon in strong interactions, which lead to the vector meson dominance picture:

$$\alpha \ (\gamma\gamma \to \text{hadrons}) \sim \frac{1}{\alpha^2} \ \sigma \ (\rho\rho \to \text{hadrons})$$

In this talk I want to report on γγ interactions and present results on

1) $\sigma(\gamma\gamma \to X)$, $X = e^+e^-$, $\mu^+\mu^-$ and f^0 (1270)

2) σ^{tot} (γγ → hadrons)

3) indications for large $p_\perp$ jets

2.2 THE WIDTH FOR f^0 (1270) →γγ

For this purpose two photon production of lepton and hadron pairs γγ → ee, μμ, ππ has been studied via the reaction e^+e^- → e^+e^-X, where X is a pair of charged particles (see fig. 2a). These processes are of considerable interest, because lepton-pair production and the pointlike production of pion pairs can be calculated in QED.

Fig. 2: QED diagrams for particle pair production

a) by γγ interactions

b) by virtual bremsstrahlung

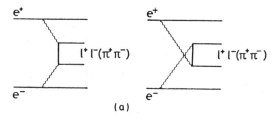

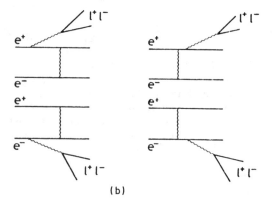

The production of C = +1 resonances, which decay into pion pairs, should show up as a deviation from the calculated QED two-prong cross section.

I will report on two-photon production of charged particle pairs [2] using the detectors PLUTO and TASSO at PETRA. The data reported in this paper have been taken at beam energies between 15 and 16 GeV ($<E_B> = 15.5$ GeV) for an integrated luminosity of 2600 nb^{-1}.

The produced pairs are identified using the central track detector of PLUTO or TASSO, respectively. For the sample where none of the scattered electrons is tagged we require two oppositely charged tracks subject to the following conditions:

a) $\Theta > 56^o$ (PLUTO), $\Theta > 35^o$ (TASSO) for each track

b) $p > 400$ MeV (PLUTO), >320 MeV (TASSO) for each track

c) acolinearity angle between the tracks $>15^o$ (PLUTO), $>10^o$ (TASSO)

d) coplanarity angle $<10^o$ (TASSO)

The polar angle Θ and the transverse momentum $p_\perp$ are measured with respect to the beam axis. Events with neutral energy not related to the tracks are rejected (PLUTO). The rather strong conditions (a) and (b) are used to ensure uniform efficiency of the central detector. Conditions (c) and (d) are needed for rejecting cosmic rays and Bhabha events.

In figs. 3a, b we plot for beam gas background subtracted no-tag events, the total energy in the central detector and the vector sum of the transverse momenta, $p_\perp sum = |\vec{p}_{1\perp} + \vec{p}_{2\perp}|$. The energy distribution peaks below 2 GeV and decreases steeply toward higher energies.

The $p_\perp^{sum}$ distribution demonstrates that the transverse momentum of the two tracks is compatible with zero. Both features strongly support the conclusion that these events originate from two-photon interactions: the energy distribution because of the bremsstrahlungsspectrum of the colliding photons, and the transverse momentum is balanced because most of the photons are radiated along the direction of the incoming beams.

518

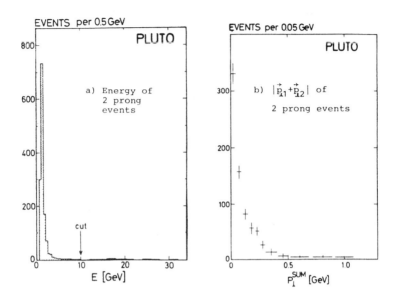

Fig. 3: Background subtracted distribution of the total energy in the central detector (a) and the vector sum of the transverse momenta $p^{SUM} = |\vec{p}_\perp^1 + \vec{p}_\perp^1|$ (b).

Fig. 4a shows the distribution of the invariant mass of the pairs, W, in the PLUTO detector. W is obtained by assigning pion masses to both tracks. The solid line is the result of the QED prediction [3] for the production of lepton pairs (ee, μμ). This prediction includes an exact calculation of the diagrams in fig. 2a plus their interference with bremsstrahlung terms (fig. 2b). Pion-pair production via the two-photon interaction is estimated [4,5] by taking the production cross section from ref. 4. The predicted rate is ∿ 15% of the lepton signal in fig. 4a.

Fig. 4b shows the 2π invariant mass distribution obtained by TASSO, after subtraction of the QED background.

The background from two photon initiated multi-hadron production with missing particles we have estimated to be smaller than 2% in the PLUTO data sample.

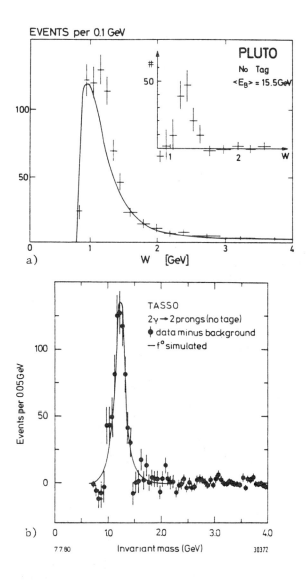

Fig. 4: Invariant 2 prong mass distribution.
The solid line in (a) shows the absolute QED
prediction for (eē, μμ̄) pairs.

The agreement of the QED calculation with the data in the bins below 1 GeV and above 1.5 GeV is very good, taking into account the systematic error of the data ($\sim 15\%$) and the fact that radiative corrections have not been included. We expect them to reduce the cross section by a few percent [6] and to have a smooth W dependence, in accordance with the good agreement of our data with the shape of the QED curve above 1.5 GeV.

For 1 GeV $<$ W $<$ 1.5 GeV there is a clear excess of the data above the QED prediction. The insert in fig. 4a shows the difference between the data and the QED prediction in the PLUTO data. It has a typical resonance behaviour. A well known candidate for such a resonance is the f^o (1270). Another possible candidate in the same mass region is the ε (1300). However, it is believed to have a larger width ($\Gamma_{tot} > 300$ MeV) [7]. We therefore reject this explanation. In principle one could distinguish the f^o from the ε hypothesis via the angular distribution of the decay pions. In practice this is not conclusive because of the limited angular acceptance and the large QED background. Also the TASSO experiment finds the same peak (fig. 4b), but their analysis has not yet arrived at a quantative result.

The width $\Gamma_{\gamma\gamma}$ was determined from the PLUTO data (fig. 4a) by fitting

$$a \cdot \sigma_{\gamma\gamma}^{QED} (W) + \sigma_{\gamma\gamma}^{f^o} (W)$$

to the mass distribution, after folding in the photon flux [4,5] and using a Breit-Wigner ansatz for the total cross section $\sigma_{\gamma\gamma}$ in the reaction $\gamma\gamma \to f^o \to \pi^+\pi^-$:

$$\sigma_{\gamma\gamma}(W^2) = 8\pi \ (2J + 1) \ \frac{\Gamma_{\gamma\gamma} \ \Gamma_{\pi^+\pi^-}}{(W^2 - M^2_{f^o})^2 + \Gamma^2 M^2 _{f^o}}$$

$\Gamma_{\pi^+\pi^-}$, Γ and M_{f^o} were taken from the standard-data compilation [7]. The f^o can be produced via three helicity amplitudes $|\lambda| = 2,1$ and 0 in photon-photon reactions. These amplitudes lead to different decay angular distributions in the pair center of mass system, and correspondingly to different acceptance corrections.

We have chosen $\lambda = 2$, because the dominance of this amplitude is predicted from widely differing theoretical approaches to the problem of radiative f^o decay [8].

The result of the PLUTO fit confirms the $\sigma_{\gamma\gamma}^{QED}$ calculation ($a = 0.95 \pm 0.07$) and yields the first determination of the radiative width of the f meson:

$$\Gamma(f \to \gamma\gamma) = 2.3 \pm 0.5 \text{ (stat.)} + 0.35 \text{ (syst.) KeV}$$

$$\text{at } <Q_\gamma^2> = .0065 \text{ GeV}^2$$

This result compares favorably with predictions (see table 1) from the non relativistic quark model with an oscillator potential 12, 13). FESR and tensor meson dominance based predictions yield larger $\Gamma_{\gamma\gamma}$ values. It should be noted that with the present value of Γ ($\omega \to \pi^0\gamma$) = 89 KeV the FESR prediction [16] changes [21] from $\Gamma_{\gamma\gamma} = 5.7$ KeV to $\Gamma_{\gamma\gamma} = 3.1$ KeV, in close agreement with the PLUTO result.

We have also looked in single tagged data for f^0 production. Fig. 4c shows the 2 prong invariant mass distribution, along with the QED prediction. The small enhancement in the f^0 mass region can be attributed to f^0 production via one almost real (as above) and one virtual ($Q^2 = 0.28$ GeV^2) photon. Due to the limited statistics we only give an upper limit of $\Gamma_{\gamma\gamma} < 2.6$ KeV (95% confidence level) again using the $\lambda = 2$ helicity hypothesis. For a ρ propagator form factor we expect a suppression by a factor of ≈ 2 here as compared to $Q^2 \simeq 0$.

Table 1   Theoretical predictions for $\Gamma(f^0 \to \gamma\gamma)$

Reference	$\Gamma_{\gamma\gamma}$ (keV)
8	8
9	28
10	0.8
11	> 1
12	1.2 - 2.3
13	2.6
14	5.07
15	7
16	5.7
17	8
18	9.2
19	11.3
20	21 ± 6
this experiment	2.3 ± 0.5

2.3 THE TOTAL CROSS SECTION FOR γγ→ HADRONS

The most obvious way of looking at total hadron production by colliding photons is the VMD picture. In this frame work, and using factorization one can infer a prediction [22] for total hadron production

$$\sigma_{\gamma\gamma}^{tot} = \frac{\sigma^2 (\gamma p)}{\sigma (pp)}$$

$$= \sigma^{diff} + \sigma^{Regge}$$

$$= 240 + \frac{270}{W} \quad (nb, GeV)$$

This parameterization includes only leading Regge trajectories. There may be also a W^{-2} term, from non leading trajectories, or from non VMD type contribution like direct photon coupling to pointlike quarks (fig. 5c).

Since we are using single tagged events here, at tagging angles from $1.2^o - 4^o$, one photon is considerably off-shell. At the two beam settings of 7.5 GeV and 15.5 GeV on the average we have $<Q^2> = 0.1$ GeV^2, and 0.25 GeV^2, respectively. The VMD picture of photons suggests a dependence of the observed cross section on Q^2 due to the ρ-form factor

$$\sigma_{\gamma\gamma}^{tot} (W, Q^2) = \sigma_{\gamma\gamma}^{tot} (W, Q^2 = 0) \cdot (\frac{M_\rho^2}{M_\rho^2 + Q^2})^2$$

The points of interest here are to see if the total cross section is as large as expected by the VMD picture, and how it varies with W and Q^2.

To measure this cross section we have used single tagged events with the following properties

a) ≥ 3 tracks in the central detector, or

b) ≥ 2 tracks plus a separate shower (PLUTO)

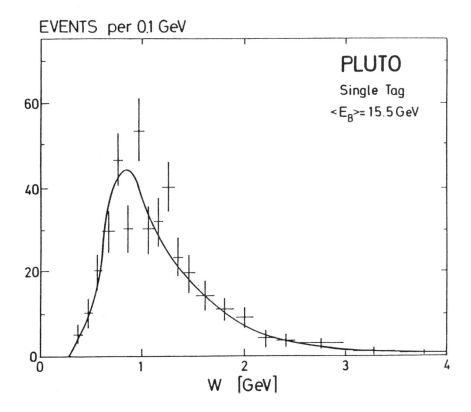

Fig. 4c: Invariant 2 prong mass distribution
of single tagged events. Curve is QED prediction
for (eē, μμ̄) pairs.

524

Fig. 5: Contributions
to the total cross section
discussed in text.

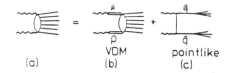

$\sigma_T + \epsilon\sigma_L$ (nb)

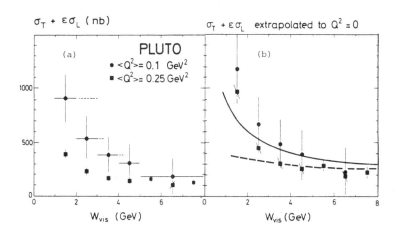

$\sigma_T + \epsilon\sigma_L$ extrapolated to $Q^2 = 0$

Fig. 6: Total cross
section $\gamma\gamma \to$ hadrons.
Corrected for acceptance.
σ as function of W_{vis} and
as measured (a) or extra-
polated to $Q^2 = 0$ (b).
σ as function of W from
fit described in text (c).
The solid curve in (b)
indicates the contribution
of the pointlike diagram
in Fig. 5c, evaluated
with m_q = 300 MeV.

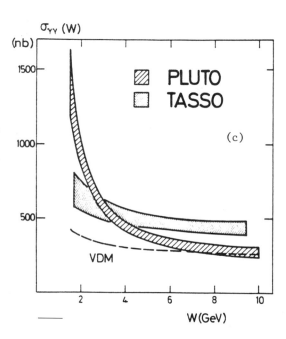

The event numbers have been corrected for acceptance losses
with simulated events which reproduce the observed distribution of
(i) the transverse momentum with respect to the beam, (ii) the
charged multiplicity and (iii) the energy observed in the central
detector. There is a further problem in arriving at the cross
section $\sigma_{\gamma\gamma}$ (W) from the fact that the luminosity of the colliding
photons is known as a function of $W_{\gamma\gamma}$, $L(W_{\gamma\gamma})$, but in the observed
events because of missing particles we only reconstruct $W_{vis} < W_{\gamma\gamma}$.
With simulated events we have studied the transition $W_{\gamma\gamma} \to W_{vis}$
leading to an effective luminosity L_{eff} as a function of W_{vis}.
This way we can calculate

$$\sigma_{\gamma\gamma} (W_{vis}) = \frac{\text{No. of events } (W_{vis})}{L_{eff} (W_{vis})}$$

This cross section can be compared to any prediction for
$\sigma_{\gamma\gamma}$ (W) by using $\sigma_{\gamma\gamma}$ (W) as starting point for a simulated experiment yielding $\sigma_{\gamma\gamma}$ (W_{vis}). Unfolding of $\sigma_{\gamma\gamma}$ (W) is not
possible at present statistics without "smoothing"
assumptions for $\sigma_{\gamma\gamma}$ (W).

In fig. 6a we show $\sigma_{\gamma\gamma}$ (W_{vis}, Q^2) for $<Q^2> = 0.1$ [23] and
$<Q^2> = 0.25$ GeV^2. The larger $<Q^2>$ cross section is systematically smaller. Assuming that the reaction $\gamma\gamma\to$ proceeds along
the VMD picture we can use the ρ form factor to do a model
dependent extrapolation of both cross sections to $Q^2 = 0$.

Fig. 6b shows this cross section. Both measurements now are
consistent. We also show the mentioned prediction from VMD and
factorization for $\sigma_{\gamma\gamma}$ (W) as $\sigma_{\gamma\gamma}$ (W_{vis}) (hatched curve) We
have made a fit to the data allowing for an additional term $\propto 1/W^2$.

$$\sigma_{\gamma\gamma} (W) = (.97 \pm .16) (240 + \frac{270}{W}) + \frac{2250 \pm 500}{W^2} \quad (nb, \text{ GeV})$$

The $1/W^2$ term can be due to non leading Regge trajectories, as
well as to a Born term type quark exchange diagram (fig. 5c).
Besides the statistical errors shown we estimate a 25% systematic
uncertainty mainly from acceptance corrections.

In fig. 6c $\sigma_{\gamma\gamma}$ (W) from the PLUTO fit, and from a TASSO fit

$$\sigma_{\gamma\gamma}(W, Q^2 = 0) = 380 + \frac{520}{W_{\gamma\gamma}} \quad (nb, \text{ GeV})$$

are shown, together with the VMD expectation.

In conclusion, the present measurement by PLUTO and TASSO indicate that there may be more contributions to $\sigma(\gamma\gamma\rightarrow$ hadrons) than expected from the VMD prediction, at $W \leq 3$ GeV.

2.4 INDICATIONS FOR LARGE $p_\perp$ PROCESSES

The Born term quark exchange diagram (fig. 5c) can lead to large $p_\perp$ events. At energies of W_{vis} >4 GeV it might actually lead to 2 coplanar quark jet events. We have found such events. Fig. 7 shows an example.

The magnitude of the rate for $\gamma\gamma \rightarrow q\bar{q}$ can be related [24] to $\gamma\gamma \rightarrow \mu\bar{\mu}$

$$R_{\gamma\gamma} = \frac{\sigma(\gamma\gamma\rightarrow q\bar{q})}{\sigma(\gamma\gamma\rightarrow\mu\bar{\mu})} = 3 \sum_q e_q^4 \left(1+0\left(\frac{\alpha_s}{\pi}\right)\right) \cong \frac{34}{27}$$

Here the quark charge e_q enters with the 4th power. We have searched for such two-jet events by

a) selecting events with > 4 tracks, and $4 < W_{vis} < 10$ GeV

b) maximizing a non collinear thrust (twoplicity) with ≥ 2 tracks per jet

c) requiring for the two jets

$$p_{t1}, \ p_{t2} > 1.25 \text{ GeV}$$

$$\theta_1 > 51^{\circ}, \quad \theta_2 > 36^{\circ}$$

relative to the e^+e^- beam axis.

In a preliminary analysis the following number of large $p_\perp$ jets and large $p_\perp$ lepton pairs with corresponding final state kinematic were found:

$q\bar{q}$ jets:	7
$\mu\bar{\mu}$ pairs:	4 (5.5 expected)
$e\bar{e}$ pairs:	5 (5.5 expected)

These numbers are consistent with the predicted $R_{\gamma\gamma}$.

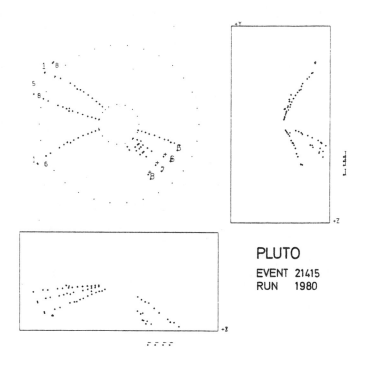

PLUTO

EVENT 21415
RUN   1980

Fig. 7:   Single tagged γγ 2 jet event.
Projections perpendicular and along the beam.
W_{vis} = 8 GeV.

Finally, we show in fig. 8 the single particle $p_\perp^2$ distribution,
from all events (tag or no tag) with 3 < W < 9 GeV.   A $p_\perp^2$ calculation for γγ → q$\bar{q}$, q = u, s, c, d with quark fragmentation is
shown. The cross section for this contribution in the Born
approximation is

$$\sigma_{\gamma\gamma} = \frac{4\pi\alpha}{W^2} \cdot \sum_{\substack{color \\ flavor}} e_{qi}^2 \cdot \ln \frac{W}{m_q^2}$$

m_q here is the quark mass.

528

Obviously the predicted reaction $\gamma\gamma \rightarrow q\bar{q}$ is consistent with the data. This process, however, is not the only source for large $p_\perp$ particles [24].

In conclusion, indications for the hard scattering process $\gamma\gamma \rightarrow q\bar{q}$ have been observed in terms of 2 jet events, and further implications like large p_t^2 particles and a $1/W^2$ term in the total cross section are consistent with the data.

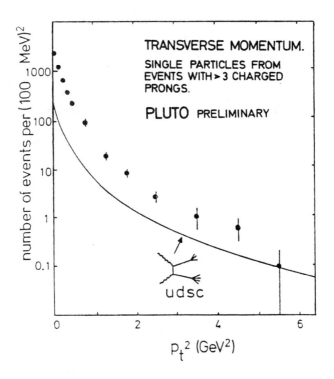

Fig. 8: Single track momentum square transverse to the beam, from events with $3 < W_{vis} < 9$ GeV, and more than 3 tracks. The curve is the expectation for Born term $q\bar{q}$ production.

3.   PROPERTIES OF GLUONS

Here I want to summarize which properties of gluons have been observed in the PLUTO experiment. At the Υ resonance and in the continuum at energies up to W = 32 GeV the data satisfying the requirements that

1)   gluons are colored

2)   gluons have spin 1

3)   gluons are radiated by quarks as bremsstrahlung

are observed.

3.1   THE COLOR CHARGE OF GLUONS

The Υ decay into hadrons provides evidence for the fact that gluons do not come as color singlet states. Fig. 9a compares the thrust distribution [25] of the non electromagnetic decays of Υ (9.46) into hadrons, to the one of hadron events, produced in the nearby continuum. The latter one, exhibiting the properties of quark pair production, is significantly different from the Υ distribution: the Υ does not decay into quark pairs. This demonstrates that gluons do not mediate a transition between 2 color singlet states (fig. 9b). The color octet nature of the gluons inhibits such a transition.

3.2   THE SPIN OF GLUONS

In fig. 9a we also compare the Υ thrust distribution to a phase space model (same multiplicity as Υ) and a 3 gluon QCD decay model with subsequent gluon fragmentation. The Υ thrust distribution is significantly different from phase space, and agrees with the 3 gluon decay model prediction. We consider this as evidence for the 3 gluon decay of the Υ. It is then meaningfull to compare the thrust axis angular distribution (relative to the e^+e^- beams) of the Υ decays (fig. 10) to the 3 gluon decay model expectations. These expectations depend on the spin assigned to the gluons. In fig. 10 we show the spin 1 distribution [26] which compares well with the data. Spin 0 gluons [27] would lead to a significantly different distribution: hence for spin 1 we find a non trivial agreement, and spin 0 gluons are ruled out.

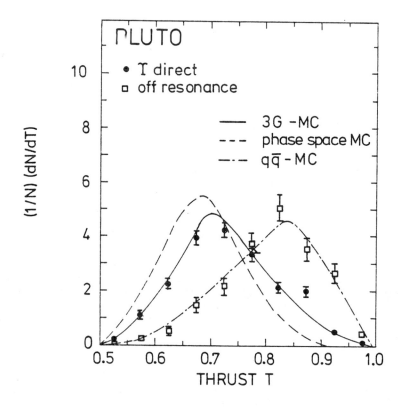

Fig. 9a: Thrust distribution for T non-electromagnetic decays, and for continuum e⁺e⁻ annihilation events. Expectations for 3 vector gluon decay, phase decay and q$\bar{q}$ jet model are shown.

Fig. 9b: T decay mode forbidden for colored gluons.

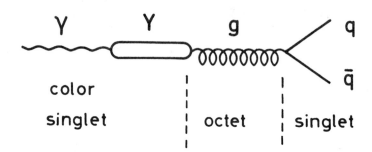

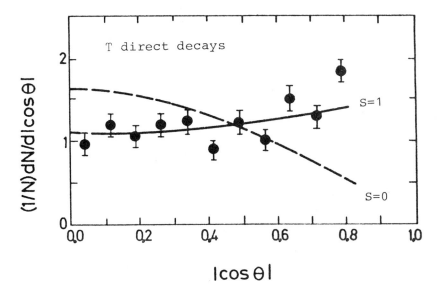

Fig. 10: Thrust axis angle (versus beam) distribution, of T non-electromagnetic decays. Spin 1 (solid line) and spin 0 (broken line) expectations are shown.

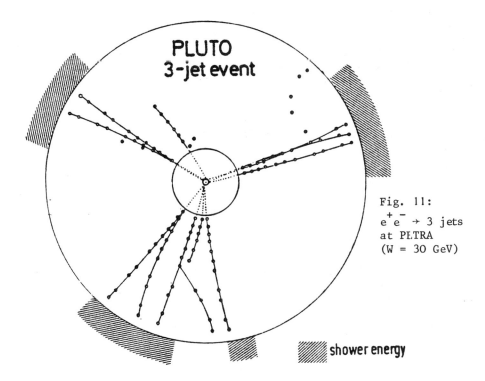

Fig. 11:
$e^+e^- \rightarrow 3$ jets
at PETRA
(W = 30 GeV)

There is a further discrepancy of the spin 0 gluon hypothesis with the data. For an T decay into 3 spin 0 gluons [27], with the same quarklike fragmentation as used for the spin 1 gluon calculations, we find a thrust distribution which is close to the one for qq events, and in clear disagreement with the experiment. This discrepancy, however, depends on the fragmentation assumption, whereas the thrust axis angular distribution provides a fragmentation independent test of the gluon spin.

3.3   EFFECTS FROM GLUON BREMSSTRAHLUNG

In the following I will summarize briefly the analysis of single hard gluon emission effects, and present a comparison of the data with the predictions from all order calculations in the leading logarithmic approximation, for parton jets.

3.31   HARD GLUON EFFECTS

The most direct demonstration of single hard gluon emission is the observation at high energies of 3-jet events: 2 quark jets and 1 hard gluon jet. Fig. 11 shows a nice example. But also the production of planar, asymmetric ("slim - fat") 2-jet events, with increasing cm energy, is evidence for the radiation of one hard gluon. In general hard gluon radiation does not always lead to the formation of a separated, third jet.

The asymmetric character of the events is demonstrated in the seagull plot [28], $<p_\perp^2>$ vs x_p, in fig. 12. It is made separately for the "slim" and the "fat" jet in each event. Comparing energies of 13 <W< 17 GeV and 27 <W < 32 GeV shows that at the higher energies the average $p_\perp^2$ cannot be understood in terms of "symmetric" $q\bar{q}$ 2-jet event production, as generated along the Field-Feynman jet model [29]. However with an admixture of "asymmetric" $q\bar{q}g$ events produced along the Hoyer et al model [30] (with α_s = 0.17) we get perfect agreement with the seagull plots for the slim and the fat jet.

The planar character of the 30 GeV events becomes obvious from a comparison of $<p_{in}>^2$ and $<p_{out}>^2$ at W = 12 - 13 GeV and W = 27 - 32 GeV in fig. 13. The squares of the average track momenta in a reconstructed event plane grow on the average by a factor 2.4, while those out of the plane grow by a factor 1.6 only. The increase of p, with respect to thrust, mainly spreads events into a plane. Again, the hard gluon emission model calculation describes this feature of the data quantitatively.

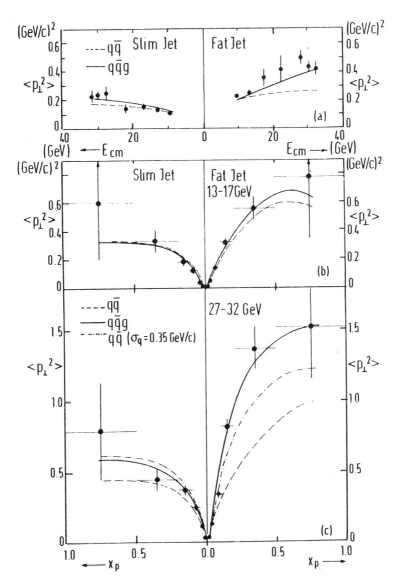

Fig. 12: $\langle p_\perp^2 \rangle$ vs E_{CM}, and seagull plots.
Curves for $q\bar{q}$ and $q\bar{q}g$ models.

534

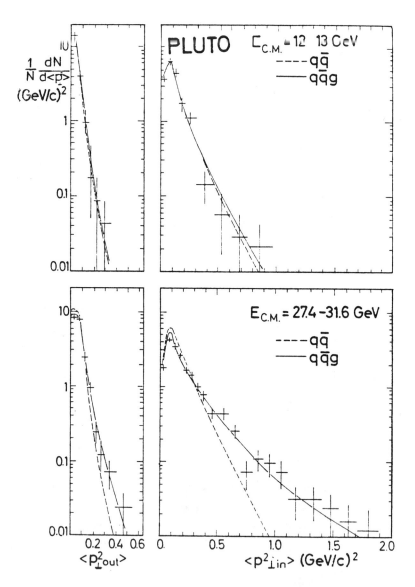

Fig. 13: Momentum components (squared) perpendicular to sphericity axis, in and out of planarity plane.

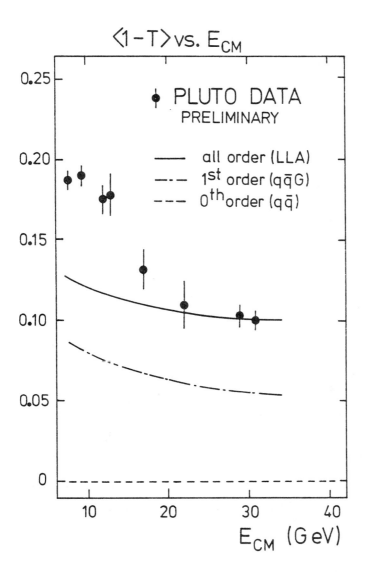

Fig. 14: Average of 1-thrust, versus energy. Curves for 0th, 1st and all order (LLA) QCD predictions (no fragmentation) are shown.

Using the momentum tensor [31] eigenvalues λ_i, or the normalized Q_i parameters

$$Q_i = 1 - \frac{?\lambda_\perp}{\lambda_1+\lambda_2+\lambda_3} = \frac{\sum n_\perp^2}{\sum p^2} \qquad i = 1,2,3$$

one can define cuts for planar events [32]. The cuts used here [28] are $Q_3 > 0.83$ and $Q_1 > 0.03$. The expected fraction of events within these cuts depends on the value of α_s in the $q\bar{q}g$ model calculations. Good agreement with the observed fractions of

$$3.9 \pm 1.3\% \quad \text{at} \quad W = 12, \ 13 \ \text{GeV}$$

$$5.7 \pm 0.8\% \quad \text{at} \quad W = 27 - 32 \ \text{GeV}$$

could be achieved with the $q\bar{q}g$ Monte Carlo simulation [3o], but not with $q\bar{q}$ jet events [29] only, even after variation of the $p_\perp$ width of the jets [28].

From these various hard gluon effects we get [33] a value for the strong coupling constant of

$$\alpha_s = .16 \pm .03 \pm .03$$

$$\text{statist., syst.}$$

3.32 COMPARING JET EVOLUTION WITH QCD PREDICTIONS

Instead of understanding events in terms of jets as the entities representing hard QCD quanta, and describing the properties of jets in some phenomenological models unrelated to QCD, one can also try to understand jets as the result of multiple gluon emission processes initiated by an energetic confined quark. There are QCD treatments for the development of such a quark-gluon cascade in the leading logarithmic approximation (LLA). They ignore however the final parton-hadron transition process which is considered to be soft. With growing jet energy these confinement processes are expected to have a decreasing effect on the overall jet evolution, so that jet properties as measured with hadrons can be compared to predictions for jets of partons.

The question is whether the present energies are high enough for this picture to be valid. For this purpose we show first in fig. 14 the average of $1 - T$ at energies of $9.4 < W < 32$ GeV. It is calculated here from charged particles only and corrected with simulated events [3o]. We compare it to 3 levels of pure QCD predictions. 0th order ($q\bar{q}$) trivially predicts $\langle 1 - T \rangle = 0$. First order [34] ($q\bar{q}g$) already gives a large improvement, but disagrees with the

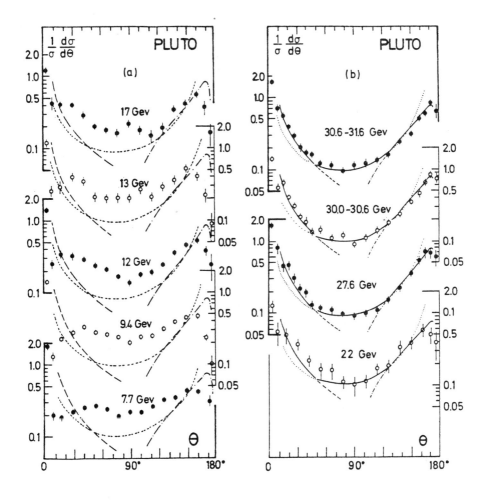

Fig. 15: 2-particle energy weighted angular correlations.
Curves see text. Solid sections indicate regions used for
α_s fit.

538

thrust observed after hadronization at all energies. The all order summation in the leading logarithmic approximation [35, 36] however merges with the experimental result, at W > 30 GeV. The LLA calculation [35] was done with a value of Λ = 500 MeV for the QCD scale parameter.

Second we show in fig. 15 the energy weighted two-particle angular correlation over the full angular range from $0°$ to $180°$ [37]. For each particle a of an event all 2-particle combinations with the other particles b (including b=a) are formed and at the angle θ (a,b) the weight

$$W_{ab} = \frac{E_a}{E_{Beam}} \cdot \frac{E_b}{E_{Beam}}$$

is plotted. If all final state particles are observed, we have

$$\sum_{a,b} W_{ab} = 1$$

This sum rule is needed for integration in deriving the theoretical prediction [38]. In case of missing particles or when using only charged particles as in these data this relation can be restored in the events by replacing 2 x E_{Beam} by the observed energy. The shapes of the experimental distributions are corrected for effects from particle losses using simulated events [30].

The theoretical predictions from LLA calculations are valid for forward correlations [39, 38] ($10° < \theta < 50°$) and for backward correlations [40] ($120° < \theta < 170°$). To achieve a coverage of the full angular distribution at the large angle correlations around $90°$ we used the results of a first order calculation [41]. The comparison of these calculations with the data shows:

1) At W ≥ 27.6 GeV, there is a remarkable agreement over the full angular range (fig. 15b). For all 3 angular sections a common Λ has been used in fitting the curves to the data. Table 2 shows the resulting χ^2 and α_s values at the various energies. The χ^2 are acceptable, and the α_s values are higher than from the hard gluon effects discussed before. With independent Λ values the backward region gives a smaller α_s (0.21 ± 0.01), together with a significantly better χ^2. Using $\alpha_s(p_\perp^2)$ in the fit rather than $\alpha_s(Q^2)$, as suggested in other papers [38,42,43], then $\alpha_s(900\ GeV^2)$ comes down to $\alpha_s \approx 0.15$.

Table 2: Values of the strong
coupling constant α_s from
2 particle correlations

	COMMON	Λ
Energy (GeV)	α_s	χ^2/ND
27.6	0.25±0.01	18/ 26
30.0–30.6	0.27±0.01	34/ 26
30.6–31.6	0.26±0.01	38/ 26
All together α_s for Q = 30	0.26±0.01	140/107

2) At W < 17 GeV the LLA
predicts much stronger
forward-backward peaks
than observed. This
corresponds to a more pronounced jet character in
the LLA treatment than observed at these energies.
Λ has not been fitted
here. A value of Λ =
500 MeV was used for
calculating the curves.
A too strong jettiness
was also predicted in the
LLA calculation for
<1 – T> in fig. 14, at
the lower energy points.

In the backward direction, the position of the peak of the
opposite side correlation is reasonably well described. In contrast
the independent 2-jet fragmentation model [29,30], without multiple
soft gluon emission, fails to reproduce the observed opposite side
acollinearity distribution [44] at very backward angles. In fig. 16
we compare the corresponding (uncorrected) acollinearity distributions from simulated jet events [30] to the data. The simulated
events predict a much stronger back-to-back correlation effect. The
same conclusions have been found in independent calculations by
Kajantie and Pietarinen [45].

In conclusion, the LLA predictions for the properties of
e^+e^- annihilation events in terms of parton jets, as expressed by
the average thrust and by the 2-particle energy correlation merge
with the gross features of the data at energies of W > 27 GeV. At
the same time, the broadening of the opposite side acollinearity
distribution, which is expected as a consequence of multiple soft
gluon emission is unaccounted for in q$\bar{q}$ phenomenological jet models.
For determining a consistend value of α_s from all three angular
regions an improved understanding of the details of the theoretical
treatment is obviously needed.

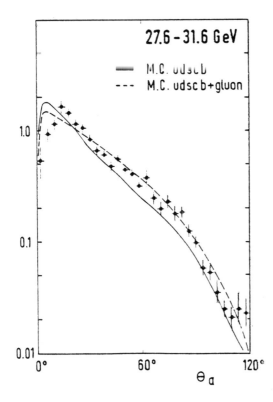

27.6 – 31.6 GeV

M.C. udscb
M.C. udscb+gluon

1.0

0.1

0.01

0°   60°   120°

Θ_a

Fig. 16. Opposite side ('acollinearity') correlation (observed, as from charged particles). Curves are from simulated events from $q\bar{q}$ and $q\bar{q}g$ models.

I greatly appreciate the help of numerous colleagues at DESY in preparing this talk, and the interesting talks and conversations at the EMS '80.

REFERENCES

1. Euler and Kockel, Naturwissenschaften 23 (1935) 246.
2. PLUTO collaboration, Ch. Berger et al., Phys. Lett. 94B (1980) 254, TASSO collaboration, E. Hilger, DESY report 80/75.
3. J.A.M. Vermaseren, private communication and R. Bhattacharya, J. Smith, G. Grammer, Phys. Rev. D15 (1977) 3267.
4. S.J. Brodsky, T. Kinoshita, H. Terazawa, Phys. Rev. D4 (1971) 1532.
5. C. Carimalo, P. Kessler, J. Parisi, Phys. Rev. D21 (1980) 669, N.M. Budnev, I.F. Ginzburg, Phys. Rep. C15 (1974) 181. We used the formula given in ref. 4 which in our range differs by less than 5%.
6. De Fries, private communication and discussion session at the International Workshop on $\gamma\gamma$ Interactions, Amiens (1980).
7. Particle data group C. Bricman et al., Phys. Lett. 75B (1978).
8. J. Babcock, J.L. Rosner, Phys. Rev. D14 (1976) 1286.
9. G.M. Radutzkij, Soviet Journal of Nuclear Physics 8 (1969) 65.
10. A. Bramon, M. Greco, Lett. Nuovo Cimento 2 (1971) 522.
11. D. Faiman, H.J. Lipkin, H.R. Rubinstein, Phys. Lett. 59B (1979) 269.
12. S.B. Berger, B.T. Feld, Phys. Rev. D8 (1973) 3875.
13. V.M. Budnev, A.E. Kaloshin, Phys. Lett. 86B (1979) 351.
14. N. Levy, P. Singer, S. Toaff, Phys. Rev. D13 (1976) 2662.
15. H. Kleinert, L.P. Staunton, P.H. Weisz, Nucl. Phys. B38 (1972) 87.
16. B. Schremp-Otto, F. Schrempp, T.F. Walsh, Phys. Lett. B36 (1971) 463.
17. B. Renner, Nucl. Phys. B30 (1971) 634.
18. M. Greco, Y. Srivastava, Nuovo Cimento 43A (1978) 88.
19. G. Schierholz, K. Sundermeyer, Nucl. Phys. B40 (1972) 125.
20. V.N. Novikow, S.I. Eidelman, Soviet Journal of Nuclear Physics 21 (1975) 529.
21. F. Schrempp, private communication.
22. J.L. Rosner, Resonance physics at very high energies, in Isabelle Physics Prospects, ed. by R.B. Palmer, Brookhaven National Laboratory Report BNL 17522, Upton, N.Y., (1972) p.316-350. T. Walsh, J. Physique C2 Suppl. 3 (1974).
23. PLUTO collaboration Ch. Berger et al., Phys. Lett. 89B (1980) 120.
24. S.J. Brodsky, T. De Grand, J. Gunion, J. Weis, Phys. Rev. D19 (1979) 1418.
25. PLUTO collaboration, presented by S. Brandt at the 1979 EPS Conference, DESY Report 79/42 (1979)

542

26. K. Koller, H. Krasemann and T.F. Walsh, Z. Physik C1 (1979) 71.
27. K. Koller and H. Krasemann, Phys. Lett. 88B (1979) 119.
28. PLUTO collaboration, Ch. Berger et al., Phys. Lett. 86B (1979) 418.
29. R. Field and R.P. Feynman, Nucl. Phys. B136 (1978) 1.
30. P. Hoyer, P. Osland, H.G. Sander, T.F. Walsh and P.M. Zerwas, Nucl. Phys. B161 (1979) 349.
31. J.D. Bjorken and S.J. Brodsky, Phys. Rev. D1 (1970) 1416.
32. G. Knies, Proc. XIVth Rencontre de Moriond (1979), vol. 2, p. 99; edited by Tran Than Van.
33. PLUTO collaboration, forthcoming paper.
34. A. de Rújula, J. Ellis, M.K. Gaillard and E. Floratos, Nucl. Phys. B138 (1978) 387.
35. G. Schierholz, DESY report 79/71 (1979).
36. P. Binétruy, preprint CERN/TH/2807.
37. PLUTO collaboration, forthcoming paper.
38. Yu. L. Dokshitzer, D.T.D'yakonov, Fizika Elementarnykh Chastits, (Leningrad 79) translated as DESY-L-Trans 234 (79), Yu. L.Dokshitzer, D.T. D'yakonow and S.T.Troyan, Phys. Rep. 58C (80) 269.
39. K. Konishi, A. Ukawa and G. Veneziano, Phys. Lett. 80B (79) 259, and Nucl. Phys. 157B (79) 45.
40. Yu. L. Dokshitzer, D.T. D'yakonov and S.T. Troyan, Phys. Lett. 78B (1978) 290.
41. C.L. Basham, L.S. Brown, S.D. Ellis and T.S. Love, Phys. Rev. Lett. 41 (1978) 1585.
42. G. Parisi, R. Petronzio, Nucl. Phys. B154 (1979) 427.
43. W. Marquardt and F. Steiner, DESY 80/24 (unpublished).
44. PLUTO collaboration, Ch. Berger et al., Phys. Lett. 90B (1980) 312.
45. K. Kajantie, E. Pietarinen, DESY 80/19 (unpublished).

AN OVERVIEW OF QCD IN RELATION TO E$^+$E$^-$ ANNIHILATION, AND
BRIEF VIEW OF SINGLE PARTICLE SPECTRA

Harvey L.Lynch

Deutsches Elektronen-Synchrotron DESY, Notkestrasse 85,
2000 Hamburg 52, Germany

ABSTRACT
A review of the relation of e$^+$e$^-$ annihilation experiments to
QCD is given, discussing evidence for gluon bremsstrahlung and
measurements of the coupling constant.
Single particle spectra with identified $\pi^\pm$, $K^\pm$, p, $\bar{p}$, and K_s
are presented.

I have a couple of topics I wish to cover today. The first is
an overview of our current knowledge of QCD, especially in light
of recent e$^+$e$^-$ annihilation experiments. The second is a brief look
at recent TASSO results on single particle inclusive spectra with
particle identification. Since this talk only lasts 12 minutes I'll
ask your understanding for being rather brief. As a member of the
TASSO Collaboration[1], extensive use of their results will be made.
When not otherwise cited results are from this source.

Let us begin with a user's guide to QCD covering what we have
learned in the past few years and the conclusions we can draw from
the various pieces of information. This guide is given in Table 1
where various pieces of experimental evidence are grouped under a
category, and the conclusion from this category is given.

This is a meson conference, and I am not competent to discuss
the details of a large body of work in lepton scattering. I do,
however, wish to recognize the fact that the observation of scaling
violation in lepton scattering[2] and the production of high p_T
lepton pairs[3] historically are early pieces of experimental information pointing towards QCD.

Let us now turn to e$^+$e$^-$ annihilation and treat several related
subtopics. The first is the subject of jet broadening. By this we
mean that as the center of mass energy is increased the jets produced become broader. The conclusion from such an observation is
that one is not dealing with a simple q$\bar{q}$ pair production model. The
first, and most dramatic, evidence for jet broadening is the production and decay of the Υ[4]. This has been discussed by other speakers, so I shall not dwell further on the topic. At higher energies we have the W (center of mass energy) dependence of the momentum component perpendicular to the jet direction. Fig. 1 shows
the mean value <$p_\perp^2$> of the square of single particle momenta with
respect to the jet direction as a function of W. There is a clear
increase with W. A similar conclusion[5] can be drawn from mean values of sphericity or thrust vs W; in this case these values fall
more slowly with W than a simple q$\bar{q}$ model would predict; see
Fig. 2. A simple q$\bar{q}$ model predicts the mean sphericity to fall

Table I — QCD Milestones

Scaling violations in lepton scattering

High p_T lepton pairs

Jet broadening → NOT (simple $q\bar{q}$)

   Υ decay
   $<S>$, $<T>$ vs W
   $d\sigma/dp_T^2$ for different W
   "Seagull" plot

Energy Flow → NOT (simple $q\bar{q}$ OR phase space)

Planar events → 3-body kinematics

   $<p_T^2>_{in}$, $<p_T^2>_{out}$ vs W
   Fox-Wolfram moments
   A-S , $T-T_3$ correlations
   Oblateness

Three jet events → Gluon bremsstrahlung

   Charged particles only
   p_T^2 with respect to 3 jets
   Charged + neutral particles

Measurements of coupling constant

   ψ decay
   Lepton scattering
   W = 30 GeV "model dependent"
   W = 30 GeV "model independent"

Gluon spin (favor J = 1)

   Lepton scattering
   Υ decay

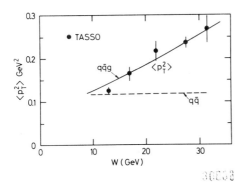

Fig. 1

$\langle p_T^2 \rangle$ vs. center of mass energy. The solid line shows the QCD prediction and the dotted line for simple $q\bar{q}$.

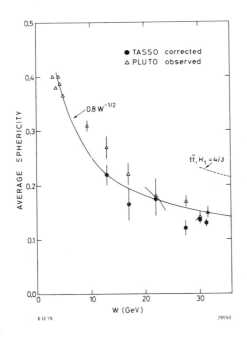

Fig. 2

Mean sphericity vs. center of mass energy. The solid line is a pheno-minological fit to the data.

roughly like 1/W. An important question is "How does this increase come about ?" Figure 3 shows the differential p_T^2 distribution[6] for two different W groupings of data. The interesting thing is that at low values of p_T^2 the two different data samples are quite similar, but at higher p_T^2 values there is a tail which develops in the higher energy data. While the low energy data may be reasonably well represented by a Field-Feynman type model having a characteris-

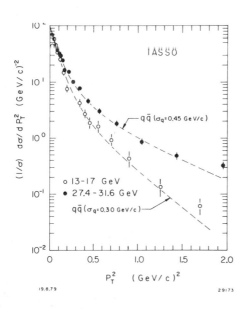

Fig. 3

$(1/\sigma)\ d\sigma/dp_T^2$ at low and high energies. The dotted lines show the prediction of a simple $q\bar{q}$ model for different values of the characteristic transverse momentum σ_q.

tic transverse momentum σ_q = 300 MeV/c, the higher energy data require σ_q = 450 MeV/c to obtain a reasonable representation of the data over the region shown. At higher p_T^2 values even σ_q = 450 MeV/c is too small to well describe the data. Another quantity, the so-called "seagull plot" gives similar information. This is a plot of the mean value of p_T^2 as a function of $z = p_{hadron}\ /\ p_{beam}$. In addition one divides the event into two halves and makes the plot for the broader jet and the narrower jet separately. In this way it could be possible to see jets preferentially broader on just one side if gluon bremsstrahlung is responsible for the broadening. Such a plot[6] is shown in Fig. 4. By virtue of the selection process the two halves are different. This is shown by the superposed lines of a simple $q\bar{q}$ model with σ_q = 300 and 450 MeV/c. Unfortunately within the available statistics the data cannot exclude a simple change of σ_q; this plot is therefore demoted to the category of jet broadening.

The next item in the user's guide is energy flow. This was first used in T decay[7], and later it has been used for e^+e^- annihilations at higher energies[8]. This was discussed by another speaker. The conclusion of this analysis is that one excludes simple (non broadened) $q\bar{q}$ or phase space models.

We continue next to "planar events". These are particularly important because the existence of events in a plane suggests three-body kinematics. This is what one wants to see if gluon bremsstrahlung is important. There are various pieces of information sug-

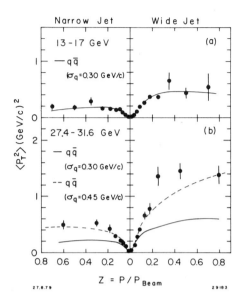

Fig. 4

"Seagull Plot" :

$\langle p_T^2 \rangle$ vs. z = p/p_{beam}
for particles of narrow
and wide jets. (a) and
(b) show TASSO data with
expected behavor of simple
$q\bar{q}$ models. (c) and (d)
show similar data from
PLUTO where the QCD pre-
diction is also shown.

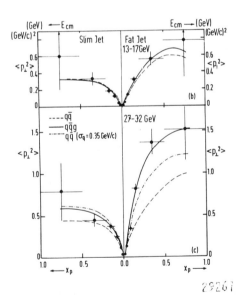

gesting planar event structure. One simple one is the distribution
of momentum transverse to the jet direction (a) in and (b) out
of the event plane. If the gluon bremsstrahlung picture makes sense,
one expects $\langle p_T^2 \rangle_{in}$ to broaden with W, while $\langle p_T^2 \rangle_{out}$ should not

548

grow with W. Figure 5 shows the data[6], where the event plane is defined by the eigenvectors of the momentum tensor having the 2 largest eigenvalues. The low energy data are well represented by the

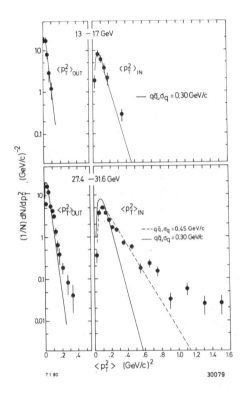

Fig. 5

Mean squared momentum with
respect to jet axis in
and out of the event plane.
Also shown are the ex-
pectations from simple $q\bar{q}$
models.

simple $q\bar{q}$ model. Note that the selection criteria naturally cause
$\langle p_T^2\rangle_{in}$ to be broader than $\langle p_T^2\rangle_{out}$, but this bias is completely explained at low energies by a simple $q\bar{q}$ model. The higher energy
data, however, cannot be adequately represented this way even if
σ_q = 450 MeV/c is used. A large excess of $\langle p_T^2\rangle_{in}$ events remain.
Therefore even a broadened but cylindrically symmetric $q\bar{q}$
cannot describe the data. Similar conclusions can be reached using
other quantities, such as Fox-Wolfram moments. This method has the
advantage of giving information on event shape without explicitly
computing jet directions or plane orientation. Another useful quantity to study planar structure is a correlation plot[9] between sphericity and aplanarity, or similarly thrust and triplicity. Still
another measure of planar structure is oblateness[8]. For the Spheri-
city-Aplanarity plot one uses the normalized eigenvalues of the mo-
mentum tensor, Q_i, where $Q_i = \Lambda_i / (\Lambda_1 + \Lambda_2 + \Lambda_3)$, and Λ_i are the

eigenvalues. The eigenvalues have simple physical interpretation: Λ_1 is the sum of squares of momenta perpendicular to the jet axis, and perpendicular to the event plane; Λ_2 is the sum of the squares of momenta perpendicular to the jet axis and in the event plane; and Λ_3 is the sum of the squares of momenta parallel to the jet axis. In terms of the Q's, the sphericity $S = (3/2)(Q_1 + Q_2)$ and the aplanarity $A = (3/2) Q_1$. Data at low and high energies from TASSO are shown in Fig. 6. Quantitatively there are 125 events having $A < 0.05$ and $0.2 \leq B < 1$ (B is an axis perpendicular to A) observed in the high energy data. If each of the particles in an event is rotated at random around the jet axis to destroy any natural planar structure and then analyzed, only 35 events remain the (A, B) region. Thus there is strong evidence that the events tend to be planar.

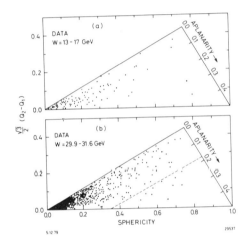

Fig. 6

Sphericity - Aplanarity triangle plot for data at low and high energies.

Clearly, what one would really like to have is what John Ellis described as the "smoking jet of gluon bremsstrahlung"[10]. That means that one wants to see nice clean three-jet events. Fig. 7 shows typical 2 and 3-jet events using charged particles only. While this 3-jet event is particularly pretty visually, it is in fact typical of three-jet events in that such events have about 4 tracks per jet on the average. If such three-jet events are to be due to gluon bremsstrahlung, then the distribution of momenta transverse to each of the three jet axies should be about the same as the p_T^2 distribution to two-jet events analyzed with a single jet direction. That this is true can be seen[11] in Fig. 8 where $(1/\sigma) \, d\sigma/dp_T^2$ is plotted for 3-jet events at $W \sim 30$ and compared to the $W = 12$ GeV data, which are primarily 2-jet events.

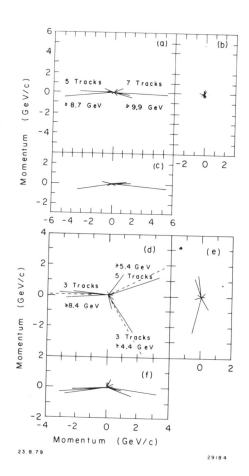

Fig. 7

Events in momentum space. (a,d) in the plane of the event; (b,e) viewed along the jet axis; (c,f) viewing event edgewise. Parts (a-c) show typical 2-jet structure, and (d-f) show 3-jet structure.

The two distributions are clearly very similar. Both the PLUTO[12] and JADE[13] Collaborations have found 3-jet events using both charged and neutral particles. Figure 9 shows three such events from JADE[13]. This is, once again, a display of particles in momentum space in the event plane. Note in particular the outer periphery of the events, which shows a histogram of the energy flow for each event.

Now let us turn to measurement of the strong coupling constant α_s. Unfortunately we are isolated from the physics we are trying to study by the hadronization and all its associated misery. This fact complicates life rather substantially. So in order to find α_s must to through some Monte Carlo type calculations[14,15] to represent the world one can observe. These models have several parameters in addition to the interesting quantity α_s. Most importantly there is a parameter σ_q, which describes the transverse momen-

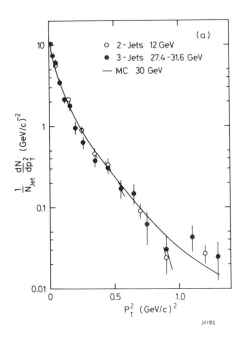

30185

Fig. 8

$(1/\sigma)$ $d\sigma/dp_T^2$ for 12 GeV data analyzed as 2-jet events and for 30 GeV data analzed as 3-jet events. The line shows the prediction of QCD.

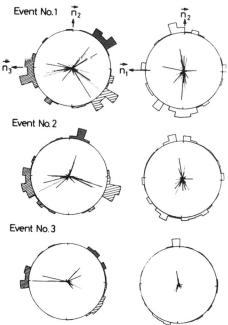

30209

Fig. 9

Events having both charged and neutral particles observed displayed in the event plane and viewed along the jet axis. The outer periphery shows the energy flow.

tum distribution, and which is inextricably coupled to α_s in nearly all observables one can think of. In a Field-Feynman type model[16] there is a quantity "a" which parametorizes the fragmentation itself. This is an important parameter but not very interesting to the question we wish to address. Similarly the relative probability of producing pseudoscalar mesons vs. vector mesons is important, but not very interesting to the question. Lastly there are some totally uninteresting parameters, such as the transverse momentum distribution for quarks from gluons and a thrust cut off to avoid infrared divergences; we have checked that no reasonable variation of these parameters affects the determination of α_s significantly. The real problem is that in general the first 4 "important" parameters are coupled. This means that one cannot simply use the values of "σ_q" and "a" from Field and Feynman and try to fit σ_s, because these people used $\alpha_s = 0$ to determine these parameters; therefore one would be inconsistent. There are various methods which have been used[17,18] to determine α_s. Generally they count events in the tail of some distribution; e.g. sphericity, thrust, planarity, oblatenes, and $\langle p_T^2 \rangle_{in}$ - $\langle p_T^2 \rangle_{out}$ distributions, use estimates of the other parameters and then fit α_s. The fundamental question which one must answer is, "What is the effect of the ignorance of all these other parameters on the final value of α_s?" The ideal solution to the problem is to repeat all the fits of Field and Feynman but including QCD effects. Then one gets a value for α_s, among other things. The difficulty is that this is an extremely ambitious, and I might add, hard program. It is harder than it looks because the model is not quite good enough to handle all the subtleties of the observed data; there are some inconsistencies which must be carefully resolved in order to obtain a believable result. There is an easier method which the TASSO people have used[19]; that is to emulate the methods of Edison[20] and find a physical observable which is nearly independent of the "uninteresting" parameters; and still depends critically upon α_s. The result of this research is to choose the "triangle plot" of sphericity vs. aplanarity and fit the population density for S > 0.25. One obtains a value of $\alpha_s \sim 0.16$ for any reasonable choices of the parameters a and P/(P/V). Figure 10 shows the values of α_s obtained from 25 separate fits to α_s and σ_q with fixed values of these other parameters. Clearly α_s is quite stable; in fact the r.m.s. scatter of all the results is only 0.01 while the statistical error is 0.04. Another way of saying this is that the χ^2 at the minimum in α_s - σ_q space ranges from 1.2 to 5.4 for 3 degrees of freedom over the entire range of a and P/(P/V). Therefore we have succeeded in decoupling the "uninteresting" parameters. It suffers from one problem, namely that the statistical error of 0.04 is rather large.

One can try to reduce the statistical error by using 2-jet events to determine the other parameters, in particular σ_q. To do this events having S < 0.2 were used, and fits were made to the mean charged multiplicity, $\langle p_T^2 \rangle_{out}$, and s d$\sigma$/dx to obtain estimates σ_q = 0.32 $\pm$ 0.04 GeV/c, a = 0.57 $\pm$ 0.20, and P/(P/V) = 0.56 $\pm$ 0.15.

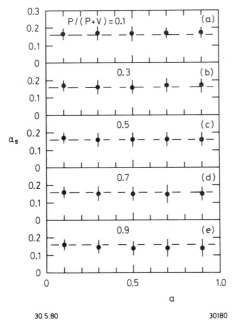

Fig. 10

Fitted values of α_s vs. a for different values of P/(P/V). The dotted lines show an average value of 0.16. The errors shown are statistical.

30.5.80   30180

It should be emphasized that these represent a self consistent set of parameters, but they should not be taken too seriously as absolute physical quantities. Using this value of σ_q one can return to the S > 0.25 data and obtain α_s = 0.17 $\pm$ 0.02.[9] We have by this method cut the statistical error in half but at the cost of heavier reliance upon the model. We estimate that the systematic un-certainty within the framework of the model is $\pm$ 0.03. The TASSO re-sults are in reasonable agreement with the other PETRA experiments: MARK-J[17], α_s = 0.23 $\pm$ 0.02 $\pm$ 0.04; JADE[13], α_s = 0.17 $\pm$ 0.04; and PLUTO[18], α_s = 0.17 $\pm$ 0.06.

All of this analysis was done using the Ali et al.[15] represen-tation of QCD. This model is an extension of work by Hoyer et al.[14] whereby higher order gluon terms have been computed by Ali et al. The analysis has been repeated using the Hoyer et al. model; the result is that α_s increases from 0.17 to 0.19. The larger va-lue of α_s is to be expected in the Hoyer et al. model because only lowest order gluon terms were computed.

How well does QCD represent the e^+e^- annihilation data ? To answer this question we have chosen 19 different distributuions, among them are multiplicity, thrust, sphericity, aplanarity, $<p_T^2>_{in}$, $<p_T^2>_{out}$, s dσ/dx, seagull plot, and the triangle plot, and compared them with Monte Carlo calculations using the fitted parameters. Representative plots are shown in Fig. 11. The result in a χ^2 of about 500 for 350 degrees of freedom. In terms of a χ^2-probability this is very poor, but if one allows that the mo-del may not be perfect this means an average of only 1.2σ per point instead of the expected 1.0. For a 4-parameter theory this is impressive agreement indeed !

554

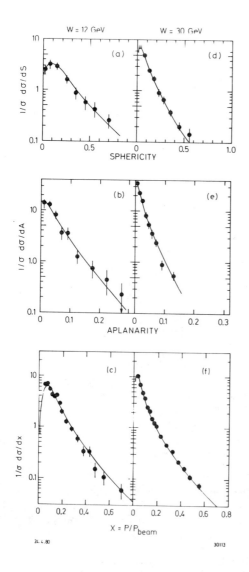

Fig. 11

Comparison of measurement with QCD predictions using 12 and 30 GeV data. (a,d) phericity, (b,e) aplanarity, (c,f) Bjorken scaling.

Let me briefly turn to another topic, viz. single particle inclusive spectra with particle identification. TASSO has two separate time of flight systems. The first is at a distance of 130 cm from the interaction point and covers about 80% of 4π in solid angle; it has an r.m.s. time resolution of 0.4 ns. The second system is about 550 cm from the interaction point and covers about 20% of 4π; it has an r.m.s. time resolution of 0.5 ns. Using time of flight one can separate pions from kaons for momenta up to 1.1 GeV/c and π/K from protons up to 2.2 GeV/c. Figures 12 a

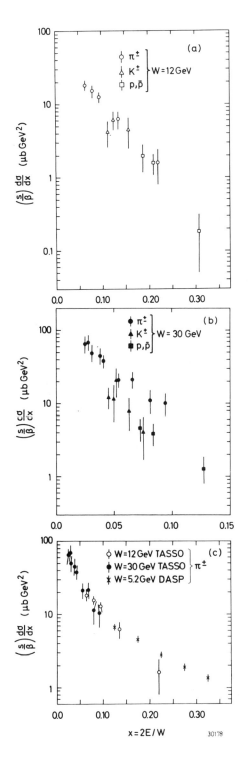

Fig. 12
Bjorken scaling cross sections
as a function of x for
different particle types.

and b[21] show the Bjorken scaling cross sections as a function
of x, where x = F_{hadron} / F_{beam} (i.e. with mass included). These
yields are relatively similar at a given x
regardless of the particle type, for both the low and high energy
data. Figure 12c compares data on pions with the lower energy
DASP data[22]; within the region of overlap the agreement is good.
An interesting result is that in the 30 GeV data (39 ± 11)% of
the events contain a p, p̄, or n, n̄, assuming the neutron yield
is the same as the proton yield. This is noteworthy to model buil-
ders, since existing fragmentation models in current use produce no
nucleons at all !

Finally K^0's have been identified using the channel
$K_s \to \pi^+ \pi^-$. This involves finding a charge zero vertex separated
from the event vertex and pointing back to the event vertex. Making
an invariant mass cut gives a sample having very little background.
This method can identify K^0's over a much wider range of momenta
than time of flight could identify $K^\pm$. The Bjorken scaling cross
section for kaons is shown in Fig. 13a[23]. The agreement with the
low energy data[24] is good. The ratio of the inclusive K^0 cross
section to the μ-pair cross section is shown in Fig. 13b as a
function of center of mass energy. The rapid rise in this ratio
follows the total charged multiplicity of the events[5]. At 7 GeV the
mean number of K^0 + $\bar{K}^0$'s per event is about 0.5, while at 30 GeV
the number is (1.5 ± 0.3) / event. The total charged multiplicity
goes from 5 to 13 from W = 5 to 30 GeV.

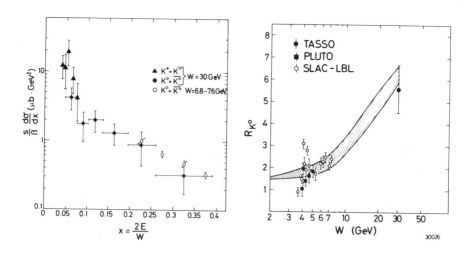

Fig. 13 - (a) Bjorken scaling cross sections for $K^\pm$ and $K^0 + \bar{K}^0$ as
a function of x. (b) The inclusive cross section for
$K^0 + \bar{K}^0$ production divided by the μ-pair cross section
as a function of center of mass energy.

REFERENCES

1. TASSO Collaboration
   R.Brandelik, W.Braunschweig, K.Gather, V.Kadansky, K.Lübelsmeyer,
   P.Mättig, H.-U.Martyn, G.Peise, J.Rimkus, H.G.Sander, D.Schmitz,
   A.Schultz von Dratzig, D.Trines, W.Wallraff,
   I.Physikalisches Institut der RWTH Aachen, Germany
   H.Boerner, H.M.Fischer, H.Hartmann, E.Hilger, W.Hillen, L.Koepke,
   H.Kolanoski, G.Knop, P.Leu, B.Löhr, R.Wedemeyer, N.Wermes,
   M.Wollstadt,
   Physikalisches Institut der Universität Bonn, Germany
   H.Burkhardt, D.G.Cassel, D.Heyland, H.Hultschig, P.Joos, W.Koch,
   P.Koehler, U.Kötz, H.Kowalski, A.Ladage, D.Lüke, H.L.Lynch,
   G.Mikenberg, D.Notz, J.Pyrlik, R.Riethmüller, M.Schliwa, P.Söding,
   B.H.Wiik, G.Wolf,
   Deutsches Elektronen-Synchrotron DESY, Hamburg, Germany,
   R.Fohrmann, M.Holder, G.Poelz, O.Römer, R.Rüsch, P.Schmüser,
   II. Institut für Experimentalphysik der Universität Hamburg,
   Germany
   D.M.Binnie, P.J.Dornan, N.A.Downie, D.A.Garbutt, W.G.Jones,
   S.L.Lloyd, D.Pandoulas, J.Sedgbeer, S.Yarker, C.Youngman,
   Department of Physics, Imperial College London, England
   R.J.Barlow, I.Brock, R.J.Cashmore, R.Devenish, P.Grossmann,
   J.Illingworth, M.Ogg, B.Roe, G.L.Salmon, T.Wyatt,
   Department of Nuclear Physics, Oxford University, England,
   K.W.Bell, B.Foster, J.C.Hart, J.Proudfoot, D.R.Quarrie,
   D.H.Saxon, P.L.Woodworth
   Rutherford Laboratory, Chilton, England
   Y.Eisenberg, U.Karshon, D.Revel, E.Ronat, A.Shapira,
   Weizmann Institute of Science, Rehovot, Israel
   J.Freeman, P.Lecomte, T.Meyer, Sau Lan Wu, G.Zobernig,
   Department of Physics, University of Wisconsin,
   Madison, Wisconsin, USA
2. P.H.Perkins, Proc. 16th Int.Conf. on High Energy Physics,
   Chicago-Batavia 1972, p. 189;
   H.Deden et al., Nucl.Phys. B 85, 269 (1975)
   E.M.Riordan et al., SLAC PUB 1639 (1975);
   H.L.Anderson et al., Phys.Rev.Lett. 37, 4 (1976);
   D.L.Fox et al., Phys.Rev.Lett. 37, 1504 (1976);
   P.C.Bosetti et al., Nucl.Phys. B 142, 1 (1978);
   J.G.H.de Groot et al., Zeit.Phys. C1, 143 (1979),
   Phys.Lett. 82B, 456 (1979);
   H.L.Anderson et al., FNAL Preprint PUB-79/30-EXP (1979)
3. J.K.Yoh et al., Phys.Rev.Lett. 41, 684 (1978)
4. PLUTO Collaboration, C.Berger et al., Phys.Lett. 82B, 449 (1979)
5. TASSO Collaboration, R.Brandelik et al., Phys.Lett.
   89B, 418 (1980)
6. TASSO Collaboration, R.Brandelik et al., Phys.Lett.
   86B, 243 (1979)
7. PLUTO Collaboration, C.Berger et al., DESY 79/43
8. MARK-J Collaboration, D.P.Barber et al.,
   Phys.Rev.Lett 43, 83P (1979)

558

9. S.L.Wu and G.Zobernig, Particles and Fields
   (Z.Phys.C) 2, 107 (1979)
10. J.Ellis, CERN TH-2817
11. TASSO Collaboration, R.Brandelik et al., DESY 80/40
12. PLUTO Collaboration, C.Berger et al.,
    Phys.Lett. 86B, 418 (1979)
13. JADE Collaboration - DESY 79/80
14. P.Hoyer et al., Nucl.Phys. B 161, 349 (1979)
15. A.Ali et al., DESY 79/86
16. R.D.Field and R.P.Fenyman, Nucl.Phys. B136, 1 (1978)
17. MARK-J Collaboration, D.P.Barber et al.,
    Phys.Lett. 89B, 139 (1979)
18. PLUTO Collaboration, Contrubution to this conference
19. TASSO Collaboration, R.Brandelik et al., DESY 80/40
20. T.A.Edison, The Electric Light Bulb (1879)
21. TASSO Collaboration, R.Brandelik et al., DESY 80/49
22. DASP Collaboration, R.Brandelik et al., Nucl.Phys. B148,189 (1979)
23. TASSO Collaboration, R.Brandelik et al., DESY 80/39
24. V.Lüth et al., Phys.Lett. 70B, 120 (1977)

RECENT RESULTS FROM THE JADE COLLABORATION* AT PETRA ON e+e- ANNIHILATION TO MULTIHADRONS

L. H. O'Neill, Jr.

Deutsches Elektronen-Synchrotron, Hamburg

* W. Bartel, T. Canzler 1), D. Cords, P. Dittmann, R. Eichler, R. Felst, D. Haidt, S. Kawabata, H. Krehbiel, B. Naroska, L.H. O'Neill, J. Olsson, P. Steffen, W.L. Yen 2)

Deutsches Elektronen-Synchrotron, Hamburg, Germany

E. Elsen, M. Helm 3), A. Petersen, P. Warming, G. Weber

II. Inst. Experimentalphysik Univ. Hamburg, Germany

H. Drumm, J. Heintze, G. Heinzelmann, R. D. Heuer, J. von Krogh, P. Lennert, H. Matsumura, T. Nozaki, H. Rieseberg, A. Wagner

Physikalisches Inst. Univ. Heidelberg, Germany

D.C. Darvill, F. Foster, G. Hughes, H. Wriedt

University of Lancaster, England

J. Allison, A. H. Ball, I. Duerdoth, J. Hassard, F. Loebinger, H. McCann, B. King, A. Macbeth, H. Mills, P.G. Murphy, H. Prosper, K. Stephens

University of Manchester, England

D. Clarke, M.C. Goddard, R. Marshall, G.F. Pearce

Rutherford Laboratory, Chilton, England

M. Imori, T. Kobayashi, S. Komamiya, M. Koshiba, M. Minowa, S. Orito, A. Sato, T. Suda 4), H. Takeda, Y. Totsuka, Y. Watanabe, S. Yamada, C. Yanagisawa 5)

Lab. of Internatl. Collaboration on Elementary Part. Physics and Dept. of Physics, Univ. of Tokyo, Japan

1) Present address: Siemens AG., Munich, Germany
2) Present address: Purdue University, Indianapolis, USA
3) Present address: Texaco AG., Hamburg, Germany
4) Present address: Cosmic Ray Lab., Univ. Tokyo, Japan
5) Partly supported by the Yamada Science Foundation

ABSTRACT

A search for production of a new quark flavor in
multihadronic states from e+e- annihilation has been made
up to an energy of 35.8 GeV in the center of mass. No
evidence is seen for such production. A new statistical
analysis by the JADE collaboration of the combined data
of four PETRA experiments from a fine energy scan in the
region 29.90 to 31.46 GeV in the center of mass sets new
upper limits on the integrated cross section for a bound
state consisting of a new flavor quark and antiquark. The
ability of the JADE detector to measure dE/dx provides
new upper bounds on the production of fractionally
charged particles such as free quarks, or of heavy,
integrally charged states such as long-lived B mesons.
Finally the fractions of the final state energy carried
by gamma rays and by neutral particles of all kinds are
measured at center of mass energies from 12 to 35 GeV.
The gamma ray and neutral energy fractions are
approximately 26% and 38% respectively, while the
fractional energy carried by neutrinos is less than 15%.

INTRODUCTION

This report on results from the JADE collaboration
at PETRA will be in four parts as follows: Part I will
briefly introduce the detector and cover results on the
total cross section for the reaction e+e- -->
multihadrons at the highest energies achieved as of the
date of this conference. Part II will present a new
analysis of combined results from the four PETRA
experiments which were taking data in the autumn of 1979
as a fine energy scan in search of narrow resonances was
made. Part III will deal with a search, using the JADE
dE/dx measurement, for fractionally charged and heavy,
integrally charged particles. Part IV will treat
measurements of the total energies carried by gamma rays
and by neutral particles of whatever type in
multihadronic final states.

PART I DETECTOR, TOTAL CROSS SECTION

Extant publications describe the JADE detector in
greater detail than is possible here[1,2]. Its central
tracking device is a cylindrical drift chamber providing
48 coordinate measurements on charged tracks in the
region of polar scattering angle $(\theta) |\cos(\theta)| < 0.83$. Its

wires run parallel to the beam axis at radii from
0.21 to 0.78 m so that, in a polar coordinate system in
the plane perpendicular to the beam axis and with origin
at the beam, the wire positions give the radii (r) and
the drift times give the azimuths (ϕ) of points along the
charged tracks. The coordintes of these points in the
direction parallel to the beam (z) are given by the
current division method through the signal amplitudes
measured at both ends of each wire. The same amplitudes
make possible a measurement of the ionization density
(dE/dx) along charged tracks. With each event the data
for up to eight tracks passing through the drift space of
the same wire can be recorded.

The drift chamber is surrounded by an array of 2712
lead glass Cerenkov counters covering the polar angle
regions $|\cos(\theta)| < 0.82$ ("barrel" section) and $0.89 <
|\cos(\theta)| < 0.97$ ("end caps"). The barrel counters are 12
and the end cap counters 10 radiation lengths thick at
normal incidence. The lead glass array is in turn
surrounded by a large muon detector, consisting of an
iron-loaded concrete hadron filter interspersed with 622
single-wire drift chambers arranged in five layers. The
outermost layer is separated from the beam by some 6 pion
interaction lengths of material. Muons at normal
incidence to the rectangular hadron filter must have 1.3
GeV of energy to reach the outermost layer.

Fig. 1 is a section of the detector in a vertical
plane containing the beam. Fig. 2 shows a graphics

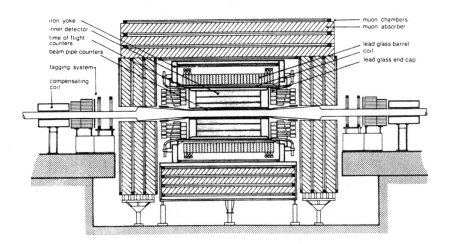

Fig. 1. A vertical section of the JADE detector

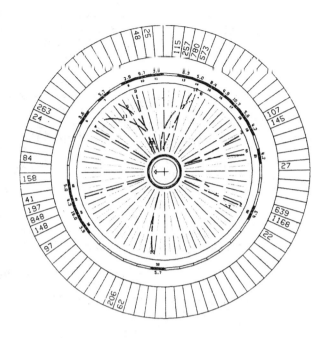

Fig. 2. Graphics display of a multihadronic event

representation of a typical e+e- --× multihadrons event
in a plane normal to the beam axis. The jet chamber and
the lead glass barrel, but not the muon detector, are
shown. Inside the chamber the measured coordinates along
the various charged tracks are represented by crosses.
The numbers within the lead glass segments represent the
total energies, in MeV, in the rows of lead glass
counters running parallel to the beam.

The JADE detector has been used to measure the total
cross section for the process e+e- --> multihadrons at
the highest center of mass energies reached by any
electron storage ring as of the date of this conference.
The event selection criteria and method of analysis are
essentially identical to those used previously at lower
energies and have been described in an extant
publication . Table I gives, for various center of mass
energies between 33.0 and 35.8 GeV, the numbers of
multihadronic events recorded, the integrated
luminosities and the values of the quantity "R". The
latter is defined as the ratio of the total cross section
for the process e+e- --> multihadrons to the total cross
section for the process e+e- --> μ+μ-. The data of Table
I are preliminary.

Table I The quantity "R" for three energies

C.M. Energy (GeV)	NO. Evts.	Int. Lum. (nb-1)	R
33.00	26	78.1	3.88 ± 0.76
35.23	422	1440.3	3.89 ± 0.19
35.80	23	88.7	3.55 ± 0.74

The table entry for 35.23 GeV represents a mean energy and a mean R for all the data collected between 35.00 and 35.56 GeV inclusive, a region in which a fine energy scan was made. The errors quoted are statistical only. A 10% systematic uncertainty, highly correlated for the three different energies, is estimated to apply to the same data. These results are in good agreement with those recently published by the MARK-J collaboration[3]. Fig. 3 shows the JADE and MARK-J results, with combined systematic and statistical error bars, along with various measurements of R at lower energies.

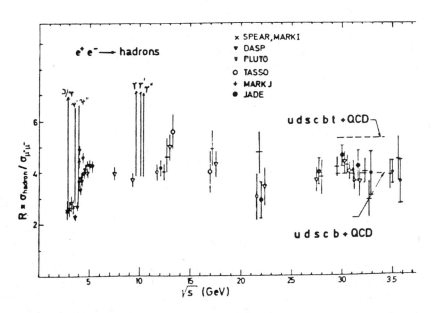

Fig. 3. R versus center of mass energy

If a threshold for production of a charge 2/3 quark were crossed, one would expect an increase in R of about 4/3. Fig. 3 indicates no such step. The lower dotted line on the same figure indicates the value of R predicted by model calculation" assuming that only the established u, d, s, c and b quarks, and gluons, contribute. This result is seen to be in excellent agreement with the experimental data. The upper dashed line indicates the R value, quite inconsistent with the data, which would be given by the same model calculation should a "t" quark with charge 2/3 and mass below 14 GeV be taken to exist. The R measurement does not exclude the existence of a new, charge 1/3 quark beyond the b.

PART II RESONANCE SEARCH

Bound states consisting of a heavy, new-flavored quark and its antiquark would be observed in e+e- collisions as narrow resonances in the total cross section for annihilation to multihadrons and would lie at masses 1 to 2 GeV below the threshold for production of particles bearing the new flavor quantum number. In the autumn of 1979 a fine scan was made at PETRA in search of such resonances with masses between 29.20 and 31.46 GeV. The center of mass energy was varied in steps of 20 MeV, which is nearly equal to the r.m.s. variation in the same quantity due to the energy spread of the stored beams. At each energy an integrated luminosity of, on average, 24 nb was acquired at each interaction region. This luminosity corresponds to an average of some 8 events per energy point per experiment. Each of the four PETRA experiments taking data at the time in question, PLUTO, TASSO, MARK-J and JADE has published results on the 29.90 to 31.46 GeV resonance search[5]. Fig. 4 shows the values of R obtained by combining the data of all four experiments, plotted versus center of mass energy. No strong evidence of a resonance is seen.

In order to make this observation more quantitative the data of Fig. 4 have been fitted by functions of the form:

$$R(W) = R0 + S*\exp(-(W-M_R)^2/(2*\sigma_W^2)) \qquad (1)$$

where a fit is done for each value of "M " in 10 MeV steps from 29.90 to 31.46 GeV, "R0" and "S" are parameters of the fit, "W" is the abscissa of Fig. 4 and σ_W is the r.m.s. variation in the center of mass energy due to the energy spread of the colliding beams. The highest value of S, 1.57 ± 0.65, occurs for the M_R = 29.92 GeV fit. The significance of this

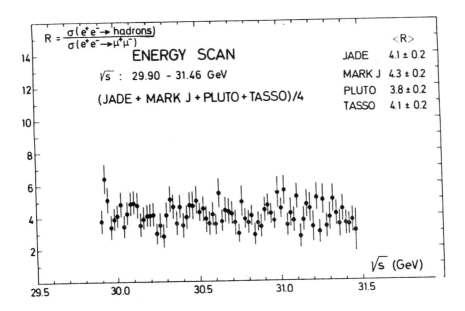

Fig. 4. Combined energy scan data of four experiments

maximum may be evaluated with the aid of a Monte Carlo procedure which tests the hypothesis that no resonance is present and that R is constant over the region of interest. For each experiment and for each of the 79 values of W for which points appear on Fig. 4, a value of R is chosen at random from a certain gaussian distribution. The mean of this distribution is the value of R obtained by the experiment in question, averaging its data over the entire range of energies scanned. The width of the distribution is given by the statistical uncertainty of the value of R obtained by the same experiment at the particular value of W in question. Then for each of the W's the R's of the four experiments are combined with statistical weighting factors.

We now have in hand a list of 79 pseudo data points, entirely analogous to Fig. 4, from a simulated experiment measuring an R which in truth has the same value at each of the 79 points. The gaussian fitting procedure described above for the real data is now repeated for the pseudo data, and the highest value of the parameter S for any value of M_R is selected. Then the entire procedure of generating the 79 pseudo data points is repeated a large number of times, each time with a different sequence of random numbers, and for each repetition the maximum value of S for any value of M_R is selected.

566

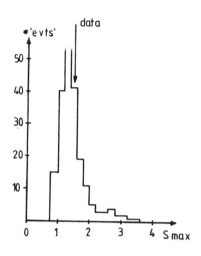

Fig. 5. Distribution
of maximum "S's" as
as defined in text.

Fig. 5 shows the
distribution of these maximum
S's. It is seen that the
maximum S obtained for the real
data, 1.57, is merely a typical
value. Thus there is no
statistically significant
evidence of a peak in the data
of Fig. 4.

If there should
nevertheless be a resonance at
29.29 GeV too weak to be
observed with statistical
significance, what upper limit
can be set on its integrated
cross section (Σ)? Taking the
90% confidence level upper
limit on S at M_R = 29.92 GeV,
assuming any resonance to be
narrow compared to the energy
spread of the colliding
beams and taking radiative
corrections into account we
obtain Σ < 17 nb*MeV. Then using
the relation:

$$\Sigma = \frac{6\pi^2}{M_R^2} \Gamma_{ee} \frac{\Gamma_h}{\Gamma_{tot}}$$

(2)

where Γ_{ee}, Γ_h and Γ_{tot}
represent respectively
the partial width of
the resonance for
decay to e+e-, the
partial width for
decay to multihadrons
and the total width,
we obtain a limit of
0.7 KeV (90%
confidence level) on
the quantity
Γ_{ee} * Γ_h/Γ_{tot}. Under
the rather pessimistic
assumption that only
50% of resonance
decays are to
multihadrons we arrive
at Γ_{ee} < 1.4 KeV
(90% C.L.).

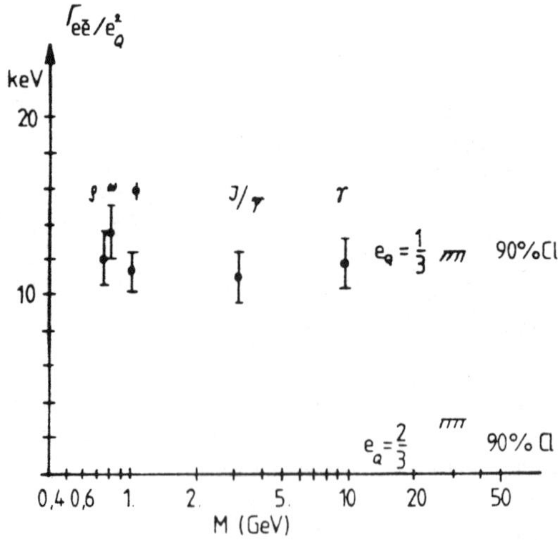

Fig. 6 shows Γ_{ee}
divided by the quark
charge squared for the
known ground state

Fig. 6. Γ_{ee} divided by quark
charge squared

vector mesons. This quantity is seen to be approximately 12 KeV for all of the resonances. If this value holds for a new ground state vector meson we may exclude the existance of such a particle consisting of charge 2/3 quarks at 29.92 GeV. $(1.4 \text{ KeV}/(2/3)^2 = 3.15 \text{ KeV} << 12 \text{ KeV}.)$ A system of charge 1/3 quarks would be compatible with the data. Our limits are of course stronger for masses other than 29.92 GeV.

PART III SEARCH FOR FRACTIONAL CHARGES AND LONG-LIVED MASSIVE PARTICLES

The JADE detector is capable of measuring the ionization density along charged tracks, using the same amplitude measurements at the ends of each wire with which the z coordinates of hits are determined. Individual amplifiers are calibrated with test pulsers, and the gas gain for each wire is determined with the help of Bhabha-scattering electrons. Corrections to the amplitudes associated with individual tracks are made to allow for the path length of the track in each wire's drift space, for saturation effects in the gas amplification and for absorption of drift electrons over long drift paths. Because tracks at small scattering angles leave the end of the drift chamber before reaching the outermost wire and because tracks inside jets are sometimes partially or wholly shaddowed by tracks with similar drift times – the two-track resolution is 7 mm – it is not possible to sample the ionization density (dE/dx) as many as 48 times for each track. A minimum of 25 samplings is required before it is considered that dE/dx has been meaningfully measured for a given track. This requirement results in a nearly momentum-independent efficiency of 65% for the dE/dx determination. The actual determination for a given track is made by averaging the lowest 60% of the dE/dx measurements at individual wires (method of the truncated mean). This procedure reduces the sensitivity to large Landau fluctuations in the single-wire measurements and results in a dE/dx resolution (FWHM) of 22% for tracks within jets. (For well-isolated tracks such as those from Bhabha scattering the dE/dx resolution is 14% (FWHM).)
Fig. 7 is a scatter plot of dE/dx versus "apparent momentum" (momentum divided by charge) for 8500 tracks, taken from 1668 multihadronic events recorded at center of mass energies between 27 and 35 GeV. Fig. 7 also shows the (solid) curves near which tracks of the particles e, π, K, p, deuteron and triton are expected to lie. The dotted curves correspond to hypothetical particles of mass 5 GeV and charge 2/3 and 1.

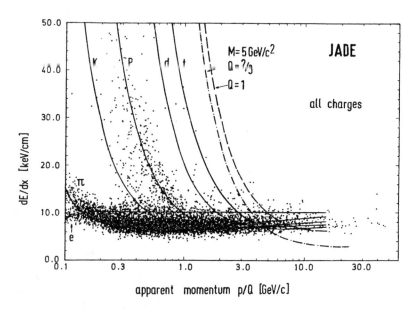

Fig. 7 dE/dx versus apparent momentum

Nuclear interactions in the ∿ 3 gm/cm of material between the beam line and the first wire of the jet chamber account for the large number of protons seen in Fig. 7.

Of the 8500 points on Fig. 7 65 are found more than 2.5 standard deviations from any of the curves for e, π, K or p. All can be explained without resorting to new particles. 14 lie in regions of the scatter plot consistent with their being deuterons or tritons. A visual scan confirms that the corresponding tracks originate in the beam pipe or inner wall of the jet chamber pressure vessel. The remaining tracks with anomalous ionization density can be explained as pairs of tracks of which one track is completely shaddowed within the two-track resolution of the other or as extreme statistical fluctuations in the dE/dx measurement. Thus no evidence is found for the production of any previously unknown type of particle.

An upper limit on the production cross section for a new type of particle with some hypothesized charge and mass can be obtained so long as the curve corresponding to a particle with the assumed properties passes through an event-free region of Fig. 7. Specifically we require that for some momentum interval or intervals a band 2.5 standard deviations in dE/dx to either side of the curve be free of events. Then an upper limit for production of the particle in question, within the momentum interval or intervals described above, is obtained by setting the expectation value for the number of observable particles to 2.3. This implies that the Poisson probability of observing 0 is 10%.

To obtain the corresponding upper limit on the total cross section for production of a new type of particle, i. e. irrespective of momentum, it is necessary to make some hypothesis about the momentum spectrum of the hypothetical particles, since in general they could not be distinguished from known particles over the entire range of momenta. We choose two drastically different invariant momentum spectra, the "soft spectrum":

$$E/(4\pi p^2)*d\sigma/dp \quad \exp(-3.5*E(GeV)) \qquad (3)$$

where "p" and "E" are respectively the particle momentum and energy, and the "hard spectrum":

$$E/(4\pi p^2)*d\sigma/dp = constant \qquad (4)$$

The soft spectrum is characteristic of multipion production in e+e- collisions at SPEAR-DORIS energies[6], while the hard spectrum is merely a very different alternative, which will indicate the sensitivity of the result to the assumed spectrum.

Fig. 8 shows, as a function of the assumed quark mass, the 90% C.L. upper limit on the total cross section, "Rincl", for producing pairs of charge 2/3 quarks within multihadron events. Rincl is given for both the hard and soft momentum spectra specified above and, like the other results on Fig. 8, is normalized to $\sigma_{\mu\mu}$, the total cross section for producing pairs of muons. The figure also gives the analogous limit, "Rexcl", for production of pairs of charge 2/3 quarks without other particles in the final state. The dotted curve indicates the production cross section quark pairs would have if they consisted of free, pointlike, spin 1/2 particles with 2/3 charge.

Although quarks of charge 1/3 would have a reasonably high ($\sim$60%) efficiency for producing signals at individual wires of the jet chamber, the many gaps in their tracks would cause the present pattern recognition codes to have a low efficiency for seeing the tracks as a whole. For the time being no limits will be presented for charge 1/3 quarks.

The same techniques used to search for free quarks can be applied to look for heavy particles of unit charge if these are sufficiently long-lived to reach the jet chamber. In particular, it has been suggested[7] that hadrons containing heavy quarks, such as the b quark, might be rather long-lived. We hypothesize then that pairs of point-like, spin 1/2 b quarks are produced according to the quantum electrodynamic cross section (well above threshold) and then decay into massive particles of unit charge. The cross section for this process is just $3*(1/3)^2 * \sigma_{\mu\mu}$, where the coefficient of 3 arises from the existence of three different quark colors and the factor of $(1/3)^2$ accounts for the 1/3 charge of the b quark.

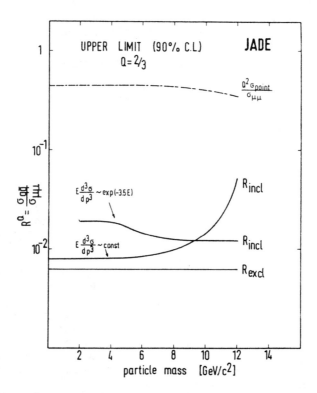

Fig. 8. Limits on production cross section
of charge 2/3 particles

Fig. 9 shows the (90% C.L.) upper limits on the lifetime of a massive, unit-charged particle produced with this cross section. Limits are shown as a function of the hypothesized particle mass for both the "soft" spectrum defined above and for a constant (non-invariant) momentum spectrum, "dσ/dp". It is seen that for either spectrum the lightest particles that could contain the b quark (∿5 GeV in mass) cannot have lifetimes greater than $2*10^{-9}$ seconds.

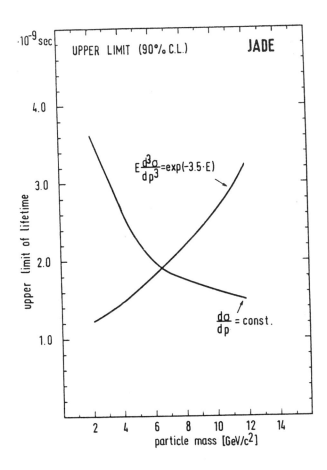

Fig. 9. Limits on the lifetime of a massive,
unit-charge particle

PART IV GAMMA RAY AND NEUTRAL ENERGY

The JADE detector covers 90% of the solid sphere
with lead glass Cerenkov counters and is well suited to
detect the total electromagnetic energy content of
multihadronic final states. At the same time, by
measuring the momenta of nearly all charged particles
with the jet chamber, it can derive the total energy
carried by neutral particles of whatever type. The total
energy of all gamma rays in a final state divided by the
center of mass energy be called the "gamma ray energy

fraction. " The analogous quantity for all neutral particles will be called the "neutral energy fraction."

The difference between the gamma ray and neutral energy fractions is of considerable interest because a model [8] in which the colliding e+ and e- produce free, unit-charged quarks which then decay to mesons and neutrinos would predict that 20% to 30% of the center of mass energy is carried away by neutrinos alone, so long as the colliding beams are energetic enough to produce the free quarks. Furthermore early data from SPEAR [9] seemed to indicate the neutral energy fraction was increasing with increasing center of mass energy. The neutral energy fraction itself is of some interest because certain theoretical bounds [10,11] apply to it in the approximation that multihadron final states consist only of pions and have isospin 0 or 1.

The neutral and gamma ray energy fractions change with time through the decay of particles in the final state and thus require careful definition. The procedure followed here will be to correct the fractions in question for the decay of K_S^0 and Λ particles, but not for others. This approach simply arises from practical considerations: The K_S^0 and Λ multiplicities and energy spectra can be estimated from available data and the corresponding corrections are small enough that their uncertainty does not dominate the systematic error of the final results.

The data for the present analysis were taken at center of mass energies of $\sqrt{s}$ = 12 GeV, near 30 GeV and near 35 GeV. The 30 and 35 GeV samples were not accumulated at fixed energies, but rather obtained by collecting all events between 27.4 and 31.56 GeV and between 33.0 and 35.8 GeV. The trigger conditions and the criteria for selection of hadronic events were the same as applied previously to measure the total cross section for annihilation to multihadrons , except that one additional acceptance cut is made for this analysis, as follows: In order to minimize loss of energy due to particles which escape down the beam pipe or the gaps between the lead glass barrel and the end caps, events are selected for which the jet axis points well inside the region covered by the barrel. Specifically $|\cos(\theta)| < 0.7$ is required. As an estimator of the jet axis direction the sphericity axis [12] is used.

In deriving the gamma ray energy fraction from raw measurements it has to be considered that energy clusters from photons and charged tracks, particularly in jet events, often overlap in the lead glass counters. Thus in general clusters cannot uniquely be attributed to charged particles or to photons. Therefore we have adopted the following procedure to determine the photon energy fraction: For each event an expectation value is computed for the total amount of Cerenkov light generated in the lead glass by all of the charged particles belonging to that event. The Cerenkov light due to charged particles, converted to an equivalent shower energy, is subtracted from the total energy measured in the lead glass arrays to obtain the gamma ray energy fraction. As a first approximation all charged particles are treated as pions and the data of Barber et al. [13] are used to obtain an expected pulse height for each track. These data are for pions of 1.8 to 3.9 GeV momentum, incident normally upon 40 cm thick lead glass counters. The main uncertainties in our treatment arise from interpolating between 1.8 GeV and Cerenkov threshold, and in scaling from 40 cm to other thicknesses of lead glass. (The lead glass barrel is 30 cm thick at normal incidence.) Averaging over all events the Cerenkov light associated with charged particles is equivalent to 23% of the center of mass energy at 12 GeV and 18% in the vicinity of 30 GeV.

Fig. 10 shows the distributions of gamma ray energy fraction, corrected for charged particle contribution as described above, for the three center of mass energies.

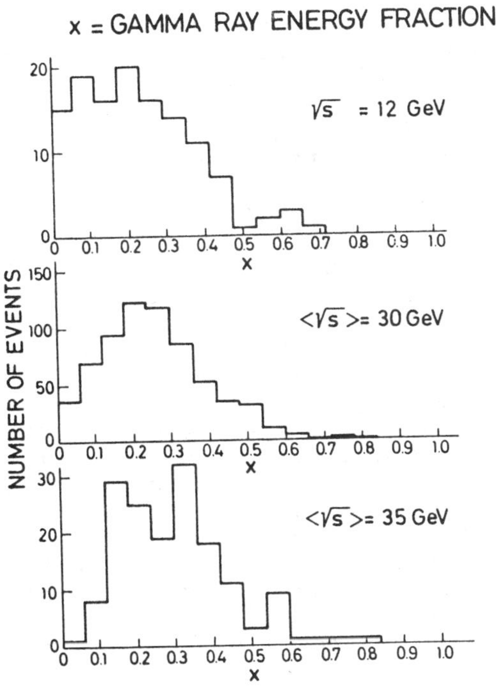

Further corrections are applied to the data of Fig. 10 for several effects which are estimated only in the mean over all events, not event by event. Approximately 14% of the photons in the events selected for this analysis convert before reaching the jet chamber. The resulting electrons and positrons are treated as pions and some fraction of their energy is subtracted from the total registered in the lead glass. The size of this effect can be estimated in a straight-forward way with Monte Carlo simulations. The same is true of the small losses due to gaps in the lead glass system.

Fig. 10. Distribution of the gamma ray energy fraction

The latter two effects are combined into an overall efficiency for detecting gamma ray energy, which also includes the effect of "radiative corrections", i.e. the reduction of the total energy of the visible final state because the initial e+ and e- radiate gamma rays in directions close to the beam axis.

Additional corrections have been made for gamma rays generated by nuclear interactions of hadrons in the material before the jet chamber and for protons ejected from the same material with momenta below Cerenkov threshold. These protons are initially treated as pions, so that on their account energy is incorrectly subtracted from the total recorded in the lead glass. The lead glass energy is further corrected for the interactions of the neutral hadrons K_L^0 [14], n and $\bar{n}$ [15]. The extra energy released by annihilating p's on the one hand, and the unavailability of the p rest mass for creating light on the other, are taken into account[15]. The higher Cerenkov thresholds of charged K's, p's and p's in comparison to pions are also corrected for with the help of the available data on kaon[14] and proton[15] production. A hand scan of a small sample of the data at each energy has been made to search for and estimate the size of all effects not simulated in transporting Monte Carlo events through the detector or otherwise taken into account. Each of the individual corrections discussed in this paragraph changes the final result for the mean energy carried by gamma rays by 2% of the center of mass energy or less. Adding these corrections together, with the appropriate signs and estimated uncertainties, we obtain -1.2% ± 2.5%, -1.9% ± 1.7% and -1.0% ± 1.7% of the center of mass energy at 12, 30 and 35 GeV respectively.

Finally the mean energy per event of gamma rays from π^0's originating in the strange particle decays $K_S^0 \longrightarrow \pi^0\pi^0$ and $\Lambda(\bar{\Lambda}) \longrightarrow n(\bar{n})\,\pi^0$ is subtracted from the mean total energy of all photons. The K_S^0 correction is estimated at -1.1% ± 0.6% of the center of mass energy at 12 GeV and -1.2% ± 0.6% at the higher energies. The Λ effect is negligible (<0.1%) everywhere.

The corrected mean gamma ray energy fractions at the various center of mass energies are listed in Table II. It is observed that the energy fraction which is carried by photons does not change outside experimental uncertainty between 12 and 35 GeV center of mass energy, and that its value is about 26%. The indicated uncertainties are predominantly systematic and are mainly due to the treatment of nuclear interactions by pions in the lead glass at energies below those covered by the data of Barber et al. [13].

Table II The corrected gamma ray energy fraction

C.M. Energy (GeV)	Gamma Ray Energy Fraction
12.00	0.213 ± 0.070
30.35	0.261 ± 0.059
34.93	0.307 ± 0.060

The neutral energy fraction, i. e. the fractional energy carried away photons, K_S^0's, Λ's, long-lived neutral hadrons and neutrinos, can be obtained from the charged tracks by calculating the complement to the charged energy fraction. The following method has been adopted: The charged energy is calculated from the measured momenta by assuming as a first approximation that all particles have the pion mass.

Fig. 11 shows the distributions, for the three different center of mass energies, of the neutral energy fraction computed in this approximation.

Just as with the gamma ray energy fraction several corrections are applied to the neutral energy fraction only in the mean rather than event by event. Geometrical acceptance and losses due to the finite two-track resolution of the jet chamber are corrected for with the help of Monte Carlo simulations of meson jets[4]. With these simulations an efficiency for observing charged particle energy is computed so that radiative corrections are taken into account. The efficiency for charged particle energy also contains a correction for

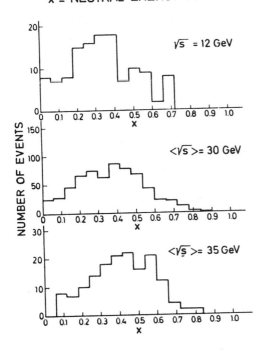

x = NEUTRAL ENERGY FRACTION

$\sqrt{s}$ = 12 GeV

$\langle\sqrt{s}\rangle$ = 30 GeV

$\langle\sqrt{s}\rangle$ = 35 GeV

NUMBER OF EVENTS

Fig. 11. Distribution of neutral particle energy

the e+e- pairs from photon conversion, which must not contribute to the charged particle energy. Because the Monte Carlo meson jets include kaons, the effect of attributing the pion mass to kaons is contained in the same efficiency.

Proton (and antiproton) mass effects are estimated from the available data[15] on p and p̄ production. They require a correction of -1.4% ± 0.7% to the final neutral energy fraction at 12 GeV in the center of mass, and -0.7% ± 0.4% at the higher energies. Nuclear interactions, almost all by charged hadrons, in the material before the jet chamber cause an overestimate of the neutral energy fraction by releasing π^0's and neutrons. This effect requires a correction of -1.0% ± 1.0% at all energies. At the same time these interactions give rise to an effect in the opposite direction by ejecting low energy protons. It is correct to regard the kinetic energy of such protons as belonging to the charged particle energy of the final state since this kinetic energy comes originally from charged hadrons. However the protons in question are counted as having the total energy of pions with the same momenta. This rather large and uncertain effect requires correction of the neutral energy fraction by +2.3% ± 2.3% at 12 GeV in the center of mass and by +1.5% ± 1.5% at the higher energies. A hand scan of a small sample of the events at each energy has been made in search of effects not otherwise accounted for. This scan leads to a correction to the neutral energy fraction of +2.3% ± 1.2% at 35 GeV center of mass energy and to smaller changes at the lower energies. The scan corrections are mainly due to charged particles ejected backwards into the jet chamber by nuclear interactions in the lead glass.

Finally the computed neutral energy fraction is increased by 3.2% ± 1.2% at 12 GeV in the center of mass and by 3.2% ± 1.3% at the higher energies in order to account for the charged decays of the neutral strange particles K_S^0 and Λ[14,16]. The finally corrected mean neutral energy fractions for the three center of mass energies are listed in Table III. The indicated uncertainties are mainly systematic. They are largely due to momentum measurement errors for the highest energy tracks and to the various effects of nuclear interactions in the material between the jet chamber and the beam axis.

Table III The corrected neutral energy fraction

C.M. Energy (GeV)	Neutral Energy Fraction
12.00	0.312 ± 0.041
30.35	0.375 ± 0.037
34.93	0.438 ± 0.041

It is observed that the neutral fraction also remains fairly constant over the energy range under investigation, although a slight increase with energy cannot be excluded by the present measurement. Fig. 12 shows the data of Tables II and III in graphical form.

It should be noted that the values of Table III are well below the theoretical upper limit of 0.77 on the energy fraction carried by π^0's in all-pion states of isospin 0 or 1 [10]. The arguments of Ref. 10 provide no lower limit on the π^0 energy fraction. However lower (and upper) limits are available on the mean multiplicity of π^0's, once more restricted to all-pion final states of isospin 0 or 1 [11]. For isospin 0 this multiplicity is exactly 1/3 the total number of pions, while for isospin 1 the analogous fraction is bounded from below by a limit which tends to 1/5 as the number of pions becomes large. The limit is lower for finite numbers. If we make the hypothesis that all pions have the same mean energy and neglect non-pions and low multiplicity final states, we obtain a rather soft lower limit of 20% for the fractional energy carried by

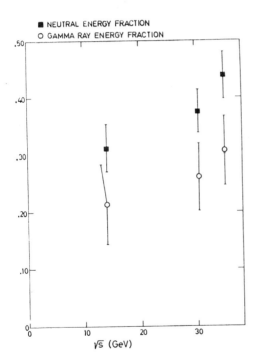

■ NEUTRAL ENERGY FRACTION
O GAMMA RAY ENERGY FRACTION

$\sqrt{s}$ (GeV)

Fig. 12. Energy dependence of neutral and gamma ray energy fractions

π^0's. Our results are entirely consistent with this lower limit.

The high energy data of this experiment may be compared to measurements performed at SPEAR by the Crystal Ball experiment [17] and the SLAC-LBL collaboration [9,18]. The more recent data from the Crystal Ball experiment indicate that the observed neutral energy fraction is about 0.3 between 4 and 5.2 GeV. This indicates practically constant neutral fractions from charm threshold up to 35 GeV.

A comparison between the neutral and the gamma ray energy fractions provides an upper limit on the amount of energy carried by neutrinos. If we remove the K_S^0 and Λ corrections explained in the third paragraph of this part, which are irrelevant to the neutrino energy problem and correlate the errors on the neutral fraction and the gamma ray fraction [19], we conclude that $5.6\% \pm 8.0\%$, $7.0\% \pm 6.8\%$, and $8.8\% \pm 7.2\%$ of the center of mass energy is carried by neutrinos and long-lived neutral hadrons at 12, 30 and 35 GeV respectively. Then using the available data on K^0_L and n and $\bar{n}$ production we can derive neutrino energy fractions of $-2.2\% \pm 8.4\%$, $-1.4\% \pm 7.3\%$ and $+1.6\% \pm 7.6\%$ respectively for the same three center of mass energies. Therefore 15% corresponds approximately to a two-standard-deviation upper limit on the fraction of the total center of mass energy carried by neutrinos in e+e- --> multihadrons final states at PETRA energies. If, however, we assume that liberated quarks with unit charge are produced before they fragment into hadrons [8], the neutrino energy fraction is expected to lie between 20% and 30%, a result strongly disfavored by the present experiment.

REFERENCES

1. JADE collaboration, W. Bartel et al., Phys. Lett. 88B 171 (1979).

2. J. Heintze, Nucl. Instr. and Meth. 156 227 (1978).
   H. Drumm et al., IEEE Trans. Nucl. Sci. 26.1 81 (1979).
   H. Drumm et al., DESY Report 80/38.

3. MARK-J collaboration, D. P. Barber et al., Phys. Rev. Lett. 44 1722 (1980).

4. P. Hoyer et al., Nucl. Phys. B161 349 (1979).

5. JADE collaboration, W. Bartel et al., Phys. Lett 91B 152 (1980).
   MARK-J collaboration, D. P. Barber et al., Phys. Rev. Lett. 44 1722 (1980).
   PLUTO collaboration, Ch. Berger et al., Phys. Lett. 91B 148 (1980).
   TASSO collaboration, R. Brandelik et al., Phys. Lett. 88B 199 (1979).

6. R. Brandelik et al. DASP collaboration, Nucl. Phys. B148 189 (1979).

7. F.N. Cahn, Phys. Rev. Lett. 40 80 (1978).
   H. Fritsch, Phys. Lett. 78B 611 (1978).
   E. W. Lee, and S. Weinberg, Phys. Rev. Lett. 38 1237 (1977).

8. J.C. Pati and A. Salam, Nucl. Phys. B144 445 (1978)

9. It now appears that this "energy crisis" is mainly not real and can be accommodated by a plausible model of multihadron final states consistent with the existent data but not verified in detail. See. D. Scharre, SLAC-PUB-2315 April 1979 and Proceedings of the 14th Recontre de Moriond (1979), J. Tran Thanh Van editor, p. 205. It remains of interest to observe the neutral fraction far above SPEAR energies, where a trend might lead to a large effect.

10. J.S. Bell, G. Karl and Ch. Llewellyn Smith, Phys. Lett. 52 363 (1974).

11. Ch. Llewellyn Smith and A. Pais, Phys. Rev. D6 2625 (1972).

12. The sphericity axis is that principal axis of the sphericity tensor which corresponds to the largest eigenvalue. The sphericity tensor is defined in: JADE collaboration, W. Bartel et al., Phys. Lett. $\underline{89b}$ 136 (1979) and references therein.

13. D.P. Barber et al., Nucl. Instr. and Meth. $\underline{145}$ 453 (1977).

14. At 30 and 35 GeV center of mass energy the K_L^0 (and K_S^0) energy spectrum and multiplicity are taken from the data of R. Brandelik et al., DESY Report 80/39 and DESY Report 80/49, with the assumption that charged and neutral kaons have the same inclusive cross section. For 12 GeV center of mass energy the multiplicity and energy spectrum of the K's is estimated by interpolation between the PETRA data just cited and the SPEAR results of V. Lüth et al., Phys. Lett. $\underline{70B}$ 120 (1977). The Cerenkov light created by K_L^0-nucleus interactions is estimated from the data of Reference 13 after subtracting the contribution of the incident charged pions before interaction.

15. p, $\bar{p}$, n and $\bar{n}$ are assumed all to have the same multiplicity and energy spectrum. The invariant cross section is taken to be an exponential function of particle energy, with a slope which is estimated, both for 12 GeV and the higher center of mass energies, from the data of R. Brandelik et al., DESY Report 80/49. With the shape of the energy spectrum so determined, the JADE detector is used to measure the number of $\bar{p}$'s. $\bar{p}$'s are identified below 1 GeV/c momentum by their large dE/dx. Various estimates are made for the amount of Cerenkov light produced by nucleons and annihilating antinucleons.

16. The ratio of $\Lambda + \bar{\Lambda}$ to $p + \bar{p}$ multiplicity is taken from the SPEAR data of G.S. Abrams et al., Phys. Rev. Lett. $\underline{44}$ 10 (1980), and assumed independent of the center of mass energy. The Λ energy spectrum is taken to be the same as that of protons.

17. E. Bloom, SLAC-PUB-2425 (1979).

18. D.L. Scharre et al., Phys. Rev. Lett. $\underline{41}$ 1005 (1978).

19. It is evident that the assumed energy of K_S^0's has no influence whatever on the finally computed K_L, n, $\bar{n}$ and ν energy, i. e. on Neutral Energy - Gamma Ray Energy - K_S^0 Energy - $\Lambda(\bar{\Lambda})$ Energy. This is because all K_S^0's decay entirely into charged particles or gamma

rays. Therefore the uncertainty in the K_S^0 energy
should not be allowed to affect the estimate of the
uncertainty in the neutrino and long-lived neutral
hadron energy. Λ's decay into charged particle,
gamma rays or neutrons, but our neutron spectrum is
inferred from the experimental proton spectrum
(Reference 15), which does not distinguish between
protons from Λ decay and other protons. We thus
regard neutrons from Λ decay as included in the
neutron spectrum estimated in Reference 15 . Thus
uncertainty as to the energy should also not be
permitted to affect the estimated uncertainties
assigned to the neutrino and long-lived neutral
hadron energy. In effect we are simply using, for our
present purpose, neutral and gamma ray energy
fractions at a time after the K_S^0's and Λ's have
decayed.

D AND E MESONS AND POSSIBLE KKK ENHANCEMENTS[*]

Presented by Jean O. Dickey
Representing the Caltech-Fermilab-Illinois-Indiana Collaboration

C. Bromberg[a], J. Dickey, G. Fox, R. Gomez, W. Kropac[b],
J. Pine, S. Stampke
California Institute of Technology, Pasadena, California 91125

H. Haggerty, E. Malamud
Fermi National Accelerator Laboratory, Batavia, Illinois 60510

R. Abrams, R. Delzenero, H. Goldberg, F. Lopez,
S. Margulies, D. McLeod, J. Solomon
University of Illinois at Chicago Circle, Chicago, Illinois 60680

A. Dzierba, F. Fredericksen, R. Heinz, J. Krider,
H. Martin, D. Petersen
Indiana University, Bloomington, Indiana 47401

ABSTRACT

Results are presented from the study of multikaon final states
in the reactions $\pi^- p \rightarrow K^0 K^\pm \pi^\mp \tilde{X}$, $K^0 K^+ K^- \tilde{X}$ at 50 and 100 GeV/c. Here $\tilde{X}$
is semi-inclusive, because a specific forward topology and an inter-
action registering in the counters surrounding the target are required. The D(1285) meson is seen in the $\delta\pi$ mode, while the E(1420)
meson is observed in both the $\delta\pi$ and the $K^* K$ modes. In addition,
there are two possible enhancements in the three kaon final state.
The first (M ~ 1840 MeV) is associated with the $K^0 \phi$ mode and is consistent with being the charmed D^0. The second, a K^* state (M = 2003
± 14 MeV, Γ = 87 ± 43 MeV), decays into KA_2.

I. INTRODUCTION

Results are presented from the study of multikaon final states
in the reactions $\pi^- p \rightarrow K^0 K^\pm \pi^\mp \tilde{X}$, $K^0 K^+ K^- \tilde{X}$, where $\tilde{X}$ indicates a semi-
inclusive trigger. The data are from the experiment E110 using the
Fermilab multiparticle spectrometer (MPS - see Figs. 1 and 2) in the
M6W beamline with beam energies of 50 GeV (7% of the data) and 100
(93% of the data) GeV. The trigger was designed to select events

---

*Work supported in part by the U.S. Department of Energy under Con-
tract Nos. DE-AC-03-79ER0068 (Caltech) and DE-AS02-76-EY-02009 (Indiana); and the National Science Foundation under Grant No. PHY-78-
07452 (Illinois).
(a) Presently at Michigan State University, East Lansing, Michigan
(b) Presently at Hughes Aircraft Corporation, Los Angeles, California

584

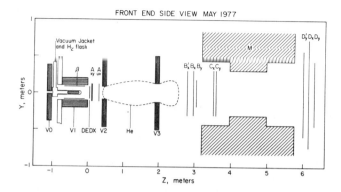

Fig. 1. Multiparticle spectrometer plan view, May 1977.

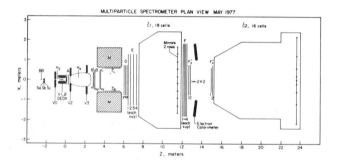

Fig. 2. Front end side view of the multi-particle spectrometer, May 1977.

with a neutral vee (decaying into two forward charged particles) plus two charged particles that trace back to an interaction vertex. In addition, a recoil registering in the counters surrounding the target is required. The identification of the outgoing charged particles is done using two Cerenkov counters. A more detailed discussion of the apparatus, event reconstruction and particle identification is given in Ref. 1.

II. $K^0 K^\pm \pi^\mp$ FINAL STATES

The mass spectra of the $K^0 K^\pm \pi^\mp$, $K^0 K^- \pi^+$ and $K^0 \pi^- K^+$ systems shown in Fig. 3 have no obvious resonance signals. The $K^0 K^\pm \pi^\mp$ mass spectrum requiring a "δ cut" ($M(K^0 K^\pm) \leq$ 1.04 GeV) is shown in Fig. 4. A D(1285) and E(1420) are seen over little background. The contributions from the two charged modes are equal as expected; the two modes have been combined to increase the statistics. Two types of fits are done (see Table I for results). The first is a fit to two Breit-Wigner resonances plus a background, which underestimates the background near the E region. To obtain a better estimate of the background, a sample of $\pi^+ \pi^- K^0$ events was deliberately misidentified as being $K^\pm \pi^\mp K^0$ and the "δ cut" was made. This "background", which was smoothly varying and had no enhancements, was then subtracted from the true $K^\pm \pi^\mp K^0$ events and a fit was made to two Breit-Wigner resonances (see Fig. 5). The results, compared with the world average[2] and recent results[3-5], are shown in Fig. 6. The various resolutions have been calculated using the error matrix from the fitting program, and the results of fits presented

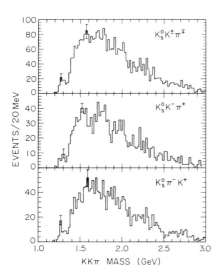

Fig. 3. KKπ mass distribution

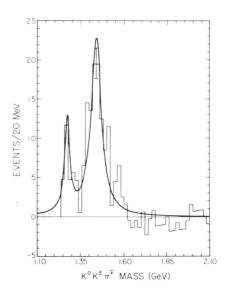

Fig. 5. $K^0 K^{\pm} \pi^{\mp}$ mass distribution with δ cut (M($K^0 K^{\pm}$) ≤ 1.04 GeV) and background subtracted. The full line refers to a resonance fit.

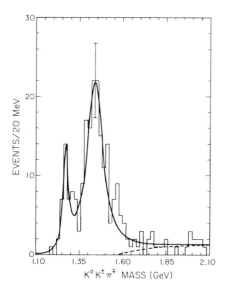

Fig. 4. $K^0 K^{\pm} \pi^{\mp}$ mass distribution with δ cut (M($K^0 K^{\pm}$) ≤ 1.04 GeV). The full line refers to a resonance plus background fit and the dashed line is the background fit.

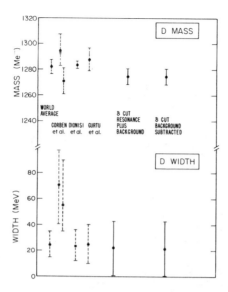

Fig. 6(a). Comparison of D mass and width results. Corden et al. reports two values; the first is the result for the $\eta^{o}\pi^{+}\pi^{-}$ decay mode and the second is for the $K^{+}K^{-}\pi^{o}$ decay mode.

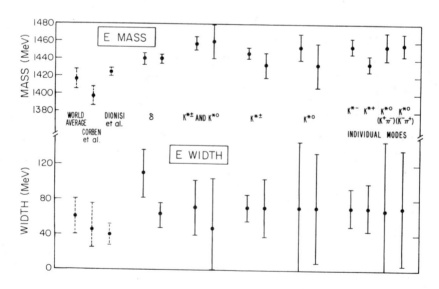

Fig. 6(b). Comparison of E mass and width results.
If two values are presented, the first is the result from a resonance plus background fit and second is the background subtracted fit result.

Table I.   E Production in $\pi^- p \to K^0 K^{\pm} \pi^{\mp} \tilde{X}$

	$\delta\pi$ Mode	$K^{*0}K^0$ and $K^{*\pm}K^{\mp}$ Modes	$K^{*\pm}K^{\mp}$ Mode	$K^{*0}K^0$ Mode
Resonance Plus Background Fit				
Mass (MeV)	1440±6	1458±7	1447±6	1454±15
Width (MeV)	110±27	71±31	71±15	71±75
Background Subtracted Fit				
Mass (MeV)	1440±5	1460±20	1433±14	1433±25
Width (MeV)	62±14	47±58	71±33	71±62

δ cut: $m(K^0 K^{\pm}) \le 1.04$ GeV

K^* cut: $(0.84 \le m(K\pi) \le 0.94$ GeV$)$

here have included the unfolding of the resolution. The agreement of the D mass and width with other results is very good. The width of the E obtained with the background subtracted fit agrees well with the others, while the E mass is slightly high (less than one standard deviation, however).

The D cuts $(1.257 \le m(KK\pi) \le 1.307$ GeV$)$ were imposed on the $K^0 K^{\pm} \pi^{\mp}$ mass spectrum with the resulting two-body mass spectra shown in Fig. 7. A "δ enhancement" is seen in $K^0 K^{\mp}$ peaking near the threshold. The corresponding E cut $(1.356 \le m(KK\pi) \le 1.476$ GeV$)$ spectra are shown in Fig. 8. An enhancement is present both in the $K^{\pm}\pi^{\mp}$ and $K^0 K^{\pm}$ indicating $K^*(890)$ and δ production.

K^* cuts $(0.84 \le m(K\pi) \le 0.94$ GeV$)$ were applied and the resulting $K^0 K^{\pm} \pi^{\mp}$ mass spectra are shown in Fig. 9. An E signal is seen. (Note that the D meson region is below K^*K threshold.) The largest signal over background ratio is obtained when we demand that both $(K\pi)^0$ and $(K\pi)^{\pm}$ be in the K^* region (see Fig. 9(a)). This cut reduces the amount of background present. Signal to background ratio is larger when one requires the charged $K\pi$ combination rather than the neutral combination to be in the K^* region (see Fig. 9(b), 9(c) and Table I). Again a sample of $K\pi\pi$ events was deliberately misidentified as $KK\pi$ and a K^* cut was imposed. This background was subtracted from the true events and the difference was fit to a Breit-Wigner resonance.

588

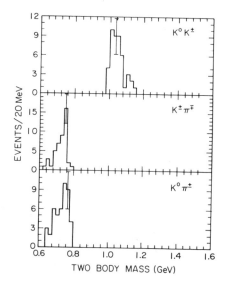
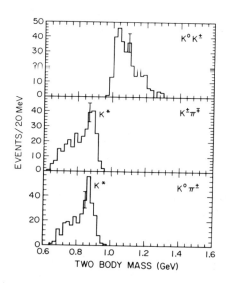

Fig. 7. Two body mass distribution with D cut (1.257 ≤ M(KKπ) ≤ 1.307 GeV).

Fig. 8. Two body mass distribution with E cut (1.356 ≤ M(KKπ) ≤ 1.476 GeV).

Both the results from the resonance plus background fit and the background subtracted fit are shown in Table I. The fits have included the unfolding of the resolutions. A resonance plus background fit was performed on the individual modes (Table II). With the limited

Table II. E Production by Individual Modes in $\pi^- p \to K^0 K^\pm \pi^\mp \tilde{X}$

	$K^{*-}K^+$ Mode	$K^{*+}K^-$	$K^{*0}(K^+\pi^-)K^0$ Mode	$K^{*0}(K^-\pi^+)K^0$ Mode
Mass (MeV)	1455±9	1435±9	1455±16	1457±13
Width (MeV)	71±22	71±27	68±79	71±66
Number of Events in Peak	99±22	90±23	59±73	70±72

δ cut: $m(K^0 K^\pm) \leq 1.04$ GeV

K^* cut: $(0.84 \leq m(K\pi) \leq 0.94$ GeV)

statistics, the number of events from each mode is equal as expected.

While the E meson signal is seen with a K^* cut, it has a better signal to background ratio with a "δ cut". Therefore, the background subtracted fit with a "δ cut" is presented as the best fit.

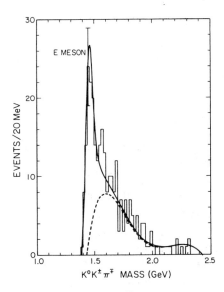

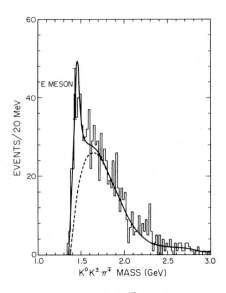

Fig. 9(a). $K^0 K^\pm \pi^\mp$ mass distribution with K^{*o} and $K^{*\pm}$ cut $(0.84 \leq M(K\pi) \leq 0.94$ GeV$)$.

Fig. 9(b). $K^0 K^\pm \pi^\mp$ mass distribution with $K^{*\pm}$ cut $(0.84 \leq M(K\pi)^\pm \leq 0.94$ GeV$)$.

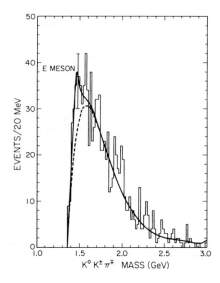

Fig. 9(c). $K^0 K^\pm \pi^\mp$ mass distribution with K^{*o} cut $(0.84 \leq M(K\pi)^o \leq 0.94$ GeV$)$.

Dionisi et al.[4] obtained as a result of a partial wave analysis a branching ratio $E \rightarrow K^{*}\bar{K} + c.c./E \rightarrow [\delta\pi + (K^{*}\bar{K} + c.c.)] = 0.86 \pm 0.12$, where $\delta \rightarrow K\bar{K}$. Taking the ratios of cross section times branching ratio[1], we determined this branching ratio to be 0.76 ± 0.06 which is in agreement with their determination.

III. $K^{o}K^{+}K^{-}$ FINAL STATE

Some exciting physics are indicated in low statistics three kaon results where two possible mass enhancements are seen. Because of the kinematics ($M_{resonance} \gtrsim 3 M_{k}$) this is a particularly good channel for studying high mass strange states. The first of these enhancements (M = 1842 ± 9 MeV), associated with a $K^{o}\phi$ decay mode, is consistent with being the charmed D^{o}. Its observed width (35 ± 30 MeV) agrees with the resolution of the spectrometer for this final state and has a semi-inclusive cross section times branching ratio of 0.05 ± 0.02 μb[1]. The second enhancement, a K^{*} state (M = 2003 ± 14 MeV, Γ = 87 ± 43 MeV) decays into KA_{2} and has a semi-inclusive cross section times branching ratio of 12 ± 2 μb[1]. The $K^{o}K^{-}K^{+}$ mass spectrum is shown in Fig. 10.

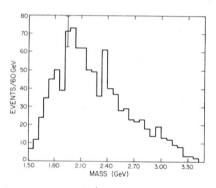

Fig. 10. $K^{o}K^{+}K^{-}$ mass spectrum.

$K^{o}\phi$ FINAL STATE

A well-defined ϕ signal is observed in the $K^{+}K^{-}$ mass spectrum (see Fig. 11). A ϕ cut (1.01 ≤ $m(K^{+}K^{-})$ ≤ 1.03 GeV) is imposed and the resultant $K^{o}K^{+}K^{-}$ mass spectrum is shown in Fig. 12. A low-statistics enhancement is seen in the 1.82-1.86 GeV bin. The ϕ's in our data are produced preferentially at high total momentum (or high x, where $x = \dfrac{P_{tot}(K^{o}K^{+}K^{-})}{P_{incident}}$).

For x ≥ 0.9, the ϕ cut (1.01 ≤ $M(K^{+}K^{-})$ ≤ 1.03) produces a data sample which is almost pure resonance (82%), while for x < 0.9, the cut produces a less pure (65%) signal. The resultant $K^{o}K^{-}K^{+}$ spectra (Fig. 13) show the corresponding increase in the resonance signal to background. The 20 MeV binning shows more events are centered towards the higher half of bin. A fit to a resonance plus background gives a mass of 1842 ± 9 MeV and a width of 35 ± 30 MeV. Since the number of events is small, Poisson statistics should be applied. A relevant measure of the significance of these results is the probability of such an event occurring. The probability of obtaining a signal of 10

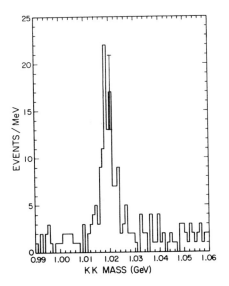

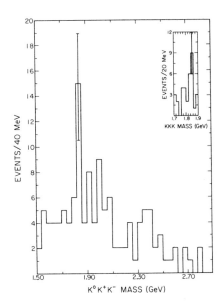

Fig. 11. K⁺K⁻ mass distribution.

Fig. 12. $K^0K^+K^-$ mass distribution
with ϕ cut $(1.01 \leq M(K^+K^-) \leq 1.03$
GeV).

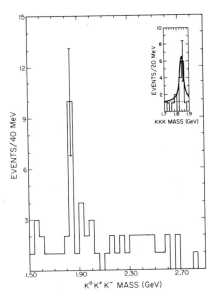

Fig. 13. $K^0K^+K^-$ mass distribution with ϕ $(1.01 \leq M(K^+K^-) \leq 1.03$ GeV)
and X $(X \geq 0.90)$ cuts.

when one expects a signal of 1.4 (the background - see Fig. 13) is
2×10^{-6}. As a check, the $K\pi\pi$ and $KK\pi$ systems have been deliberately
misidentified as the 3K system and the "ϕ" cut has been made. No
peak occurred in either the $KK\pi$ or the $K\pi\pi$ system. As a further
check, tighter Cerenkov cuts were imposed on the interaction vertex
particles and the 3K enhancement remained. A cut was made on the 3K
mass $(1.82 \leq M(3K) \leq 1.86)$ with the corresponding two particle mass
plots given in Fig. 14. A large ϕ production is seen in the K^+K^-

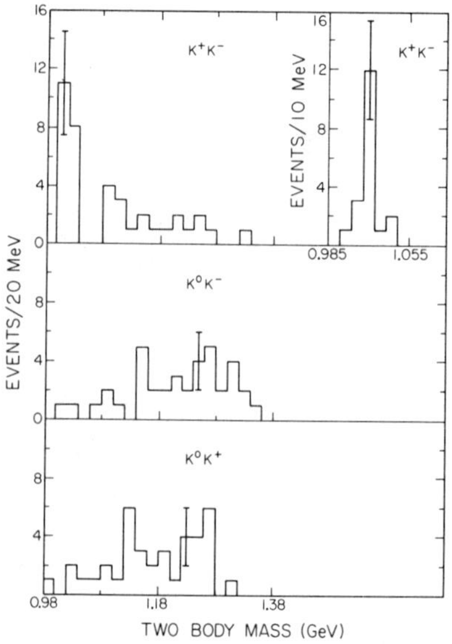

Fig. 14. Two body mass distribution with the cut $(1.82 \leq M(K^0K^+K^-) \leq 1.86)$.

spectrum. Therefore, we conclude
that this 3K enhancement has the
following decay mode: $K^* \rightarrow K^0\phi$;
$\phi \rightarrow K^+K^-$.

The widths of the K^* and ϕ
resonances are rather narrow and
are consistent with the resolution
of the spectrometer for these final
states. The various resolutions
are calculated using the error matrix from the fitting program. The
mass resolution in the ϕ mass region is small (the mean resolution
$\delta M_{mean} = 1.7 \pm 0.2$ MeV at 100
GeV/c) and thus the natural decay
width dominates in the plot of Fig.
11. The ϕ's are well resolved
since their decay products (the K^+
and K^-) have low ϕ center of mass
momenta (low Q value for the ϕ decay) and all chambers may be utilized in fitting these tracks. The
largest uncertainty comes from the
K^0 resolution ($\delta M_{mean} = 22.6 \pm 0.7$

MeV at 100 GeV/c). Here one must not only fit the tracks with fewer
chambers but also do a vertex determination. The final calculated
mean resolution of the $K^0\phi$ state is 35 ± 2 MeV at 100 GeV/c. Therefore, the width of the resonance is consistent with the resolution of
the experiment. The charmed D^0 meson, a narrow state, has a mass of
1863 ± 0.86 MeV[2]. Taking account of our limited statistics and systematic measurement errors, we feel that the enhancement is consistent with being the charmed D^0.

Such a decay of the charmed D^0 would be explained by the diagram
of Fig. 15(a) and its conjugate. Here the D^0 ($c\bar{u}$) decays by the exchange of a W and an $\bar{s}s$ pair is created. Similarly, one would expect
decays with a creation of $\bar{d}d$ and $\bar{u}u$ pair (Figs. 15(b) and (c)) with a
comparable probability. (Maybe the light quark processes are about

(a)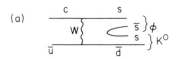

(b)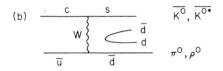

(c)

Fig. 15. Diagrams for charmed D^o decay.

twice the size of the contribution of Fig. 15(a) where the $\bar{s}s$ is created from the vacuum.) Because the vee plus two charged particles trigger discussed here also includes the state $K^o \pi^+ \pi^-$, the diagram of Fig. 15(c) and its conjugate can be studied. The $K^{*\pm} \pi^{\mp}$ mode is investigated, since the decay of the D via $K^*(890)$ has been reported in pp collisions[6]. The $K^o \pi^+$ (Fig. 16(a)) spectrum has a large K^{*+} signal, while the $K^o \pi^-$ (Fig. 16(b)) distribution has a weaker K^* signal. Making a K^{*+} cut $(0.84 \leq M(K^o \pi^+) < 0.94$ GeV), the $K^o \pi^+ \pi^-$ mass spectrum is shown in Fig. 16(c). Since we are selecting K^{+*} mode (no K^{-*}) and the $K^o \pi^+$ decay is observed only (no $K^+ \pi^o$), the ratio of the observed decay of diagram Fig. 15(a) to 15(c) would be reduced from about 1 : 2 to 1 : 2/3. Our $K^o \phi$ resonance signal at the D^o mass is 10 events (100 and 50 GeV/c data combined) with a background of 5. Hence we would expect a signal of about 8 from the $K^{*+} \pi^-$ mode. Examining Fig. 16(c), we find the data are consistent with this expectation but there is no statistically significant effect.

Charmed baryons and mesons have been observed in hadronic experiments and their results are summarized in Table III. The cross sections times branching ratios serve as a check. Since the results here are semi-inclusive and serve as a lower limit for a truly inclusive reaction, one would expect this cross section times branching ratio to be much smaller than the other results. This is indeed the situation.

KA$_2$ FINAL STATE

The mass spectrum requiring $M(K^+ K^-)$, $M(K^+ K^o)$ or $M(K^- K^o)$ to lie in the A_2 mass region (1.261 to 1.363 GeV) is shown in Fig. 17. A rather striking peak is seen at 2.0 GeV. Examining the different modes A_2^o, A_2^- and A_2^+, the contribution from each is approximately equal as expected (Fig. 18 and Table IV). A clear peak is seen with the A_2^o cut, while a shoulder on the peak exists for the $A_2^{\pm}$ modes. This is explained by kinematic reflection. When one does an $A_2^{\pm}$ cut (in this case with $(1.80 \leq M(3K) \leq 2.10$ GeV), some ϕ contribution is

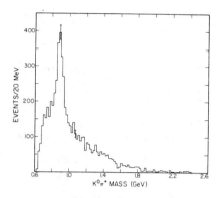

Fig. 16(a). $K^o\pi^+$ mass spectrum from the $K^o\pi^+\pi^-$ data.

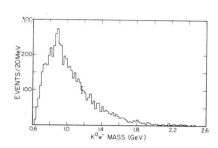

Fig. 16(b). $K^o\pi^-$ mass spectrum from the $K^o\pi^+\pi^-$ data.

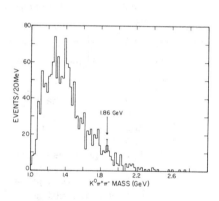

Fig. 16(c). $K^o\pi^+\pi^-$ mass spectrum with K^{+*} cut ($0.84 \leq M(K^o\pi^+) < 0.94$ GeV).

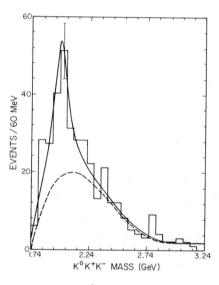

Fig. 17. $K^o K^+ K^-$ mass distribution with A_2 cut ($1.261 \leq M(KK) \leq 1.363$ GeV).

also obtained. This shoulder is associated with the $K\phi$ enhancement. If one subtracts the A_2^o signal from both the $A_2^{\pm}$ contributions, a signal (with poor statistics) is seen near 1.8 GeV. Since the A_2^o cut signal is the cleanest (being unaffected by this kinematic reflection), its fit is taken as the best value. The resonance plus background fit gives a mass of 2003 ± 14 MeV with a width of 87 ± 43 MeV.

Table III. Charmed Hadronic Production Results

Group	Mode	σ x BR (μb)
This experiment	$D^o \to \phi K^o \to K^+ K^- K^o$	0.05 ± 0.02*
D. Drijard et al.[6]	$D^+ \to \overline{K^{*o}} \pi^+ \to K^- \pi^+ \pi^+$	Range 4.0 to 62.0†
ACCDHW[7]	$\Lambda_c^+ \to \overline{K^{*o}} p$	6.2†
		3.0†
	$\Lambda_c^+ \to K^- \Delta^{++}$	6.7†
		3.3†
UCLA-Saclay[8]	$\Lambda_c^+ \to \Lambda^o \pi^+ \pi^+ \pi^-$	2.8 ± 1.0
	$\Lambda_c^+ \to K^- \pi^+ p$	2.3 ± 0.3
Giboni et al.[9]	$\Lambda_c^+ \to \Lambda^o \pi^+ \pi^+ \pi^-$	0.3 - 0.7
	$\Lambda_c^+ \to K^- \pi^+ p$	0.7 - 1.8

* Semi-inclusive result (lower bound for inclusive result)

† Results here are model dependent

Table IV. K^* (2000) Production†

Cut	A_2^{all}	$A_2^o \to K^+ K^-$	$A_2^- \to K^o K^-$	$A_2^+ \to K^o K^+$
Mass (MeV)	1986 ± 15	2003 ± 14	1968 ± 20	2009 ± 20
Width (MeV)	150 ± 68	87 ± 43	136 ± 90	150 ± 78
Number of Events	137 ± 60	50 ± 20	66 ± 44	68 ± 31

† Results of resonance plus background fits with both 50 and 100 GeV/c data.

A cut ($1.92 \leq M(3K) < 2.10$ GeV) is imposed with the resulting two-body spectra given in Fig. 19.

As a check, the Kππ and KKπ events were deliberately misidentified as KKK, and the A_2 cuts were imposed. The resulting three body spectrum shows no signs of this enhancement and looks very much like a background. Further, tight Cerenkov cuts were used and this enhancement is still clearly present.

596

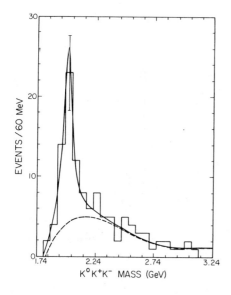

Fig. 18(a). $K^o K^+ K^-$ mass distribution with A_2^o cut ($1.261 \leq M(K^+ K^-) \leq 1.363$ GeV).

Fig. 18(b). $K^o K^+ K^-$ mass distribution with A_2^- cut ($1.261 \leq M(K^o K^-) \leq 1.363$ GeV).

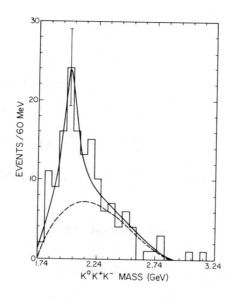

Fig. 18(c). $K^o K^+ K^-$ mass distribution with A_2^+ cut ($1.261 \leq M(K^o K^+) \leq 1.363$ GeV).

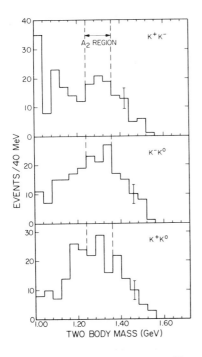

Fig. 19. Two body mass distribution with the cut (1.92 ≤ M(3K) < 2.10 GeV).

The mass resolution was calculated from the fitting program error matrix for the KA_2 final state (at 100 GeV/c, δM_{mean} = 32.7 ± 0.8 MeV). The width of this state (87 ± 43 MeV) is wider than the resolution and hence it is inconsistent with a very narrow state. The excited charmed D^{o*}, where the mass is consistent with this value, would not be expected to have a KA_2^o mode, as it decays into $D^o\pi^o$ or $D^o\gamma^o$. This enhancement associated with KA_2 would therefore be an "ordinary" K^* resonance.

CONCLUSIONS

We regard our observation of the D and E mesons - the first in any high energy experiment - as an important test of the technique and an encouraging indication that our methods could make useful contributions to meson spectroscopy. Our $K\bar{K}\bar{K}$ results are a glimpse of the possible physics that can be probed by future experiments of this type. Results were presented from the study of multikaon final states in the reactions $\pi^-p \to K^oK^\pm\pi^\mp\tilde{X}$, $K^oK^+K^-\tilde{X}$ at 50 and 100 GeV/c. Here $\tilde{X}$ is semi-inclusive and is only a fraction of a true inclusive event. In the $KK\pi$ final state, the D and E mesons are observed with following masses, widths and cross sections times branching ratios (at 100 GeV/c - see Ref. 1 for discussion):

$$M_D = 1275 \pm 6 \text{ MeV}, \quad \Gamma_D = 22 \pm 21 \text{ MeV}, \quad \sigma_D \times BR = 0.27 \pm 0.06 \text{ μb}$$
$$(\delta(\delta \to K\bar{K})\pi \text{ mode})$$

$$M_E = 1440 \pm 5 \text{ MeV}, \quad \Gamma_E = 62 \pm 14 \text{ MeV}, \quad \sigma_E \times BR = 0.63 \pm 0.06 \text{ μb}$$
$$(\delta(\delta \to K\bar{K})\pi \text{ mode}),$$

$$= 0.90 \pm 0.57 \text{ μb}$$
$$(K^{*o}K^o \text{ mode}),$$

$$= 1.11 \pm 0.16 \text{ μb}$$
$$(K^{*\pm}K^\mp \text{ mode}).$$

The low statistics three kaon results presented here indicate two possible mass enhancements. The first state, which decays via the $K^o\phi$ mode, is consistent with being the charmed D^o. Its width

(35 ± 30 MeV) agrees with the resolution of the spectrometer for this final state. Its cross section times branching ratio of 0.05 ± 0.02 µb compares favorably with those of charmed baryons and mesons produced in other hadronic reactions. The $\hat{K}\pi$ decay mode (suggested by Fig. 15) has been examined. We find that the data are consistent with this mode but there is no statistically significant effect. The second enhancement, a K^* state, (M = 2003 ± 14 MeV, W = 87 ± 43 MeV) decays into KA_2 and has a cross section times branching ratio of 12 ± 2 μb^1.

ACKNOWLEDGMENTS

We would like to thank F. J. Nagy for furnishing the resonance plus background fitting routines and for valuable discussions concerning their use. We also thank P. Schlein for providing the Cerenkov counters used in this experiment.

REFERENCES

1. C. Bromberg et al., California Institute of Technology preprint, CALT-68-747.
2. C. Bricman et al., Phys. Lett. 75B, 1 (1978).
3. M. J. Corden et al., Nucl. Phys. B144, 253 (1978).
4. C. Dionisi et al., Bergen-CERN-Collège de France-Madrid-Stockholm Collaboration, CERN/EP 80-1 (January 1980).
5. A. Gurtu et al., Nucl. Phys. B151, 181 (1979).
6. D. Drijard et al., Phys. Lett. 81B, 250 (1979).
7. D. Drijard et al., Phys. Lett. 85B, 452 (1979).
8. K. L. Giboni et al., Phys. Lett. 85B, 437 (1979).
9. W. Lockman et al., Phys. Lett. 85B, 443 (1979).

LIST OF PARTICIPANTS

Aaron, R.	Northeastern University
Aerts, A.T.	Los Alamos Scientific Laboratory
Aguilar-Benitez, M.	Cludad Universitaria
Ashford, V.	Brookhaven National Laboratory
Aston, D.	Stanford Linear Accelerator Center
Atiya, M.	Columbia University
Aubert, B.	Lab. d'Annecy-le-Vieux de Phys. des Particules
Baggett, N.	Department of Energy
Ballam, J.	Stanford Linear Accelerator Center
Bebek, C.J.	Harvard University
Bensinger, J.R.	Brandeis University
Bionta, R.M.	Carnegie-Mellon University
Block, M.M.	Northwestern University
Bloom, E.D.	Stanford Linear Accelerator Center
Bowler, M.G.	Oxford University
Brockman, P.	McGill University
Cannata, P.	Bell Labs
Carnegie, R.K.	Carleton University
Carroll, A.S.	Brookhaven National Laboratory
Cashmore, R.	Oxford University
Cason, N.M.	University of Notre Dame
Cence, R.J.	University of Hawaii
Chen, M.	Massachusetts Institute of Technology
Christenson, J.	New York University
Chung, S.U.	Brookhaven National Laboratory
Cohn, H.O.	Oak Ridge National Laboratory
Condo, G.	University of Tennessee
Creutz, M.	Brookhaven National Laboratory
Crittenden, R.R.	Indiana University
Cutts, D.	Brown University
Dankowych, J.	University of Toronto
Dickey, J.	California Institute of Technology

Donoghue, J.	Massachusetts Institute of Technology
Dover, C.	Brookhaven National Laboratory
Dowd, J.	Southeastern Massachusetts University
Dulude, R.	Brown University
Dunwoodie, W.	Stanford Linear Accelerator Center
Durkin, S.	Stanford Linear Accelerator Center
Eichten, E.	Harvard University
Eisler, F.	College of Staten Island
Etkin, A.	Brookhaven National Laboratory
Fabbri, F.L.	Lab. Nazionali di Frascati/CERN
Ferbel, T.	University of Rochester
Fernow, R.C.	Brookhaven National Laboratory
Ferrer, A.	CERN
Firestone, A.	Iowa State University
Flaminio, E.	Universita di Pisa
Foley, K.J.	Brookhaven National Laboratory
Franzini, P.	Columbia University
Galik, R.S.	University of Pennsylvania
Gettner, M.W.	Northeastern University
Glashow, S.L.	Harvard University
Godfrey, S.	University of Toronto
Goldhaber, G.	Lawrence Berkeley Laboratory
Goldman, J.H.	Florida State University
Gordon, H.	Brookhaven National Laboratory
Gottesman, S.	University of Massachusetts, Amherst
Grannis, P.	SUNY, Stony Brook
Green, D.	Fermi National Accelerator Laboratory
Guerin, F.	Brown University
Hart, E.L.	University of Tennessee
Heller, L.	Los Alamos Scientific Laboratory
Hemingway, R.J.	Carleton University
Henri, V.P.	Universite de l'Etat
Hey, T.	University of Southampton

602

LIST OF PARTICIPANTS (continued)

Hitlin, D.	California Institute of Technology
Honma, A.	Stanford Linear Accelerator Center
Hoogland, W.	NIKHEF-H and Zeeman Lab.
Inagaki, T.	National Laboratory for High Energy Physics, KEK
Jaffe, R.L.	Massachusetts Institute of Technology
Johnson, R.	Brookhaven National Laboratory
Kagan, H.	University of Rochester
Kalogeropoulos, T.	Syracuse University
Kass, R.	University of Rochester
Kayser, B.	National Science Foundation
Kern, W.	Southeastern Massachusetts University
Kirk, H.	Brookhaven National Laboratory
Kirz, J.	SUNY, Stony Brook
Knies, G.	DESY
Kramer, M.A.	City College of New York
Kumar, B.R.	Rutherford Laboratory
Kycia, T.F.	Brookhaven National Laboratory
Lanou, R.E.	Brown University
Lassila, K.E.	Iowa State University
LeBritton, J.A.	University of Rochester
Lee-Franzini, J.R.	SUNY, Stony Brook
Legacey, D.	Carleton University
Leith, D.	Stanford Linear Accelerator Center
Lindenbaum, S.J.	Brookhaven National Laboratory/CCNY
Littenberg, L.S.	Brookhaven National Laboratory
Longacre, R.	Brookhaven National Laboratory
Lynch, H.L.	DESY
Machacek, M.	Northeastern University
Magnuson, B.	Brandeis University
Mallik, U.	City College of New York
Mandelkern, M.	University of California, Irvine
Mann, T.	Tufts University

Margolis, B.	McGill University
Martin, M.	Univ. de Geneve
Marx, M.	Brookhaven National Laboratory
McPherson, A.C.	Carleton University
Meadows, B.	University of Cincinnati
Moneti, G.C.	Syracuse University
Morris, T.W.	Brookhaven National Laboratory
Nakamura, K.	University of Tokyo
Ne'eman, Y.	Tel-Aviv University
Nef, C.	Univ. de Geneve
Nicholson, H.	Mount Holyoke College
Oneda, S.	University of Maryland
O'Neill, L.H.	DESY
Ozaki, S.	Brookhaven National Laboratory
Paige Jr., F.E.	Brookhaven National Laboratory
Pakvasa, S.	University of Hawaii
Palmer, R.B.	Brookhaven National Laboratory
Patel, P.M.	McGill University
Peaslee, D.	Department of Energy
Perrin, D.	CERN
Perruzzo, L.	Universita di Padova
Polychronakos, V.	Brookhaven National Laboratory
Poster, R.	Brandeis University
Potter, D.	Rutgers University
Prentice, J.D.	University of Toronto
Protopopescu, S.	Brookhaven National Laboratory
Raja, R.	Fermi National Accelerator Laboratory
Ratcliff, B.	Stanford Linear Accelerator Center
Rau, R.R.	Brookhaven National Laboratory
Robinson, D.K.	Case Western Reserve University
Roos, C.	Vanderbilt University
Roos, M.	University of Helsinki
Ruddick, K.	University of Minnesota

LIST OF PARTICIPANTS (continued)

Russell, J.J.	Southeastern Massachusetts University
Sakitt, M.	Brookhaven National Laboratory
Samios, N.P.	Brookhaven National Laboratory
Scharre, D.L.	Stanford Linear Accelerator Center
Schouten, M.M.	University of Nijmegen
Schröder, H.	DESY
Schröder, V.	DESY
Shapiro, A.M.	Brown University
Skjevling, G.	CERN
Slaughter, M.D.	Los Alamos Scientific Laboratory
Smith, G.A.	Michigan State University
Stanton, N.	Ohio State University
Strand, R.C.	Brookhaven National Laboratory
Strauch, K.	Harvard University
Sun, C.R.	SUNY, Albany
Taft, H.	Yale University
Teramoto, Y.	Brookhaven National Laboratory
Trower, P.	Virginia Polytechnic Institute and State University
Trueman, T.L.	Brookhaven National Laboratory
Turnbull, R.M.	University of Glasgow
Wang, L.L.	Brookhaven National Laboratory
Webb, R.	Princeton University
Weilhammer, P.M.	CERN
Weinstein, R.	Northeastern University
Weisberg, H.	Brookhaven National Laboratory
Weygand, D.	Brookhaven National Laboratory
Wheeler, C.D.	Brookhaven National Laboratory
White, D.H.	Brookhaven National Laboratory
Whitmore, J.	Michigan State University
Wicklund, A.B.	Argonne National Laboratory
Willen, E.H.	Brookhaven National Laboratory
Wiss, J.	University of Illinois at Urbana-Champaign

LIST OF PARTICIPANTS (continued)

Witherell, M.   Princeton University

Yamin, P.   Brookhaven National Laboratory

COMPLETE AUTHOR INDEX

Abrams, R. 583

Aihara, H. 217

Bar Yam, Z. 170

Bebek, C.J. 365

Bensinger, J.R. 170

Berg, D. 94

Biel, J. 94

Bloom, E.D. 312

Bromberg, C. 583

Button-Shafer, J. 170

Cashmore, R.J. 1

Chandlee, C. 94

Chen, M. 421

Chiang, I-H. 415

Chiba, J. 217

Chung, S.U. 170

Cihangir, S. 94

Cleland, W.E. 55

Collick, B. 94

Delfosse, A. 55

Delzenero, R. 583

Dhar, S. 170

Dickey, J. 583

Donoghue, J.F. 104

Dorsaz, P.A. 55

Dowd, J. 170

Dzierba, A. 583

Eichten, E. 387

Etkin, A. 170

Ferbel, T. 94

Fernow, R. 170

Ferrer, A. 123

Foley, K.J. 170

Fox, G. 583

Fredericksen, F. 583

Fujii, H. 217

Fujii, T. 217

Fukuma, H. 217

Garren, L.A. 415

Gloor, J.L. 55

Goldberg, H. 583

Goldhaber, G. 223

Goldman, J.H. 170

Gomez, R. 583

Green, D.R. 152

Haggerty, H. 583

Heinz, R. 583

Heppelmann, S. 94

Hey, A.J.G. 194

Hogan, G.E. 415

Huston, J. 94

Iwasaki, H. 217

Jensen, T. 94

Johnson, R.A. 415

Jonckheere, A. 94

Joyce, T. 94

Kamae, T. 217

Kern, W. 170

Kienzle-Focacci, M.N. 55

Kirk, H. 170

Knies, G. 513

Koehler, P.F. 94

Kopp, J. 170

Kramer, M.A. 170

Krider, J. 583

607

COMPLETE AUTHOR INDEX (continued)

Kropac, W. 583
Kumar, B.R. 69
Kwan, B. 415
Kycia, T.F. 415
Lee-Franzini, J. 375
Lesnik, A. 170
Li, K.K. 415
Lichti, R.L. 170
Lindenbaum, S.J. 170
Littenberg, L.S. 415
Lobkowicz, F. 94
Lopez, F. 583
Love, W. 170
Lynch, H.L. 543
Makdisi, Y. 94
Malamud, E. 583
Mancarella, G. 55
Margulies, S. 583
Marshak, M. 94
Martin, A.D. 55
Martin, H. 583
Martin, M. 55
McDonald, K.T. 415
McLaughlin, M. 94
McLeod, D. 583
Morris, T.W. 170
Morris, W. 170
Muhlemann, P. 55
Nakamura, K. 217
Nef, C. 55
Nelson, C.A. 94
O'Neill Jr., L.H. 559

Oshima, T. 94
Ozaki, S. 170
Pal, T. 55
Petersen, D. 583
Peterson, E. 94
Pine, J. 583
Platner, E. 170
Prentice, J.D. 297
Protopopescu, S.D. 170
Ratcliff, B.N. 37
Ruddick, K. 94
Rutschmann, J. 55
Saulys, A. 170
Scharre, D.L. 329
Schröder, H. 356
Shupe, M. 94
Slattery, P. 94
Smith, A.J.S. 415
Smith, G.A. 186
Solomon, J. 583
Stampke, S. 583
Sumiyoshi, T. 217
Takada, Y. 217
Takeda, T. 217
Takeshita, T. 217
Thaler, J.J. 415
Thompson, P. 94
Wang, L.L.C. 403
Weygand, D. 170
Wheeler, C.D. 170
Wijangco, A. 415
Willen, E. 170

COMPLETE AUTHOR INDEX (continued)

Winik, M. 170
Wiss, J.E. 257
Witherell, M.S.............285
Yamauchi, M. 217
Zeidler, H. 55

AIP Conference Proceedings

		L.C. Number	ISBN
No.1	Feedback and Dynamic Control of Plasmas	70-141596	0-88318-100-2
No.2	Particles and Fields - 1971 (Rochester)	71-184662	0-88318-101-0
No.3	Thermal Expansion - 1971 (Corning)	72-76970	0-88318-102-9
No.4	Superconductivity in d-and f-Band Metals (Rochester, 1971)	74-18879	0-88318-103-7
No.5	Magnetism and Magnetic Materials - 1971 (2 parts) (Chicago)	59-2468	0-88318-104-5
No.6	Particle Physics (Irvine, 1971)	72-81239	0-88318-105-3
No.7	Exploring the History of Nuclear Physics	72-81883	0-88318-106-1
No.8	Experimental Meson Spectroscopy - 1972	72-88226	0-88318-107-X
No.9	Cyclotrons - 1972 (Vancouver)	72-92798	0-88318-108-8
No.10	Magnetism and Magnetic Materials - 1972	72-623469	0-88318-109-6
No.11	Transport Phenomena - 1973 (Brown University Conference)	73-80682	0-88318-110-X
No.12	Experiments on High Energy Particle Collisions - 1973 (Vanderbilt Conference)	73-81705	0-88318-111-8
No.13	π-π Scattering - 1973 (Tallahassee Conference)	73-81704	0-88318-112-6
No.14	Particles and Fields - 1973 (APS/DPF Berkeley)	73-91923	0-88318-113-4
No.15	High Energy Collisions 1973 (Stony Brook)	73-92324	0-88318-114-2
No.16	Causality and Physical Theories (Wayne State University, 1973)	73-93420	0-88318-115-0
No.17	Thermal Expansion - 1973 (lake of the Ozarks)	73-94415	0-88318-116-9
No.18	Magnetism and Magnetic Materials - 1973 (2 parts) (Boston)	59-2468	0-88318-117-7
No.19	Physics and the Energy Problem - 1974 (APS Chicago)	73-94416	0-88318-118-5
No.20	Tetrahedrally Bonded Amorphous Semiconductors (Yorktown Heights, 1974)	74-80145	0-88318-119-3
No.21	Experimental Meson Spectroscopy - 1974 (Boston)	74-82628	0-88318-120-7
No.22	Neutrinos - 1974 (Philadelphia)	74-82413	0-88318-121-5
No.23	Particles and Fields - 1974 (APS/DPF Williamsburg)	74-27575	0-88318-122-3

No.24	Magnetism and Magnetic Materials - 1974 (20th Annual Conference, San Francisco)	75-2647	0-88318-123-1
No.25	Efficient Use of Energy (The APS Studies on the Technical Aspects of the More Efficient Use of Energy)	75-18227	0-88318-124-X
No.26	High-Energy Physics and Nuclear Structure - 1975 (Santa Fe and Los Alamos)	75-26411	0-88318-125-8
No.27	Topics in Statistical Mechanics and Biophysics: A Memorial to Julius L. Jackson (Wayne State University, 1975)	75-36309	0-88318-126-6
No.28	Physics and Our World: A Symposium in Honor of Victor F. Weisskopf (M.I.T., 1974)	76-7207	0-88318-127-4
No.29	Magnetism and Magnetic Materials - 1975 (21st Annual Conference, Philadelphia)	76-10931	0-88318-128-2
No.30	Particle Searches and Discoveries - 1976 (Vanderbilt Conference)	76-19949	0-88318-129-0
No.31	Structure and Excitations of Amorphous Solids (Williamsburg, VA., 1976)	76-22279	0-88318-130-4
No.32	Materials Technology - 1975 (APS New York Meeting)	76-27967	0-88318-131-2
No.33	Meson-Nuclear Physics - 1976 (Carnegie-Mellon Conference)	76-26811	0-88318-132-0
No.34	Magnetism and Magnetic Materials - 1976 (Joint MMM-Intermag Conference, Pittsburgh)	76-47106	0-88318-133-9
No.35	High Energy Physics with Polarized Beams and Targets (Argonne, 1976)	76-50181	0-88318-134-7
No.36	Momentum Wave Functions - 1976 (Indiana University)	77-82145	0-88318-135-5
No.37	Weak Interaction Physics - 1977 (Indiana University)	77-83344	0-88318-136-3
No.38	Workshop on New Directions in Mossbauer Spectroscopy (Argonne, 1977)	77-90635	0-88318-137-1
No.39	Physics Careers, Employment and Education (Penn State, 1977)	77-94053	0-88318-138-X
No.40	Electrical Transport and Optical Properties of Inhomogeneous Media (Ohio State University, 1977)	78-54319	0-88318-139-8
No.41	Nucleon-Nucleon Interactions - 1977 (Vancouver)	78-54249	0-88318-140-1
No.42	Higher Energy Polarized Proton Beams (Ann Arbor, 1977)	78-55682	0-88318-141-X
No.43	Particles and Fields - 1977 (APS/DPF, Argonne)	78-55683	0-88318-142-8
No.44	Future Trends in Superconductive Electronics (Charlottesville, 1978)	77-9240	0-88318-143-6

No.45	New Results in High Energy Physics - 1978 (Vanderbilt Conference)	78-67196	0-88318-144-4
No.46	Topics in Nonlinear Dynamics (La Jolla Institute)	78-057870	0-88318-145-2
No.47	Clustering Aspects of Nuclear Structure and Nuclear Reactions (Winnepeg, 1978)	78-64942	0-88318-146-0
No.48	Current Trends in the Theory of Fields (Tallahassee, 1978)	78-72948	0-88318-147-9
No.49	Cosmic Rays and Particle Physics - 1978 (Bartol Conference)	79-50489	0-88318-148-7
No.50	Laser-Solid Interactions and Laser Processing - 1978 (Boston)	79-51564	0-88318-149-5
No.51	High Energy Physics with Polarized Beams and Polarized Targets (Argonne, 1978)	79-64565	0-88318-150-9
No.52	Long-Distance Neutrino Detection - 1978 (C.L. Cowan Memorial Symposium)	79-52078	0-88318-151-7
No.53	Modulated Structures - 1979 (Kailua Kona, Hawaii)	79-53846	0-88318-152-5
No.54	Meson-Nuclear Physics - 1979 (Houston)	79-53978	0-88318-153-3
No.55	Quantum Chromodynamics (La Jolla, 1978)	79-54969	0-88318-154-1
No.56	Particle Acceleration Mechanisms in Astrophysics (La Jolla, 1979)	79-55844	0-88318-155-X
No. 57	Nonlinear Dynamics and the Beam-Beam Interaction (Brookhaven, 1979)	79-57341	0-88318-156-8
No. 58	Inhomogeneous Superconductors - 1979 (Berkeley Springs, W.V.)	79-57620	0-88318-157-6
No. 59	Particles and Fields - 1979 (APS/DPF Montreal)	80-66631	0-88318-158-4
No. 60	History of the ZGS (Argonne, 1979)	80-67694	0-88318-159-2
No. 61	Aspects of the Kinetics and Dynamics of Surface Reactions (La Jolla Institute, 1979)	80-68004	0-88318-160-6
No. 62	High Energy e^+e^- Interactions (Vanderbilt , 1980)	80-53377	0-88318-161-4
No. 63	Supernovae Spectra (La Jolla, 1980)	80-70019	0-88318-162-2
No. 64	Laboratory EXAFS Facilities - 1980 (Univ. of Washington)	80-70579	0-88318-163-0
No. 65	Optics in Four Dimensions - 1980 (ICO, Ensenada)	80-70771	0-88318-164-9
No. 66	Physics in the Automotive Industry - 1980 (APS/AAPT Topical Conference)	80-70987	0-88318-165-7
No. 67	Experimental Meson Spectroscopy - 1980 (Sixth International Conference , Brookhaven)	80-71123	0-88318-166-5
No. 68	High Energy Physics - 1980 (XX International Conference, Madison)	81-65032	0-88318-167-3
No. 69	Polarization Phenomena in Nuclear Physics -- 1980 (Fifth International Symposium, Santa Fe)	81-65107	0-88318-168-1
No. 70	Chemistry and Physics of Coal Utilization - 1980 (APS, Morgantown)	81-65106	0-88318-169-X